GRAPHIC CORRELATION

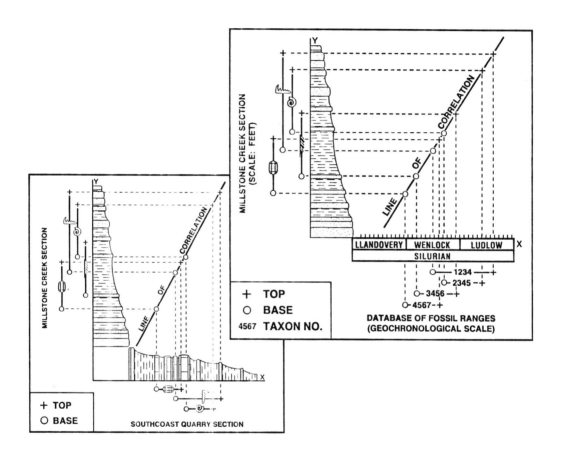

Edited by

Keith Olin Mann, *Juniata College, Huntingdon, PA*
and
H. Richard Lane, *Amoco Production Company, Houston, TX*

SEPM Society for Sedimentary Geology

Peter A. Scholle, *Editor of Special Publications*
Special Publication No. 53

Tulsa, Oklahoma, U.S.A.

October, 1995

A PUBLICATION OF
SEPM SOCIETY FOR SEDIMENTARY GEOLOGY

ISBN 1-56576-023-9

© 1995 by
SEPM (Society for Sedimentary Geology)
1731 E. 71st Street
Tulsa, Oklahoma 74136
1-800-865-9765

Printed in the United States of America

PREFACE

Alan Shaw developed graphic correlation during the late 1950s to solve a problem that he could not address using traditional biostratigraphic zonations. Later, Shaw (1964) introduced graphic correlation in the book *Time in Stratigraphy*. Although *Time in Stratigraphy* and subsequent publications presented the technique, most geologists did not immediately embrace it; however, an increasing number of geologists has begun to use graphic correlation because they find this robust technique provides finer stratigraphic resolution and better accuracy and precision than traditional zonations. Recently, geologists have turned to graphic correlation to help them address many questions that require the finest resolution possible.

Graphic correlation has proved its validity and merit in various contexts. When researchers tested graphic correlation on a number of hypothetical data sets, for which they knew true taxon ranges, graphic correlation performed exceptionally well. Graphic correlation not only performs well on theoretical data, but it also continually performs well on real data in practical appplications. Amoco has devoted many people and innumerable hours to biostratigraphy based on graphic correlation, because the technique provides an advantage in a competitive market.

This volume presents the graphic correlation method, recent methodological developments, and a number of technical papers exemplifying the technique. Geologists from academia, industry, and government have cooperated to produce this volume. Besides advancing the use of graphic correlation, this synthesis should also promote and enhance high-resolution stratigraphy in general.

This special publication resulted as a consequence of three professional meetings. While attending the National Geological Society of America Meeting in Dallas in 1990, I realized that many colleagues were unfamiliar with graphic correlation. In fact, one colleague mentioned that although he had heard of graphic correlation and saw it presented in a number of recent stratigraphy textbooks, he was unfamiliar with this new technique. I responded by telling him that Shaw developed graphic correlation before he was born. To put the age of graphic correlation in a geological perspective, since its inception, 2.4 m of new sea-floor has formed on each side of the East Pacific Rise! Later while attending the National Conference on Undergraduate Research in 1991, I mentioned to Charlotte Mehrtens (the University of Vermont), co-organizer of the 1993 Northeastern Section Meeting of the Geological Society of America, that graphic correlation would make an excellent topic for a theme session. Several months later, she asked me to organize a symposium on graphic correlation. I set forth three goals for that symposium: to expose the uninitiated to graphic correlation, to enhance discussion among practitioners of the technique, and to publish a volume dedicated to graphic correlation. The symposium accomplished two of the goals and all three are realized through the publication of this volume. This special publication synthesizes and summarizes the current work with graphic correlation; it is not intended to be the definitive work on graphic correlation. Instead this collection of papers presents a summary of the technique as currently practiced and it should provide a starting point for those interested in high resolution stratigraphy through graphic correlation. Graphic correlation continues to develop and spread as more geologists use this important and innovative technique. Its potential is only beginning to become realized.

Besides requesting contributions from the participants of the Burlington symposium, we sought contributions from other researchers as well. Because an increasing number of geologists use the technique and practitioners of graphic correlation no longer form a small group, we assuredly, although unintentionally, failed to contact all practitioners of graphic correlation. We apologize for this omission; however, the increase in practitioners using graphic correlation clearly demonstrates the blossoming of this powerful technique.

ACKNOWLEDGMENTS

I want to acknowledge Alan Shaw for developing the technique and having the tenacity to stick to his idea. I owe a special debt of gratitude to two teachers: Rex Crick who first introduced graphic correlation to me in the graduate course "Regional Stratigraphy of the World" and Gilbert Klapper who rekindled my interest in the technique during my doctorate studies at the University of Iowa. I would like to thank the scientists who contributed to this volume and the reviewers for conscientiously reviewing the manuscripts. I also wish to express my gratitude to Dana Ulmer-Scholle, SEPM Special Publication Technical Editor, for her rapid responses over the internet and good humor to my numerous questions. I would also like to thank Norman MacLeod, Jeffery Stein and Kate Bucklen for their assistance. Several institutions provided supplementary funding for this project: Amoco provided funds for several editorial meetings, and Juniata College (Huntingdon, Pa) covered my postage. I am especially grateful to Susan Rose and Kyle Mann.

Keith Olin Mann
Editor

CONTENTS

PREFACE

I. INTRODUCTION TO GRAPHIC CORRELATION

GRAPHIC CORRELATION: A POWERFUL STRATIGRAPHIC TECHNIQUE COMES OF AGE........................*Keith Olin Mann and H. Richard Lane* 3

EARLY HISTORY OF GRAPHIC CORRELATION ....*Alan B. Shaw* 15

II. THE GRAPHIC CORRELATION METHOD

GRAPHIC CORRELATION AND COMPOSITE STANDARD DATABASES AS TOOLS FOR THE EXPLORATION BIOSTRATIGRAPHER..*John L. Carney and Robert W. Pierce* 23

GRAPHIC CORRELATION: SOME GUIDELINES ON THEORY AND PRACTICE AND HOW THEY RELATE TO REALITY...*Lucy E. Edwards* 45

ESTIMATING THE LINE OF CORRELATION... *Norman MacLeod and Peter Sadler* 51

EXTENDING GRAPHIC CORRELATION TO MANY DIMENSIONS: STRATIGRAPHIC CORRELATION AS CONSTRAINED OPTIMIZATION...*William G. Kemple, Peter M. Sadler And David J. Strauss* 65

EVALUATING THE USE OF AVERAGE COMPOSITE SECTIONS AND DERIVED CORRELATIONS IN THE GRAPHIC CORRELATION TECHNIQUE... *Kenneth C. Hood* 83

INTEGRATION OF THE GRAPHIC CORRELATION METHODOLOGY IN A SEQUENCE STRATIGRAPHIC STUDY: EXAMPLES FROM NORTH SEA PALEOGENE SECTIONS .... *Jack E. Neal, Jeff A. Stein, and James H. Gamber* 95

III. TECHNICAL APPLICATIONS OF GRAPHIC CORRELATION

WORLDWIDE AND LOCAL COMPOSITE STANDARDS: OPTIMIZING BIOSTRATIGRAPHIC DATA ....*Richard W. Aurisano, James H. Gamber, H. Richard Lane, Edward C. Loomis, and Jeff A. Stein* 117

HIGH-RESOLUTION BIOSTRATIGRAPHY IN THE UPPER CAMBRIAN ORE HILL MEMBER OF THE GATESBURG FORMATION, SOUTH-CENTRAL PENNSYLVANIA .... *James D. Loch and John F. Taylor* 131

GRAPHIC ASSEMBLY OF A CONODONT-BASED COMPOSITE STANDARD FOR THE ORDOVICIAN SYSTEM OF NORTH AMERICA... *Walter C. Sweet* 139

GRAPHIC CORRELATION OF MIDDLE ORDOVICIAN GRAPTOLITE-RICH SHALES, SOUTHERN APPALACHIANS: SUCCESSFUL APPLICATION OF THE TECHNIQUE TO APPARENTLY INADEQUATE STRATIGRAPHIC SECTIONS ....*Barbara J. Grubb and Stanley C. Finney* 151

A CONODONT- AND GRAPTOLITE-BASED SILURIAN CHRONOSTRATIGRAPHY....................................*Mark A. Kleffner* 159

GRAPHIC CORRELATION OF A FRASNIAN (UPPER DEVONIAN) COMPOSITE STANDARD........................*Gilbert Klapper, William T. Kirchgasser, and John F. Baesemann* 177

MEASURING THE DISPERSION OF OSTRACOD AND FORAMINIFER EXTINCTION EVENTS IN THE SUBSURFACE KIMMERIDGE CLAY AND PORTLAND BEDS, UPPER JURASSIC, UNITED KINGDOM........................*David H. Melnyk, John Athersuch, Nigel Ainsworth, and Paul D. Britton* 185

CORRELATION ACROSS A CLASSIC FACIES CHANGE (LATE MIDDLE THROUGH LATE CENOMANIAN, NORTHWESTERN BLACK HILLS): APPLIED SUPPLEMENTED GRAPHIC CORRELATION .... *Cynthia Fischer* 205

GRAPHIC CORRELATION OF NEW CRETACEOUS/TERTIARY (K/T) BOUNDARY SUCCESSIONS FROM DENMARK, ALABAMA, MEXICO, AND THE SOUTHERN INDIAN OCEAN: IMPLICATIONS FOR A GLOBAL SEDIMENT ACCUMULATION MODEL .... *Norman Macleod* 215

GRAPHIC CORRELATION OF PLIO-PLEISTOCENE SEQUENCE BOUNDARIES, GULF OF MEXICO: OXYGEN ISOTOPES, ICE VOLUME, AND SEA LEVEL.. *Ronald E. Martin and Ruth R. Fletcher* 235

APPENDIX... 249
GENERAL INDEX.. 257
STRATIGRAPHIC INDEX 260
TOXONOMIC INDEX.. 262

PART I
INTRODUCTION TO GRAPHIC CORRELATION

GRAPHIC CORRELATION: A POWERFUL STRATIGRAPHIC TECHNIQUE COMES OF AGE

KEITH OLIN MANN
Department of Geology, Juniata College, Huntingdon, Pennsylvania 16652
AND
H. RICHARD LANE
Amoco Production Company, Houston, Texas 77253

INTRODUCTION

While working as a civil engineer in England, William Smith played a key role in establishing the science of biostratigraphy when he noted that different species of invertebrate fossils characterized the successive Secondary formations in England (Rudwick, 1976). From this, Smith formulated what geologists now recognize as the principle of faunal succession about sixty years before Darwin (1859) published the *Origin of Species*. Darwin's theory of evolution, through variation and natural selection, explained why the geologic succession of organisms occurs. Augmenting the principle of superposition with the principle of faunal succession not only allowed geologists to develop a better relative geologic time scale, but it also enabled them to extend their correlations geographically.

Although biostratigraphy, paleobiogeography, and paleoecology address temporal and spatial distributions of fossils in rocks, biostratigraphers place primary emphasis on the vertical sequence of fossils and use this sequence to place strata in temporal order and to correlate rocks separated in space. Since conceptualization of the fossil zone in the 19th century, biostratigraphers have traditionally used zonal schemes to establish the relative time scale. Ideally, biostratigraphic zones are temporally confined, spatially widespread, and possess approximately synchronous boundaries. Biostratigraphic zones carry time significance and are recognized in rocks where the fossils are present (Johnson, 1992). Despite a number of factors that limit the distribution and preservation of organisms in rocks (e.g., facies control, ecologic barriers, migration, dispersal barriers, local extermination, taphonomic effects, misidentification, erosion, and metamorphism), geologists have been able to define zones and erect a relative time scale.

Since the inception of biostratigraphy, stratigraphers have continually striven to refine and improve both the geologic time scale and correlations. Lyell's quantitative analysis of Tertiary fauna represented an early attempt to produce a quantitative faunal chronometer against which other events and processes could be plotted and measured (Rudwick, 1978). Since the work of Lyell, biostratigraphers have developed a number of quantitative techniques to achieve the finest resolution possible from the rock record. By refining stratigraphic resolution, through subdividing the geologic time scale into finer and finer increments that are both widespread and consistent, geologists will be able to address and answer geologic questions that require a high degree of resolution.

A BRIEF OVERVIEW OF GRAPHIC CORRELATION

Shaw (this volume) found that traditional biostratigraphic zonations were inadequate to answer stratigraphic problems he faced, and so he developed graphic correlation. Graphic correlation plots data from measured sections, placing the bases of the sections at the origin, on a two-axes graph. Usually, first and last appearances of fossil taxa provide the majority of data. Shaw (1964) noted that the utilitarian nature of fossils rests on the premise that extinct fossil taxa divide time into three segments: the time before the organism existed, the time during the existence of the organism, and the time after the organism existed. If both sections spanned identical time periods, experienced identical rates of rock accumulation, and contained a number of time-equivalent data points, the plotted points would align along a line that bisected the axes. This simple graph adds veracity to the technique because it forces the stratigrapher to consider all data since every point is readily visible. Connecting these points with a line forms the line of correlation (LOC). This LOC represents the proposed correlation of the two sections. Using the LOC, the stratigrapher can then transfer, either graphically or mathematically, data present in one section to the other section. This procedure allows the stratigrapher to predict where in the section a fossil should occur. Clearly this predictive attribute constitutes a powerful feature of graphic correlation.

The methodology of graphic correlation is elegantly simple. However, the stratigrapher can only perform graphic correlation after completing the arduous task of data collection. This most important step consumes, by far, the majority of time and effort in any quantitative stratigraphic study. The initial step in graphic correlation requires the stratigrapher to select the standard reference section (SRS). The SRS is the "most" complete section (i.e., this section is the least likely of the measured sections to contain a fault or hiatus; represents the longest stratigraphic interval; contains many species, optimally from several phyla and kingdoms; and is well sampled) to which information is compounded from other sections. After the stratigrapher uses the LOC to integrate information into the SRS from the next best section, the SRS becomes the composite standard (CS). As the geologist incorporates many more sections into the CS, it matures, and total stratigraphic ranges (the stratigraphic interval between the lowermost specimen and the uppermost specimen) of fossil taxa are realized. As the number of data points increases and the precision and accuracy of the composite standard improves, the stratigrapher can delineate finer and finer intervals of geologic time. Shaw (1964) stated that the composite standard, containing information from many sections, allows the stratigrapher to build a precise statement concerning the overall succession of species. Shaw (1964) also demonstrated that the composite standard allows the stratigrapher to measure disturbances of stratigraphic continuity, to create "time-of-flooding" and shoreline maps, to map rates of rock accumulation, and provide time control to stratigraphic studies. Subsequently, geologists have used the composite standard as a Rosetta stone for a number of different research objectives (see the technical section of this volume for several examples). The general methodology of graphic correlation is easily understood and the ramifications of this process and its results are considerable.

The stratigrapher can subdivide the CS into a number of segments called composite standard units (CSUs). Carney and Pierce (this volume) present a full discussion of this procedure. Composite standard units are predicated on the measurements of rock thickness at the SRS. Although CSUs are of equal thickness as measured in the SRS, they are not necessarily of equal temporal duration (Fig. 1). Composite standard units would only be of equal duration if the SRS experienced constant rock accumulation rates.

Kemple and others (this volume) observed that stratigraphic correlation requires three tasks: establishing the temporal sequence of marker events, determining the relative sizes of the intervals between those events, and locating the horizons that correspond with each event. Ideally, the techniques that biostratigraphers use should be able to accomplish these functions. Additionally, the techniques should not force the biostratigrapher to make the *a priori* assumption that first and last fossil occurrences in widely-separated sections are synchronous and represent actual evolution and extinction events. Darwin (1859, p. 293) cautioned against such an assumption when he made the following statement:

> When we see a species first appearing in the middle of any formation, it would be rash in the extreme to infer that it had not elsewhere previously existed. So again when we find a species disappearing before the uppermost layers had been deposited, it would be equally rash to suppose that it then became wholly extinct. We forget how small the area of Europe is compared to the rest of the world: nor have the several stages of the same formation throughout Europe been correlated with perfect accuracy.

Later, Hedberg (1965) produced a figure, which has since appeared in many textbooks, illustrating several factors that restrict stratigraphic ranges (Fig. 2). Figure 2 clearly shows that stratigraphers should not expect widespread synchronous first and last occurrences in the stratigraphic record. Although stratigraphers do recognize that these variables (e.g., migration, nonpreservation, barriers, and local extermination) could truncate species ranges, often these variables and their effects are not fully considered or incorporated within stratigraphic correlations.

Current work on the Eocene-Oligocene boundary exemplifies this point well. Correlation of several sections to the Massignano section (the Eocene-Oligocene boundary stratotype section) near Ancona, Italy, using standard biostratigraphic techniques, has frustrated researchers. Recently, Hazel and Pitakpaivan (1994) determined the reason for the previous, unsatisfactory correlations; they used graphic correlation to show that several fossil events at the Massignano section represent unfilled ranges. Additionally, they were able to use graphic correlation to correlate other sections that overlap the Massignano section in time to the boundary stratotype.

The mode and tempo of sedimentation, often progradational and episodic, that Recent sedimentological research has documented would also inhibit the widespread preservation of synchronous fossil occurrences in the fossil record. Shaw (1964) realized this and took great care to articulate this point. For example, he concluded that "the autochthonous epeiric marine sedimentary pattern would be one showing the lateral deposition of carbonate-evaporate rock types ranging successively from limestone to evaporites." He proposed that in any case when different members of the autochthonous series are superposed without interruption in the stratigraphic sequence the boundaries between the rock types cannot be time parallel. Shaw (1964) also concluded that widespread rock units reflecting uniform depositional conditions must be diachronous. In summarizing the first section of *Time in Stratigraphy*, he made the following statement: "All laterally traceable nonvolcanic epeiric marine sedimentary rock units must be presumed to be diachronous."

Ager (1973) built upon this concept in discussing the process of sedimentation. He noted that sedimentation in the form of a "gentle rain from heaven," in which sediment rains down and preserves all contemporaneous environments simultaneously, currently seems to occur only in the oceanic depths. Ager (1973) acknowledged that even on the ocean floors there are vast areas without sediment and great gaps within the sediment that is present. He felt that the chief contribution of Recent sedimentary studies on the main part of the continental shelves was the demonstration of lateral rather than vertical sedimentation.

Dott (1983) expanded upon this theme in his 1982 SEPM Presidential Address: Episodic sedimentation—how normal is average? how rare is rare? does it matter? He concluded that recognizing the importance—if not the predominance—of episodic sedimentation matters a great deal and may necessitate revision of many depositional models. Without even considering the other processes that affect the distribution and fossilization of organisms, the punctuated mode of sedimentation alone naturally impedes the preservation of fossil ranges that exhibit widespread synchronous first and last appearances.

Given the probability that many stratigraphic sections do not possess entire species ranges, the assumption that first and last appearances do record evolution and extinction events would seriously compromise stratigraphic conclusions. Graphic correlation does not make such *a priori* assumptions. In fact,

FIG. 1.—Ten-meter thick Formation A took 2,000,000 years to accumulate, but the ten-meter thick Formation B accumulated in 1,000,000 years. The eight composite standard units (CSUs) of Formation A represent 250,000 years each, while the eight CSUs of Formation B represent 125,000 years each.

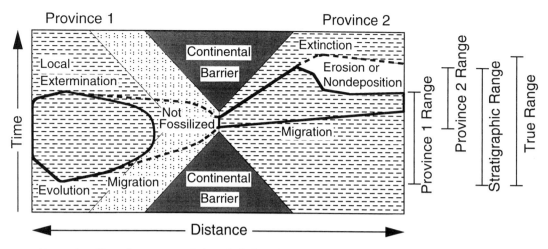

FIG. 2.—The geographic, stratigraphic, and true ranges of a hypothetical species. The first and last appearances of the species in the two provinces clearly differ temporally (modified from Hedberg, 1965 and Eicher, 1968).

graphic correlation commonly demonstrates that many sections contain ranges that are incomplete. This single attribute makes graphic correlation one of the most powerful stratigraphic techniques available.

The stratigrapher should not restrict graphic correlation databases to single taxonomic groups or only biostratigraphic data. Such restrictions hinder the development of robust databases. Shaw (1964) primarily used range data of fossil taxa, but he also suggested that geologists could incorporate other data (such as ash beds) into the technique. Stratigraphers have followed his advice and supplemented biostratigraphic databases with other unique and nonunique temporally significant data. Authors of at least one study (Prell and others, 1986) that used graphic correlation did not even use biostratigraphic data. Integrating a diversity of data (e.g., biostratigraphic, stable isotopes, magnetic, isotopic dates, ash beds, etc.) strengthens the methods and conclusions of any stratigraphic investigation.

The answers to many geologic questions are predicated on resolving the geologic record into very fine increments. Geologists have used the resolving power of graphic correlation to address a number of research queries (e.g., determining the stratigraphic ranges of fossils; determining the diachronous or synchronous nature of rock formations and taxon first and last appearances; delineating smaller segments of the stratigraphic column than possible using traditional zonations; determining the synchrony of geologic events; testing the assumption that sequence boundaries are synchronous; defining, characterizing, and identifying stratigraphic boundaries; determining durations and rates of geologic events; and measuring relative rates of rock accumulation). Recently geologists have also used graphic correlation to address topics that have captivated many scientists and the public as well: MacLeod (1991) reassessed the classic study of Malmgren and others (1983) concerning the tempo of evolution, and MacLeod and Keller (1991a, b) addressed the stratigraphic completeness of Cretaceous/Tertiary boundary sections in order to help evaluate causal mechanisms of the extinction event.

The Biostratigraphic Working Group of the International Geologic Correlation Program (IGCP Project 148) identified three biostratigraphic goals: (1) develop algorithms for quantitative biostratigraphic correlation, (2) implement computer programs, and (3) evaluate quantitative methods through case studies using both actual and simulated data sets (Brower and Millendorf, 1978). Hypothetical data sets allow researchers to test the accuracy, precision, and resolution of biostratigraphic techniques. Brower and Bussey (1985) cautioned that although simulated data sets have the advantage of possessing true answers, erecting a hypothetical data set requires careful scrutiny so that it realistically depicts a natural data set and is neither too simplified nor too complicated. In a series of papers, Edwards (1982a, b, 1984) used simulated data sets to test biostratigraphic methodologies. Edwards (1982a) erected a data set to compare four categories (probabilistic, multivariate, graphic, and relational) of quantitative biostratigraphic techniques. Subsequently, she constructed a different simulated data set to determine why graphic correlation works so well. Edwards (1984) showed that although graphic correlation did not replicate the true ranges of taxa exactly, it did reproduce them with surprisingly high fidelity. Brower and Bussey (1985) used the data set developed by Edwards (1982b) to compare five quantitative biostratigraphic techniques. They concluded that graphic correlation, along with the ranking model they tested, performed the best of the techniques they investigated in terms of overall effectiveness.

Amoco geologists have used graphic correlation successfully in the petroleum industry for close to thirty years. If graphic correlation did not perform well in the competitive petroleum industry, geologists and management would have abandoned the technique long ago. Petroleum geologists, as well as others, have demonstrated the effectiveness and utilitarian nature of the technique for years; however, only recently has the number of geologists, and hence the number of publications, using graphic correlation increased significantly. In a recent review of Paleozoic biostratigraphy, Lenz and others (1993) noted that graphic correlation shows considerable potential for refining biostratigraphic zonation and correlation. They recounted the work of Sweet (1984) to illustrate the resolving power of graphic correlation, noting that the 80 subdivisions (standard time units) he delineated corresponded to four zones and subzones in the North Atlantic Realm. More and more geologists are beginning to use graphic correlation and are finding it to be a powerful technique that provides finer precision and resolution of the stratigraphic record than achievable with traditional zonations.

At the 1994 Society of Economic Paleontologists and Mineralogists Research Conference, "Graphic Correlation and the Composite Standard: the Methods and their Application," MacLeod (1994) encouraged the use of comparative and analytical (as opposed to anecdotal and narrative) approaches to stratigraphy, because these approachs would increase the quality of biostratigraphic research. Graphic correlation allows stratigraphers, in certain instances, to use the hypothetico-deductive structure. This method dismisses narrative-type scenarios and requires alternative hypotheses that produce different results to be expressed in testable formats. Scientific disciplines that advanced more rapidly than others systematically, explicitly, and regularly used such a methodology. The graphic correlation technique permits stratigraphers to conduct research programs that systematically test hypotheses. For example graphic correlation has the ability to test whether sequence stratigraphic boundaries are synchronous. Scott and others (1993) used graphic correlation to address the question: are seismic/depositional sequences time units? Scott and others (1994) also recently used graphic correlation to test previous correlations of the Mid-Cretaceous depositional cycles of the Western Interior and the Gulf Coast.

Besides the mechanical prerequisites outlined by Carney and Pierce (this volume), graphic correlation also requires an intellectual openness and curiosity. The geologist needs to gaze beyond traditional zonal definition and correlation to explore the resolving power and ramifications that graphic correlation affords. Although biostratigraphers use fossils to define, characterize, and identify zones, stratigraphers need not restrict themselves to zonal concepts. Biostratigraphers should examine other methodologies that use biostratigraphic data differently and which may produce temporal scales with finer increments than traditional zonations. The fossils in the strata are the raw material: the factual data. It is the geologist's task to make the best use of that data and to construct a temporal metric with the most sensitive, precise, and accurate scale attainable. In scrutinizing graphic correlation, the geologist may find that it provides better resolution, precision, and accuracy than traditional zonations do and the geologist may even discontinue the use of zones. Since the most important attribute of a geologist is the ability to synthesize information from a variety of sources to address questions of earth history, the techniques the stratigrapher uses should also integrate a diversity of data. The ability of graphic correlation to integrate a variety of data (e.g., fossil ranges, key beds, stable isotope data, magnetic data, radioactive isotopic dates, well-log data) clearly exemplifies its superiority over other stratigraphic techniques.

Geologists have employed and augmented graphic correlation in ways that Shaw probably did not foresee, and in some instances would not endorse. Some geologists followed Shaw's (1964) suggestions, while others modified and amended graphic correlation to address their particular geologic problems. Augmentations, such as supplemental graphic correlation developed by Edwards (1989) and the integration of seismic and graphic correlation with sequence stratigraphy, have helped expand the use of graphic correlation. Because graphic correlation can use large databases, and in fact performs best with large data sets, the advent of the computer probably has been the single most important addition to the methodology. More recently, the ubiquitous personal computer alone has caused the most significant increase in the use of graphic correlation. Originally practitioners wrote their own computer programs (Edwards, pers. commun. 1993). Later, geologists used spreadsheets and graphic programs to execute graphic correlation. More recently two geologists have produced computer programs dedicated to graphic correlation. K. Hood developed GraphCor, a computer program for the IBM-based platform, and N. MacLeod developed two public domain programs for the Macintosh: Shaw Stack (based on Hypercard) and Stratigraphica.

A BRIEF HISTORY OF GRAPHIC CORRELATION RESEARCH

During the past thirty years, a number of groups have used graphic correlation including industry (Amoco), government (e.g., the United States Geological Survey, the Natural History Museum–London), and academia (e.g., the Ohio State University, the University of California-Riverside, Louisiana State University, the University of Iowa, the University of Delaware).

Because geologists at Amoco used and refined graphic correlation essentially since its inception, it is instructive to chronicle its history within Amoco which demonstrates the commitment needed to build a global database using graphic correlation as well as the difficulties in the implementation of graphic correlation that were faced at Amoco. The early years (1962–1968) of graphic correlation and composite standard development at Amoco occurred at its Exploration and Production Research Center in Tulsa, Oklahoma. The hiring of Shaw in 1962 brought his ideas, based on his work at Shell (Shaw, this volume) to Amoco (at that time Pan American Petroleum). The Director of Geological Research, R. Tremaine (who later became a Vice President of Exploration and a strong advocate for Shaw's methods) hired Shaw. Within the next two years, Shaw hired several Paleozoic specialists. The initial work on graphic correlation and the composite method, directed by Shaw, was limited to a few people at the Research Center. Establishing a computer database in these early years proved difficult. Certain critics charged that company computing resources were being used simply as typewriters to print species lists. Some members of the computer staff also actively resisted the development of a paleontologic database. Eventually, Shaw was denied permission to use the company computer program for the paleontology database, but he was allowed to use one developed in the Producing Department of Honolulu Oil Company (a recent acquisition of Pan American). The Honolulu production program was modified to print paleontologic data, but it was written in hexadecimal code. This caused many problems through the years because fewer and fewer computer specialists knew how to program in hexadecimal language.

Shaw instilled procedures among the staff conducive to database development. As Supervisor, he enforced fairly rigidly several important procedural rules:

1. All outcrops and wells had to be accurately located and described geographically, by latitude and longitude, and assigned an Amoco locality number. Paleontologic samples had to be accurately placed within a detailed measured sequence. A single technician assigned the locality number to ensure accuracy and consistency and to limit duplication. No

research or technical service reports would be released without these and the other numerical designations listed below. Initially, Shaw encountered great resistance to locality numbers from computer personnel because the numbers were based on latitude and longitude. Because Pan American was largely a domestic firm, it commonly used the General Land-Office System (township and range). Resistance disappeared when technical service samples from Mozambique arrived and Shaw asked the critics to assign a locality number using their system. Basically, this locality number procedure continues today with more than 16,000 localities established in the Amoco database.

2. Before entering data into the mainframe computer database, all locality and sample computer load sheets had to be completed in their entirety. This required a great deal of geographic, stratigraphic, sample, and coded information. Although Amoco has now automated data entry, the requirement for information has not changed.

3. All identified taxa had to have a unique species number; however, before a taxon number was assigned, a full description, a diagnosis, a comparison of the species, an illustration, and a type specimen were required. Each record and specimen was deposited into company files and collections. In part, this procedure reflected the primitive state of micropaleontology taxonomy at that time. With many excellent taxonomic publications now available, if judged adequately defined, taxa now are assigned numbers directly from literature.

During these early years, the Paleozoic Paleontology Staff consisted of Shaw (trilobites and brachiopods) and Supervisor until 1964; G. Verville (fusulinids) and Supervisor from 1964 until about 1972; W. Creath (ostracodes); G. Klapper (conodonts); A. Ormiston (Siluro-Devonian trilobites); K. Ciriacks (Upper Paleozoic brachiopods and Tertiary molluscs); J. Derby (Cambro-Ordovician trilobites); and H. R. Lane (a summer graduate-student assistant working on Carboniferous conodonts). In the early 1960's, J. Grayson headed a palynology group of about the same size. With growth in program size and parallel organizational style, a competitive, rather than a cooperative, relationship developed between the palynology and paleontology groups. Palynologists at the time had a particularly difficult time contributing to the developmental work in graphic correlation because relatively few taxa had been published and little was known of complete biostratigraphic ranges.

Naturally, computer technology in the sixties was primitive by modern standards. Shaw had several computer technologists assisting him. It was not uncommon to see him and his computer staff carrying large, heavy boxes of computer punch cards. Shaw, acting as the catalyst, started several Paleozoic composite standards. Amoco later largely abandoned these early standards, but geologists did capture the data and learned much about graphic correlation. Shaw also oversaw the development of a Mississippian composite standard as it helped him determine stratigraphic relations in the western United States. In addition, he also worked closely with J. Derby and W. Creath in developing an early version of the Lower/Middle Ordovician composite standard in southern Oklahoma, and he worked with G. Klapper and A. Ormiston in building an early version of the Upper Devonian composite standard.

In 1968, Shaw transferred to the Denver office where he worked on Cook Inlet, Gulf of Alaska, and Alaskan-Peninsula Tertiary composite standards. Important additional participants in this work included D. Englehardt, F. X. Miller, K. Ciriacks, A. Ormiston, and members of the Denver Paleontologic Staff. The Denver office applied graphic correlation for the first time in an operational and drilling environment supporting exploration. During this same time, the Research Palynology Staff in Tulsa, headed by C. F. Upshaw, actively began participating in developing composite standards. This probably reflected the rise of Shaw's influence in research and his location in a regional office. H. R. Lane (conodonts) and D. Mishell (palynology) developed a Late Mississippian through lower Middle Pennsylvanian composite standard and F. X. Miller began his important work on palynological-based standards in the Cretaceous of the western United States from his Denver office, while C. Albers of the Houston office and several people in the New Orleans office attempted to create a Gulf Coast foraminifera-based composite standard in the Tertiary, but "doglegs" abounded ("doglegs" represent changes in the slope of a line of correlation that record differing rates of rock accumulation). Today these "doglegs" occur generally at flooding events, where the slope of the line of correlation abruptly changes above and below the event. Generally, "doglegs" occur at levels of sediment starvation and, like unconformities, they are expressed on the line of correlation as terraces. Also like unconformities, "doglegs" require the practitioner to break the line of correlation into different computer processing intervals, known as "blocks." Shaw (1964) referred in his book to this process called "blocking"; however, in the early 1970's the concept of blocking had not matured, and what is more important, computing power and understanding were not sufficient to make such data manipulations simple. Therefore, a Gulf-Coast Tertiary composite standard languished and never really progressed until the early 1990's.

In 1976, Amoco named Shaw as the Chief Paleontologist of Amoco Production Company in Chicago and a year later he also became Chief Geologist. Because the company's management placed him in a position to oversee the entire corporate paleontologic effort, he promoted graphic correlation and the composite standard approach throughout Amoco. At the same time, C. F. Upshaw became head of a combined Paleontology/Palynology Group, succeeding G. Verville. In 1978, F. X. Miller transferred from Denver to Tulsa Research to become the overall composite standard database "construction engineer" for the company. These three (Shaw, Upshaw, and Miller) formed a dynamic trio who wove graphic correlation and the composite standard approach into the research fabric of Amoco. At this time, Shaw, as Chief Paleontologist and Geologist, and C. F. Upshaw, as Research Director, presented to Amoco management a ten-year plan for intensive paleontologic database and composite standard building. Company management approved the plan and as part of that effort, Amoco hired a large number of new and experienced geologists to the paleontologic staffs in each of the operating offices in Calgary, Denver, Houston, New Orleans, and Tulsa. Amoco recruited discipline specialists in the Mesozoic and Cenozoic especially. At this time, the Tulsa Research Paleontology Staff provided operational support for all international exploration. The Tulsa staff, along with Denver's domestic work, provided the bulk of the data for the composite standard. Amoco Paleontology also undertook the massive, and important, effort of modernizing computer systems. Driven by the tireless efforts of C. F. Upshaw, the improvements focused mainly on developing an interactive mainframe database and a composite standard system. These initial efforts set the tone for regular computer modernizations at Amoco to keep up with the saltatory evolution of computer hardware and software.

From 1977 to 1992, the graphic correlation and composite standard database building effort progressed methodically but was to various degrees enthusiastically endorsed and severely criticized by Amoco staff, inside and outside of the Research Center. Critics looked upon the paleontology program as overstaffed and the graphic correlation approach as fatally flawed because of perceived built-in range over-extensions and thus with little "real-world" exploration value. In general, critics charged that this effort caused significant financial drain on ever-decreasing corporate resources. Critics included other scientific research specialists, other paleontologists (especially in the operating offices), and company management itself. Supporters mustered every argument possible to preserve the effort. The value of the approach was always appreciated and supported whenever it was presented to Amoco executives and whenever it was used to solve a business-related problem. The support of the ten-year program persisted, and the database was built. The program survived through difficult times. During that period, numerous successful exploration applications using graphic correlation had created a loyal customer base that strongly supported the work.

Tulsa-based graphic correlation research technology began to migrate to an operational environment in Houston. In 1988, R. W. Pierce assumed the lead of the Tulsa-based paleontologic staff and began to infill the database and test new ways of displaying interpretations based on graphic correlation. H. R. Lane became the Manager of the Worldwide Paleontologic Services Group in Houston and began to apply the research database to customer-defined problems. In early 1989, C. F. Upshaw, Research Director, died suddenly, thus removing an effective stalwart for graphic correlation research. Shaw had retired earlier in 1986, and the company experienced several downsizings, leaving the paleontologic staff somewhat smaller than it had been in ten years. F. X. Miller retired in 1992, and J. Stein became the Database Coordinator in Houston. Through the late 1980s and early 1990s, Amoco consolidated its Paleontologic efforts in New Orleans and Denver into the Houston office. The final merger of paleontologists occurred in 1992 when Amoco also consolidated the Tulsa Research staff into the Houston office. Although this ended a clearly defined research effort, program consolidation offered other possibilities that otherwise would not have occurred. For example, the consolidation required a transformation from a mainframe computer system to a UNIX/Oracle computer system. The funds to accomplish this probably would have been unavailable if they had not been attached to an office consolidation budget. This reprogramming effort took three years and required a large capital budget to upgrade computing equipment and paleontologic databases. An outside major consulting firm, in a recent benchmarking of Amoco technology, judged the paleontology program as the best practice in the industry and one of the most efficiently run technologies in the company. This excellent rating results directly from the unique, widespread use of graphic correlation and composite standard technology within Amoco. Graphic correlation is the main reason Amoco continues to maintain the most viable paleontologic program in the industry.

In summary, the graphic correlation and composite standard program at Amoco succeeded for several reasons: Amoco benefited from very strong leadership which installed procedures necessary to create a consistent, high-quality database; management styles emerged that led the program through difficult times with vision of a future; and Amoco assembled a versatile and world-class paleontologic and support staff that was able to change technical directions from being basically research-oriented to one that is dominantly business-application oriented.

Although no other company or research group has mounted as sustained an effort devoted to graphic correlation as has Amoco, other companies (including both major oil companies and consulting companies), as well as many academic and governmental geologists, have used the technique. A quick perusal of *Science Citation Index* shows that since the publication of *Time in Stratigraphy*, authors have cited it in more than 200 scientific papers. Documenting the total number of articles published that used or discussed graphic correlation is difficult; however, the results of a simple literature search using *GeoRef* (American Geological Institute) and several key words (graphic correlation and composite standard) show that the number of such publications is rapidly increasing (Fig. 3). Shaw's (1960) abstract clearly represents the first graphic correlation and composite standard publication. Besides Shaw (1960, 1964), Ormiston (1973) and Mikan and Sweet (1974) published early abstracts, and Upshaw and others (1974) in all likelihood, published the first research paper that used graphic correlation. In addition to Shaw's (1964) book, Baron (1975) most likely published the first graphic correlation diagram. Although other books (Eicher, 1968; Mathews, 1974) reviewed the graphic correlation technique, only after Miller (1977) recapitulated the

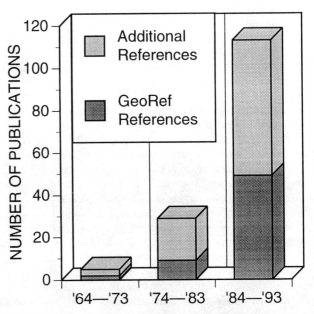

FIG. 3.—A histogram showing the increase in the number of graphic correlation publications.

technique did articles began to appear more frequently than during the first decade. In the past ten years, over one hundred articles have appeared that use graphic correlation. Even though searching *Science Citation Index* and *GeoRef* does not document all articles that used or discussed graphic correlation, these resources, supplemented with our knowledge of other papers, do document a general trend: an increasing number of scientists are using graphic correlation and publishing their results (Fig. 3).

Although some authors (Miller, 1977; Edwards, 1984) published papers devoted to explaining the methodology of graphic correlation and a number of textbooks (Eicher, 1968; Raup and Stanley, 1971, 1978; Mathews, 1974, 1984; Miall, 1984, 1990; Boggs, 1987; Brenner and McHargue, 1988; Prothero, 1990; Lemon, 1990; McKinney, 1991) have presented graphic correlation, why has it taken so long for graphic correlation to begin to blossom? Was it that geologists did not fully understand the technique? Was it the advent of the personal computer? Was it because it is different? Was it because it is a quantitative technique and most stratigraphers are more comfortable with qualitative techniques? Was it because graphic correlation calls into question techniques and assumptions that many stratigraphers commonly use (MacLeod, this volume), or was it because graphic correlation requires too great of an intellectual leap? In ranking the severity of epistemic change that occurs during conceptual changes and revolutions, Thagard (1992) proposed that adding a new concept ranked sixth in a list of nine degrees of change (the ninth being the most severe). Graphic correlation calls for the addition of the composite standard and composite standard units to biostratigraphy. In fact, some stratigraphers have even replaced biostratigraphic zones with the composite standard and composite standard units. Possibly a change of this magnitude to biostratigraphy causes many biostratigraphers to resist graphic correlation. Practitioners have offered a variety of explanations; however, as with many questions, no one answer suffices. The paucity of graphic correlation practitioners led Miall (1984) to conclude his presentation on graphic correlation with the following statement: "The value of the graphic method for correlating sections with highly variable lithofacies and no marker beds is obvious, and it is perhaps surprising that the method is not more widely used." The questions of why it took almost twenty-five years for graphic correlation to begin to bloom and why many geologists oppose the technique when first presented with it will make a lively and interesting study for the historians and philosophers of science. One thing is certain, more and more geologists are using graphic correlation because it successfully integrates many types of stratigraphic data and provides a greater level of stratigraphic resolution than do traditional zonations.

Hallam (1973) observed that precision, scope, explanatory value, and testability are the common hallmarks used to judge the quality of scientific contributions. Considering these criteria, graphic correlation continues to perform in an exemplary fashion. Graphic correlation does not represent a discovery of a new fact, but what is more important, it represents the development of a new method that leads to a better understanding of earth history than was previously attainable. It is not simply the accumulation of more facts that normally leads to major scientific advances; instead, major advances often come about when a complete change of standards and methods occurs, and facts are integrated and synthesized differently than before. For his methodological contributions, Alan Shaw deserves recognition as an important stratigraphic innovator.

VOLUME REVIEW

This special publication continues the work that has gone before it. The volume strives to present graphic correlation in its various permutations as geologists currently practice it. We envision geologists using this volume to learn how to apply graphic correlation, on their own, to their data. Naturally it is essential for the researcher to peruse the seminal work of Shaw (1964). Additionally, one should also consult the previous literature, including the general discussions of Miller (1977) and Edwards (1984), as well as the numerous technical presentations (see appendix). The reader should keep in mind that graphic correlation, as does any biostratigraphic approach, requires consistent identification, classification, and taxonomy, and it can only be implemented after the arduous and time-consuming task of data acquisition.

In the past, researchers have not always completely and explicitly explained many of the subtle aspects of graphic correlation. Biostratigraphers commonly had to discover features of graphic correlation on their own or learn subtleties from other practitioners. We requested the authors of this volume to provide complete, clear, and exact explanations concerning the methods they used. We also asked them to state explicitly how they use and weight data. We hope that the papers in this volume elucidate the many subtle details of graphic correlation.

We did not want to define numerous terms through editorial proclamation, but we did ask the authors to follow consistent word usage for several concepts. The standard reference section (SRS) is the "most" complete section (actually a section is either complete or it is not) to which information is compounded. After information is added to the SRS through the line of correlation, the SRS becomes the composite standard. The composite standard (CS) is the reference that contains information from many sections. Note that some authors in this volume use the term "composite section" to describe a section they compiled from several measured sections of rock. Please do not confuse the terms composite standard and composite section.

We divided the volume into three sections: *Introduction to Graphic Correlation, The Graphic Correlation Method,* and *Technical Applications of Graphic Correlation.* After this introduction, Shaw presents a personal account of the early history of graphic correlation. He describes the development of his innovative technique and later in the paper he discusses the early history of graphic correlation at Amoco. Shaw draws particular attention to the resistance he encountered when he introduced graphic correlation. Resistance to a new scientific idea is common. Wilfred Trotter (1942) made the following observation during a lecture he gave at the Institute of Pathology, St. Mary's Hospital, in 1939.

> The mind likes a strange idea as little as the body likes a strange protein and resists it with similar energy. It would not be too fanciful to say that a new idea is the most quickly acting antigen known to science. If we watch ourselves honestly we shall often find that we have begun to argue against a new idea even before it has been completely stated. I have no doubt that that last sentence

has already met with repudiation—and shown how quickly the defense mechanism gets to work.

Many geologists continue to resist graphic correlation and others simply dismiss the technique; however, despite the criticisms voiced against graphic correlation, no one has disproved its validity or discredited its accuracy, precision, or resolution. Conversely, when researchers (Edwards, 1984; Brower and Bussey, 1985) tested graphic correlation with hypothetical data sets, in which true taxa ranges and correlations were known, it performed very well. MacLeod (this volume) notes that some geologists simply do not fully understand the technique. In addition to Shaw (1964), Miller (1977), and Edwards (1984), this volume addresses that deficiency. MacLeod (this volume) also suggests that many stratigraphers reject graphic correlation because it challenges the very assumptions that they use in their own analyses. He notes that despite being cautioned against considering biostratigraphically defined datums as synchronous, such assumptions permeate analyses within the stratigraphic literature. MacLeod (this volume) proposes that for the geologic community to realize fully the potential of graphic correlation, a major paradigm shift within a large segment of the stratigraphic community needs to occur.

Shaw's personal account of graphic correlation is not only important for scientists, but it also holds significance for the historian. A participant and scientist records and selects different aspects of an event for discussion than does the historian. As Mayr (1982) noted, the history of science deserves inspiration, information, and methodological assistance from both science and history and in turn contributes through its findings to both fields.

The Graphic Correlation Method

Carney and Pierce (this volume) open the methods section with a brick-and-mortar approach to the technique. This paper presents the theory and practice of graphic correlation, as well as many practical procedures currently used at Amoco. Amoco geologists have used graphic correlation for thirty years and clearly have the most experience with the technique. This contribution builds upon the previous publication of Miller (1977), also an Amoco geologist, and provides an in-depth discussion of many methodological subtleties. Their lucid discussion concerning the calibration of the composite standard, a topic that has caused confusion for many geologists, will help all graphic correlation practitioners. Not only do Carney and Pierce (this volume) present the method of graphic correlation, but they also draw on their experience and use case studies to clarify many aspects, including practical application, of graphic correlation.

As editors we decided not to support any particular graphic correlation practice over another (naturally when this many researchers gather, differences of opinion certainly exist); rather, we wanted to expose readers to the various practices, and reasons for those practices, that researchers employ. We caution the reader that other papers in this volume may present slightly different views and methodological approaches to graphic correlation from the original presentation of Shaw (1964) or Carney and Pierce (this volume). Additionally, note on which axis the researcher places the composite standard: some practitioners place it on the ordinate and others place it on the abscissa. Pay close attention to how the researcher weights data and determines the line of correlation. Lastly, as you read the papers, mentally compare the accuracy, precision, and resolution of graphic correlation to traditional zonations; you will discover that graphic correlation provides superior results.

MacLeod and Sadler (this volume) present an extensive and thorough discussion of determining the line of correlation. They discuss the various methodologies used, as well as the assumptions, strengths, and weaknesses of each method. Any geologist seriously interested in graphic correlation and high-resolution stratigraphy should peruse this paper. Establishing the line of correlation is the most problematic procedure of graphic correlation and has prompted more discussions among practitioners than any other topic. Such a comprehensive discussion of this single topic is long over-due.

Kemple and others (this volume) present constrained optimization, a technique performed by a computer program they developed. Constrained optimization operates in many dimensions simultaneously, as opposed to the two-dimensional, serial procedure of traditional graphic correlation. They state that constrained optimization treats all sections at once and eliminates (constrains) the impossible solutions and then seeks the best (optimization) solution that should represent the true history of events. They applied constrained optimization to data sets already analyzed with graphic correlation by experienced geologists. Their computer program provides answers similar to, and at times better than, the solutions of the experienced geologists. Kemple and others (this volume) propose that this successful replication indicates that most of the geologist's expertise is not subjective, but in fact represents knowledge garnered from all sections. This should quell detractors who contend that graphic correlation is not a quantitative technique because of inherent subjectivity in the method.

Hood (this volume) presents two new modifications (average composite models and derived correlations) of graphic correlation. He suggests that average composite models can supplement the traditional maximum composite model. He maintains that average composite models, which are based on the average ranges of taxa, constrain the LOC better than the traditional method, because average ranges generally are similar to ranges observed in the comparison sections within a basin. Amoco geologists (Carney and Pierce, this volume; Aurisano and others, this volume) use primarily local composite standards to address the existence of locally-restricted stratigraphic ranges. Hood (this volume) also extends the graphic correlation methodology into several dimensions by introducing derived correlation, a technique that compares correlations among sections.

Neal and others (this volume) also expand the application of graphic correlation; they demonstrate how to synthesize two powerful techniques, sequence stratigraphy and graphic correlation, that many geologists have used separately. They demonstrate how to use graphic correlation to provide temporal control to seismic and sequence stratigraphy. This paper clearly shows that integrating geophysical data with biostratigraphic data permits a fuller understanding of stratigraphy than either offers alone. This integration supplies a potent research method to the stratigrapher. Certainly many geologists will use this approach in the future.

On the basis of her experience using and teaching graphic correlation, Edwards (this volume) explains many theoretical and practical elements of graphic correlation. She addresses several of the most important intricacies of graphic correlation in an easily understandable manner.

Technical Applications of Graphic Correlation

This section contains both summaries of research programs and presentations of new analyses. Geologists from academe, industry, and government present research that spans local, regional, and global correlations throughout the Phanerozoic. We have organized this section simply by geologic system. Many papers in the technical section also present important and insightful comments concerning graphic correlation methodology, and in doing so follow the advice of G. K. Gilbert (1883): "... whoever in publishing the result of a scientific inquiry sets forth at the same time the process by which it was attained, contributes doubly to the cause of science."

Aurisano and others (this volume) discuss many essential rudiments of graphic correlation. After briefly recounting the composite standard approach used at Amoco, they define and demonstrate the use of the two most important types of composite standards: worldwide composite standards and local composite standards. By drawing on their experience and using several case studies, these authors show the advantages of both composite standards. They also note the many economic and practical advantages graphic correlation has provided Amoco.

Loch and Taylor (this volume) constructed a composite standard, based primarily on trilobites, for the Upper Cambrian Ore Hill Member of the Gatesburg Formation in south-central Pennsylvania. They noted that the exposures produced a limited data set; however, graphic correlation allowed them to achieve a finer resolution of the stratigraphic record than other geologists accomplished using traditional range charts. Furthermore, graphic correlation permitted them to recognize two thin and widespread units that stratigraphers had not previously recognized in the Appalachians. This study shows that researchers with restricted data sets can benefit from graphic correlation.

Sweet (this volume) builds on his previous work and presents a lucid discussion of assembling a composite standard for the North American Ordovician Period. He uses the ranges of more than 300 conodont species in measured sections from seven areas to produce this august work. This research clearly represents a significant contribution and will be invaluable to researchers of Ordovician stratigraphy.

Grubb and Finney (this volume) present a concise report documenting the construction of a composite standard, and its use, from twenty-three sections of Middle Ordovician sediments in the Appalachian Basin. They note that their composite standard emanated from a data set that appeared to be inadequate. This successful application to a troublesome data set demonstrates the power of graphic correlation. They used the composite standard to measure the rate of basin migration by temporally tracking a basal-shale contact. Graphic correlation also allowed them to measure rates of rock accumulation within the basin.

Kleffner (this volume) expanded upon his previous (Kleffner, 1989) research and integrated twelve additional sections into his Silurian composite standard, which, with this new revision, contains over 200 taxa and has applicability as a global, high-resolution reference. He divided the Silurian composite standard into 92 subdivisions (standard time units) and observes that this resolution is, at a minimum, twice that of any previously proposed for the Silurian Period. He also shows how the Silurian composite standard relates to series, stages, zones, selected taxa ranges, and zonal indices.

Klapper and others (this volume) present the most recent results of their research program; constructing a composite standard for the Frasnian. This Frasnian composite standard currently consists of 27 sections, two of which are regional composites of 20 and 16 sections. They divided the Frasnian composite standard into 34 composite standard units, which, they note, represents a far finer stratigraphic resolution than any available through traditional conodont or ammonoid zonations. Additionally they observe that graphic correlation allowed them to correlate between the mostly mutually exclusive *Palmatolepis* and *Polygnathus* conodont biofacies, a problem that was insoluble using traditional zonal methods. In a recent summary of Paleozoic biostratigraphy, Lenz and others (1993) remarked that zones established in one biofacies commonly may be difficult or even impossible to apply to other well-differentiated biofacies. The ability of graphic correlation to transcend the limits inherent in zonations erected in well-differentiated biofacies further exemplifies its power. In addition to the technical aspect of their paper, Klapper and others (this volume) also present several insightful comments concerning graphic correlation methodology.

MacLeod (this volume) continues his research on the Cretaceous-Tertiary (K/T) boundary and expands his database by including Danian data from several new sections (high latitude and proximal to the proposed Chicxulub impact structure). He uses his global database and graphic correlation to determine the stratigraphic completeness of K/T sections. He has found that Danian sequences from very shallow inner neritic settings and the deep sea are incomplete, while Danian sequences from neritic settings through upper bathyal settings are temporally complete. Therefore, if a researcher does not take a global view, but rather investigates sections only from very shallow inner neritic settings or the deep sea, the investigator would acquire an incomplete view of the geologic record; a view biased toward abrupt and sudden patterns of change produced by unrecorded time from these settings. MacLeod (this volume) also answers previous criticisms directed at graphic correlation in general and his research specifically.

Melnyk and others (this volume) used graphic correlation to record the dispersion of ostracod and foraminifer extinction events compared to digitally filtered wireline log data in Upper Jurassic sections of the Wessex Basin, United Kingdom. After erecting a composite standard, they identify a number of taxa with short ranges, small dispersion (compared to the geophysical data), and large numbers of occurrences. They also use the amount of dispersion (compared to the geophysical data) to measure the confidence of each biostratigraphic event.

Fisher (this volume) applied the supplemental graphic correlation technique developed by Edwards (1989) to the unique and non-unique, but temporally significant, data she collected from Cenomanian strata in Montana and Wyoming. She faced a similar difficulty to that of Klapper and others (this volume); differences in physical conditions among facies often affects the distribution of taxa and makes correlation difficult. Supplemental graphic correlation enabled her to procure high-resolution correlations across a significant facies change within the

Western Interior Seaway and to delineate 23 divisions within strata representing one million years.

Martin and Fletcher (this volume) summarize the work led by Martin during the past decade. They intend their paper to serve as a heuristic aid to help stratigraphers apply graphic correlation to their own data. They recount the progression of the research program and the construction of an integrated data set. They used graphic correlation to recognize sequence boundaries, identify erosionally truncated or reworked biostratigraphic markers, and relate zonations to paleoclimate and sea-level changes. They state that graphic correlation of subzonal boundaries coupled with oxygen isotope, other biostratigraphic events, and magnetostratigraphic datums allowed them to demonstrate the utility of planktic foraminiferal assemblage zones in the subdivision of the Pleistocene epoch. Additionally, they make an important observation when they note that graphic correlation not only can be used in a predictive manner, but it also can suggest further avenues of research.

SUMMARY

Graphic correlation possesses many fundamental attributes that make it probably the most robust and powerful quantitative stratigraphic technique available. As mentioned previously, graphic correlation requires an intellectual openness on the part of the stratigrapher, and encourages the stratigrapher to look beyond traditional zonations. Finally, the geologist should heed the caution of Shaw (1964) that Edwards (1984) highlighted:

> The technique described here does not in any way relieve the paleontologist from thinking about what he is doing. The tool presented by the graph is extremely powerful, but also insensate. It cannot overcome inconclusive or inadequate data, nor can it compensate for slipshod thinking in its applications. On the contrary, by giving a result that can be expressed numerically—it may lend a spurious aura of sanctity to otherwise worthless conclusions.

ACKNOWLEDGMENTS

We would like to thank Fred Rogers, Lucinda Pitcairn, Susan Rose, and Bob Washburn for reviewing this manuscript and making suggestions that improved both the content and presentation of the manuscript. We also want to thank Chip Carney for bringing the quotation by Charles Darwin to our attention.

REFERENCES

AGER, D. V., 1973, The Nature of the Stratigraphical Record: New York, John Wiley and Sons, 122 p.

BARON, J. A., 1975, Marine diatom biostratigraphy of the Upper Miocene–Lower Pliocene strata of southern California: Journal of Paleontology, v. 49, p. 619–632.

BOGGS, S., 1987, Principles of Sedimentology and Stratigraphy: Columbus, Merrill Publishing Company, 784 p.

BRENNER, R. L. AND MCHARGUE, T. R., 1988, Integrated Stratigraphy: Englewood Cliffs, Prentice Hall, 419 p.

BROWER, J. C. AND BUSSEY, D. T., 1985, A comparison of five quantitative techniques for biostratigraphy, in Gradstein, F. M., Agterberg, F. P., Brower, J. C., and Schwarzacher, W. S., eds., Quantitative Stratigraphy: Dordrecht, Reidel Publishing, p. 279–306.

BROWER, J. C. AND MILLENDORF, S. A., 1978, Quantitative biostratigraphic correlation: biostratigraphic correlation within the IGCP Project 148: Computers in Geosciences, v. 4, p. 217–220.

DARWIN, C., 1859, On the Origin of Species, a facsimile of the first edition, 1964: Cambridge, Harvard University Press, 513 p.

DOTT, R. H., 1983, 1982 SEPM Presidential address: Episodic sedimentation—how normal is average? how rare is rare?, does it matter?: Journal of Sedimentary Petrology, v. 53, p. 5–23.

EDWARDS, L., 1982a, Numerical and semi-objective biostratigraphy: review and predictions: Third North American Paleontological Convention, Proceedings, v. 1, p. I47-I52.

EDWARDS, L., 1982b, Quantitative biostratigraphy: the methods should suit the data, in Cubitt, J. M. and Reyment, R. A., eds., Quantitative Stratigraphic Correlation: Chichester, John Wiley and Sons, p. 45–60.

EDWARDS, L., 1984, Insights on why graphic correlation (Shaw's method) works: Journal of Geology, v. 92, p. 583–597.

EDWARDS, L., 1989, Supplemental graphic correlation: a powerful tool for paleontologists and nonpaleontologists: Palaios, v. 4, p. 127–143.

EICHER, D. L., 1968, Geologic Time: Englewood Cliffs, Prentice-Hall, 149 p.

GILBERT, G. K., 1883, Review of Archibald Geikie's Geological Sketches at Home and Abroad, and Textbook of Geology: Nature, v. 27, p. 237–239.

HALLAM, A., 1973, A Revolution in the Earth Sciences: Claredon Press, Oxford, 127 p.

HAZEL, J. E. AND PITAKPAIVAN, K., 1994, Graphic correlation of the Eocene-Oligocene boundary stratotype at Massignano, Italy, in Lane, H. L., Blake, G., and MacLeod, N., eds., SEPM Research Conference: Graphic Correlation and the Composite Standard: The Methods and their Applications: Society of Economic Paleontologists and Mineralogists, p. 18.

HEDBERG, H. D., 1965, Chronostratigraphy and biostratigraphy: Geological Magazine, v. 102, p. 451–461.

JOHNSON, J. G., 1992, Belief and reality in biostratigraphic zonation: Newsletters on Stratigraphy, v. 26, p. 41–48.

KLEFFNER, M. A., 1989, A conodont-based Silurian chronostratigraphy: Geological Society of America Bulletin, v. 101, p. 904–912.

LEMON, R. R., 1990, Principles of Stratigraphy: Columbus, Merrill Publishing Company, 559 p.

LENZ, A. C., JIN, J., MCCRACKEN, A. D., UTTING, J., AND WESTROP, S. R., 1993, Paleoscene 15. Paleozoic biostratigraphy: Geoscience Canada, v. 20, p. 41–73.

MACLEOD, N., 1991, Punctuated anagenesis and the importance of stratigraphy to paleobiology: Paleobiology, v. 17, p. 167–188.

MACLEOD, N., 1994, Database Systems, in Lane, H. L., Blake, G., and MacLeod, N., eds., SEPM Research Conference: Graphic Correlation and the Composite Standard: The Methods and their Applications: Society of Economic Paleontologists and Mineralogists, p. 3.

MACLEOD, N. AND KELLER, G., 1991a, How complete are Cretaceous/Tertiary boundary sections? A chronostratigraphic estimate based on graphic correlation: Geological Society of America Bulletin, v. 103, p. 1439–1457.

MACLEOD, N. AND KELLER, G., 1991b, Hiatus distributions and mass extinctions at the Cretaceous/Tertiary boundary: Geology, v. 19, p. 497–501.

MALMGREN, B. A., BERGGREN, W. A., AND LOHMANN, G. P., 1983, Evidence for punctuated gradualism in the Late Neogene *Globorotalia tumida* lineage of planktonic foraminifera: Paleobiology, v. 9, p. 377–389.

MATTHEWS, R. K., 1974, Dynamic Stratigraphy: Englewood Cliffs, Prentice-Hall, 370 p.

MATTHEWS, R. K., 1984, Dynamic Stratigraphy (second edition): Englewood Cliffs, Prentice-Hall, 489 p.

MAYR, E., 1982, The Growth of Biological Thought: Cambridge, Belknap Press of Harvard University Press, 974 p.

MCKINNEY, F. K., 1991, Exercises in Invertebrate Paleontology: Boston, Blackwell Scientific Publications, 272 p.

MIALL, A. D., 1984, Principles of Sedimentary Basin Analysis: New York, Springer-Verlag, 490 p.

MIALL, A. D., 1990, Principles of Sedimentary Basin Analysis (second edition): New York, Springer-Verlag, 668 p.

MIKAN, F. A. AND SWEET, W. C., 1974, Graphic correlation of key Permo-Triassic sections in Kashmir, Pakistan and Iran: North-Central Section, 8th Annual Meeting, Geological Society of America Abstracts, v. 6, p. 531.

MILLER, F. X., 1977, The graphic correlation method in biostratigraphy, in Kauffman, E. and Hazel, J., eds., Concepts and Methods of Biostratigraphy: Stroudsburg, Dowden Hutchinson and Ross, p. 165–186.

MORMISTON, A. R., 1973, Advantages and limitations of graphic correlation: Geological Society of America Abstracts, v. 5, p. 759

PRELL, W. L., IMBRIE, J., MARTINSON, D. G., MORLEY, J. J., PISIAS, N. G., SHACKLETON, N. J., AND STREETER, H. F., 1986, Graphic correlation of oxygen isotope stratigraphy application to the Late Quaternary: Paleoceanography, v. 1, p. 137–162.

PROTHERO, D. R., 1990, Interpreting the Stratigraphic Record: New York, W. H. Freeman and Company, 410 p.

RAUP, D. M. AND STANLEY, S. M., 1971, Principles of Paleontology: San Francisco, Freeman and Company, 481 p.

RAUP, D. M. AND STANLEY, S. M., 1971, Principles of Paleontology: San Francisco, Freeman and Company, 460 p.

RUDWICK, M. J. S., 1976, The Meaning of Fossils: Chicago, University of Chicago Press, 287 p.

RUDWICK, M. J. S., 1978, Charles Lyell's dream of a statistical palaeontology: Palaeontology, v. 21, p. 225–244.

SCOTT, R. W., EVETTS, M. J., AND STEIN, J. A., 1993, Are seismic/depostional sequences time units? Testing by SHADS cores and graphic correlation: *in* Offshore Technology Conference, no. 25, p. 269–276.

SCOTT, R. W., FRANKS, P. C., STEIN, J. A., BERGEN, J. A., AND EVETTS, M. J., 1994, Graphic correlation tests the synchronous Mid-Cretaceous depositional cycles: Western Interior and Gulf Coast, *in* Dolson, J. C., Hendricks, M. L., and Wescott, W. A., eds., Unconformity-related Hydrocarbons in Sedimentary Sequences: Denver, Colorado, Rocky Mountain Association of Geologists, p. 89–98.

SHAW, A. B., 1960, Quantitative fossil correlations: Geological Society of America Bulletin, v. 71, p. 1972.

SHAW, A. B., 1964, Time in Stratigraphy: New York, McGraw-Hill, 365 p.

SWEET, W. C., 1984, Graphic correlation of upper Middle and Upper Ordovician rocks, North American Midcontinent Province, U.S.A., *in* Burton, D. L., ed., Aspects of the Ordovician System: Oslo, Universitetsforlaget, Paleontological Contributions from the University of Oslo, no. 295, p. 23–35.

THAGARD, P. 1992, Conceptual Rveolutions: Princeton, Princeton University Press, 285 p.

TROTTER, W., 1942, The Collected Papers of Wilfred Trotter: London, Oxford University Press, 194.

UPSHAW, C. F., ARMSTRONG, W. E., CREATH, W. B., KIDSON, E. J., AND SANDERSON, G. A., 1974, Biostratigraphic framework of Grand Banks: American Association of Petroleum Geologists Bulletin, v. 58, p. 1124–1132.

EARLY HISTORY OF GRAPHIC CORRELATION

ALAN B. SHAW
1315 Kamira Drive, Kerrville, Texas 78028–8805

As organizer of the Burlington symposium on graphic correlation of March, 1993, Dr. Mann invited me to open this volume with a review of the birth and early history of graphic correlation.

From the Summer of 1949 through the Summer of 1955, I taught invertebrate paleontology at the University of Wyoming. While there, I used the conventional zonal approach to biostratigraphy that I had learned in school. However, when I went to work in the Denver Area office of Shell Oil Company in the Fall of 1955, I was immediately faced with a problem that conventional zonation could not solve.

At that time the big play in the Rocky Mountain region was for Mississippian oil in the Williston basin of North Dakota and adjacent Montana. Production from the basin had sparked interest in Mississippian rocks elsewhere, and Shell had recently supported Union Oil in the coring of the Union Otter Woman Morning Gun No. 1, drilled in the foreland just east of the Folded Belt in western Montana. My first assignment at Shell was to study the Morning Gun core to verify the correlations that had been made between it and the Williston basin.

The Williston basin Mississippian is naturally divisible into three parts: the upper, oil producing dolomites and evaporites of the Charles Formation, the massive limestones of the Mission Canyon Formation in the middle, and the thin bedded limestones of the Lodgepole Formation (Fig. 1).

The three major stratigraphic units, the Charles, Mission Canyon and Lodgepole Formations, had been correlated on the assumption that rock units are time-parallel with the three lowest divisions of the Mississippian Period in the Mississippi Valley, the Meramecian, Osagean and Kinderhookian, respectively.

The Morning Gun core had cut a small interval of Jurassic rock before penetrating Mississippian dolomite, which was called Sun River Dolomite after a dolomite exposed along the nearby Sun River. The limestones of the lower part of the core were called Madison Limestone. I was asked to sample the core for fossils and to confirm the correlation of the Sun River Dolomite with the presumably Meramecian Charles Formation and the Madison Limestone with the presumably Osagean Mission Canyon of the Williston basin.

In 1955, there was no workable Mississippian fossil zonation. There were the classic descriptions of the Mississippi Valley brachiopods by Weller (1914) and much older descriptions of the Kinderhookian assemblage from the "Lake Valley Beds" of New Mexico. Neither of these works had adequate information on the stratigraphic distribution of the taxa. Work by Girty (1899) gave some rudimentary stratigraphic information from Yellowstone Park, but too few taxa were identified to be generally useful. Studies by Moore (1928) and Crickmay (1955) were documented stratigraphically, but most of their taxa were not present in the faunas I was studying.

In addition to the paucity of published stratigraphic data, there was much dispute over the correlation of these widely separated areas themselves, so I was in the position of making correlations among problems rather than correlating against any established standard.

My first approach was to try to set up a zonation from the available literature. I had all of the plates in the works of Weller (1914) and Moore (1928) copied photographically. Then I cut out pictures of each taxon and pasted them on a wall-sized sheet of linen that I had divided into the standard Mississippi Valley Series. Because there was little detailed information on the position of many of the taxa within each series, the chart was only marginally useful. But it did show one thing— the Morning Gun fossils were mainly Kinderhookian with a few possibly Osagean additions. I was already in trouble!

Figure 2 illustrates the basic information that was available about the Morning Gun core. The top of the core cut a few inches of Jurassic clastics before entering Mississippian dolomite at 8960 ft. Jurassic clastics filled Mississippian solution channels, showing that the contact itself was a normal sedimentary surface and not faulted.

From 8960 to 9163 ft, the rocks were dolomitized and were correlated lithically with the Sun River Dolomite. From 9163 ft to total depth at 9559 ft, the rocks were primarily limestones and were called Madison Limestone. Three faults were also visible in the core: (1) At 9000 ft, flat dips above the fault changed to 15° dips below; the dips below the fault were probably due to fault drag and died out a short distance below, making it unnecessary to correct the stratigraphic position of fossils from the footwall block; (2) A middle fault, visible at 9094 ft, was marked by an abrupt shift in the orientation of internal structures such as joints. There was only slight change in dip below the fault, but the strike of the bedding was very different;

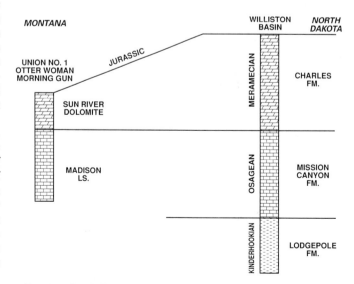

FIG. 1.—Correlations of Mississippian rock units in Early 1955. At the height of the Mississippian play in Montana and North Dakota, the major rock units were correlated on a gross lithologic basis. It was assumed that these rock units were time-parallel so that the names of the lower three Stages recognized in the Mississippi Valley could be transferred to the Williston basin on a one-to-one basis; fossil evidence was not considered in these assignments.

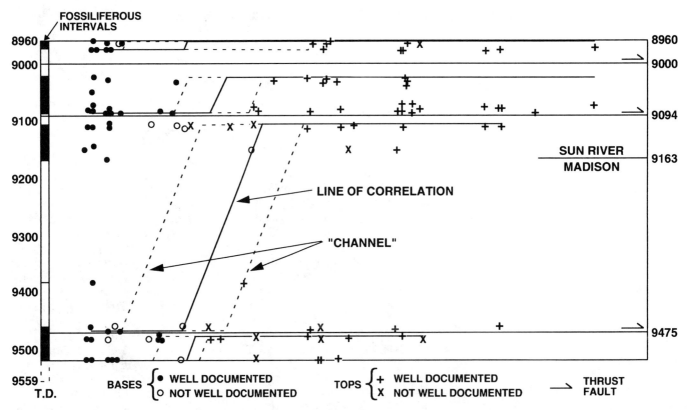

FIG. 2.—Graph of Union Morning Gun #1 well against a composite standard developed in Montana, North and South Dakota and Iowa. The Y-axis of the graph is a schematic representation of the Union Morning Gun #1 well. The entire Mississippian section was cored, but only those intervals shown in solid black yielded identifiable fossils. The dotted interval at the base of the diagram indicates that the actual total depth (T.D.) of the core was at 9559 ft, but that the true stratigraphic position of the bottom of the core had to be corrected for steep dip in the lowest block to 9538 ft as explained in the text. The X-axis is based on a current composite standard; the scale is uniform, but the numerical values are irrelevant to this presentation. The solid lines of correlation are those deemed most reliable, but the wide channels are an effect of the paucity of control in the well; the channels are delimited by well documented taxa whose range in the composite standard is unlikely to be extended by the scattered occurrences in the core. Faults shown at 9000, 9094 and 9475 were observed in the core and are not inferred from fossil control. The secondary Sun River dolomite (8960–9163 ft) cuts across the two upper fault planes. Below 9163 ft the core is predominantly limestone.

and (3) The lower fault lay at 9475 ft and was strikingly apparent by a change from flat beds above the plane to 50° dips below. The high angle of dip, which continued to the bottom of the core, made it necessary to adjust the stratigraphic position of fossils recovered from the bottom block. The true stratigraphic position of the total depth (T.D.) at 9559 ft was 9538 ft after correction.

These stratigraphic details were all known before I sampled the core. I did not discover any of them. The faults had been recognized but were interpreted as due to minor movements of little significance because of the lithic correlations between the Sun River-Madison sequence in the core to the Charles-Mission Canyon sequence of the Williston basin.

By the time I had identified such fossils as I could find, I found myself in an uncomfortable position. I had been expected to confirm the Meramecian age of the Sun River Dolomite, but the very first fossil I found at the top of the core (8961.5 ft) was the Kinderhookian brachiopod *Punctospirifer solidirostris*. However, the unfamiliar paleontologic evidence, set against the generally accepted interpretation, was not persuasive.

Lacking acceptance of the fossil evidence, I hoped instead that I could demonstrate that the Sun River Dolomite was a secondary dolomite, unlike the evaporitic dolomites of the Charles Formation. I reasoned that if the four fault blocks in the core were of essentially the same age, repeated by thrusting, then I thought it would be obvious that the only way the upper blocks could be dolomite while the lower blocks of the same rock were still limestone was if dolomitization had occurred *after* the faulting. But this solution threw me back on the fact that I had no sufficiently refined zonation within the Kinderhookian section against which I could correlate the individual blocks. The solution had to be internal to the core.

I surmised that where the same fossils were present in different blocks, their physical separation might reflect in a general way the stratigraphic throw of the faults so I calculated the separation in feet of all pairs of taxa present in more than one block. Unfortunately, the small number of common occurrences plus the differing thicknesses of the blocks gave me a jumble of numbers that I could not interpret.

To eliminate the effects of the different thicknesses of the four fault blocks and to assist me in seeing which occurrences were from comparable stratigraphic positions, I thought of plotting the common taxa from each block against the other three on X-Y graphs. The slope of any line on such plots would necessarily be 45° because the same rock was present in each block. The resulting graphs were not all that impressive to anyone except an enthusiast, but they led me naturally to the next idea of plotting separate sections against one another and ultimately to combining data from many sections by projecting them arithmetically into a single section, creating what became a composite standard.

During 1956 and 1957, with the assistance of Del Davis, Pat Tidwell and Delia Wilson, I had studied other surface collections from Montana and cores from the Williston basin. By the beginning of 1958, I had enough documented data so that on February 4 (according to my diary) I could plot the first of what would be recognizable today as a true graphic correlation between two surface sections in Montana.

All of this work did not take place as tidily as this recital may make it seem. I had other business to occupy me besides the Morning Gun core, which had become less interesting to Exploration as the Mississippian play ebbed from western Montana, but I went on to test the new method with fossil data from the Cambrian, Ordovician, Pennsylvanian and Upper Cretaceous sections. I also plotted Upper Cretaceous bentonites. The graphs of Pennsylvanian fusulinids and Upper Cretaceous bentonites were important because they showed that, where control is good enough, graphs are rectilinear and not doglegged.

Returning to Figure 2, the Y-axis is a plot of the Morning Gun core against a current composite standard (X-axis) built from surface sections and cores from Montana, North Dakota, South Dakota and Iowa. It is not a tight graph because there are not enough well documented taxa in the core, but it does confirm that there was enough movement along the three thrust faults to repeat what is basically the same section of rock in each block. The solid lines are my best estimate of the correlation of each block; the dashed lines mark the channels around those lines of correlation. The channels are defined by the most firmly established taxa in the composite standard; it seems unlikely that the rare specimens in the Morning Gun core are likely to extend the ranges of these taxa that have been documented extensively in other sections. But this is not the place to argue the correlation of the Morning Gun core. The importance of the Morning Gun is that it gave rise to the whole idea of plotting fossil occurrences on a graph.

To return to history, I had stirred up trouble by calling into question both the correlation of the Sun River Dolomite with the Charles Formation and the lithic similarity of the two dolomites. My conclusion that the faulting was important also irritated the geophysicists, who could not see it.

Following the first graphic correlation of surface sections in February, 1958, I added more surface sections, and by April, 1958, I had drawn a cross section from central Montana to the Folded Belt of western Montana. This showed that the Lodgepole and Mission Canyon Formations were diachronous. At that time there were no published studies that had reached such a conclusion so the implications of the cross section were dismissed as too bizarre to be credible. An acrimonious dispute arose with those who rejected the idea that rock units (or, by extension, electric log picks) could be diachronous.

In late 1958, I presented a talk on the theory of graphing to the Denver paleontology group. That group was understandably more receptive. The paleontologists did not have difficulty with the idea that fossils could define stratigraphic time, but some saw no advantage in being able to make more accurate correlations. When the presentation was over one of my academic colleagues simply said, "I can't see any reason to try it." In 1988, I attended Ray Christopher's SEPM seminar on methods of correlation, and near the end of the discussion of graphing, one of the participants said, "I can't see any reason to try it." In 30 years not even the words had changed.

Most surprising was the frequently voiced objection that graphing of fossil data simply could not be done. It was as though I was breaking some sort of law by even trying it. I never really understood this objection, but it was sufficiently widespread to cause me, when writing *Time in Stratigraphy*, to adopt the device of saying that I was plotting a section against *itself* in my illustrations to sidestep automatic rejection of the argument if I had said I was plotting one section against another.

A second common objection to graphing was that it would be impossible to discover the earliest and latest specimens — which the critics felt would mean that the range limits in the composite standard would never be correct and would therefore never provide points against which to plot. It is ironic nowadays to see graphs in which the line of correlation zigzags from one fossil limit to another, implying that only the limiting occurrences have been found.

One reaction took me completely by surprise. I wrote to a friend of many years, whom I had accompanied in the field and whose stratigraphic work I knew to be of the best. I asked him to supply me with data from a section that he was still working on but that I had not seen. I already had a composite standard for that part of the section, and I offered to try to predict from it what fossils he would find at levels that he had not yet sampled. I thought this would be comparatively easy to do and would be a good test of the strength of the composite standard idea. In due course he sent me a set of data, but after a long effort I could not make any sense of it. Years later, he admitted to me that he had deliberately falsified the data to make the experiment fail.

Plotting the work of others also turned out to be hazardous; my relations with some authors turned abruptly frosty when the plots showed overlooked faulting or mismeasured sections. And when graphs of environmentally controlled assemblages that had been correlated as time equivalents showed that these assemblages were really of different ages the graphs were not well received.

Meanwhile, back at the office, the debate over the existence of diachronism and the validity of fossil correlations continued, and it began to upset management. To resolve the issue, they asked a senior research paleontologist to review my work. I met with him in the Fall of 1959 and presented all of the existing examples. As soon as I had finished, he said to me that since rocks are time-parallel I had simply proved that fossils are diachronous and that none of the correlations had any meaning. Hearing this, management decided that paleontology had no place in the Exploration effort and told me to shut down all further work. Early in 1960 my father died, and I took a leave of absence to settle his affairs. Shortly after I returned to work I was told to resign.

After my dismissal in October, 1960, I made two decisions. First, I would write an exposition of my ideas on diachronism and the graphic method. Second, I would not spend the rest of my career trying to get unwilling colleagues to accept either idea. I wrote the first draft of *Time in Stratigraphy* during the months of unemployment before joining Pan American Petroleum (now Amoco Production) in 1961, but except for the book itself, I have not written about it further. I have, of course, delivered many talks and demonstrations as part of my jobs with Amoco, but I have not tried to convince the profession at large.

I knew from my own work and from the experiences of the few who would try it for themselves that graphic correlation was much more accurate than other methods. But I realized, too, that it would not be widely accepted until a new generation of biostratigraphers came along. Keith Mann's symposium in Burlington in March, 1993, was held 35 years and one month after the first successful graphic correlation and 29 years after the appearance of *Time in Stratigraphy* — very close to the traditional generation of thirty years.

After I joined Pan American Research in July, 1962, I was given the job of organizing a research program in paleontology. Pan American had long had a functioning program of Gulf Coast biostratigraphy based on Foraminifera. It had been set up in the 'thirties' by Julius Garrett, one of the true pioneers of Gulf Coast paleontology. My job was made easier by the existence of that program and the attitude of a management that accepted paleontologic input routinely. At the time, too, Pan American already had under way a fledgling program in the then-new field of palynology.

My job was to see that all forms of paleontology that are relevant to petroleum exploration became routinely operational. Palynology was at first too expensive and had to be made cost effective, but it was important to introduce it as a balance to the exclusive reliance on Foraminifera. Invertebrate fossils were needed, too, because their significantly different environmental and taphonomic meanings provide a useful check on the more abundant microfossils. My long term goal was to get paleontologists and palynologists out into exploration offices, using adequate bodies of data based on composite standards, to become a routine part of the Company's exploration effort.

My approach in Pan American was to take data from those who were doing the research and to make the graphs myself. If an individual showed interest in the method I would encourage him. If he did not, and continued to use other less accurate techniques, I could at least keep track of what he was doing and prevent any egregious errors from being sent out to Operations.

By the time I moved from Research to Operations in 1968, there was a sufficient acceptance of the value of graphic correlation so that I felt it would survive. Not all of the staff used graphs, but composite standards had been started and enough was being done so that I was satisfied that the program was viable.

I leave to others in this volume to show where those early efforts have led, but I would like to mention in closing a few in Amoco who early on began to use graphs and who contributed greatly to making graphic correlation an acceptable part of Amoco's effort. When I joined Amoco Research the first two plays that involved paleontology were those in the Tertiary of the Cook Inlet, Alaska, and the Devonian Swan Hills reef play in Alberta. The Cook Inlet correlations were based on palynomorphs and were in the hands of Don Engelhardt. The graphs correlating the fiordlike fillings of Cook Inlet with the much thinner onshore sections put graphic correlation to as severe a first test as could be imagined. Amoco's first composite standard was based on a section sampled for palynomorphs from Kaliakh Mountain, Alaska. The original palynology-based standard was later augmented by the invertebrate studies of Ken Ciriacks and Allen Ormiston.

The Devonian Swan Hills reefs presented an even more intractable problem in attempting to establish time planes from the normally fossiliferous strata outside the reefs into the specialized biota dominated by stromatoporoids behind the reef. This effort occupied Allen Ormiston for several years, but he ultimately solved it, producing the first formal Amoco report to operations on the use of a composite standard.

The late Charlie Upshaw was also a pioneer in Amoco, although I never actually saw him make a graph. His contribution was to dedicate himself to the education of Amoco personnel about the functions of paleontology and its role in exploration.

There were several others who were in Pan American Research in those early days when graphing was new, but most of them moved on to other things. One who moved but kept his interest in graphing was Gil Klapper. His influence through his teaching at the University of Iowa has probably been greater than my own.

And, one of the early recruits, F. X. Miller, came to us from geophysics. After serving a stint in research, he was one of the first palynologists to move into Operations. Ultimately, he returned to research to manage the Amoco database.

There were others throughout Amoco who became involved with graphing in later years, but this history covers only the early efforts before 1968, when I moved from research into exploration. To all of those who have assisted in making graphic correlations a routine part of Amoco's business, I am most truly grateful. My goal was from the beginning to make biostratigraphic control available to stratigraphers on a scale at least a refined as their subdivisions of the rock column, and that appears to have happened .

Years ago I read a comment by an author whose name I have forgotten that ideas, like living organisms, go through a period of "natural selection" after they are born. Like a new organism, a new idea departs from the familiar and therefore does not "fit in." It is therefore forced to overcome rejection and must fight to make a niche that has not existed before. Some, like Sorby's facies, Wegener's continental drift and Ma's daily growth rings in Paleozoic corals must endure decades of rejection, while other plausible but erroneous ideas, like Wernerism and layer cake stratigraphy, persist and are accepted long after they should have been laid to rest. I am most grateful to Dr. Mann, who has made it possible for me to see that my simple idea has survived its period of selection and appears to be prospering.

Disclaimer: The views expressed in this paper are solely those of the author and do not necessarily reflect the opinions of the editors, SEPM or its members.

REFERENCES

CRICKMAY, C. H., 1955, The Minnewanka Section of the Mississippian: Calgary, Imperial Oil Company, 15 p.

GIRTY, G. H., 1899, Devonian and Carboniferous Fossils, *in* Hague, A., Iddings, J. P., Weed, W. H., Walcott, C. D., Girty, G. H., Stanton, T. W., and Knowlton, F. H., Geology of the Yellowstone National Park: Part II, Descriptive Geology, Petrography, and Paleontology: Washington, D. C., United States Geological Survey Monograph 32, Part 2, p. 479–599.

MOORE, R. C., 1928, Early Mississippian formations in Missouri: Rolla, Missouri Bureau of Geology and Mines, second series, v. 21, 283 p.

SHAW, A. B., 1964, Time in Stratigraphy, McGraw-Hill, 365 p.

WELLER, S., 1914, Mississippian Brachiopoda of the Mississippi Valley basin, Illinois Geolological Survey Monograph 1

PART II
THE GRAPHIC CORRELATION METHOD

GRAPHIC CORRELATION AND COMPOSITE STANDARD DATABASES AS TOOLS FOR THE EXPLORATION BIOSTRATIGRAPHER

JOHN L. CARNEY AND ROBERT W. PIERCE

Exploration and Production Technology Group, Amoco Corporation, 501 Westlake Park Boulevard, Post Office Box 3092, Houston, Texas 77253–3092

ABSTRACT: Graphic correlation, based on a Cartesian coordinate system, is a tool which is used to derive precise, consistent, and highly resolved biostratigraphic correlations among wells and/or measured outcrop sections thereby reducing technical risk in hydrocarbon exploration. Non-paleontological data may also be correlated using this method. A database of fossil ranges, the composite standard, is created by graphic correlation. Graphic correlation and composite standard construction are readily handled by computer software. Large quantities of biostratigraphic data can be compiled, organized, retrieved, and applied to future correlation studies. The enhanced resolution of this method facilitates recognition of unconformities and condensed sections, resulting in interpretations well suited for sequence stratigraphic basin studies. Used in conjunction with log and seismic data, this technique improves the explorationist's ability to identify and exploit subtle stratigraphic traps. It can also facilitate recognition of ancestral structural features masked by unconformities and overlying flat lying beds.

The technique is a mainstay in Amoco's approach to biostratigraphy. It is a powerful tool for evaluating and interpreting paleontological data generated in-house as well as that acquired from vendors, through data trades, in acreage bid round data packages purchased from foreign governments, and from the literature. This method does not replace traditional paleontological techniques; it is a tool for data handling and display designed to enhance traditional methods.

SCOPE OF THIS PAPER

Although the idea of correlating two stratigraphic sections by plotting each one on the perpendicular axes of a two-dimensional graph is elegant in its simplicity, the day-to-day application of this method requires familiarity with a number of concepts whose nature and interrelationships are not necessarily immediately apparent. The purpose of this paper is to introduce the terminology and techniques of graphic correlation. To provide the reader with a sense of graphic correlation in action, several basic principles will be illustrated using a series of simplified examples from marine Paleogene units of the northern Gulf of Mexico. The present paper is not intended to be an exhaustive treatment of graphic correlation. The most in-depth description of all aspects of this approach to biostratigraphy is still Shaw's original work (Shaw, 1964). Miller (1977) is a helpful reference as well.

Various practitioners of graphic correlation assign the two graphic axes in different ways: the composite standard database against which less well known sections will be plotted can be placed either on the X-axis or the Y-axis. This paper follows the Amoco convention of placing the composite standard database on the X-axis and the new stratigraphic section to be evaluated in terms of that database on the Y-axis. The opposite layout of axes is equally effective and is used widely.

INTRODUCTION TO GRAPHIC CORRELATION AND COMPOSITE STANDARD DATABASES

General Description

Graphic correlation is a process; the composite standard (CS) is the database of fossil ranges that the geologist creates using that process. Graphic correlation, based on a Cartesian coordinate system, was introduced by Alan Shaw. As Shaw (1964) indicated, it is important to bear in mind that points of correlation between any two time-equivalent stratigraphic sections (in wells or outcrops) exist in nature. It is the job of the paleontologist to recognize them. Graphic correlation is simply a means of recognizing these points which allows biostratigraphers to make their interpretations more consistently, in a more unbiased mode, and based on a larger number of individual pieces of evidence. This method deals primarily, though not exclusively, with fossil range data (tops and bases), which, as Shaw (1964) argued convincingly, provide the best available means of paleontological correlation. Other types of temporally distributed data which can be related to a measured vertical scale can also be used. Radiometric ages, isotope stages, magnetic reversal patterns, electric log correlations, and seismic reflectors which can be tied to wells are all commonly graphed with paleontological data.

Graphic correlation may be used in two ways. The observed ranges of fossils from two stratigraphic sections of similar age may be compared to each other for the purpose of recognizing correlations between them (Fig. 1, left side). Also, the observed ranges from a single stratigraphic section may be compared to a database of composited fossil ranges scaled in chronostratigraphic units (Fig. 1, right side). By means of the latter type of comparison, the section can be calibrated to the scale of the database. Once two or more stratigraphic sections have been so calibrated, correlations among them will be obvious.

To effect both types of comparisons, the fossil range data from one stratigraphic section (well or outcrop) are plotted parallel to one axis of the graph (generally the Y-axis), and the fossil range data from another section or from a composite database of ranges are plotted parallel to the other axis (generally the X-axis). The range tops and bases of fossils in common between the two axes are then projected horizontally from the Y-axis and vertically from the X-axis and cross plotted in the field of the graph, presenting the paleontologist with all of the possible fossil range correlation points between the two axes. Some of these *potential* points of correlation will be useful; some will not. Based on conventional paleontological skills and experience and on an understanding of the local geology, the interpreter selects the cross-plotted points which are most reasonable and designates them as *points of correlation* between the two axes. Presented with an array of such points of correlation on the graph, the paleontologist can then finalize the interpretation by drawing a *line of correlation* (LOC) based on these points (Fig. 1). Again, the interpreted LOC represents the paleontologist's best effort at *discovering* and *expressing* the relationship which *naturally exists* between two stratigraphic sections or between a stratigraphic section and the database.

It is important to stress that graphic correlation is not a "Black Box" which accepts data, digests it, and extrudes an-

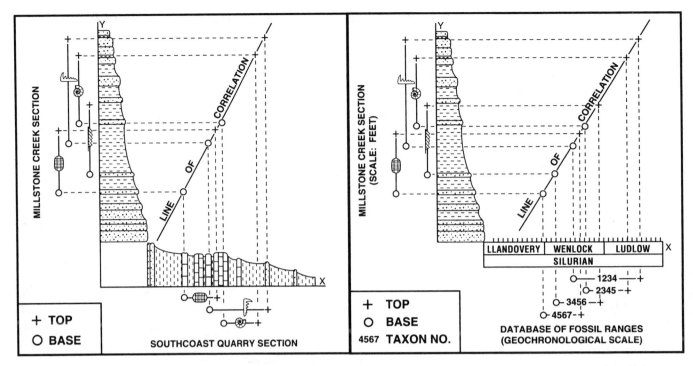

FIG. 1.—Graphic correlation can be used to compare two stratigraphic sections (left) and to compare a single stratigraphic section to a database (right).

swers. It is, rather, an excellent system for compiling, storing, organizing, and retrieving large volumes of data and for displaying those data in such a way that the paleontologist can make the most informed decisions possible as to the biostratigraphic interpretation of a well or an outcrop section. All factors the biostratigrapher considers in using traditional correlation methods are equally important when using graphic correlation. Faulting, unconformable relationships, reworking, the influence of shifting environments, varying degrees of fossil preservation, the redistribution of death assemblages, and sample quality all must be taken into account. Because this method allows one to interpret correlations based on the entire observed fossil assemblage rather than just on certain "index" species, it makes the effects of these often confusing factors easier to recognize.

A CS can potentially be enhanced each time it is used to evaluate data from a new, at least partially coeval stratigraphic section. Each additional section can provide new insights into the worldwide ranges of fossils already in the database, can add new taxa, and can extend the stratigraphic coverage of the database. Eventually, after many iterations of application and refinement, a *mature* CS becomes a continuous geochronologic scale expressed in terms of an ordered sequence of fossil bases and tops. Thus, the process of comparing fossil range data from a given stratigraphic section to a composite standard database of fossil ranges by graphic correlation can be thought of as calibrating rock thickness to a geochronologic scale. The importance of this is that, if there are significant chronostratigraphic intervals not represented by strata or represented by a very small thickness of strata in the section being evaluated, horizontal offsets in the LOC make this obvious (Fig. 2). This enhanced ability to *identify* and *measure* stratigraphic discontinuities (hiatus or condensed section) makes the interpretations

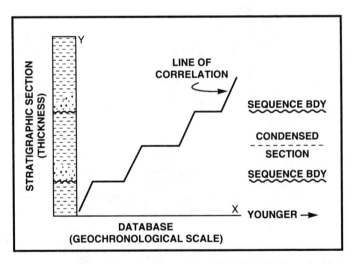

FIG. 2.—Graphic correlation interpretations of stratigraphic discontinuities may be applied directly to sequence stratigraphic studies.

from graphic correlation particularly well suited to sequence stratigraphic studies.

Advantages of the Graphic Correlation and Composite Standard Approach

There are a number of significant advantages to biostratigraphic correlation based on graphic correlation and CSs. First, the cross-plotted points on the graph provide easy visual comparison of all fossil range data in common between two stratigraphic sections. All *potential* points of correlation are visible.

If reevaluation and revision of correlation interpretations between two sections is warranted, the points originally honored as well as those originally ignored are available for easy comparison in light of new data or further insights into the geology of the area.

Second, graphic correlation is an excellent computerized bookkeeping system. During the CS building process, range tops and bases from any number of wells and/or outcrops can be compared to each other graphically to establish a stable, consistent database. The stratigraphic ranges of tens of thousands of fossil taxa can thus be compiled, refined, stored and retrieved for reference when needed to evaluate new stratigraphic sections.

Third, using a mature CS against which to graphically correlate increases the likelihood that any new section will be able to be evaluated successfully. The composited ranges in the database eventually approach the true worldwide ranges of species. Also, as the ranges of more and more fossils, representing a diversity of ages and a variety of depositional environments, are added to the database, it becomes an increasingly flexible and universal correlation tool. Regardless of the age or depositional environment of the new (unknown) stratigraphic section, it will likely contain bases and tops in common with the database thus making graphic correlation possible.

Fourth, the process of graphic correlation, the interpretations derived from that process (mathematically definable lines of correlation), and the CSs created, all lend themselves to being handled by computer software. Indeed, considering the volume of data involved as a database matures, management of by computer becomes a virtual necessity.

Lastly, graphic correlation greatly streamlines the work of the biostratigrapher. Traditional methods may require the comparison of fossil range data from a given stratigraphic section to that from several other sections and/or to range data from a number of reference range charts representing various fossil groups and/or the work of various authors. This is an often confusing, multi-step process. On the other hand, the evaluation of the same section by comparing it, via graphic correlation, to a mature CS is a single-step process which presents all of the evidence for correlation from all fossil groups on a single piece of paper or on a single video display.

The Components of the Process

The process of graphic correlation, its use to create a CS database, and the application of that database to the biostratigraphic evaluation of stratigraphic sections may be broken down into several fundamental components. Each component is described briefly and, where appropriate, later developed more fully in this paper.

Raw Data.—

The first step is to obtain suitable raw data. Such data necessarily consist of a record of fossil occurrences observed at measured intervals over a vertical span of sedimentary strata. Typically, a dataset contains a listing of fossils identified in each of a series of samples. A checklist of fossils from a well or measured outcrop section is the easiest type of data to use, but sample-by-sample listings of observed fossil assemblages (typical in oil company files) are also acceptable. Literature descriptions of measured outcrop sections which note the fossils occurring at successive sampling points may also be used. The key requirement of the raw data is that there must be a measured vertical scale to which the samples are related. Sample spacing directly controls the degree of resolution and precision obtainable. Typically, sampling intervals in ditch cuttings samples range from 10 to 60 ft. In a core or outcrop, sampling may be carried out to any desired degree of detail.

Data Capture.—

Once obtained, the raw data must be entered into a computer file, one sampling interval at a time. There are various software packages used to accomplish this (Bug-In, BugWare, and various proprietary software packages).

Fossil Range Charts.—

The creation of fossil range charts for each well or measured outcrop section is generally accomplished by the same software used to build the original computer file. There are, however, additional software systems (e.g., RagWare) which can do this. These software packages extract from the sample by sample records of fossil occurrence in a well or outcrop the lowest (oldest) and the highest (youngest) occurrence for each species. These extracted ranges are the data required for input into graphic correlation software and the range charts are themselves useful in hardcopy form.

Graphic Correlation.—

The process of graphic correlation compares the fossil range data from a given well or outcrop either to the fossil range data from another well or outcrop, or to the fossil ranges in a CS. Geologists at Amoco use a proprietary UNIX-based software system specifically designed for graphic correlation. Certain P.C. software packages, both commercial (GraphCor by Hood Software) and public domain (Shaw Stack and Stratigraphica by Norman MacLeod), also perform graphic correlation.

Composite Standard Building and Refinement.—

Graphic correlation of a stratigraphic section against a CS is a two-stage process. First, a LOC is established using tops and bases of fossils common to both the stratigraphic section and the CS. Second, tops and bases of fossils which are present in the stratigraphic section, but not in the CS, can be projected into the CS using the LOC derived in the first stage. The first stage provides a chronostratigraphic interpretation and calibration of the stratigraphic section; the second stage serves to enrich the CS.

It is important to consider whether it is indeed advisable to proceed beyond the first stage. If the data from the well or outcrop being analyzed are thought to be of questionable quality, it is better to stop the process after the first stage. It is important to protect the developing CS from contamination by poor quality fossil range data. Thus, in the case of certain stratigraphic sections, the full benefits of graphic correlation may be derived (Stage 1), but the fossil ranges from those sections need not be added to the CS (Stage 2). A CS should be built using only the highest quality data available.

Thus, by successive comparison of fossil range data from various wells and/or outcrops to each other, a CS is developed. This database may then be used in later correlation studies.

Through continued use, the database is enhanced; additional species are added to the database; an improved understanding of individual fossil ranges is obtained; and the stratigraphic coverage of the database is extended. The construction and refinement of a database is done using the same software which performs the graphic correlation operation.

Application and Further Refinement.—

Once constructed, a CS becomes a powerful tool for evaluating fossil range data from wells or measured outcrop sections. Each new stratigraphic section can be compared to and calibrated in terms of the database. Once calibrated in this manner, each section can be correlated to other calibrated sections precisely and consistently. A CS never becomes truly finished or complete; with each use, there is the potential for further refinement and expansion.

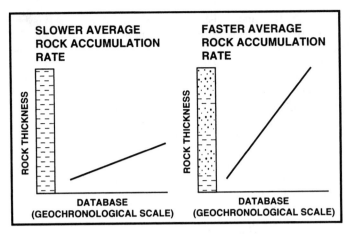

FIG. 3.—The slope of the LOC can be used as an indication of relative rate of rock accumulation.

FUNDAMENTAL TERMS AND CONCEPTS

Rate of Rock Accumulation

It is tempting to consider measured stratigraphic thickness versus the magnitude of the chronostratigraphic interval represented by that thickness as *rate of sedimentation*. However, a direct relationship between the original rate of sedimentation and the observable thickness of rock cannot easily be made because of the effects of geological processes (e.g., compaction, redistribution, partial removal, diagenesis, and metamorphism) acting on those sediments before and during lithification. Miller (1977, p.166) states "In a stratigraphic section, we observe only the sediment that was preserved and lithified, not the total amount of sediment that was originally deposited. The immeasurable effects of processes such as the original rate of sedimentation, subaerial and subaqueous erosion prior to burial, and compaction after burial are compensated for in the thickness of rock that we observe and measure. Shaw defines the (rock thickness that is actually observed in a stratigraphic section versus the chronostratigraphic interval represented) as rate of rock accumulation as opposed to rate of sedimentation . . ."

The LOC represents a comparison of observed rock thickness in two stratigraphic sections, or a comparison of observed rock thickness in one stratigraphic section to a geochronologic scale expressed in terms of the fossil ranges observed in many sections (Fig. 1). Thus, there exists a relationship between the slope of the LOC and the average rate at which rock accumulated in the stratigraphic sections involved. In the case of graphing two stratigraphic sections against each other, if the average rock accumulation rate was identical at both localities, the LOC representing any interval of stratigraphic age in common between the two sections would lie at 45° to the X-axis. Any difference in average rock accumulation rate between the two sections would result in a LOC at an angle greater or less than 45°. Likewise, if the CS, representing a scaled sequence of chronostratigraphic events, is plotted on the X-axis, variations in the inclination of the lines of correlation, derived as one stratigraphic section after another is graphed against the CS, provide some indication of relative average rates of rock accumulation among the various localities (Fig. 3).

The history of any single stratigraphic section represents deposition in varying environments and undoubtedly includes many changes in rate of sedimentation. Further, it is reasonable to assume that after a body of sediment is deposited it may be partially or completely removed before any subsequent sediments are deposited. Thus, there are many small intervals of time not represented by rock. If reasonably reliable interpretations can be made as to the magnitude of the time interval represented by a given rock thickness, an average rate of rock accumulation can be calculated. It is this average rate of rock accumulation that is represented by the long, straight LOC. A LOC representing the details of changing rock accumulation rates, if the required resolution were available, would be broken into many small "doglegs" and would contain many small horizontal terraces.

Calculations of average rock accumulation rate provide a bottom-line figure that encompasses the additions and subtractions of all geological processes acting on a body of sediment from the time of deposition to the time of lithification and later diagenesis. From this bottom-line figure and from an understanding of observable present-day depositional processes, one can attempt to make reasonable inferences on average sedimentation rates over limited stratigraphic intervals. Rates of sedimentation, however, cannot be measured in the rock record and there are many obstacles in the path of calculating them for any significant interval of geologic time. Calculations of relative *rate of rock accumulation* can be used to help *estimate* relative *rate of sedimentation*, but it is important to bear in mind the differences between the two and the shortcomings of such estimates.

Fossil Ranges

As mentioned above, the observable stratigraphic ranges of fossils are the primary data used in the graphic correlation process. For an extinct species, the base and top which can be determined in a measured outcrop section or well bore (disregarding for the moment downhole caving of cuttings samples) define its *local stratigraphic range*. The local stratigraphic range represents an interval of geologic time which is less than or equal to the *total stratigraphic range* of that species. The total stratigraphic range of an extinct species is derived from the study of many stratigraphic sections distributed over a broad geographic area and may be defined as the interval of geologic time between the oldest *observed* appearance of that species worldwide and its *observed* worldwide disappearance. The stratigraphically oldest *documented* appearance of a species may be referred to as its *first appearance datum* (FAD); the *documented* disappearance of that species may be referred to as its *last appearance datum* (LAD). Thus, the FAD and LAD define the *total stratigraphic range,* also referred to as the *composite standard range.*

The *true total stratigraphic range* of an extinct species is the interval of geologic time between the first evolutionary occurrence of that species and its *worldwide* extinction. The *true total stratigraphic range* is a theoretical concept and may or may not be observable; all that can be said about its extent is that it is greater than or equal to the total stratigraphic range.

It is a basic premise of stratigraphic paleontology that the base and top of the true total stratigraphic range of a given extinct species represent unique events in earth history. It is reasoned that the FAD and LAD of that species in a mature CS database approximate the base and top of its true total stratigraphic range as closely as is possible within the limits of the paleontologist's ability to observe. Thus, in graphic correlation theory, the total stratigraphic range or composite standard range of a fossil, defined by its FAD and LAD, is the best available tool for establishing time stratigraphic correlations between geographic localities, no matter the distance separating them.

The stratigraphic record is incomplete. At any given locality it is to be expected that significant intervals of geological time are not accounted for by the strata observed. Also, intervals of unrecorded time differ in geological age and magnitude from section to section. If the base or top of the true total stratigraphic range of a given taxon falls within a hiatus at a given locality, an observed lowest occurrence or an observed highest occurrence which does not represent the true total stratigraphic range will result.

Sediments at any given locality may record, completely or in part, deposition in an environment which the species of interest could not tolerate. Thus, even though the time represented by those sediments overlaps or includes the true total stratigraphic range of that species, observations regarding its range at that locality will be distorted; a base younger than the true total stratigraphic range base and/or a top older than the true total stratigraphic range top will be recorded.

Even in a locality where the stratigraphic record is essentially complete and where the environment of deposition was suitable, fossil preservation and/or sample quality may limit observations. A *local stratigraphic range* that is abbreviated as compared to the true total stratigraphic range may still be observed.

The foregoing discussion paints a bleak picture of ever being able to arrive at an approximation of the true total stratigraphic ranges of fossil taxa. Indeed, such considerations should underscore the need for caution in assuming the synchroneity of "tops" or "bases" from one locality to another. It was in order to overcome such difficulties that Shaw (1964) developed the graphic correlation method.

Graphic correlation provides a systematic method of compositing information from many localities, adding gradually to a CS to develop a composite standard range for each species, and documenting each step along the way.

Graphic correlation software creates a table of occurrence records which lists, for every species included in a CS database, all localities at which each species was ever observed. For each observed occurrence of each species, the table indicates the chronostratigraphic extent of the local stratigraphic range. This is possible because the localities have been calibrated to the chronostratigraphic scale of the CS by means of their respective lines of correlation. If the local stratigraphic range of a given extinct species at a new locality is found to have a younger (but not reworked) top than had been previously recorded, the CS range of that species can be extended in the database so that its LAD moves to a stratigraphically younger position. Thus, by adding information one locality at a time, the LAD of a species can be extended in the database increment by increment. Eventually an LAD will be defined which approximates the true total stratigraphic range top of the species. By the same process the FAD can be gradually extended to older and older levels in the database, eventually arriving at a level approximating its true total stratigraphic range base.

The composite standard base (FAD) and top (LAD) of a species in the database define its *composite standard range,* also referred to as its total stratigraphic range. The CS base for a given species is the oldest local base seen so far; its CS top is the youngest local top seen so far. Thus a CS range for a given fossil may be thought of as the sum of all observed *local stratigraphic ranges* of that fossil. In the early stages of compositing data, tops and bases may be extended significantly as new localities are graphed. Later, as the database matures through the addition of data from many broadly distributed localities, fossil range extensions become smaller and the range end points eventually stabilize. This suggests that the observable CS range is approaching the theoretical *true total stratigraphic range.* With use, the CS becomes a powerful tool for objectively evaluating the local stratigraphic ranges observed in new localities.

Stratigraphic Expansion of a Composite Standard Database

In addition to providing a means of extending CS ranges, graphic correlation also allows the gradual expansion of the CS database into progressively older and younger stratigraphic intervals.

To begin constructing a CS database, the available data must first be inventoried. These data will be in the form of local stratigraphic range observations from a number of localities. The process is started by selecting a *standard reference section* (SRS): the single best stratigraphic section available. Assuming that the sample quality and fossil abundance in all sections is acceptable, the criteria for choosing the SRS include the following: (1) it should be the most complete section (i.e., the section with the least likelihood of containing faults or significant unconformities); (2) it should be the section representing the longest estimated chronostratigraphic interval; (3) it should be a section made up of sediments deposited in paleoenvironments suitable for the fossils to be used; and (4) it should be the section which has been sampled in the greatest detail. The SRS will be placed on the X-axis of the first graphic correlation plot and forms the nucleus of the CS.

Figure 4 (left side) shows five hypothetical outcrop sections from localities A–E. They have been measured and sampled; local stratigraphic range data have been recorded for each. A hypothetical succession of stratigraphic units, 1 (oldest) through 10 (youngest), represents the total, overlapping chronostrati-

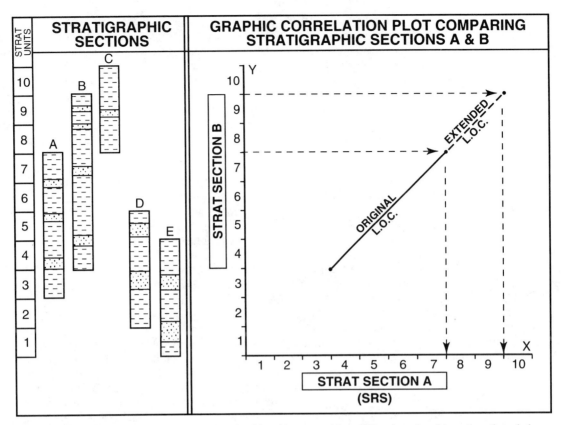

Fig. 4.—Extension of the stratigraphic coverage of the database is achieved by compositing additional stratigraphic sections through the process of graphic correlation.

graphic extent of sections A–E. The right side of Figure 4 shows the first in a series of graphs. This graph illustrates the procedure for stepwise incremental building and extension of a CS to gradually encompass additional stratigraphic intervals.

Locality A was chosen to be the standard reference section. This section represents stratigraphic units 3 through 7. In the first graphic correlation plot, Locality A is placed on the X-axis and Locality B on the Y-axis. Localities A and B are coeval over stratigraphic intervals 4 through 7, thus a LOC may be derived which expresses the correlative relationship between these two localities over that interval ("Original L.O.C."). Locality B, however, contains additional local stratigraphic range data from intervals 8 and 9. It is desirable to capture that information so that the database will have utility over a greater stratigraphic interval than represented by section A alone. The LOC between Localities A and B represents a comparison of average rates of rock accumulation between the two sections. If, through an understanding of the local geology, the safe assumption can be made that similar average rates of rock accumulation would likely have persisted, the LOC can be extended ("Extended L.O.C.") to express a projected correlative relationship between the upper part of the section at Locality B and the section not represented above the exposure at Locality A. This extended LOC allows the projection of the additional information on the Y-axis into the SRS on the X-axis (Fig. 4, right side). The concept of projecting points of correlation from one axis to the other is further discussed in a later section of this paper. By this process, the stratigraphic interval covered by the SRS has been extended and now includes stratigraphic intervals 3 through 9. As the process of compositing data into the SRS begins, it is appropriate to refer to the X-axis as the *composite standard reference section* (CSRS) or, simply, the composite standard.

Locality C may then be graphed against the CS. In this second graph (not figured), a LOC can be derived which ties Locality C to the CS over stratigraphic intervals 8 and 9. Using an extension of this LOC as before, the additional information from stratigraphic unit 10 at Locality C may be projected into the CS, once again extending its coverage into younger section. In a like manner, Localities D and E may be graphed against the CS (third and fourth graphs) and the local stratigraphic range information from older sections may be added to the database so that it now includes stratigraphic intervals 1 and 2 as well. Thus, by completing a sequence of four graphs, the stratigraphic coverage of the CS has been greatly extended and it has been done in a careful, systematic manner. The four graphic correlation plots themselves provide detailed records of exactly how the database was constructed and which evidence was used.

The *ultimate* goal of graphing is to assemble a CS that contains fossil ranges observed in many, widely distributed, stratigraphically overlapping sections which, together, represent the entire geologic column. Such a mature database may be thought of as a geochronologic scale expressed in terms of the overlapping ranges of many fossil species. The ideal and mature database is a chronostratigraphic model built on many observations. It contains, for any interval of geologic time, the *total stratigraphic ranges* of any fossils likely to be encountered in new localities, no matter the environment of deposition represented. As a model, the CS predicts the sequence in which fossil bases and tops will be seen in stratigraphic sections yet to be studied.

Additional Rounds of Graphing

At any point in the process of building a CS, data from localities already composited may be regraphed. The reasons for doing this are that new species are constantly being added to the database, the ranges of species in the database are constantly being adjusted, and the stratigraphic coverage of the database is ever increasing. It is thus advantageous to reevaluate data from some of the original localities by comparing them to the more mature database now available.

General practice is to build an initial CS by graphing the data from a group of localities in sequence. This process is referred to as the *first round* of graphing. After the last locality of the initial group has been composited, the process is repeated, graphing each locality once again against the CS. This recorrelation process is referred to as the *second round* of graphing. As described above, each section graphed may or may not be used to contribute range information to the CS. Prior to regraphing any section it is important to remove from the CS the range information derived from that section, if any. This avoids the effect of a section simply being graphed against itself.

A minimum of three rounds of graphing should be carried out for the stratigraphic sections used in establishing a database. For a given stratigraphic section, it may be noted that the interpreted lines of correlation for each early round of graphing lie in somewhat different positions. This is to be expected since each round of graphing involves a slightly different (more mature) database. As rounds of graphing continue, changes in the lines of correlation for a given section should become smaller as CS ranges eventually stabilize. When this stabilization occurs, there can be reasonable confidence that the best possible interpretations have been achieved based on the present data. It has been found that, after three to six rounds of graphing, fossil ranges are generally stabilized and fixed to the scale of the CS database.

Scaling of the Composite Standard

The SRS on the X-axis of a graph is no more than a stratigraphic section which contains the local stratigraphic ranges of certain fossils. It represents rates of rock accumulation and an interval of geologic time which can be known only in general terms. It is appropriate to scale the SRS in units of rock thickness (feet or meters). The mature CS database which evolves from the SRS, however, serves as a geochronologic scale (Harland and others, 1990) against which to compare additional stratigraphic sections. Such a database contains properly sequenced and proportionally spaced events (in the sense of Shaw, 1964) compiled from many stratigraphic sections. These events are the bases (FADs) and tops (LADs) of the CS ranges of fossils. Using these events in the CS, the database can be related to the worldwide scheme of chronostratigraphic units (zones, stages, etc.) and/or to a chronometric scale (Harland and others, 1990) in years.

As described previously, building a CS begins with a graphic correlation of the first two, at least partially coeval, stratigraphic sections. The section chosen as the SRS (an outcrop or a well) is placed on the X-axis. Thus, at this stage, the scale of the X-axis is expressed in terms of feet or meters. For the purposes of establishing a LOC in the first graph, an X-axis scale in units of rock thickness is acceptable; the derived correlations between the two sections can be expressed in a satisfactory manner. However, as soon as the interpreted LOC is used to project additional fossil species into the SRS, to extend fossil ranges in the SRS, or to extend the stratigraphic coverage of the SRS, measurements in feet or meters on that axis would cause confusion. At this point in the process, the X-axis becomes a compilation of information from multiple stratigraphic sections in which a given increment of rock thickness could represent a broad range of rock accumulation rates and time intervals. If graphic correlation is to be used to composite many additional stratigraphic sections and to build a CS representing a sequence of chronostratigraphic events over a significantly long stratigraphic interval on the X-axis, a more universal scale is desirable.

The most obvious first choice would be to simply scale the CS in years. However, in the early days of developing CSs (i.e., the 1960s and 70s), scaling in years was viewed as an unattainable alternative. At that time, the database of available radiometric dates in stratigraphic sections with good fossil assemblages was relatively meager, and the technology for obtaining those dates was much less precise than it is today. There are also certain considerations with regard to database maintenance and management which bear on this issue. Interpretations in the literature of the ages (in years) of biozone and stage boundaries vary depending on the fossil group studied and change from time to time as new insights are gained. This is a source of confusion if the CS is scaled in years, because the ages given in the literature for chronostratigraphic boundaries would always be in some degree of disagreement with the ages of those boundaries in the CS. In the past, therefore, the general feeling was that a nonannual scale which could remain stable and independent of the dynamics of the literature was needed as a substitute for a scale in years.

Today the database of radiometric dates has grown significantly and the accuracy with which ages are derived has improved (Harland and others, 1990). Biostratigraphers might now be in a position to make a strong case for building a CS scaled to absolute time. Certainly the nonannual scales typically used for CSs can be equated to annual scales with some degree of confidence and the process of equating the two is in many ways similar to establishing a scale in years for the CS. However, considerations with regard to controversy over the absolute ages of chronostratigraphic boundaries, both within the literature and between the literature and the CS database, remain significant.

This paper expresses the conservative point of view in which a nonannual scale is preferred. The scale chosen serves several purposes. It allows the bases and tops of fossil ranges to be expressed numerically; it keeps bases and tops in their proper relative position as new data are added to the database; it provides for the assignment of stable, numerical values to the boundaries of biozones and stages; and it permits correlation interpretations to be quantified.

Though the actual choice of numbers for the scale is arbitrary, certain considerations should be born in mind. The scale should be designed from the outset so that all Phanerozoic time will be represented by a single, continuous numerical sequence with no repetition of values. It is also helpful to arrange the scale in ascending order with zero at the oldest point in the database to avoid confusion with an annual scale.

Biostratigraphers at Amoco use a numerical CS scale which ranges from zero (at the base of the Cambrian) to 23000 (representing Holocene deposition). In Figure 5, a hypothetical outcrop section is used to illustrate the development of this scale by the division of the original SRS into equal increments. In this section, 500 ft of stratigraphic thickness have been measured and sampled. It has been determined, by means of planktonic foraminifera and calcareous nannofossils, that the section pertains in general to the Early and Middle Miocene Epoch. The section is subdivided so that each 100 ft of stratigraphic thickness equals 50 units of a numerical scale. The specific numbers used to calibrate this section were chosen so that sufficient additional units would be available for assignment to older and younger Phanerozoic sections. Any sections younger than the top of this outcrop would be assigned numbers ascending from 21850; sections older than the base of this outcrop would be assigned numbers descending from 21600. If the base of this outcrop is, in round figures, at 20 Ma, approximately 555,000,000 years of Phanerozoic time remain to be accounted for below it. Thus the 21,600 units available between basal Miocene and basal Cambrian time would each represent an average of 25,000 to 30,000 years. Approximately 550 units are required above the top of this outcrop to provide a similar number of years per unit for younger sections. This should be sufficient to express any degree of resolution likely to be obtained.

The units thus defined are referred to as composite standard units (CSU). Each CSU has thickness and duration in time. The boundaries between the CSUs are points with neither thickness nor duration and are referred to as *composite standard time equivalents* (CSTE). The numerical scale pertains to the CSTE at the base of each CSU.

It is important to recognize that CSUs, as *defined*, are units of equal *rock thickness*, at least insofar as the initial SRS is concerned. In Figure 5, each 100 ft was divided into 50 CSU, giving each CSU a *thickness* of 2 ft. The CSU scale, however, is used as a representation of geochronology. This may appear to be a contradiction; a brief statement of pertinent considerations should provide clarification.

Each CSU in the scale does not represent the same number of years, but the *age* of the CSTE bounding a given CSU, often defined by a fossil FAD or LAD in the CS, remains constant from place to place. As discussed earlier, the mature CS FAD and LAD are believed to closely approximate true total stratigraphic bases and tops, unique events in earth history, and may be regarded for practical purposes as dependable for chronostratigraphic correlations between geographic localities. In a mature CS, a constant relationship is maintained between the FADs and LADs of the *composite standard ranges* and the CSTE scale which defines the boundaries of CSUs. Many CSUs, then, not unlike biozones and stages, have bases which are fixed and defined by unique biostratigraphic events (FAD or LAD). This suggests that each CSU might reasonably be thought of as a chronostratigraphic unit.

Calibrating the Composite Standard Scale to Absolute Time

The goal of equating a nonannual scale (i.e., a CSU scale) to a scale in years is so that chronostratigraphic boundaries in the CS can be related to published information on worldwide phenomena such as coastal onlaps, sea-level fluctuations and plate tectonics events. Calibration of correlations in Ma is also valuable in geohistory/source rock maturation studies which require input in units of absolute age.

Absolute time, by the sound of the term, should be a robust and dependable scale which can be referred to confidently by the geoscience community. On closer observation, however, absolute time is found to be a somewhat rickety construct based on certain physical evidence which has accumulated over the years. Individual bits of evidence suggest that earth history amounts to a imprecisely definable number of "years" and that certain events in earth history (volcanic eruptions, crystallization of magmatic bodies, reversals of the poles of the earth's magnetic field, etc.) took place a certain number of "years ago." These events are arranged in a chronological sequence and thought of as a framework of "fixed" dates on which to base geochronology. Beyond the actual samples which yield radiometric dates, however, the scale of absolute time, like the CSU scale, falls within the realm of projection, interpolation, correlation, superposition, assumptions as to the uniformity of var-

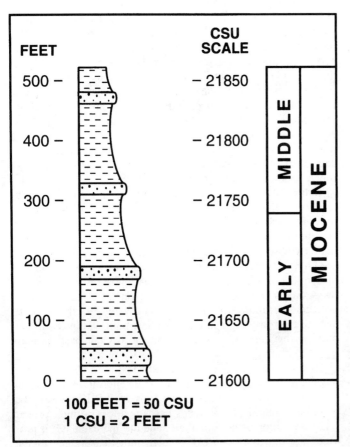

FIG. 5.—A CSU scale for the SRS is established by subdividing the measured section into segments of equal CSU value.

ious processes, and, statistics. Harland and others (1990) provide an excellent review of the absolute time scale and it's construction. The point of these comments is not to berate the absolute time scale; it is a marvel of painstaking ingenuity and is of tremendous value to geoscience in general. The point is that the absolute age of most sedimentary rocks cannot be measured directly with today's technology. Some can be measured, but they are few and the cost is high. There is nothing mystic, unduly precise, or sacred about the methods by which the stratigraphic column, between radiometric age control points, is calibrated to time.

Thus, in considering the question of equating CSUs to absolute time, the approach need not be made timidly with the conviction that the CSU scale is to be compared to a more precise or sophisticated scale. The process of calibrating the CS to absolute time admittedly involves projection, interpolation, correlation, and assumption, but these are concepts not foreign to the absolute time scale itself. In reality, two rather imperfect scales are being compared one to the other.

There are two possible approaches to calibrating a CSU scale to an annual scale. Each has advantages and disadvantages and the two approaches are not necessarily mutually exclusive. In the first approach, stratigraphic sections containing both abundant fossils in common with the CS and strata, such as volcanic ash beds with established radiometric dates, can be graphed against the CS. The radiometric ages can then be projected into the CS using the LOCs of the graphs, thus establishing the relationship between absolute age data points and events already in the CS. Once this is done, the intervals in the CS between radiometric age data points can be scaled to time by interpolation. Biozone and stage boundaries already exist in the CS by definition, since they are based on biostratigraphic events (FADs and LADs). Chronostratigraphic boundaries will thus be assignable, independent of the literature, to specific points on an annual scale. The weakness of this approach is that interpolation between radiometric dates assumes a constant rate of rock accumulation over significant intervals. This probably was not the case. This drawback can be minimized by graphing as many sections as possible. The advantage of this method is that the radiometric data points have been placed in the database by graphing, a procedure less subject to inconsistencies than correlation by fossil zones.

Another method involves inserting the published absolute ages of biozone and stage boundaries directly into the CS database at the appropriate points in the succession of FADs and LADs. The only requirements are that the fossils whose bases and tops define the biozone or stage boundaries must already be in the database and that their CS ranges must be reasonably close to their true total stratigraphic ranges. An important weakness of this approach is that the published absolute ages of biozone and stage boundaries rely on traditional methods of correlation to relate the sections containing radiometric dates with the boundary stratotype sections for the biozones or stages, assuming that stratotypes have been established. In addition to this, once the sections are correlated, interpolation is used to assign a specific age to the chronostratigraphic boundary. This interpolation, as in the previous approach, relies on assumptions of uniform rates of rock accumulation. The weakness of this approach can be reduced if the correlations among sections are based on paleomagnetic and/or oxygen isotope data as well as on fossil zones. Statistical evaluation of boundary ages such as error function versus scanning age *chronograms* (Harland and others, 1990) can also add strength to published boundary ages.

Any absolute age calibration of the CSU scale, regardless of how it was derived, results in a non-linear relationship with respect to the radiometric time scale. In certain intervals each CSU will have an absolute annual duration which differs from the duration of individual CSUs in other intervals. It would be virtually impossible to avoid this because rock accumulation rates vary among the sections used to build the database.

Blocking of a Graphic Correlation Interpretation

The interpretation of a graphic correlation plot amounts to an analysis of all of the possible points of correlation suggested by cross-plotted bases and tops. Based on a number of criteria, certain points will emerge as having a greater likelihood of reliability than others in establishing time correlations between the stratigraphic section represented on the Y-axis and the CS on the X-axis. These high-grade points of correlation are used to draw the LOC which defines the relationship between the two axes. The LOC may be a single straight line; it may be a series of "doglegs" (straight line segments attached but at some angle to each other) indicating changes in average rate of rock accumulation; and/or it may be a series of straight line segments offset horizontally from each other indicating intervals of geologic time not recorded in the stratigraphic section on the Y-axis.

Once the best LOC has been determined, the interpretation must be quantified so that it may be stored in the computer file for that locality. The overall interpretation is first separated into blocks where each block takes in one continuous straight-line segment. If the entire interpretation consists of a single straight line, there will be only one block for the graph. If the LOC changes slope at some point, that point will be a boundary between blocks. And, if the line is offset horizontally because of a fault or unconformity in the section being analyzed, the level of the offset will form a boundary between blocks. Typical graphs comprise one to several blocks.

When the blocking of a graph is established, each straight line segment may then be uniquely characterized by the algebraic solution for "A" and "B" in the linear regression formula for that segment, $X = A + BY$. The "A-value" (Y-intercept) relates to the age of the section under study and the "B-value" defines the slope of the LOC. By the blocking process, unique numerical values of A and B are calculated and stored in a computer file for each interpreted graphic correlation plot.

GRAPHIC CORRELATION AND MARINE PALEOGENE UNITS OF THE
NORTHERN GULF OF MEXICO

The graphic correlation examples in this section of the paper, based on marine Paleogene planktonic and benthic foraminifera of the northern Gulf of Mexico, illustrate the basic concepts of graphic correlation and CS database building in action. Each species is represented in the graphs by a taxon number (unique computer identification code) or a taxon number along with the taxon name. In the text, species are generally referred to by number. The CS ranges used in these examples generally reflect the status of the Amoco database in April 1993; some have been modified to better illustrate various points. The local stratigraphic range data from "outcrops A–F" and "well 1" are fictitious. Range bases are indicated by "o"; range tops are shown by "+."

Graphic Correlation Compared to Traditional Methods

Figure 6 is an example of traditional correlation methods applied to two measured outcrop sections, A and B. The taxon numbers and observed ranges of several fossils have been plotted for each outcrop. To correlate Outcrop A to Outcrop B, only the tops and bases of fossil species present in both sections (i.e., fossils 5 and 9, may be used). Assuming that the bases and tops are synchronous between outcrops, a reasonable correlation may be established, but the observed local stratigraphic ranges of many fossils are of no use (i.e., 1, 6, 7, 10, 11 and 12). In all likelihood, this information will go unrecorded and forgotten and will be of little or no future benefit. In addition, no correlation information beyond the simple observation of interval thickness will be derived for the strata between correlation points and above and below the highest and lowest correlation points.

In Figure 7, the same two outcrops, A and B, are compared using graphic correlation. The ranges of fossils observed in Outcrop A, along with their taxon numbers, are plotted along the Y axis of the graph. The fossil ranges and pertinent taxon numbers from Outcrop B, chosen as the SRS, are plotted along the X-axis. Once again, fossils 5 and 9 are the first to be used because they occur in both outcrops. Thus (Fig. 8), the top of fossil 5 in Outcrop A is cross-plotted against the top of fossil 5 in Outcrop B to derive the Cartesian coordinates of the graphic correlation point (5_+) for this top. The same process is employed to derive the coordinates of the graphic correlation points for the base of fossil 5 (5_o), the top of fossil 9 (9_+) and the base of fossil 9 (9_o). Using these cross-plotted points, a LOC may be drawn (Fig. 9) which defines the correlative relationship between Outcrops A and B.

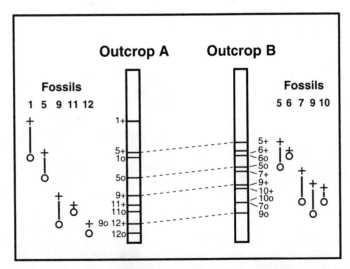

FIG. 6.—Using conventional correlation methods, time-equivalent tops and bases of fossils present in both outcrops can be used to correlate; tops and bases present in only one of the sections are of no use.

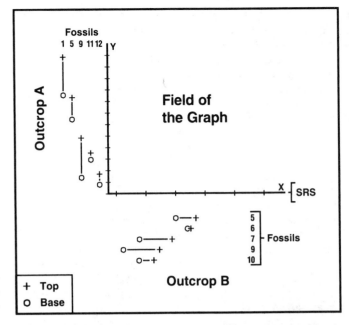

FIG. 7.—Using graphic correlation, two outcrops to be compared are placed on the X and Y axes of a graph. The older ends of each section are placed toward the graph's origin.

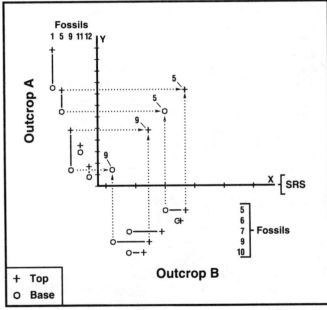

FIG. 8.—Time-equivalent bases and tops of fossils present in both outcrops are projected into the field of the graph to define cross-plotted points of correlation.

Development of the Paleogene Northern Gulf of Mexico Composite Standard

Figure 10 illustrates an important feature of graphic correlation. Once the LOC is established, an infinite number of proportionally interpolated or extrapolated correlative horizons may be drawn between the two graphic axes. Thus, any point

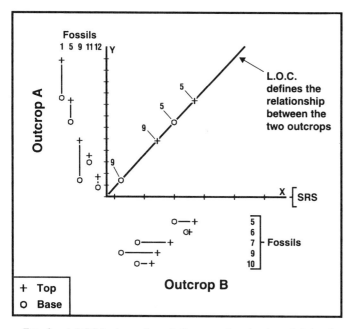

FIG. 9.—A LOC is drawn through the cross-plotted points, defining the correlative relationship between outcrops A and B.

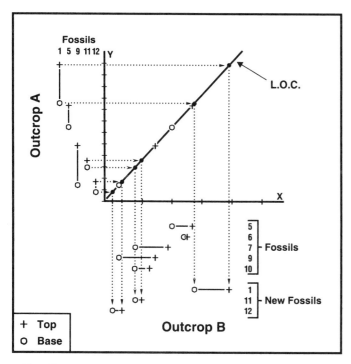

FIG. 10.—Using the LOC based on time-equivalent fossils common to both outcrops, the bases and tops of fossils present only in outcrop A are projected into sequentially and proportionally appropriate positions among the bases and tops of the fossils in outcrop B, thus compositing the range information from outcrops A and B.

in Outcrop A may be projected horizontally to the LOC, and vertically to Outcrop B, to determine the correlation between the outcrops at that point. Using this capability, the range data for fossils not used for correlation between the two outcrops can, nevertheless, be captured and recorded for use in future correlation. Based on the LOC, the bases and tops of fossils 1, 11 and 12 in Outcrop A can be transferred into the dataset for Outcrop B. Projection of these points through the LOC assures that their stratigraphic positions are sequentially and proportionally appropriate with respect to the stratigraphic positions of the bases and tops of fossils 5, 6, 7, 9 and 10, which were originally observed in Outcrop B.

Scaling of the Composite Standard

From this point forward in the process, the X-axis of the graph can no longer be properly thought of as "Outcrop B" or "SRS"; it is now referred to as the CSRS, or, simply, the CS (Fig. 11). At this point the evolution of the database has begun; the X-axis becomes a geochronologic scale, expressed in terms of fossil bases and tops, for the interval between the base of fossil 12 and the top of fossil 1. The scale of the X-axis should no longer be measured in linear units such as feet or meters. Beginning with this stage of the process, the X-axis will be scaled in CSU. Since there are certain fossil species with established stratigraphic ranges present in the CS, the chronostratigraphic interval represented can be recognized as lower Paleocene through Middle Oligocene time. It is significant to recall, however, that, if these "index" species were not present, the correlation process would not be impeded. It has been suggested earlier in this paper that CSUs be considered chronostratigraphic units (Fig. 12). The base of CSU 21738 is defined by the FAD of the index planktonic foraminifera, *Praeorbulina*

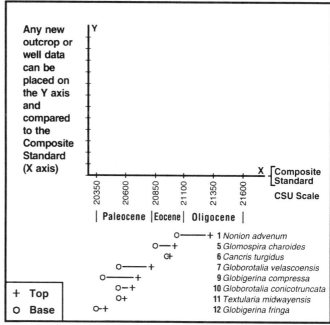

FIG. 11.—Upon completion of the first compositing step, the combined range information from outcrops A and B on the X-axis becomes the composite standard and CSUs replace linear units of measure for that axis.

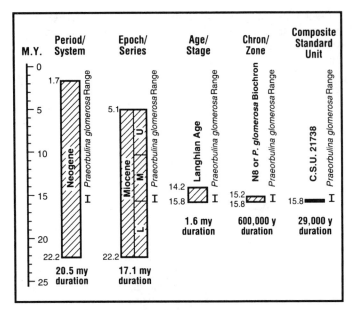

FIG. 12.—A hierarchy of chronostratigraphic units: the relative duration of each unit is shown and all are compared to the duration of a typical species of planktonic foraminifera. The worldwide stratigraphic base of *Praeorbulina glomerosa* defines the base of CSU 21738, N8 (*Praeorbulina glomerosa* Biozone), the Langhian Stage, and Middle Miocene time.

glomerosa. In the literature this taxon also defines the base of the Langhian Stage. Figure 12 shows how CSU 21738 may be compared to various traditional units of chronostratigraphy, to geologic time, and to the stratigraphic range of *Praeorbulina glomerosa*. Figure 13 places the entire sequence of Phanerozoic CSUs, which ranges from CSU 1 to CSU 23000, in context as compared to absolute time and gross chronostratigraphic subdivisions.

Expansion and Refinement of the Composite Standard Database

The reason for creating a CS such as the one now established on the X-axis of the graph is so that it can be used to evaluate additional sets of fossil range data from new localities. In Figure 14, a third stratigraphic section, Outcrop C, is compared graphically to the CS. There are two fossils (1 and 9) in common between Outcrop C and the database; their bases and tops may be cross-plotted to establish a LOC. Note that one of these, fossil 1, is a part of the CS database only by virtue of the use of the graphic correlation. If in the previous correlation, Outcrop A had simply been compared to Outcrop B and then Outcrop B to Outcrop C using traditional correlation techniqus, there would only be one fossil range (fossil 9) in common between Outcrops B and C. Such a correlation interpretation would be less complete than attainable with graphic correlation. This illustrates the point that, as additional stratigraphic sections are compared to the CS and the local stratigraphic ranges of their contained fossils are projected into that database, the chances of being able to derive biostratigraphic correlations with each new dataset are increased. With the LOC established for Outcrop C, the tops and bases of fossils 2 and 4 can be projected into appropriate positions in the CS (Fig. 15). The database has once again been enriched.

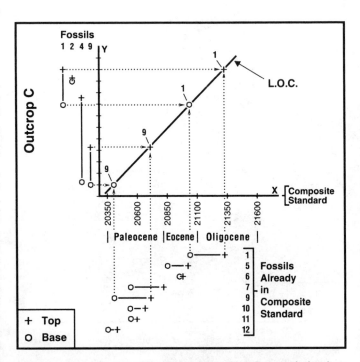

FIG. 13.—The Amoco scale of CSUs compared to a Ma scale and to gross chronostratigraphic terms. The Amoco scale was initially planned by dividing the estimated 575 my of Phanerozoic time into 23,000 equal 25,000 yr segments.

FIG. 14.—Local stratigraphic range data from outcrop C are graphed against the CS. An LOC is drawn based on the cross-plotted bases and tops of fossils 1 and 9.

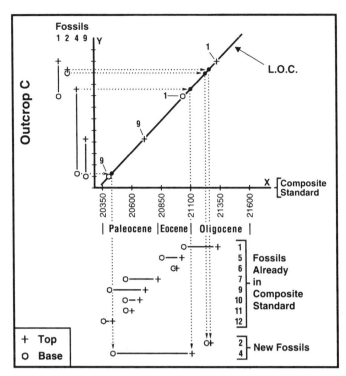

Fig. 15.—The LOC for outcrop C based on fossils 1 and 9 is used to project the tops and bases of fossils 2 and 4 into the CS. It is by repeating this process that the CS is enriched.

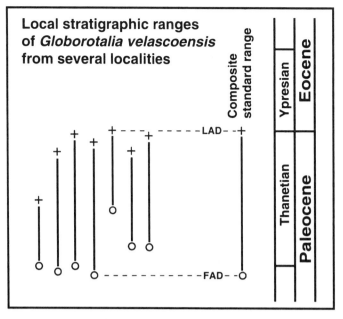

Fig. 16.—The CS range for *Globorotalia velascoensis* is defined by the oldest base and youngest top among the local stratigraphic ranges observed for that species at various localities.

The foregoing discussion illustrates once again that graphic correlation of a stratigraphic section against a CS is a two-stage process. First, an LOC is derived which serves to calibrate the stratigraphic section to the geochronologic scale of the CS. Second, the LOC may (or may not) be used to project information from the stratigraphic section into the database, thus enriching it.

Graphing Data with Incomplete Fossil Ranges

For the purpose of clear illustration, the local stratigraphic range data used in examples thus far have been idealized so as to produce perfect alignment of cross-plotted points of correlation. In datasets from outcrops or wells, observed local stratigraphic ranges are often not fully developed. For various reasons (paleoenvironmental control, imperfect preservation, sample quality, etc.), the local stratigraphic range bases of some species tend to be somewhat above (younger than) their worldwide FADs and their local stratigraphic range tops tend to be somewhat lower than (older than) their worldwide LADs. By systematically compositing all observed local stratigraphic ranges the total stratigraphic range or CS range of a given fossil species is derived. Figure 16 uses the Lower Tertiary index fossil *Globorotalia velascoensis* to illustrate this concept by comparing a number of observed local stratigraphic ranges from various outcrops to the CS range for that species.

Because graphic correlation is based on the entire fossil assemblage, a LOC can generally be derived even if some of the local bases are a bit high and some of the local tops are a bit low. Assuming the maturity of the CS range of a given fossil in the database, if the lower part of its range is not developed, its local stratigraphic base will fall to the left of a properly interpreted LOC. Likewise, if it disappeared locally earlier than the time of its worldwide LAD, its local stratigraphic top will cross plot somewhat to the right of a properly interpreted LOC (Fig. 17). Elevated local bases and depressed local tops tend to

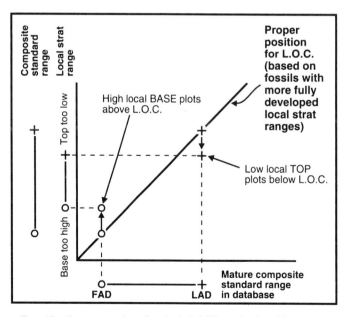

Fig. 17.—Due to a number of geological, drilling-related, and human-error factors, typical local stratigraphic range data from a given locality contain some fossil ranges which are shorter in duration than the worldwide ranges of the fossils. Assuming a mature CS range in the database, a local base which is too high cross plots to the left of a properly positioned LOC; a local top which is too low cross plots to the right of a properly positioned LOC.

form a "channel" parallel to and bisected linearly by a correctly-positioned LOC. Indeed, in some cases this channel may provide additional evidence for proper positioning of the LOC. These considerations have evolved into guidelines to assist the interpreter in deciding where to place the LOC. Simply stated, for normal (non-overturned) stratigraphic section and for taxa with mature CS ranges, non-reworked local tops should fall *on or to the right* of the LOC. Local bases, if not caved, should fall *on or to the left* of the LOC. The stipulation that the CS range must be mature is significant. In the case of taxa with immature CS ranges, the guidelines do not hold. A consideration of the CS range maturity of all cross-plotted taxa is an essential part of the interpretation of any graphic correlation plot.

In Figure 18 a fourth stratigraphic section, Outcrop D, with some incomplete local stratigraphic ranges, is in position on the Y-axis with its observed tops and bases cross-plotted against the tops and bases in the CS. The distribution of the cross-plotted points on the graph in Figure 19 shows the moderate scatter typically observed. Following the guidelines mentioned above, the LOC shown in Figure 20 may be drawn. This interpretation is based on the assumption that the tops of fossils 2 and 7 are more or less fully developed, while tops 1, 4 and 9 are somewhat depressed stratigraphically. Likewise, bases 2 and 4 appear fully developed, while the lowermost portion of the total stratigraphic ranges of fossils 1, 7 and 9 are not seen in this outcrop. The bases of species 1, 7, and 9 and the tops of species 1, 4, and 9 form a channel.

Stratigraphic Sections with Unconformities or Faults

Graphic correlation is an excellent tool for recognizing chronostratigraphic intervals recorded in the CS but not repre-

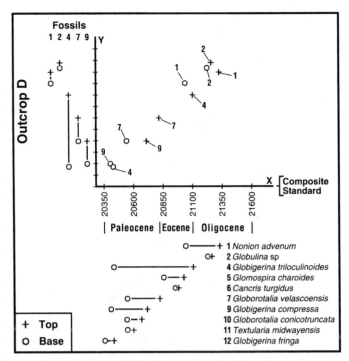

Fig. 19.—The typical scatter of cross-plotted points from a set of local stratigraphic range data containing both completely and incompletely developed bases and tops.

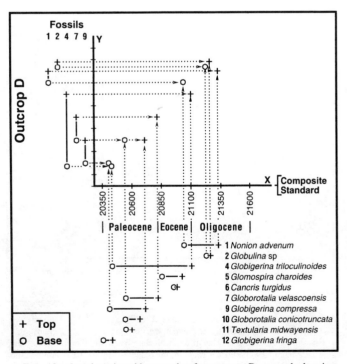

Fig. 18.—Local stratigraphic range data from outcrop D are graphed against the CS.

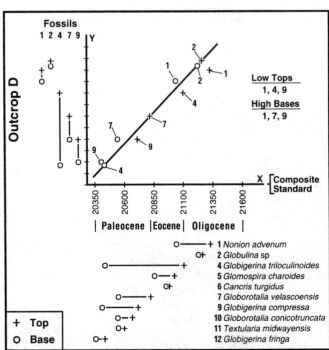

Fig. 20.—An interpreted graph for outcrop D. The LOC is drawn so that all bases lie on or to the left of the line, and all tops lie on or to the right of the line.

sented by the rock record in a well or outcrop. Figure 21 shows a dataset of local stratigraphic ranges from a stratigraphic section in which there is an unconformity representing much of the Thanetian Stage. It is evident that the ranges observed for fossils 7, 8, 9 and 10 will be significantly affected. Figure 22 shows the local stratigraphic range data from a fifth locality, Outcrop E, in position on the Y-axis with tops and bases cross-plotted against the CS. The distribution of cross-plotted points seen in Figure 23 is typical for a stratigraphic section containing an unconformity. The hiatus represented by the unconformity in this section is similar to that depicted in Figure 21. Note that the tops of fossils 7, 9 and 10, and the base of fossil 5, all appear in the same sample in this section, although previous observations recorded in the database suggest that each of these events represents a separate chronostratigraphic level in the Upper Paleocene section. In order to interpret this data, the same guidelines are followed: all bases should fall on the LOC or to the left of it, and all tops should fall on the LOC or to the right of it. To accommodate these guidelines, a two-part LOC must be drawn, offset by a short horizontal segment drawn immediately above the tops of fossils 7, 9 and 10 and immediately below the base of fossil 5 (Fig. 24). This horizontal line segment is referred to as a terrace and is an indication that a certain chronostratigraphic interval is not represented by rock thickness in Outcrop E, compared to the present CS (Sections A–D). To derive a quantitative estimate of the magnitude of the hiatus, the two end points of the horizontal line segment may be projected downward to the X-axis; the magnitude of the hiatus may then be interpreted as the chronostratigraphic interval between these two points on the X-axis. It can be seen in this case that a hiatus representing most of Late Paleocene (Thanetian) time is indicated.

In all examples prior to Figure 24, the LOC has been a single straight line, thus representing a single block with a unique solution for A and B in the linear regression equation. In the interpretation shown in Figure 24 there are two blocks: one for the Early and Middle Paleocene LOC segment and a second for

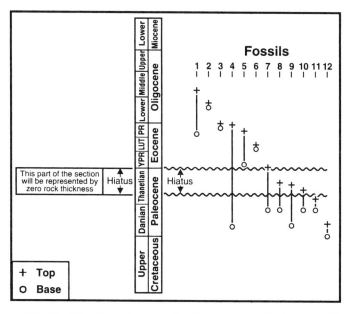

FIG. 21.—The effect of an unconformity on observed local stratigraphic ranges. Some ranges are truncated; some may be eliminated entirely (not shown in this figure).

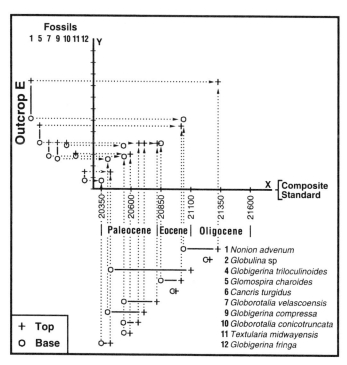

FIG. 22.—The local stratigraphic range data from outcrop E are graphed against the CS.

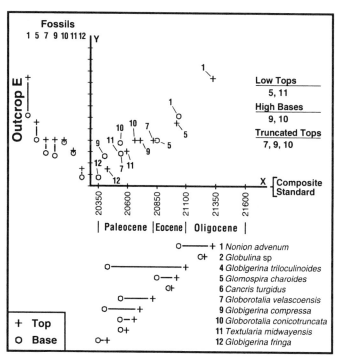

FIG. 23.—Typical data scatter for a stratigraphic section containing an unconformity. Note the horizontal alignment of the tops of fossils 7, 9, and 10 and the base of fossil 5.

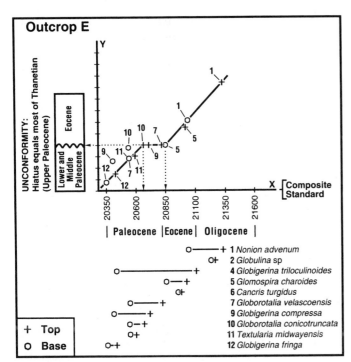

FIG. 24.—An interpreted graph for outcrop E. A stratigraphic section containing an unconformity is properly interpreted by a LOC made up of multiple (in this case, two), offset segments linked by horizontal terraces representing the hiatuses. The approximate duration (in CSUs) of a hiatus may be shown by projecting the end points of the terraces to the X-axis.

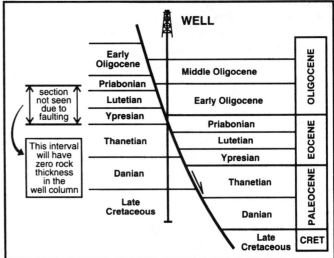

FIG. 25.—A well bore intersecting the plane of a normal fault. The stratigraphic interval not seen due to faulting would be represented by a terrace on a graphic correlation plot.

the LOC segment related to Eocene and Oligocene time. A horizontal terrace separates the two blocks.

In the case of an extremely condensed section with a significant amount of geologic time represented by the thickness of a single sampling interval, an LOC similar to that in Figure 24 would be obtained. A somewhat thicker condensed section would produce an LOC segment with a slight slope rather than a horizontal terrace. In the latter case, the condensed interval would constitute an additional block in the graph.

A faulted section, often obvious in outcrop, is frequently penetrated by a well without the geologist being aware of it until biostratigraphic analysis is undertaken. In Figure 25, a Early Tertiary/Late Cretaceous section containing a normal fault with considerable vertical displacement is shown. As can be seen in this figure, a well penetrated the downthrown block of this fault until Late Eocene sediments were reached. At that point, the wellbore intersected the fault plane and entered the upthrown block. The youngest sediments encountered in the upthrown block were Late Paleocene units. Biostratigraphic study of the samples from this well would show that most of the Eocene section is missing. Were graphic correlation applied to the fossil range data from this well, the LOC would consist of upper and lower line segments (time represented by some thickness of rock in the wellbore) offset by a horizontal terrace accounting for most of Eocene time. Thus a terrace, because it represents an interval of geologic time not represented by rock thickness at a *given locality*, is also the signature of a faulted section. In the overall analysis of a large geographic area by means of graphic correlation, faults will be distinguishable from unconformities and condensed sections. The latter two will be represented in many geographically scattered localities by a terrace spanning subequal intervals of geologic time. The former will be represented by a terrace with a stratigraphic interval unique to a single locality or a to small group of localities arrayed parallel to the strike of a fault plane.

Reworked Fossil Tops

As older sediments are eroded and the products of that erosion are transported and redeposited, fossils typical of the older section often become incorporated in sediments which are much younger. The process of reworking results in observed local stratigraphic ranges with tops which are much higher (younger) than would normally be expected. Graphic correlation, since it is based on all of the fossils in a sample, is an excellent means of calling attention to anomalously high, reworked tops. Even if the top which is reworked is that of an established index fossil, graphic correlation provides a means of checking the accuracy of that top as compared to the correlation suggested by the total fossil assemblage. This results in significantly more accurate interpretations than would be derived by methods which depend solely on the correlation indicated by index fossils.

Figure 26 shows an example in which sediments of Late Paleocene and earliest Eocene age have been eroded and redeposited along with Middle Eocene sediments. The top of *Globorotalia velascoensis*, typically occurring at the Paleocene/Eocene boundary, is extended upward into early Middle Eocene sediments. In Figures 27–29, it can be seen that graphic correlation will call attention to the fact that such reworking has occurred. Figure 27 shows data from a sixth locality (Outcrop F) cross plotted against the CS. Figure 28 shows the typical scatter of data points for a stratigraphic section with reworked fossils. Upon attempting to place a LOC among the cross-plotted tops and bases from this outcrop, it proves impossible to do so according to the basic guideline of keeping all tops on or to

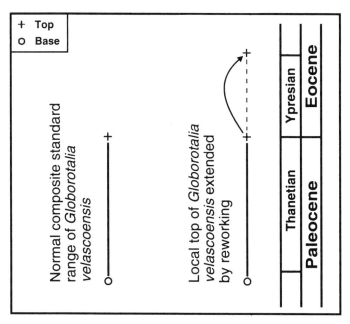

Fig. 26.—The observed local stratigraphic range of a fossil can be extended into younger section by reworking.

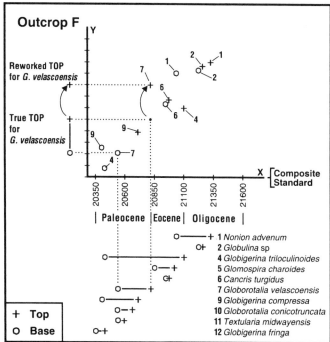

Fig. 28.—Typical data scatter for a stratigraphic section containing a reworked fossil is shown. Note occurrence of fossil 7 in an anomalously high stratigraphic position in the outcrop data.

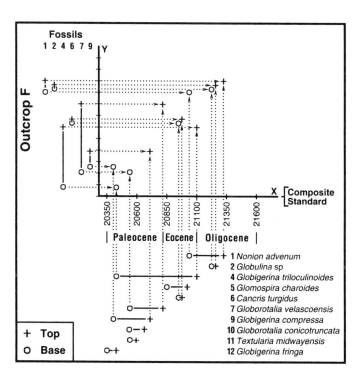

Fig. 27.—Local stratigraphic range data from outcrop F are graphed against the CS.

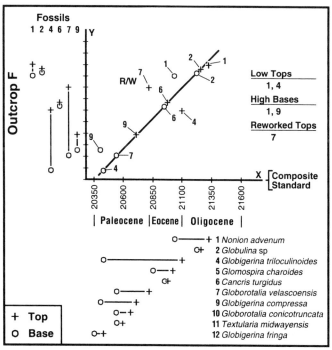

Fig. 29.—An interpreted graph for outcrop F. It is impossible to position an LOC so that all tops are to the right. This is typical of data from stratigraphic sections containing reworked fossils. Note that the top for fossil 7, *Globorotalia velascoensis*, has been designated as reworked (R/W) for future reference.

the right of the LOC and all bases on or to the left of the LOC. Indeed, the line that best fits most of the data (Fig. 29) leaves the top of fossil 7 (*Globorotalia velascoensis*) conspicuously to the left of the LOC. Though *Globorotalia velascoensis* is an established index fossil, evidence from several other fossils suggests that it be designated as "reworked" and that it be ignored in deriving an LOC for Outcrop F. In all likelihood, if traditional correlation methods had been used to analyze this data, the top of *Globorotalia velascoensis* would have been honored, and the top of the Paleocene section would have been placed in an erroneously high position. Again it is stressed that a consideration of the maturity of CS ranges in the CS is a critical part of graphic interpretation. A taxon with an immature CS top could also cross plot to the left of a properly interpreted LOC and yet not suggest reworking.

Graphic Correlation used with Data from Well Cuttings

The use of local stratigraphic range data derived from well cuttings (ditch samples) with graphic correlation requires special consideration. Because of caving of rock particles from the walls of the wellbore during drilling, the observed bases of fossils often appear somewhat extended (too old). Tops, however, are unaffected. It has been found that graphic correlation can be carried out with success using data of this type by concentrating attention on tops. The aforementioned guidelines for placing the LOC must be born in mind. It would be difficult to apply these rules successfully if the fossil bases observed in cuttings samples were simply cross plotted against the bases in the CS. Thus, the convention has evolved at Amoco of cross plotting fossil tops from a well against the corresponding tops in the CS and of cross plotting the tops in the well data against the bases for the same fossils in the CS (Fig. 30). Using this procedure, the cross-plotted top and base for a given fossil species will fall on a horizontal line on the graph. The distance separating them will depend on the CS range of the fossil as recorded in the CS on the X-axis. The cross-plotted bases resulting from the application of this convention still provide a certain amount of constraint as to the placement of an LOC. If the CS range of the fossil in the database is fairly short, the constraint can be of considerable help. As can be seen in Figure 30, a satisfactory graph can be obtained from the use of such data. In this example a terrace similar to the one depicted in Figure 24 is indicated.

PATTERNS OBSERVED IN THE LINE OF CORRELATION

As has been discussed earlier, the comparison of a stratigraphic section to the mature CS via graphic correlation may be thought of as plotting rock thickness against a geochronologic scale. Thus, graphing provides an opportunity to estimate relative average rates of rock accumulation (related to, but not the same as, rates of sedimentation). Perhaps more importantly, it also allows the recognition of intervals of geologic time not represented by significant rock thickness in any given section. Again, because identifying unconformities and condensed sections is of paramount importance to sequence stratigraphic analysis, it is easy to visualize the critical role that graphic correlation plays in these studies. Certain patterns observed in the LOC are more common than others, while some, though seldom seen, are signatures for certain geologic provinces (e.g., fold or overthrust belts). Some of the more common LOC patterns have been discussed previously in this paper. Figure 31 (A–F) presents a review of some of the possible LOC patterns which graphic correlation could produce.

Figures 31A and 31B show two lines of correlation with significantly different slopes. Both LOCs suggest normal (non-overturned) section. The LOC in Figure 31A represents a greater average rate of rock accumulation than the LOC in Figure 31B.

In Figure 31C, there are two LOC segments offset by a horizontal terrace. The terrace implies a stratigraphic discontinuity. Such a pattern of LOC and terrace indicates normal (non-overturned) section with a certain interval of time not represented by a significant thickness of rock and with similar rates of rock accumulation below and above the missing section. This pattern is generally attributed to a fault, an unconformity, or an extremely condensed section. If a condensed section is sufficiently thick, it can produce an LOC segment with a very gentle slope rather than a horizontal terrace.

Figure 31D shows the LOC derived for a stratigraphic section with a repeated interval. This is typical of reverse faulting. Such a pattern is often difficult to recognize when using data from cuttings samples due to the phenomenon of sample caving.

Figure 31E shows a dogleg pattern which indicates a normal (non-overturned) section with an early phase of rapid rock accumulation followed by a period of slower rock accumulation. Such a pattern is typical of expansion faulting provinces. The downthrown block of an active expansion fault is characterized by thick, rapidly accumulated sediment. As the locus of deposition migrates and the fault becomes inactive, the sedimentation rate on the downthrown block decreases. In sections such as this, special care must be exercised when assuming a straight line projection of the LOC for the purpose of extending the database into older or younger stratigraphic intervals.

Figure 31F shows the pattern seen when the stratigraphic section graphed is overturned. The upper part of the section is

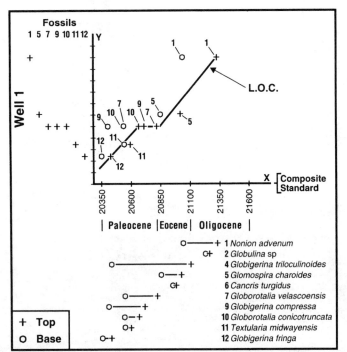

FIG. 30.—Local stratigraphic range data from well 1 are graphed against the CS. Typical well data scatter for a stratigraphic section containing an unconformity is shown. Note that only the tops in the well data have been used; down hole caving of ditch cuttings samples renders local stratigraphic bases unreliable.

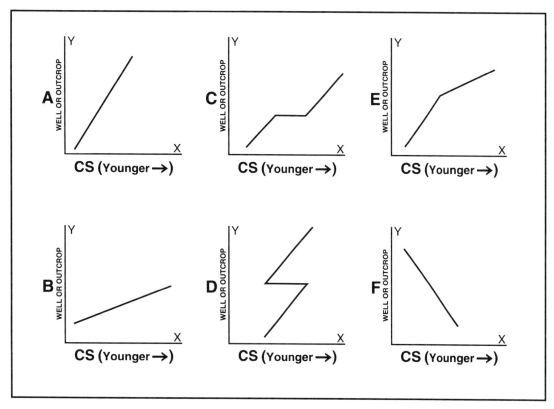

Fig. 31.—The most common LOC patterns produced by graphic correlation and the geological interpretations they suggest. (A) A LOC representing continuous, relatively rapid rate of rock accumulation; (B) A LOC representing continuous, relatively slow rate of rock accumulation; (C) Two LOC segments separated by a horizontal terrace indicating a stratigraphic discontinuity (i.e., a fault, an unconformity, or an extremely condensed section); (D) Two LOC segments separated by a horizontal terrace in a pattern typical of reverse faulting; (E) A "Dogleg" LOC typical of down thrown fault blocks in expansion faulting provinces; (F) A LOC with reversed slope seen when an overturned stratigraphic section is graphed.

older than the lower part. Interestingly, in well cuttings samples the first downhole appearances of species occur in order of decreasing age of local stratigraphic bases; tops will be subject to downhole caving.

DISPLAYING GRAPHIC CORRELATION INTERPRETATIONS

The interpreted graphic correlation plot or "Shaw diagram" is an extremely useful display, both for interpretation and communication. The LOC clearly portrays the stratigraphic interpretation and the cross-plotted tops and bases show the evidence upon which it is based. The sole limitation of the graph is that it represents only one well or outcrop section. There are additional types of displays which allow the visual comparison of interpretations from multiple wells.

Stratigraphic Nomographs

A stratigraphic nomograph is essentially a graphic correlation plot which shows the LOCs of multiple wells or outcrops superimposed (Fig. 32). The vertical scale is in feet or meters; the horizontal axis is a geochronologic scale. This display allows the comparison of relative average rates of rock accumulation and makes it easy to see the stratigraphic alignment of terraces. In Figure 32 it can be seen that the overall rates of rock accumulation in wells A and B are more rapid relative to the rates for well C. It is also clear that the stratigraphic discontinuities in all three wells tend to cluster in the Middle Oligocene and Middle Miocene Epochs. This suggests that these terraces represent an event of regional extent. It should be stressed that the *vertical proximity* of the terraces is not significant; this is controlled by the depth at which each well encountered the unconformity. It is the *vertical alignment* (above a similar chronostratigraphic interval on the X-axis) that implies a genetic relationship between the terraces seen in various wells. It is interesting to note that the middle terrace in well A does not align vertically with any terraces in the other two wells. The implication is that this middle terrace represents a feature unique to well A, possibly a fault. The stratigraphic nomograph can also be used to test the isochroneity of any other temporally distributed events which can be related to well depth (e.g., seismic reflectors used for subsurface mapping).

Chronostratigraphic Diagrams

A chronostratigraphic diagram is another means of displaying graphic correlation results for multiple wells on a single illustration. It is a bar graph with a vertical scale in chronostratigraphic units or geologic time (Fig. 33). The chronostratigraphic intervals represented by rock in each well or outcrop are shown by a vertical bar; intervals of hiatus are left blank. Like the nomograph, it allows visual comparison from one stratigraphic section to another of intervals of geologic time

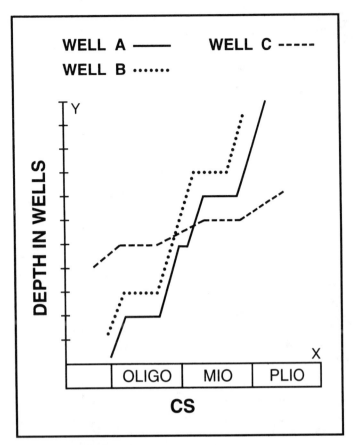

Fig. 32.—Stratigraphic nomograph showing the LOCs of three wells plotted on a single graph. The comparison of relative rates of rock accumulation and relative ages of hiatuses is facilitated by such a diagram.

represented by rock versus intervals not represented. It is a useful tool for sorting out regional and local events. Lithostratigraphic symbols and boundaries can be plotted within the vertical bars in order to study facies relationships and the diachroneity of such boundaries. In Figure 33, the lithostratigraphic unit "E," an evaporite "bed," was formerly thought to be continuous and to be a reliable chronostratigraphic datum. Analysis by graphic correlation and chronostratigraphic diagramming, however, show that there are actually several, discontinuous evaporite bodies situated within various genetically separate, unconformity-bound sediment packages.

Graphic Correlation/Seismic Displays

In view of the importance of recognizing unconformities and condensed sections in sequence stratigraphic studies, it is often instructive to plot the vertical position of graphic correlation terraces on seismic profiles. Such displays allow direct comparison of paleontological and seismic evidence for stratigraphic discontinuities. This can be done where wells which have been graphed are located on seismic lines which are to be interpreted. To insure accuracy, there should be some reasonably good means of correlating the well depth to the seismic data (e.g., a vertical seismic profile (VSP) or a synthetic seismogram based on the sonic log). In interactive seismic interpretation (e.g., using a Landmark Workstation), it is particularly helpful to load the terrace information derived from graphic correlation into the workstation database as a component of well data. This can be done by creating a computer flat file for each well showing the terraces and the depths at which they were encountered. These data are then displayed on screen superimposed on the seismic profile as a vertical series of color-coded tick marks at each well location.

CONCLUSIONS

1. Graphic correlation facilitates the recognition of unconformities and condensed sections, thus interpretations can be applied directly to sequence stratigraphic studies.
2. A composite standard database is a tool used, via graphic correlation, to calibrate the fossil range data from outcrop sections or wells to chronostratigraphy and geologic time. Such calibration facilitates correlation and provides an interpretation of geologic age.
3. Graphic correlation is primarily a method for displaying all potential points of correlation between two stratigraphic sections on a single piece of paper or computer screen. This allows the biostratigrapher to make correlations and reevaluations based on the maximum amount of evidence.
4. The graphic correlation method results in correlations derived from all fossils from all fossil groups which can be identified in a given set of samples, not just established "index" fossils or "zonal" fossils. Such correlations are less dependent on paleoenvironmental control and tend to call attention to reworked and environmentally depressed tops as well as incompletely developed, elevated bases.
5. A composite standard database improves with use. Continued graphing of additional stratigraphic sections against the database results in a better definition of the fossil ranges in the database, in extended stratigraphic coverage of the database, and in more varied representation of paleoenvironments in the database. A mature database can be applied successfully on a worldwide scale.
6. Graphic correlation and the composite standard are excellent tools for keeping accurate records of and displaying together the many fossil ranges observed during the course of a correlation study. It enables the paleontologist to benefit from all past observations.
7. Graphic correlation produces correlations which may be expressed mathematically and, thus, which may be readily recorded, stored and reproduced by computer.
8. Graphic correlation is not intended to replace traditional paleontological methods. It is meant to enhance them and add a measure of consistency to them. Conventional skills in paleontology and biostratigraphy as well as the paleontologist's regional experience are equally important in both methods.

ACKNOWLEDGMENTS

Grateful acknowledgment is made to F. X. Miller (retired, Amoco) whose years of mentoring and collegial project work while an Amoco employee are appreciated. The authors have benefited from his numerous internal proprietary reports and

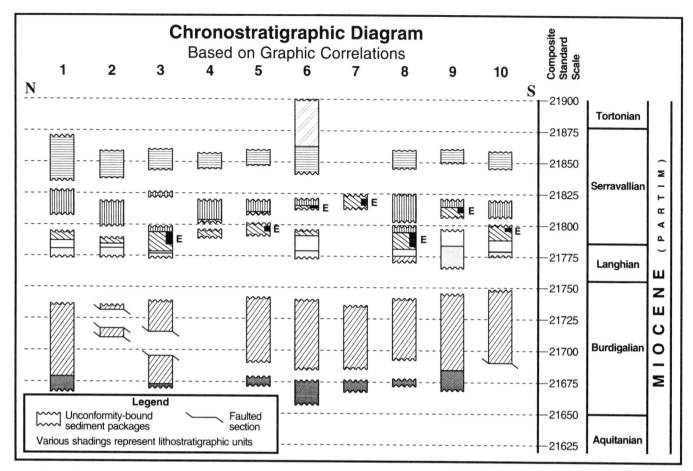

FIG. 33.—Chronostratigraphic diagram based on graphic correlations. This interpretive display arranges various graphed sections from a basin in geographic order; in this case 10 sections have been arrayed from north to south (left to right). Vertical scale is in CSUs; any horizontal line on the diagram is isochronous. Individual stratigraphic sections are shown as columns on a bar graph. Shaded and patterned intervals of columns represent lithostratigraphic units; blank intervals represent hiatuses whose duration in CSUs has been taken from the terraces on interpreted graphic correlation plots. Blank intervals which align horizontally across a number of sections (e.g., between 21750 and 21775 CSU) suggest regional events such as unconformities.

the wisdom expressed in those writings. Jeffrey A. Stein, Amoco, also provided much valuable technical advice and assistance over the course of this work. Keith Mann (Juniata College) and Dana Ulmer-Scholle (SMU) made numerous editorial suggestions which materially improved the content and organization of the paper. Their help and patience are much appreciated. Acknowledgment is expressed to Pennie Fullbright for her painstaking work drafting the final figures (on Corel Draw 4.0). The authors express their thanks to Cynthia Costello, Verda Kenworthy, and Jane Leighty for assistance with word processing (MS Word 6.0). And, lastly, appreciation is offered to Amoco for support in the writing of this paper.

SELECTED REFERENCES

HARLAND, W. B., ARMSTRONG, R. L., COX, A. V., CRAIG, L. E., SMITH, A. G., AND SMITH, D. G., 1990, A Geologic Time Scale 1989: Cambridge, Cambridge University Press, 263 p.

MILLER, F. X., 1977, The Graphic Correlation Method in Biostratigraphy, in Kauffman, E. and Hazel, J., eds., Concepts and Methods of Biostratigraphy: Stroudsburg, Dowden, Hutchinson, and Ross, p. 165–186.

SHAW, A. B., 1964, Time in Stratigraphy: New York, McGraw Hill Book Co., 365 p.

GRAPHIC CORRELATION: SOME GUIDELINES ON THEORY AND PRACTICE AND HOW THEY RELATE TO REALITY

LUCY E. EDWARDS

United States Geological Survey, 970 National Center, Reston, Virginia 22092

ABSTRACT: Graphic correlation is a geometric construction in which the locations of correlatable horizons in two sections plot as points on an xy-graph, and the line of correlation (LOC) is formed from the infinite number of points representing these horizons. In theory, the LOC between any two correlatable sections must exist; the geologist must weigh all the evidence at hand to deduce or to approximate where the LOC must be, and there is no way to "look up the answers in the back of the book." The LOC can have as many doglegs as are needed, but it cannot have a negative slope. In practice, not all the lowest and highest stratigraphic occurrences of taxa have time-significance. Furthermore, one must always check the axis labels and the scaling of any graph used in graphic correlation because there are few standard conventions in graphic correlation, and many computer programs are designed around the size of the screen or printer paper, not around the user's convenience. In reality, the hardest part of graphic correlation is generating reliable data. Graphic correlation is a means to an end; a tool to solve problems in stratigraphy, basin analysis, paleobiogeography, or evolutionary history. One should also use graphic correlation to find out where further investigation is needed, and then one should investigate further.

INTRODUCTION

A. B. Shaw developed graphic correlation as a means to establish the total ranges of fossil taxa and to correlate among stratigraphic sections by placing all sections in a standardized framework. The various steps in the graphic correlation procedure were described in detail in Shaw (1964). Updates on the procedure were given in Miller (1977) and Edwards (1984, 1989), and a brief summary with a simple example and a procedure checklist was given in Edwards (1991). Further information is given in several of the chapters of this book.

Although it may take years of experience to become totally familiar with the intricacies of graphic correlation, a few simple rules (and observations) exist. The purpose of this chapter is to emphasize some of the most important aspects of graphic correlation and to highlight some of its finer points. Armed with these insights and with knowledge of procedural details given in other chapters of this book, the reader should be adequately prepared to apply graphic correlation and to see what it can do.

THEORY

Graphic correlation depends upon locating time-equivalent levels within pairs of sections (Fig. 1A, B). The line of correlation (LOC) is the aggregation of every time-equivalent level (Fig. 1C, D).

The LOC allows information from one section to be related to information in a second section (Fig. 2). Information from individual stratigraphic sections can be combined to form a composite standard, which is a hypothetical section representing a synthesis of all the data at hand.

Rule 1. The Line of Correlation (LOC) Between Any Two Correlatable Sections Must Exist.

The xy-graph in graphic correlation is simply a geometric construction in which time-equivalent levels in the two sections plot as points on the graph. In geometry, an infinite number of points constitutes a line. In graphic correlation, the infinite number of time-equivalent levels is the LOC. If two sections do not overlap at all (for example if one is Devonian and one is Jurassic), no LOC exists. If the two sections overlap, the LOC must exist.

Rule 2. It is the Geologist's Task to Weigh All the Evidence at Hand to Deduce or to Approximate Where the LOC Must Be. The Best Deduction is the Correlation Hypothesis.

In rare cases, such as sections where continuous ash beds are found, time-equivalent levels are immediately discernible. In most real-world situations, one seeks to find correlatable levels that approach time equivalency. Stratigraphic levels containing the lowest and highest occurrences of taxa are the most common, potentially correlatable horizons used in graphic correlation. Other potentially correlatable horizons include changes in relative abundances of species (Sweet, 1979), distinctive geophysical log patterns (Edwards, 1989), paleomagnetic reversals (Dowsett, 1989), and microspherule layers (Hazel, 1989). No horizon should be used to infer the location of the LOC unless the geologist has reason to believe it has time significance.

My favorite quote on graphic correlation is appropriate here:

"The technique described here does not in any way relieve the paleontologist from thinking about what he is doing . . . It cannot overcome inconclusive or inadequate data, nor can it compensate for slipshod thinking in its application. On the contrary, by giving a result that can be expressed numerically — as we shall see — it may lend a spurious aura of sanctity to otherwise worthless conclusions." (Shaw, 1964, p. 154)

Rule 3. Until Someone Invents a Time Machine, There is No Way to "Look up the Answers in the Back of the Book." Some Correlation Hypotheses, However, Are More Geologically Reasonable than Others.

Any available information about correlatable levels or potentially correlatable levels should be weighed in the evaluation of the position of the LOC. One should not exclude available information that would help find the best answer simply because it is from a different fossil group. Some potentially correlatable data represent nonunique events or levels (log-type data). The beauty of the graph is that potentially correlatable levels can be plotted and tested. Those that are consistent with the correlation hypothesis should be retained and used to refine the correlation hypothesis. Those that are inconsistent should be excluded from any conclusions about the correlation hypothesis (they may be very interesting paleoecologically, just useless biostratigraphi-

LUCY E. EDWARDS

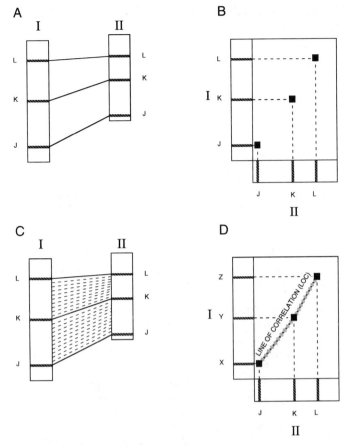

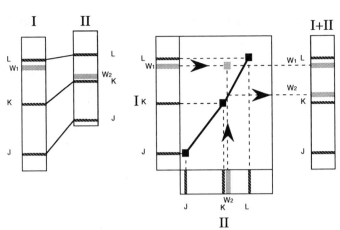

FIG. 1.—Side-by-side comparisons and graphic correlation. (A) Three synchronous horizons, J, K and L, are shown in each of two sections. (B) The same horizons and sections are shown on an xy-graph. Each horizon plots as a point (solid box) on the graph. The distance from a point to each axis corresponds to the stratigraphic position of the horizon in the section. By convention, stratigraphically higher positions are to the top and to the right. (C) The same horizons and sections as above are shown with intervening synchronous horizons. (D) The intervening horizons are plotted as points (patterned boxes) on the xy-graph; the aggregation of all synchronous points forms the line of correlation (LOC).

FIG. 2.—Side-by-side comparisons, graphic correlation, and the combining of information from two sections. In the side-by-side comparison on the left, two horizons (W1, W2) are added to the three synchronous horizons (J, K, L). The right side of the figure shows how the xy-graph and the LOC can be used to form a composite section (I+II) in which the three synchronous horizons and horizons W1 and W2 are shown. The point which represents the horizons W1 and W2 (patterned box) lies away from the LOC, demonstrating that the two horizons are not synchronous. The positions of these two horizons in the composite are found by projecting their x and y locations from the LOC onto the composite. Here, the information from sections I and II has been combined and placed in the framework of section I.

cally). Of course, if lines of evidence are inconsistent with the correlation hypothesis, one should consider revising the hypothesis.

Rule 4. Time Does Not Go Backwards, and the LOC Cannot Have Negative Slope.

The axes on the xy-graph are based on measured stratigraphic thickness assuming that stratigraphic thickness relates in some direct way to time.

Structurally overturned sections can be correlated if one can tell which way used to be up. If the sections do not obey the law of superposition and there is no way to tell older from younger stratigraphically, graphic correlation cannot be applied.

Rule 5. The LOC Can Have as Many Doglegs as Needed. See Rules 2 and 4.

The goal of graphic correlation is to deduce time-equivalency. A straight-line LOC does not imply constant sedimentation rates, but it does imply constant *relative* rates of rock accumulation for the sections being plotted. If the rate of rock accumulation in one section changes relative to the rate in another, the LOC will have a bend or dogleg in it. Figure 3 shows two common patterns involving bends in the LOC.

PRACTICE

A good guide to the positioning of the LOC is what Shaw (1964, p. 254–257) termed "economy of fit"; the best LOC is that which causes the minimum net disruption of the best established ranges. One cannot determine the position of the LOC without first identifying the points that represent the potential correlatable horizons. Once all the points are identified, each should be individually evaluated. Shaw (1964) discusses this evaluation at length; many subsequent users seem to omit this step and may be disappointed when they obtain less than desirable results.

Rule 6. Not all Potentially Correlatable Horizons are, in fact, Time Equivalent.

The most commonly used horizons in graphic correlation are the lowest and highest stratigraphic occurrences of fossil taxa. It would be convenient if the lowest observed occurrence of every species represented its evolutionary first occurrence, but this is seldom the case. The lowest observed occurrence can also represent its immigration, the appearance of a favorable environment for preservation, or simply the lowest level at which this species was sampled and correctly identified. Similarly, the highest observed occurrence may represent its extinction, its local extermination, or its emigration, or may be the result of preservation, sampling, or identification. In the case of a lowest occurrence, the "true" evolutionary first occurrence

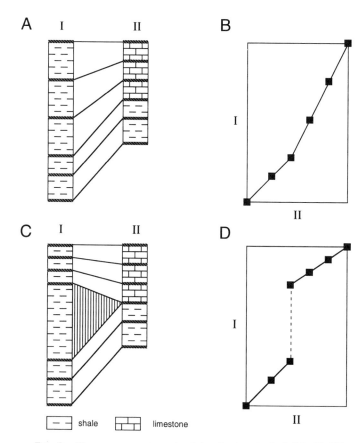

Fig. 3.—Two common patterns involving dog-legs in the LOC. (A) Side-by-side comparisons showing six synchronous horizons in each of two sections. The relative rate of rock accumulation between the two sections changes at the shale-limestone boundary in section II. (B) The same horizons and sections are shown on an xy-graph. Note the bend in the LOC. (C) Side-by-side comparisons showing seven synchronous horizons in each of two sections. The relative rate of rock accumulation between the two sections changes at the shale-limestone boundary in section II where the lithologic contact also marks an unconformity. (D) The same horizons and sections are shown on an xy-graph. The vertical (dotted) segment of the LOC indicates that there is no sediment remaining in section II that is time-equivalent to the sediment in that portion of section I.

should be at or stratigraphically below the observed occurrence. Unless there is contamination, downward mixing, or misidentification, the evolutionary first occurrence cannot be above the observed occurrence. Conversely, for a highest occurrence, the "true" extinction may be at or stratigraphically above the observed occurrence, but it cannot be below it unless there is contamination, downward mixing, or misidentification. Graphic correlation exploits these differences.

Graphic correlation can help one determine which fossil taxa do or do not have lowest or highest stratigraphic occurrences that are approximately synchronous in most of the sections studied. From an evolutionary or biogeographic standpoint, it is exciting and enlightening to explore the possible meanings and consequences of those that are not synchronous. It is not proper to use demonstrably nonsynchronous events in correlation!

As originally set forth by Shaw (1964), the beginnings and endings of ranges plot as points on the xy-graph. Since the early 1980's, I have found it more appropriate to plot range endpoints as boxes. The areas of the boxes represent one source of uncertainty; the sides of the boxes represent the unsampled intervals. When I plot first occurrences, I plot the lowest sample that contains a given taxon as well as the next lower sample that could have contained the taxon but didn't (barren samples do not count because one cannot tell what they *could* have contained). The true lowest occurrence most likely occurs somewhere between these two samples (Fig. 4). On an xy-graph, unsampled intervals exist on each axis, so first-occurrence events plot as boxes with the symbol for the observed event in the upper right-hand corner. Similarly, the highest sample that contains a taxon and the next higher sample that could have contained the taxon but didn't form boxes with the observed last-occurrence event in the lower left-hand corner of the box.

One might think that establishing the absolutely highest last occurrence or the absolutely lowest first occurrence is simply a matter of taking closely spaced samples. Sometimes, obtaining additional samples is merely inconvenient or time-consuming; other times, it is prohibitively expensive. Other times it is impossible; no amount of sampling can turn dolomite back into limestone or put back material that has since been eroded away. The sample-interval boxes that I plot represent only part of the uncertainty of the "true" locations of events. The reader should consult Strauss and Sadler (1989) and Marshall (1990) for additional discussion.

Any potentially correlatable type of data can be plotted on the xy-graph (see Rule 2, above). Data that can be expressed as magnitude (log-type data) are a good example. Some log-type data are continuously recorded, such as electric logs. Other log-type data have fixed sample locations, such as species-abundance curves. These other potentially correlatable events also have sample-interval error associated with them (Fig. 5). For example, it is very straightforward to convert paleomagnetic data to magnetic-reversal events with their associated intervals of uncertainty. The boxes that represent this uncertainty extend from the first sample with one polarity to the next sample with the opposite polarity.

In practice, I find colored pencils and a light table to be essential. First, I color-code my points: blue for most reliable, red for less reliable, and no color for unreliable. An easy way to approximate reliability is to go back to the original data sheets. Does this species, in this section, occur in most samples near its range endpoints? Frederiksen (1988) assigned numerical rankings to the reliability of range end-points in each section and summed these rankings to aid in the positioning of the LOC.

Although the line of correlation must exist (Rule 1), the available data seldom place sufficient constraints on precisely where it must lie. When one draws the LOC, it is a hypothesis, summarizing what appear to be the best correlations. Scientists should be prepared to face multiple working hypotheses. After one has drawn the LOC, one must ask: Do I like this line? This LOC mandates the following correlations and revisions of known ranges — do I still like this line? Every first-occurrence event that falls between the LOC and the axis representing the composite section, and every last-occurrence event that falls between the LOC and the axis representing the individual section, requires an extension of the known range. I find that moving and rotating the xy-graph on a piece of paper on a light table allows me to envision (and count) the range revisions each

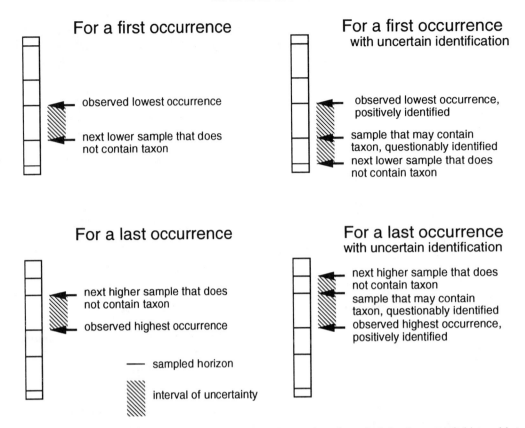

FIG. 4.—Intervals of uncertainty for first and last occurrences. The interval of uncertainty always includes the unsampled interval between the lowest or highest observation and the next lower or higher sample (left examples). It can also include sampled intervals in which the material does not permit positive identification (right examples).

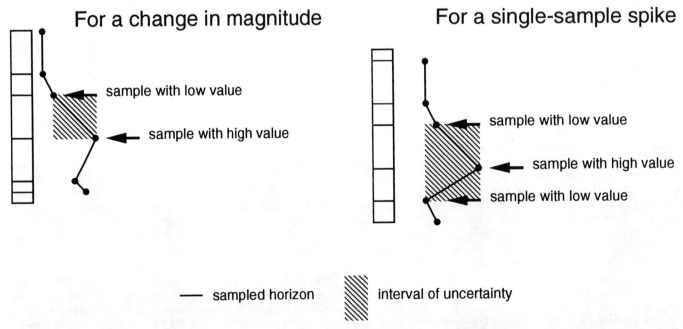

FIG. 5.—Intervals of uncertainty for quantitative data. The interval of uncertainty includes the unsampled interval between quantitative changes. This interval may extend in one direction (for an abrupt change in magnitude) or in two directions (for a single prominent value).

potential location of the LOC indicates. I can then look for a LOC that is both "economical" and geologically reasonable.

Rule 7. One Must Always Check the Axis Labels and the Scaling of any XY-graph Used in Graphic Correlation.

When more than two sections are being compared, some people choose to plot the composite section on the horizontal axis and others plot it on the vertical axis (Fig. 6). Some people work on drill holes and others work on outcrops; so high numbers can mean older sediments or younger sediments. Most computer programs are designed around the size of the screen or printer paper, not around the geologist's convenience.

In graphic correlation, the total range of each taxon is computed from its ranges in the various component sections that go to form the composite standard. Inherent to the method is the fact that slight errors in estimating the positions of the LOCs are more likely to lengthen ranges than to shorten them. Any individual error is equally likely to shorten or lengthen the range of a taxon in the compositing of a single section. In order for the range of a taxon to be artificially shortened in the composite, it must be artificially shortened in all individual sections. In order for the range of a taxon to be artificially lengthened in the composite, it need only be artificially lengthened in one of the individual sections. Thus, imprecision in the positioning of LOCs early in the compositing process can lead to artificially stretched ranges.

In order to reduce the effects of imprecise positionings of the LOCs, the graphic correlation procedure is an iterative process. The best or most complete section is selected as the standard reference section; all other sections are compared to it, and the data from them are added (composited) to it. When all sections have been compared and composited, the first round of compositing is complete. Compositing should continue until the positions of the LOCs stabilize. Early in the first round, ranges have not yet stabilized, and the correlation hypotheses are not the best that can be made. By continuing with additional rounds of compositing, more precisely positioned LOCs should improve the correlation hypotheses. Additional rounds of compositing should reduce ranges that have been extended artificially by slight errors in correlation.

A section should never be correlated against any part of itself, so one must always remember to remove any points in the composite that are derived from the section being composited. The second-lowest, lowest occurrence and the second-highest, highest occurrence are used when the lowest occurrence is based on the section being composited. Calculation of the sample-interval boxes contribute additional bookkeeping to the graphic correlation procedure, but computers are made to handle additional bookkeeping. If one thinks of the composite as containing all the samples from all the sections converted into the same standardized framework, it is easy to envision how the interval of uncertainty can be determined in the composite (Fig. 7).

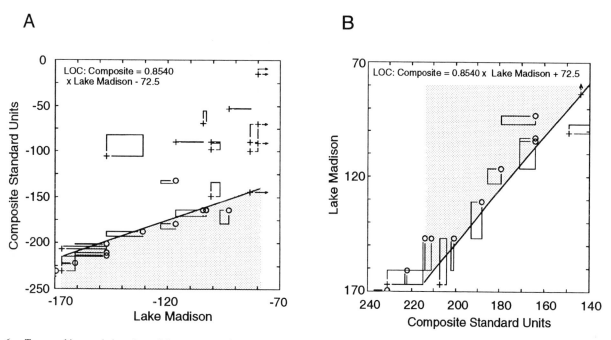

FIG. 6.—Two graphic correlation plots of the same core data (see Rule 7). On both graphs, stratigraphic up is to the top and to the right; first-occurrence events are plotted as "o," and the sample-interval boxes extend down and to the left; last-occurrence events are plotted as "+," and the sample-interval boxes extend up and to the right; and the equation for the LOC is given. (A) The composite is on the vertical axis, depths are expressed as negative numbers, and because the Lake Madison core represents only a portion of the composite section, the horizontal scale is different from the vertical scale. Here first-occurrence events that represent shorter ranges in the Lake Madison core than in the composite fall below the LOC (patterned area). (B) The composite is on the horizontal axis, depths are expressed as positive numbers with low positive numbers (shallow depths, stratigraphically younger material) to the top and right, the horizontal and vertical scales are the same, and the stratigraphically younger part of the composite has been cut off. Here, first-occurrence events that represent shorter ranges in the Lake Madison core than in the composite plot above the LOC (patterned area). For simplicity, labels of all events have been removed; these labels would be necessary for positioning of the LOC. The data are dinoflagellate-cyst occurrences in Virginia and Maryland; the Lake Madison core is in King George County, Virginia (Edwards, unpubl. data).

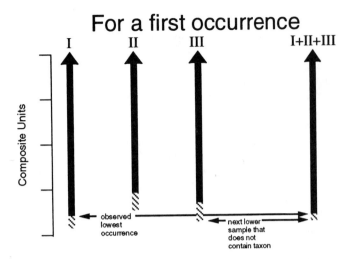

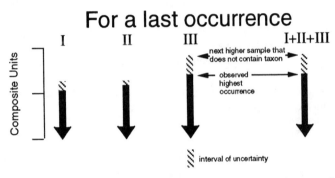

FIG. 7.—Intervals of uncertainty in the composite. The composite range of any taxon is built up from the stratigraphically lowest composite value for its first occurrence (here, section I) and the stratigraphically highest composite value for its last occurrence (here, section III). The interval of uncertainty for a first occurrence event extends from the lowest composite value for a sample that contained the taxon to the next lower sample in any section that could have contained the taxon but didn't (section III); the interval of uncertainty for a last occurrence event extends from the highest composite value for a sample that contained the taxon to the next higher sample in any section that could have contained the taxon but didn't (section III). To be conservative, I choose to exclude sample intervals in sections where the taxon does not fill its range (lower example, sections I and II). If the next lower or next higher sample is on a different segment of the LOC, so be it; what matters is where the correlation hypothesis places the sample in the composite.

REALITY

Rule 8. The Hardest Part of Graphic Correlation is Generating Reliable Data.

Sections have to be measured accurately. Samples have to be collected, processed, and examined. Taxonomy has to be accurate and consistent. The second hardest part is compiling the data. Gremlins seem to live in all keyboards. As a general rule, it takes as long to proofread a data set as it does to type it in. Remember Rule 2 above. If an occurrence plots conspicuously away from the LOC, it is important to recheck the taxonomy and the identification. It is important to take more samples, to look at more slides, to revise one's taxonomic concepts, and to do whatever it takes to get good points to use in correlation.

Rule 9. Graphic Correlation is a Tool, a Means to Solve Specific Scientific Problems.

The goal of graphic correlation is not to see how many rounds of correlation one can complete. In science, one poses questions and one seeks answers. One uses graphic correlation to solve problems in: stratigraphy, basin analysis, paleobiogeography, or evolutionary history.

Rule 10. One Should Use Graphic Correlation to Find Out Where Further Investigation is Needed, and Then One Should Investigate Further.

If graphic correlation (or any other analytical technique) produces unexpected results, do not accept them uncritically (see Rules 2 and 3).

ACKNOWLEDGMENTS

I thank Keith Mann for inviting me to write this paper and for not taking "no" for an answer. I thank N. O. Frederiksen and J. A. Stein for valuable comments on earlier versions of the manuscript, and I thank the many colleagues with whom I have discussed various aspects of graphic correlation over the last 20 years.

REFERENCES

DOWSETT, H. J., 1989, Application of graphic correlation method to Pliocene marine sequences: Marine Micropaleontology, v. 14, p. 3–32.
EDWARDS, L. E., 1984, Insights on why graphic correlation (Shaw's method) works: Journal of Geology, v. 92, p. 583–597.
EDWARDS, L. E., 1989, Supplemented graphic correlation: a powerful tool for paleontologists and nonpaleontologists: Palaios, v. 4, p. 127–143.
EDWARDS, L. E., 1991, Quantitative biostratigraphy, in Gilinsky, N. L. and Signor, P. W., eds., Analytical Paleobiology, Short courses in Paleontology, no. 4: Knoxville, The Paleontological Society, University of Tennessee, p. 39–58.
FREDERIKSEN, N. O., 1988, Sporomorph biostratigraphy, floral changes, and paleoclimatology, Eocene and earliest Oligocene of the eastern Gulf Coast: Washington, D.C., United States Geological Survey Professional Paper 1448, 68 p., 18 pl.
HAZEL, J. E., 1989, Chronostratigraphy of upper Eocene microspherules: Palaios, v. 4, 318–329.
MARSHALL, C. R., 1990, Confidence intervals on stratigraphic ranges: Paleobiology, v. 16, p. 1–10.
MILLER, F. X., 1977, The graphic correlation method in biostratigraphy, in Kauffman, E. G. and Hazel, J. E., eds., Concepts and Methods in Biostratigraphy: Stroudsburg, Dowden, Hutchinson and Ross, p. 165–186.
SHAW, A. B., 1964, Time in Stratigraphy: New York, McGraw-Hill, 365 p.
STRAUSS, D. J. AND SADLER, P.M., 1989, Classical confidence intervals and Bayesian probability estimates for the ends of local taxon ranges: Mathematical Geology, v. 21, p. 411–427.
SWEET, W. C., 1979, Late Ordovician conodonts and biostratigraphy of the western midcontinent province, in Sandberg, C. A. and Clark, D. L., eds., Conodont Biostratigraphy of the Great Basin and Rocky Mountains: Brigham Young University Geology Studies, v. 26, p. 45–85.

ESTIMATING THE LINE OF CORRELATION

NORMAN MACLEOD

Department of Palaeontology, The Natural History Museum, Cromwell Road, London SW7 5BD, England

AND

PETER SADLER

Department of Earth Sciences, University of California, Riverside, California, 92521

ABSTRACT: Accurate estimation of the line of correlation (LOC) is an important goal of graphic correlation. Since wide variety of both qualitative and quantitative methods are currently available to guide LOC estimation, it is crucial that the procedure used to infer a working LOC be completely specified and, thus, reproducible. The most popular LOC estimation technique, termed "splitting tops and bases," takes advantage of inherent differences in the chronostratigraphic implications of biostratigraphic first and last appearance datums (FADs and LADs, respectively). This method takes, as its starting point, a Standard Reference Section (SRS) that contains occurrences of a large number of species distributed over a long stratigraphic interval with no obvious gaps, hiatuses, or structural complications. If the SRS is well-chosen, a large number of its constituent taxa may occur at horizons corresponding to their global FADs and LADs. In such instances, graphic comparison of the SRS with datum positions from other sections or cores will result in the grouping of FADs and LADs on opposite sides of the true LOC. Using this relationship, the geometry of the LOC can be inferred by finding that line which divides FADs from LADs in the most efficient manner. Adjunct criteria that may be useful in establishing a qualitatively-defined LOC or distinguishing between alternative LOC geometries include consideration of key beds, datum weighting, and parsimony.

Regression analysis can also be used to quantitatively estimate LOC positions. Unlike qualitative LOC estimation, regression-based techniques make no assumptions with respect to differences between biostratigraphic datum types. Least squares linear regression has a long history of use in graphic correlation as a LOC estimation technique. However, least-squares methods make a necessary distinction between independent and dependent variables (= sections or cores). This distinction, along with its implicit corollary of a cause and effect relationship between variables, seems out of place in the context of a graphic correlation problem. Reduced major axis and major axis regression models provide a much closer match between the assumptions of the regression model and the nature of stratigraphic data, in addition to providing a better estimate of the bivariate linear trend. Reduced major axis regressions have the added advantage of being much easier to calculate than either least squares or major axis regressions. These alternative regression algorithms can be modified to take key beds and weighted data into consideration. Significance tests are available for least squares and major axis regressions, but these are, at best, indirect tests of the quality of the fit between the estimated LOC and the underlying data. Analysis of residual deviations from the LOC provides the best indication of this fit.

Linear regression methods minimize standard measures of fit between the LOC and the underlying data. While minimization of these parameters usually involves straightforward computations, these numerical recipes often fail to respect the stratigraphic implications of the minimization procedures, specifically those involving the inferred extension of local stratigraphic ranges. The sum of all local range extensions implied by a LOC has been termed the "economy of fit." No deterministic formula leads directly from the raw stratigraphic data to the LOC exhibiting the best economy of fit. However, iterative constrained optimization procedures are available to efficiently search for LOCs with very good economy of fit characteristics. In most instances, LOCs with the best economy of fit or smallest net range extension exhibit a piecewise linear geometry. Constrained optimization search strategies can also be combined with a multivariate representation of raw stratigraphic data to locate highly economical LOCs for all sections simultaneously.

INTRODUCTION

Placement of the line of correlation (LOC) is the single most important element in graphic correlation, for it is this procedure that defines the temporal relationship between two (or more) stratigraphic sequences. Because the true correlation between any two sections is never fully knowable, alternative LOC estimation procedures must be judged by their respective abilities to provide an adequate fit to the observed data in a manner that extracts the maximum amount of useful information and, at the same time, respects the chronostratigraphic implications of different datum types. In addition, any procedure used to infer the LOC must be both explicit and reproducible. It is therefore odd that few detailed descriptions of this procedure exist and that many of these treat LOC estimation in only the most general terms. Most previous discussions of the graphic correlation method (Shaw, 1964; Miller, 1977; Edwards, 1984) focus on the description of a particular LOC estimation technique rather than providing a comprehensive overview of all qualitative and quantitative procedures that either have been or might be useful in constraining LOC geometries. In this chapter, we seek to address the problem of LOC estimation from a data analytic point of view, emphasizing both the similarities and differences between various qualitative and quantitative procedures and describing the assumptions that follow from the selection of particular methods.

To facilitate our discussion of alternative LOC estimation strategies within bivariate graphic correlation analyses, we will employ the data listed in Table 1. These data were taken from Miller (1977, p. 170) who used them to illustrate a qualitative approach to LOC placement. They consist of first or last appearance datums (FADs and LADs) for twelve hypothetical taxa. [Note: Miller (1977) probably based his discussion on real rather than purely hypothetical biostratigraphic data, but the origin of his example dataset was not discussed.] In addition to these biostratigraphic data, we have added a key bed (position of the base of a hypothetical bentonite layer) for illustrative purposes.

A wide variety of stratigraphic data can be used to constrain the LOC. These include biostratigraphic datums (FADs and LADs at any hierarchical rank, including species, metaspecies, subspecific morphotypic variants, morphologs, cladistic characters, and character complexes), key beds, data from physical or chemical logs, magnetozone boundaries, etc. The only requirements necessary for using these data to infer stratigraphic correlations are that each observation be independent and located within the measured sequences by reference to a common unit of distance. Since each observation is made independently, each contributes a unique source of information to the overall correlation. The requirement of datum independence, however, in no way precludes intra-datum comparisons as a means of evaluating alternative LOC geometries.

TABLE 1.—EXAMPLE DATA*

Taxa	Section A FAD	Section A LAD	Section B FAD	Section B LAD
1	450.0	538.5	471.1	500.0
2	400.0	517.9	428.6	500.0
3	184.6	500.0	168.8	500.0
4	130.8	466.7	85.7	500.0
5	—	—	319.5	459.7
6	269.2	407.7	129.9	415.6
7	92.3	348.7	54.5	352.6
8	61.5	300.0	0.0	381.8
9	30.8	230.8	0.0	213.0
10	—	220.5	—	184.4
11	141.0	425.6	155.8	275.3
12	102.6	384.6	142.9	246.8
Bentonite		310.0		327.7

*Biotic data estimated from Miller (1977, Text-Figure 3)

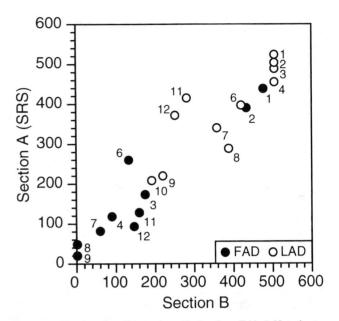

FIG. 1.—Bivariate plot of biostratigraphic data from Table 1. Note the manner in which the first appearance datums (FADs) of taxa 8 and 9 along with the last appearance datums (LADs) of taxa 1–4 line up perpendicular to the Section B axis. This geometry suggests the presence of two hiatuses in Section B bounding an interval of net positive sediment accumulation in both sections. The relative duration of these hiatuses can be estimated by determining the distance covered by these two vertical offsets along the Section A axis.

Some authors (e.g., Olsson and Liu, 1993; Ivany and Salawitch, 1993) have proposed that biostratigraphic data must include both first *and* last appearances of all taxa used in order to serve as the basis for graphic correlation analysis. While most example datasets, including the one used herein, do consist of first and last appearances for the majority of taxa, this is only a matter of descriptive convenience facilitating shorter tables and simpler discussions. No one should interpret use of these example datasets to mean that both the FAD and LAD must be present for each taxon within the sampled interval. Miller (1977, p. 169) states that, "the graphic method of correlation uses the total range of *as many fossils as possible* to locate the LOC as accurately as possible" (emphasis added), whereas Edwards (1984, p. 596) describes the notion that biotic data in a graphic correlation analysis must include each species' FADs and LADs as "unrealistic."

Once a stratigraphic dataset has been assembled and the various constituent data types identified, comparisons of sequences can be made by constructing a pairwise series of bivariate plots as in Figure 1. On these plots, data types are typically distinguished from one another, and the biotic data points identified to taxon. At this point in the analysis, the only constraint that applies to the LOC geometry is that it be a non-decreasing function passing somewhere through the cloud of data points. This constraint leads to the first and perhaps most fundamental assumption of LOC placement, which is that the graphed data bear some deterministic relationship to the actual line of correlation. It is here that the task of LOC estimation begins and the various techniques for LOC estimation diverge.

QUALITATIVE ESTIMATION PROCEDURES

Currently, the most popular method of LOC estimation for biostratigraphic or combined biostratigraphic-lithostratigraphic data involves the qualitative assessment of LOC geometries according to a principle loosely termed, "splitting tops and bases" (Miller, 1977). This phrase refers to biostratigraphic FADs (bases) and LADs (tops). Typically, the thickest, most fossiliferous, and least structurally complex succession is designated the Standard Reference Section or Standard Reference Sequence (SRS). The SRS serves as an initial estimate of the global sequence of events and provides the spatial metric for all subsequent comparisons. As biostratigraphic information from other sections or cores is added to the SRS it becomes a composite of the best information present in many different sections. To reflect this change in the nature of the SRS, once its original sequence has been modified in any way it becomes a Composite Standard (CS) or Composite Standard Reference Section (CSRS).

When comparing another section/core to the SRS on a bivariate plot, all FADs that are synchronous or that occur temporally later (above) their corresponding positions on the SRS will plot either on or below the true LOC, while all LADs that are synchronous or that occur temporally earlier (below) their corresponding positions on the SRS will plot either on or above the true LOC (Figs. 2A, B). [Note: these geometric conventions assume that the SRS is plotted on the *y*-axis. For graphs in which the SRS is plotted on the *x*-axis, exchange the terms LAD and FAD in the preceding and following sentences.] Biostratigraphic datums that are potentially out of their proper position in the global sequence (as represented by the SRS) will be identified as FADs and LADs (respectively) that plot above and below the bivariate trend of the data (Fig. 2C). Miller's (1977, Fig. 3) LOC for the example data (Fig. 3A) groups them into three categories (Table 2): (1) those datums that are synchronous in both sequences, (2) those datums that are "in sequence" with respect to the SRS but underestimate their global maxima in Section B (within the Section B sequence these datums are sometimes referred to as "unfilled local ranges," Hazel, pers. commun.), and (3) those datums that are "out of sequence" with respect to the SRS (= FADs above, LADs below the LOC). Working backward from the assumed reliability of the SRS, qualitative LOC estimation via the traditional "splitting tops and bases" criterion involves a search for the LOC geometry that

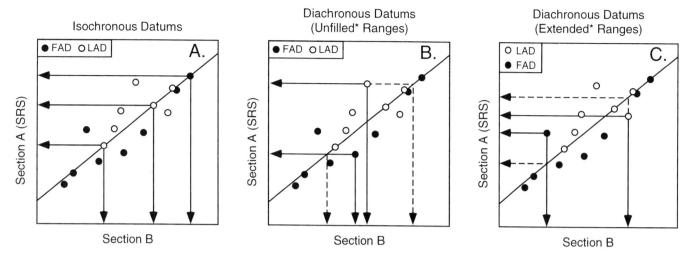

FIG. 2.—Implications for different datum geometries for chronostratigraphic inference and qualitative graphic correlation. Within any bivariate plot of stratigraphic data, a line of correlation (LOC) may be drawn. In all three diagrams, the LOC is shown as a diagonal line with a positive slope. By definition, the LOC represents the coordinate position of all isochronous data points in both sections. By using the LOC as a mapping function, a stratigrapher may project any point in one section into its temporally equivalent position in the other section. (A) Datums lying on or very close to the LOC occupy temporally equivalent positions in both sections. (B) FADs lying below the LOC and LADs lying above the LOC represent datums whose observed first and last appearances in Section B (solid lines) occur later and earlier (respectively) than their time-equivalent positions in Section A (dashed lines). Since Section B underestimates the globally maximum first and last appearances of these taxa relative to Section A (the SRS), these Section B datums constitute submaximal or unfilled ranges. (C) FADs lying above the LOC and LADs lying below the LOC represent datums whose observed first and last appearances in Section A (solid lines) occur earlier and later (respectively) than their time-equivalent positions in Section B (dashed lines). Since, in this case, the SRS underestimates the globally maximum first and last appearances of these taxa, the information provided by Section B can extend each species' maximum stratigraphic range so that they more closely approximate the global sequence of FADs and LADs. Using these relationships, a LOC geometry can be inferred via location of the line that places as may FADs on or below the LOC and as many LADs on or above the LOC as possible. This qualitative strategy for LOC estimation is informally termed "splitting tops and bases." In all of these graphs the SRS is plotted along the ordinate rather than the abscissa in order to conform to the geologically standard manner for representing the axis of time running vertically up the page. * = Relative to Section A.

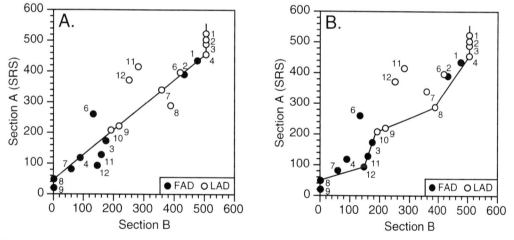

FIG. 3.—Two different qualitatively-derived LOCs drawn through the example dataset. (A) A three-segment LOC suggesting the presence of a long central period of net positive sediment accumulation at a constant rate bounded by two hiatuses within Section B. (B) A six-segment LOC suggesting that the central time interval contains evidence for at least four changes in the sediment accumulation rate. See Table 1 for subdivision of the datums into chronostratigraphic groups.

places as many biostratigraphic datums as possible into categories 1 and 2 while minimizing the number assigned to category 3.

Provided the data are densely distributed within the two sections/cores and most datums on the SRS lie close to their global maxima, there will be little doubt as to the general shape of the LOC. Unfortunately, such unambiguous correlations are exceptional. More often, a series of alternative LOC geometries that more or less equally satisfy the "splitting tops and bases" criterion can be found. For example, Figure 3B presents an alternative LOC for the Miller (1977) data that contains four changes in sediment accumulation rate relative to the SRS in the lower portion of Section B. These changes are suggested by the linear sequence of FADs for taxa 3, 11, 12; LADs for taxa 8, 9, 10; and nearly linear sequence for the LADs of taxa 4 and 8 and FADs of taxa 1 and 2. If this alternative LOC is accepted, the allocation of datums to the various categories changes as shown in the middle of Table 2.

Given that multiple LOC geometries are almost always possible for any graphic comparison between sections/cores, which LOC (= correlation model) should be preferred? The graphic correlation literature dealing with qualitative LOC placement procedures is frustratingly vague on this point. Most treatments suggest that final decisions be based on the "judgement and experience of the stratigrapher." Of course, there is no substitute for a personal knowledge of the faunas and lithologies involved. However, such gestalt-like appeals hardly constitute a reproducible procedure. Some LOC geometries will imply the existence of particular stratigraphic phenomena (e.g., hiatuses) that can be checked against other observations or that might be identified using additional analyses (e.g., data from other biotic groups, physio/chemical logs). In this way, a cycle of positive analytic feedback can often be used to improve the original correlation. In the absence of such feedback cycles, though, it is all too often the case that other, less explicit, criteria are employed to resolve discrepancies between alternative qualitatively-derived correlation models. Several of these have received some brief discussion in the graphic correlation literature while many other "rules of thumb" are applied by individual stratigraphers on a case-by-case basis. Some of the more prominent of these adjunct qualitative LOC estimation criteria are discussed below.

Key Beds

As noted by most commentators, the existence of unique lithostratigraphic units with clear chronostratigraphic significance (e.g., bases of bentonite layers, volcanic ash flow tuffs) and that are unambiguously traceable on a regional scale can provide important constraints on LOC placement. By definition, the LOC must pass through the coordinate positions of such units. This principle applies to all LOC placement procedures regardless of whether they are based on qualitative or quantitative forms of inference. Key beds, if fortuitously placed, may furnish critical data in areas where biostratigraphic control is lacking due to wide sample spacing or ambiguous due to intradatum diachrony. A plausible scenario is supplied by the example dataset (Table 1). In this example, presence of a bentonite layer within both successions resolves one aspect of the two correlation models proposed in Figure 3 by providing information in a biotically unconstrained region of the LOC (Fig. 4).

Linearity of the LOC

Both Shaw (1964) and Miller (1977) state that linear LOCs should be preferred to LOCs composed of several differently oriented segments or "doglegs." These authors accept the proposition that non-linear sediment accumulation rates should be expected at all levels of stratigraphic resolution but argue that short-term departures from linearity of modest magnitude (>10%) may be ignored because, relative to other factors, these either introduce little error (Shaw, 1964; Miller, 1977) or greatly complicate the calculations required for a complete analysis (Miller, 1977).

Our example dataset poses a problem of this general type. In this data array we contrast two correlation models, one of which (Fig. 4A) represents a single linear LOC bounded both above and below by hiatuses, while the other (Fig. 4B) consists of three LOC segments bounded, as before, by the two hiatuses. After correction of the LOCs using the key bed, both correlation models are identical for the interval >200 units on the SRS but differ substantially in the more information-rich portion of the sampled sequence between 0 and 200 units. Two different trends appear to characterize this interval. One strictly linear trend is defined by the FADs of taxa 4, 7, and 8 along with the LADs of taxa 9 and 10. The other is a doglegged trend consisting of two LOC segments defined by the FADs of taxa 3, 11, and 12 along with the LAD of taxon 10 and by the FADs of taxa 7, 8, and 12. Both of the doglegged segments of Figure 4B differ from the bivariate trend of the data (represented by Fig. 4A) by more than 10 % of arc. Certainly, if we restrict our universe to the interval between 0 and 250 units on the SRS, there is little doubt that the proper LOC consists of the two doglegged segments.

Weighting of Datum Events

Another way of attempting to resolve differences between competing qualitative LOC models is to prefer those models whose LOCs run close to the coordinate position of datums the stratigrapher considers more reliable. In many instances, these may be the zone-defining datums of traditional zone-based biostratigraphy. Any LOC that fails to maintain the integrity of well-established biozones must be treated with skepticism. Another informal rule of thumb that falls into this general category is to regard with suspicion any LOC that implies the existence of a species in a zone in which it has never been independently observed (Klapper, pers. commun.). This is not to say that the results of a graphic correlation can never point out problems with a biozonation or improve global taxon range estimates. Rather, this principle simply states that LOC geometries calling for fundamental revisions in our current understanding of taxon ranges must be very carefully evaluated and should not necessarily be preferred over alternative correlation models.

Differential weighting of particular stratigraphic datums seems a natural extension of traditional index fossil biostratig-

TABLE 2.—CHRONOSTRATIGRAPHIC DATUM TYPES

LOC Model	Datums	Isochronous Datums (= on LOC)	Diachronous Datums	
			Unfilled Ranges* (= FAD's Below, LAD's Above)	Extended Ranges* (= FAD's Above, LAD's Below)
Miller (1977):	FAD's	1,4,8,9	2,3,7,11,12	6
Fig. 3A,4A	LAD's	1,2,3,4,7,9	6,10,11,12	8
	Percent	47.62	42.86	9.52
	Cum. Percent	47.62	90.48	100.00
Six Segment				
LOC:	FAD's	3,8,9,11,12	—	1,2,6,7,4
Fig 3B	LAD's	1,2,3,4,8,9,10	6,7,11,12	—
	Percent	57.14	19.05	23.81
	Cum. Percent	57.14	76.19	100.00
Five Segment				
LOC:	FAD's	1,3,8,9,11,12	2	4,6,7
Fig 4B	LAD's	1,2,3,4,7,9	6,10,11,12	8
	Percent	57.14	23.81	19.05
	Cum. Percent	57.14	80.95	100.00

*Relative to the Standard Reference Section.

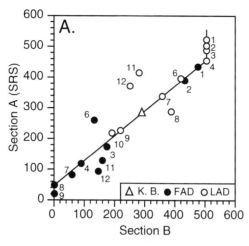 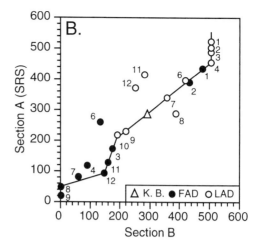

FIG. 4.—Alternative correlation models supported by the example biostratigraphic data along with a key bed (K.B. = bentonite layer in Table 1). The presence of lithostratigraphic units with an unambiguous chronostratigraphic interpretation can often aid in positioning the LOC and/or discriminating between alternative correlation models. (A) A three-segment LOC suggesting the presence of a long central period of net positive sediment accumulation at a constant rate bounded by two hiatuses within Section B. (B) A five-segment LOC suggesting that the central time interval contains evidence for at least three changes in the sediment accumulation rate. See Table 1 for subdivision of these datums into chronostratigraphic groups.

raphy into the realm of graphic correlation. Of course, different workers may assign different weights to the same datum and, on the basis of those weights, produce different results. For a rigorous quantitative procedure, this is a desirable feature. Differentially weighted data should lead to different LOC geometries. Nevertheless, when used in conjunction with qualitative LOC estimation procedures, this approach gives the individual stratigrapher a great deal of latitude in the manner in which the weights are used while at the same time failing to require explicit specification or justification of the weighting scheme. It is therefore incumbent on any stratigrapher using differentially weighted datum events in conjunction with a qualitative LOC estimation procedure to provide adequate discussion of the relative importance he or she is assigning to different biostratigraphic datums, the consistency with which differential datum weights are maintained throughout the analysis, and the manner in which they are taken into account during the LOC placement process. These recommendations are quite general and constitute little in addition to the explicit discussion of methods required of every scientific report. Failure to provide this basic information with respect to the manner in which the results were obtained risks misunderstanding on the part of the intended audience and consequent lack of confidence in those results.

We suspect that the differences between Miller's (1977) original graphic correlation of the example dataset (Fig. 4A) and our alternative correlation (Fig. 4B) stem largely from differential datum weighting. If the FADs of taxa 3, 11, and 12 are regarded as having little chronostratigraphic significance there is no reason to seriously consider the more complex LOC (Fig. 4B). However, this discrepancy points out the need for explicit discussion of assigned datum weights in evaluating alternative LOC geometries and in providing justification for the overall correlation model. In the absence of such a discussion, there seems no reason to suppose that the more complex doglegged LOC of Figure 4B is not the more reasonable model (see also discussion below). If we are to avoid needless and largely counterproductive controversy over the merits of particular graphic correlation results, the chain of reasoning that led to those results must be made clear, especially when based on qualitative analytic procedures.

Parsimony

The most general of the informal "rules" for qualitative LOC placement involves a preference for the simplest LOC geometry consistent with the available data as a whole. Aspects of this concept are touched on by Shaw (1964) under the heading, "Economy of Fit" (p. 254–257). As it applies to graphic correlation, the principle of parsimony is best understood by remembering that alternative LOC placements represent alternative explanatory models, all of which seek to account for as much of the observed data as possible. Datums falling on the LOC are identified as being time-equivalents. Thus, individual correlation models are able to "explain" or "account for" datums that lie on or close to their LOC segments. On the other hand, datums whose coordinate positions lie off the LOC are diachronous. Since diachrony may arise as the result of a large number of causes (e.g., non-uniform patterns of range expansion/contraction, environmental change, preservational factors), the reason for their diachrony remains unknown. Most LOCs will identify a greater or lesser proportion of the available data as being diachronous and thus unaccounted for by the correlation model other than through *ad hoc* or *a posteriori* means.

The potential utility of parsimony in evaluating alternative correlation models is well-illustrated by our example dataset. The alternative LOCs shown in Figure 4 differ in their complexity and their ability to explain the observed data. The three-segment LOC (Fig. 4A) passes through 10 of the 21 datum coordinates whereas the more complex six-segment LOC (Fig. 4B) passes through 12 datums. Consequently, the latter model explains a larger proportion of the data and might be preferred. Another way to assess the relative parsimony of these two correlation models is to use the spatial metric of the graphic correlation axes to summarize the total amount of range adjustment implied by each hypothesis. This method provides a measure of the overall fit of each model to the data. If, in our example, this summation is referenced to the SRS axis alone, the totals are 625 and 607 for the three-segment and six segment models, respectively. This results suggests, once again, that the six-segment correlation model provides a better fit to the data.

Note that the parsimony approaches outlined above do not distinguish between unfilled ranges and range extensions (see Table 2) and attempt to scale the parsimony indices with measures of LOC complexity. Such additions to the evaluation procedure are possible and have played a role in many previous qualitative graphic correlation analyses (see Shaw, 1964; Miller, 1977). We have not employed these additional constraints here because we see no logical reason to distinguish between these two different types of diachrony. Unfilled ranges are not "less incompatible" with particular correlation models than range extensions; they merely represent a different type of incompatibility. In addition, we find it difficult to objectively defend what seems an arbitrary preference for less complex LOCs. As LOC complexity increases, the amount of data explained will increase up to the point where the only data points remaining off the LOC are those whose positions cannot be accounted for under any conceivable geometry. Beyond this point there is nothing to be gained by increasing LOC complexity still further while any simplification of the LOC below the point of maximum parsimony unavoidably degrades the model. Since graphic correlation readily lends itself to the development of alternative correlation models, its practitioners must be explicit as to the criteria they use to evaluate alternative correlations for the same data. To the detriment of graphic correlation's image within the larger stratigraphic community, this important aspect of graphic correlation procedure has been consistently neglected. One approach to this problem is to extend Miller's (1977) concept of the correlation channel by developing and presenting both maximally parsimonious and least complex or "best case" and "worst case" correlations for the same data (see MacLeod, 1991, MacLeod and Keller, 1991, MacLeod, in press, MacLeod, this volume). Under this convention, the "true" correlation can be regarded as existing somewhere within the range specified by these end member models.

QUANTITATIVE ESTIMATION PROCEDURES

In the original presentation of graphic correlation, Shaw (1964) advocated LOC estimation via least squares linear regression. Though, at least in terms of popularity, this approach has largely been superseded by more qualitative methods, several prominent practitioners continue to employ regression methods to obtain LOC-based correlation models. Like all quantitative techniques, LOC placement via regression analysis has several advantages over more qualitative methods including: reliance on explicit and consistently applied assumptions, reproducibility, sensitivity to data weights, and in some cases, recourse to tests designed to evaluate the statistical significance of results. On the other hand, application of regression-based strategies to LOC placement involves radically different assumptions with respect to the relationship between the observed data and the true correlation. In addition, the possible existence of non-linearities in the correlation between two stratigraphic sequences greatly complicates the computational aspects of LOC estimation via regression analysis.

Prior to the advent of inexpensive but powerful computers and appropriate software, it was impractical to employ any but the simplest regression techniques to the problem of LOC placement on a routine basis. Fortunately, the current availability of powerful numerical data analysis packages have all but eliminated the computational burdens associated with such methods, thereby providing an unprecedented opportunity to explore the applicability of quantitative LOC estimation methods. In this review, we will discuss methods of bivariate linear regression first, and then introduce a multi-variate, non-linear method explained more fully by Kemple and others (this volume).

Least Squares Regression

Due to its widespread utility in many different contexts, the least squares approach to regression is by far the most popular. Indeed, owing to its treatment in most elementary statistics courses and near ubiquitous inclusion within even very low level graphing and data analysis software, many assume the least squares model to be synonymous with the concept of numerical regression analysis or "curve fitting." This is a gross oversimplification. Least squares is simply one among many different approaches to regression analysis (e.g., Myers, 1986), each of which should be evaluated for its appropriateness to particular problems.

In a bivariate context, the least squares model assumes the existence of an independent variable that can be measured without error and a dependent variable that is observed or measured with unsystematic error. A linear least squares regression finds the line that minimizes the sum of the squared deviations between itself and the observed data points *along the dependent variable axis* (see Fig. 5). Given the traditional association be-

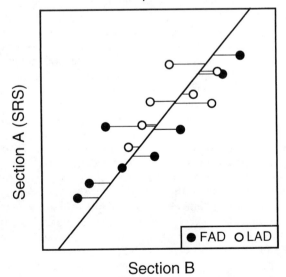

FIG. 5.—Minimization criteria used by least squares regression models to estimate LOC positions. Thin lines parallel to the abscissa = residual deviations from regression model. This procedure finds the line that minimizes the squared deviations between itself and the observed data along the dependent variable axis (Section B).

tween least squares regression and graphic correlation, it is disconcerting that the distinction between independent and dependent variables required of numerical and statistical applications of the least squares model has no detailed correspondence to the situation presented by most graphic correlation problems where the datum locations on neither axis can be assumed error (= diachrony) free. In addition, the entire rationale underlying a least squares problem that variation in the independent variable is logically associated with or "causes" variation in the dependent variable (Sokal and Rohlf's, 1981 Model I regression) seems dubious from the standpoint of graphic correlation because of the inherently complex dynamics of multi-species ecological systems coupled with the vagaries of preservation, diagenesis, and sampling.

The least squares concept, like all regression models, also forces the user to adopt a completely different stance with respect to the manner in which the bivariate distribution of datums relate to the LOC. Instead of making a distinction between FADs and LADs and fitting a line that best separates these datum subgroups on opposite sides of the plot (see above), the regression approach assumes that the error terms forcing individual datum coordinates off the LOC be randomly distributed without distinction as to datum type. Shaw's (1964) discussion contains several examples of actual data for which the least square regression line does effectively split the biostratigraphic tops and bases (see his Figs. A7-A9), but this will not necessarily be the case.

A least squares analysis of the central portion of the example dataset (excluding those datums located along the bounding hiatuses) illustrates this point (Fig. 6). [Note: since the equations for calculating a least squares regression can be found in almost any introductory statistical text they will not be repeated here.] Because the example dataset contains several datums exhibiting substantial diachrony under any linear interpretation, the least squares algorithm is forced to find the line that minimizes the sum of the squared deviations along the Section B axis (= dependent variable) for these as well as for datums that lie closer to the bivariate linear trend. This variation within the data has the effect of rotating the regression line about the centroid of the data (A = 273, B = 247) to a skewed position with respect to the main trend. In this orientation, the least squares LOC fails to lie close to many of the datums. Consequently, the least squares LOC identifies relatively few datums as being isochronous. Even these identifications might be artifactual though, in that most lie close to the data's centroid and will always occupy positions close to any LOC whose method of placement forces it to pass through this point (see below).

Reduced Major Axis Regression

It has long been realized that for data with an appreciable amount of scatter, a least squares regression often results in a line that differs markedly from one that might be "fit" by simple visual inspection. This results from the fact that the least squares model seeks to minimize deviations between the fitted line and the observed data along only one of the two variable axes. Although this minimization criterion makes perfect sense whenever a fundamental distinction between independent and dependent variables exists, in a large number of situations (such as graphic correlation) this distinction seems artificial. As an alternative, reduced major axis regressions (Kermack and Haldane, 1950; Imbrie, 1956; also known as geometric mean regressions, Ricker, 1973, and standard major axis regressions, Jolicouer, 1975) attempt to minimize the sum of the areas of the right triangles formed by deviations between the regression line and the observed data points along both *x*- and *y*-axes (Fig. 7). This minimization criterion assumes that error may be present on both axes, thus negating the distinction between independent and dependent variables. Rather than attempting to find the linear equation that best predicts the value of *y* in terms of *x*, reduced major axis regressions attempt to summarize the functional relationship between these variables (= sections or cores).

Calculation of the slope (*a*) for a reduced major axis regression is simple, being the ratio of the variables' standard deviations (S_x, S_y), as follows:

$$a = \frac{S_Y}{S_x} \quad (1)$$

For all valid applications of reduced major axis regression in graphic correlation, the sign of the slope must, of course, be positive. However, owing to the fact that equation 1 is ambiguous with respect to the sign of the slope, this parameter should be checked. The sign of the reduced major axis slope is equal to that of the corrected sum of products as calculated in the standard manner:

$$sp = \Sigma xy - \frac{(\Sigma x)(\Sigma y)}{n} \quad (2)$$

Once this slope has been determined, the *y*-intercept of the regression (*b*) may be calculated as follows:

$$b = \bar{Y} - (a \cdot \bar{X}) \quad (3)$$

Where: (X,Y) = the bivariate centroid.

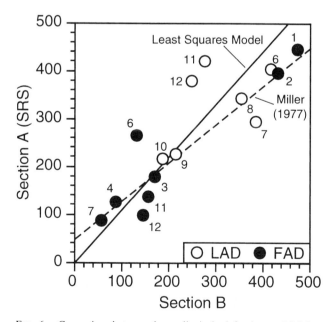

FIG. 6.—Comparison between the qualitatively-defined central LOC segment of Miller (1977) and a linear least squares LOC for the example dataset (exclusive of datum points lying along the bounding hiatuses; see Fig. 3A). Note the pronounced angular offset of the least squares LOC from the main linear trend of the data.

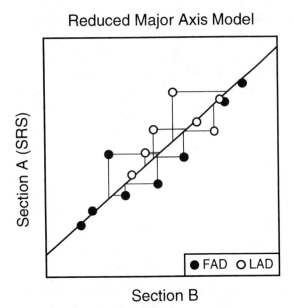

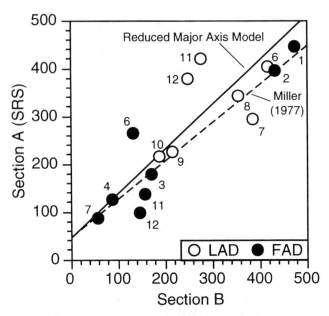

FIG. 7.—Minimization criteria used by the reduced major axis linear regression model to estimate LOC positions. Thin lines parallel to the abscissa and ordinate = residual deviations from regression model. This procedure finds the line that minimizes the sum of the areas of the triangles formed by the residual deviations along both the *x* and *y* axes. Since deviations from the model along both axes are used to estimate the LOC, this method does not require recognition of any inherent differences between variables (= stratigraphic sections) in terms of measurement accuracy or the logical causation of one by the other.

FIG. 8.—Comparison between the qualitatively-defined central LOC segment of Miller (1977) and a linear reduced major axis LOC for the example dataset (exclusive of datum points lying along the bounding hiatuses; see Fig. 3A). Note the much closer correspondence of the reduced major axis LOC to the main linear trend of the data. Compare with Figure 6.

Results of a reduced major axis regression for our example data are shown in Figure 8. In this instance the reduced major axis method yields a line much closer to Miller's (1977) linear LOC based on qualitative criteria than was achieved by the least squares approach.

Major Axis Regression

A more direct method of estimating a line that best represents the functional relationship between two variables is to minimize the perpendicular distance between the line and the individual data points (Fig. 9). This is known as a major axis regression, and it is identical to the determination of the first principal component for a matrix of bivariate data. Fortunately, calculation of the major axis slope for bivariate data can be made via a single (albeit complex) equation, given in Sokal and Rohlf (1981, p. 596–597). The major axis regression for our example data (Fig. 10) provides a marginally closer fit to Miller's (1977) qualitatively-determined LOC than does the reduced major axis approach and a substantially better fit than the corresponding least squares regression.

A comparison between the equations determined by the least squares, reduced major axis and major axis, along with an estimate of Miller's (1977) qualitatively-determined LOC equation are provided in the upper portion of Table 3. Both reduced major axis and major axis regressions converge on Miller's (1977) qualitative solution to a much higher degree than did the least squares model. Unfortunately, the fact that each regression used different optimization criteria prevents determination of whether or not these two regression models are significantly

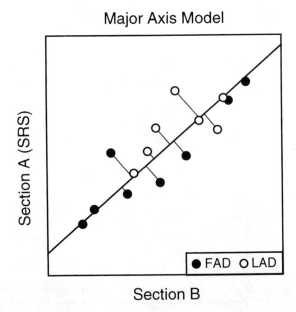

FIG. 9.—Minimization criteria used by the major axis linear regression model to estimate LOC positions. Thin lines perpendicular to the LOC = residual deviations from regression model. This procedure finds the line that minimizes the sum of the residual deviations perpendicular to the regression line itself. Since deviations from the model along both axes are used to estimate the LOC, this method, like reduced major axis regression, does not require recognition of any inherent differences between variables (= stratigraphic sections) in terms of measurement accuracy or the logical causation of one by the other.

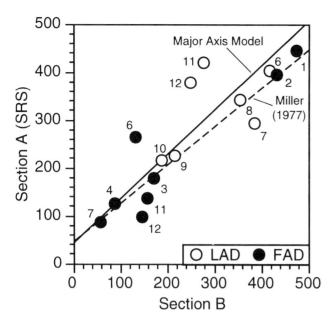

Fig. 10.—Comparison between the qualitatively-defined central LOC segment of Miller (1977) and a linear major axis LOC for the example dataset (exclusive of datum points lying along the bounding hiatuses; see Fig. 3A). Note the close correspondence of the major axis LOC to the main linear trend of the data. Compare with Figures 6 and 8.

different from Miller's (1977) LOC. Nevertheless, given the scatter in the observed data, as well as the assumption of equal weighting among the various data points, it is very doubtful that substantial differences exist. Reduced major axis and major axis regressions locate LOCs near the bivariate linear trend for these data and conform more closely to the assumptions of the graphic correlation method. Such will not always be the case, however. As indicated by equation 3, all three regression models are calculated in such a way as to pass through the bivariate centroid. While this is a desirable feature for most applications of regression models, the bivariate centroid has no particular significance in a graphic correlation problem. Thus, any quantitative LOC placement method that forces the LOC to pass through this point introduces a potentially artificial constraint on the resultant correlation model.

Weighted Regression Analysis

As with qualitative LOC placement procedures, some of the inherent difficulties in applying regression analysis to LOC es-

TABLE 3.—SUMMARY OF LOC ESTIMATION VIA LINEAR REGRESSION ANALYSIS

Regression Model	Equation
Miller (1977)	$y = 1.234x - 54.830$
Unweighted	
Least Squares	$y = 0.914x - 2.356$
Reduced Major Axis	$y = 1.065x - 43.077$
Major Axis	$y = 1.076x - 42.092$
Weighted*	
Least Squares	$y = 1.003x - 6.935$
Reduced Major Axis	$y = 1.067x - 24.398$
Major Axis	$y = 1.072x - 25.631$

x = Section A, y = Section B.
*See text for description of weighting model.

timation may be operationally minimized through the use of various data weighting schemes. The weighting of data in a regression analysis is most easily achieved via differential repetition of more highly weighted data points within the bivariate data matrix, after which regression analysis is carried out according to the methods described above (Pollard, 1977). For example, if we believe that the FADs of taxa 1 and 7, and the LAD of taxon 9 are all approximately three times more reliable than any other datum point, this information can be taken advantage of by repeating each of these datum coordinates nine times within the data matrix and then carrying out the regression of choice. Results for the least squares, reduced major axis, and major axis models are shown in Figure 11. In each case data weighting greatly improves the fit of the regression line to the bivariate linear trend.

The problem with weighting data in a regression analytic approach to LOC estimation, then, lies not in the computational mechanics, but rather in the justification for regarding particular datums as inherently more reliable than others and in the translation of this justification into a numerical weight. The points we chose to weight in our example were, quite literally, plucked out of thin air and serve no purpose other than to illustrate the potential effects of using weighted data. In an actual analysis, it would be inappropriate to arbitrarily decide which datums were to be weighted and by how much purely as a function of their proximity to a trend that "looks right." Instead, estimation of the relative reliability of stratigraphic datums should be approached in as independent, explicit, and quantitative a manner as possible. Ironically, one of the results of a graphic correlation analysis is a quantifiable analysis of each datum's chronostratigraphic reliability (see Dowsett, 1988, 1989; MacLeod and

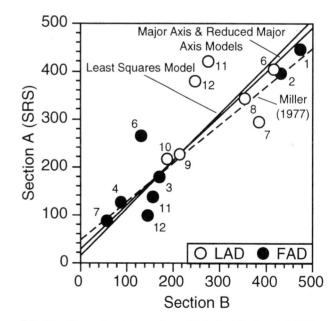

Fig. 11.—Comparison between the qualitatively-defined central LOC segment of Miller (1977) and a weighted least squares, reduced major axis, and major axis LOCs for the example dataset (exclusive of datum points lying along the bounding hiatuses; see Fig. 3A). Note the improved correspondence of all three regression models to the main linear trend of the data. Compare with Figures 6, 8, and 10. See text for a description of the weighting model and discussion of these results.

Keller, 1991). This anticipated outcome is of little use during the course of the analysis though. As an alternative, various authors have formulated "relative biostratigraphic value" (RBV) indices (see Brower, 1985 for a review) that compare datums on the basis of vertical (= stratigraphic) range, facies independence, and geographical persistence. Such indices might be useful in providing objectively determined and quantifiable weights for datums in a graphic correlation analysis despite the fact that these indices presently ignore such important criteria as ease of identification and susceptibility to diagenesis. Needless to say, the area of data weighting in graphic correlation, whether from a qualitative or quantitative point of view, needs much more attention.

Key Beds

In LOC estimation via regression, a key bed can be regarded as a data point so highly weighted that the LOC must be constrained to pass through it. So long as there is only one key bed within the study interval (or one key bed per independently analyzed subinterval), the quantitative techniques involved are simple. As originally noted by Shaw (1964), the key bed concept can be accommodated under LOC placement via regression analysis by placing it at the centroid of the data. This is operationally accomplished by subtracting the position of the key bed from the coordinate positions of all other datums. Once this procedure has been completed, regression analysis proceeds as described above with the deviations from the key bed substituting for the observed data. Since all three regression models are constrained to pass through the data's centroid, regression lines calculated in this manner will pass through the coordinate position of the key bed.

If more than a single key bed is present within a study interval or independently analyzed LOC segment, the appropriate regression may be determined entirely by these points, provided they are co-linear. If this is not the case, a piecewise linear LOC must be estimated by the appropriate qualitative (see above) or quantitative (see below) techniques.

Significance Tests

There appears to be some degree of confusion surrounding the question of statistical significance tests for regression-based LOCs. This confusion appears to stem from a misunderstanding of the distinctions between regression analysis and correlation analysis (see Sokal and Rohlf, 1981 for a particularly well-written discussion). Shaw (1964, p. 172–175) argued that the appropriate test of the significance of a linear least squares regression involves a significance test of the product-moment correlation coefficient for the same data. This is misleading. The correlation coefficient for a pair of variables measures the degree to which they covary (= exhibit similar patterns of linear variation) along a standardized scale with 1.0 representing perfectly positive covariation and −1.0 representing perfectly negative covariation. While a correlation coefficient does exist for the types of bivariate data scatters typically encountered in a graphic correlation problem, this parameter actually measures the scatter of the observed data perpendicular to the main linear trend (= the major axis regression line). Therefore, calculation of a correlation coefficient can be used to measure the degree of linear covariation existing between two variables but not to test whether a least squares model (which minimizes squared deviations on only one of the two axes) provides a compelling approximation. A far more appropriate use of the correlation coefficient test would be to determine whether or not there is any statistical basis for believing the major axis regression line exists. In addition, the correlation coefficient test is well known to be highly sensitive to deviations from the bivariate normal distribution. Since most graphic correlation data appear to depart strongly from bivariate normality, stratigraphers wishing to apply a correlation coefficient test to their data would be well advised to employ the non-parametric Spearman's ρ or Kendall's t statistics.

With respect to significance testing for least-squares regression models, statisticians favor an analysis of variance approach that involves partitioning the observed variation into components that can be attributed to the regression model itself and residual variation about the regression line. These sums of squares are then scaled by the appropriate degrees of freedom, compared with one another, and then indexed to the theoretical F-distribution to determine whether or not the residual variation is sufficiently small to preclude the possibility of the data having no trend. This analysis of variance strategy for testing the significance of a least-squares regression has been the standard method presented for evaluating least squares regressions for many years. Unfortunately, there is no comparable test for determining the significance of a reduced major axis regression.

When considering the outcome of either the correlation or analysis of variance tests, it is of the utmost importance to remember that both are actually examining the null hypothesis that the regression slope equals zero, not whether the selected regression model represents a "good" match to the underlying data. A more direct way of determining the adequacy of this match is through an examination of residual deviations from the regression model itself. Such a procedure is often recommended as an additional check to determine whether or not a specified regression model is appropriate to the data. According to the overall regression model, observations have two components, the value itself and an associated error term. For any particular regression model, the magnitude of this error is given by the difference between the predicted value based on the model and the observed value. If the regression is valid, a plot of the residuals should be randomly distributed along both sides of the regression line (= the slope of a regression through the residual values should be zero) with constant variance throughout the model. Residual deviation plots showing a pattern of points that either cluster to one side or another along the length of the model or that exhibit funnel-shaped distributions with decreased scatter at either end of the model are indications of an inappropriate model whether or not analysis of variance or correlation coefficient tests prove significant. This obtains because complex, nonlinear patterns within the observed data may retain sufficient linearity within their structures to be approximated, though not adequately modelled, by a linear function. If analysis of the regression residuals uncovers such patterns, a better fit may be obtained by eliminating outliers that, in terms of graphic correlation, most likely represent seriously diachronous datums, or by employing non-linear methods of quantitative LOC estimation.

An Approach to Non-Linear LOC Estimation

Stratigraphers who use qualitative approaches to LOC estimation often generate piecewise linear LOCs. For many of these practitioners, quantitative techniques such as those discussed above hold little appeal because they seem synonymous with linear LOCs. The assumption that true, time-averaged LOCs are essentially linear seems little more than a mathematical convenience. Strictly adhered to, the doctrine of linear LOCs implies that if the sediment accumulation rate changes in one region, rates everywhere must change by the same factor and at the same time! This implication is clearly unrealistic. Although Shaw (1964) and Miller (1977) found examples for which this assumption yielded reasonable results, their analyses were restricted to relatively closely-spaced sections and were not subjected to independent significance/residual tests. Time correlations based on magnetic polarity, isotope stratigraphy, and radiometric dating show that the assumption of linear sediment accumulation rates usually breaks down when correlation is attempted between geographically remote sections or over long intervals of geologic time. Sequence stratigraphic models also stress that the rock record is dominated by prograding wedges of sediment and that the site of the active wedge tends to oscillate between onshore and offshore positions. Large-scale non-linearity is, thus, a predicted feature of LOC geometries.

So long as we focus on a regression-based approach, it seems inevitable that quantitatively-inferred curvilinear LOCs will become more difficult to interpret as their complexity increases. But we need not be forced into the confines of regression-based LOC estimation. In order to appreciate how a very simple quantitative approach can yield a piecewise linear LOC, it is necessary to refocus attention directly on the nature of solutions of the correlation problem.

Linear regression chooses a straight line as a model solution and then computes the slope and intercept for the line that best fits local observations. We have seen how these techniques define "best" for computational convenience rather than out of any fundamental respect for the complexities of stratigraphic data or stratigraphic correlation. Traditionally, once the "best" LOC has been computed (by whatever means), all elements of the real solution to the correlation problem (e.g., the temporal sequence of events, the magnitude of the time intervals between those events, the location of horizons in each section that correspond in time with each event) are extracted. We believe that the future of quantitative approaches to LOC estimation lies in searching for the "best" sequence, spacing, and location of event horizons across all possible alternative solutions rather then attempting to directly compute a single deterministic solution. In this approach, Shaw's (1964) concept of economy of fit defines what "best" means (see also the application of parsimony criteria to graphic correlation discussed above). The best solution requires the minimum cumulative range adjustment when all locally observed event horizons are moved to positions consistent with the "true" sequence of events implied by the solution (= parsimony criterion described above). The shape and position of the LOC then emerges as a more or less passive consequence of attempting to preserve the spatial sequence of all local events.

Conceptually, this revised version of the traditional quantitative approach to LOC estimation is straightforward; simply search through all possible sequences, spacings, and locations of datum events (= LOC-based correlation models) to find the one that best conforms to the economy of fit criterion. Kemple (1991) and Kemple and others (1990; this volume) present formulae that can be used to calculate the economy of fit for any solution. Numerical algorithms can carry out the actual search. The difficulty lies in finding a procedure fast enough to consider all possible LOC geometries. As one can readily imagine, the total number of possible solutions is almost incomprehensively large. Fortunately, the field of operations research has developed procedures to address this class of numerical problems in which it is not feasible to search through every possible solution (Dell and others, 1992).

Many of these procedures are heuristic and may be most easily understood as sets of instructions one might give a short-sighted person for searching out the lowest point in a complex and unknown landscape. In this landscape metaphor, the location of each point can be thought of as representing a particular LOC-based correlation model and its elevation the corresponding economy of fit index. A good search strategy should not always progress downhill or else it will too often find only local minima. Instead, some mechanism must be provided to allows the search to "climb" out of small depressions in the economy of fit landscape and continue the search for the globally lowest point. In addition, efficient search procedures should be adaptive in the sense that they are able to find the lowest or at least an extremely low point without having to evaluate every point on the surface.

Kemple and others (this volume) explain one possible search procedure in detail. This procedure, termed "constrained optimization," eliminates impossible solutions first and only conducts the search through the landscape of feasible LOC geometries. Obviously impossible solutions specify negative sediment accumulation rates, for example. The procedure begins with a solution based on one of the local sections and generates new solutions by experimenting with random incremental changes in the LOC geometry. Changes that improve the economy of fit are always accepted; changes that worsen the economy of fit are sometimes accepted in order to escape the trap of local minima. The probability of accepting a worse economy of fit index decreases as the search progresses and is always greater for smaller as opposed to larger "uphill" steps. The progressively changing probabilities are governed by a simulated annealing procedure (Kirkpartrick and others, 1983).

For the purpose of comparison with other methods, we have applied constrained optimization, using the simulated annealing procedure, to Miller's (1977) data (Figs. 12, 13). Figure 12 illustrates that the greatest economy of fit will always correspond to a piecewise linear LOC in which the number of segments is allowed to be as large as the number of data points (if necessary) and many LOC segments are allowed to lie parallel to a section axis. Figure 13 illustrates how the search procedure quickly identifies very good solutions but may make slower progress in its convergence to a single solution exhibiting the best economy of fit.

The constrained optimization procedure allows considerable flexibility but is always subject to prior specification. Reproducibility is thus ensured. If all data from one section are weighted heavily enough (Fig. 12B), the approach can be made to mimic the qualitative strategy of splitting tops and bases.

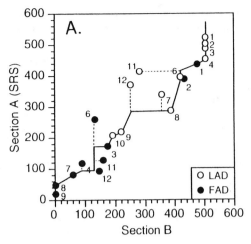
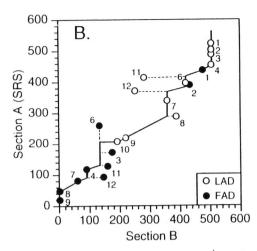

Fig. 12.—Two piecewise linear LOCs drawn through the example dataset to minimize the cumulative range extension, as explained by Kemple and others (this volume). Dotted lines are the minimum range extensions required to adjust both section to the same sequence of events. (A) Both sections equally weighted. (B) Section A weighted twice as heavily as section B. The unequal weighting allows section A to assume the role of standard reference section (SRS) and the resulting LOC tends to split tops and bases.

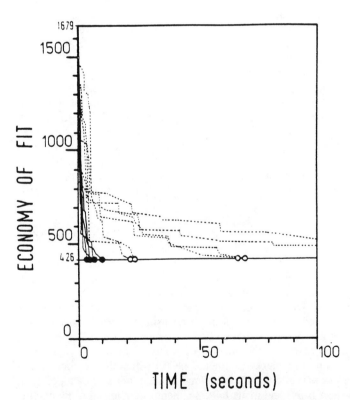

Fig. 13.—Rate of approach to the solution in Figure 12A by the constrained optimization procedures described by Kemple and others (this volume). Economy of fit is the total under-representation of ranges (if feet) implied by a solution. The final value (426) is the most economical LOC for this dataset. The initial value (1679) arises for a solution based on section A, with the FAD and LAD for the missing taxon (5) placed at the beginning and end of the sequence respectively. Solution times given for runs on an IBM 486/33MHz PC. Solid paths and filled circles show typical solution times for well-adjusted search parameters. Dotted paths and open circles show solution times for poorly adjusted search parameters; the dotted paths resemble those for much larger correlation problems (more sections and many more taxa).

Figures 12A and 12B employ only one level of parsimony, minimization of the implied misfit of local event horizons. Additional weighting terms can be added to the economy of fit formula that increases the penalties assigned to most complex LOCs (larger numbers of segments). If this second level of parsimony is weighted heavily enough (Kemple and others, 1990), the result can be made to resemble a linear regression in which the splitting of tops and bases criterion is respected.

Another important advantage of this quantitative approach arises in situations where more than two sections must be considered; this procedure easily extends to as many dimensions as there are sections. Although linear regression models might be similarly extended (e.g., multiple regression – a least squares technique), we know of no previous multivariate implementation of graphic correlation. Because all of the information contained in multiple sections can be used at once, the order in which sections are composited into the CSRS (see Edwards, 1984) becomes irrelevant, and data points do not need to be differentially weighted to reflect information from sections that cannot be represented on a single bivariate plot. Also, if a description of a new section becomes available, the entire analysis can be run again from the beginning. This avoids correlating the new section against the previously determined CSRS and facing the decision of how heavily the latter should be weighted in order to preserve its overall integrity. Eliminating the tyranny of the standard section from all stages of the process allows more precise estimates of CSRS stability to be made and holds the promise of allowing direct error estimates as well as statistical hypothesis tests to be made via jackknifing or bootstrapping of the relevant LOC parameters (see Sokal and Rohlf, 1981; MacLeod, 1991 for discussions of these statistical methods).

SUMMARY

Graphic correlation represents an extremely powerful and multi-purpose inferential tool that has the capability to transform the process of stratigraphic correlation. At present, most alternative approaches to stratigraphic correlation are based on outmoded assumptions (e.g., isochrony of biostratigraphic datums) and employ techniques (e.g., zone-based biostratigraphy) that are inherently low-resolution and fail to take full advantage of the available biotic and physical data. Since these data are often acquired at considerable cost, it is necessary that the information they provide be completely extracted and properly integrated to infer the correlation. Shaw's (1964) graphic correlation method dramatically extends the concept of stratigraphic correlation and provides a quantitative framework for making direct comparisons between different types of stratigraphic data. This, in turn, allows the stratigrapher to bring these different information sources simultaneously to bear on the problem of chronostratigraphic inference.

Among graphic correlation's many strengths are its ability to give precise expression to traditional correlation strategies, as well as its power to suggest a variety of new approaches that are peculiar to the graphic presentation of stratigraphic data. Curiously though, neither the availability of these newer methods nor the more exact description of traditional approaches has resulted in a thorough reevaluation of the inferential basis for stratigraphic correlations. Thus, many quantitative and qualitative methods for estimating the line of correlation continue to be employed, all of which make different assumptions with respect to the nature of different types of stratigraphic data, their relationship to one another and their relationship to the true line of correlation, and all of which produce different results.

The methods used for estimating the LOC in a graphic correlation problem can be subdivided according to whether or not they are based on qualitative and quantitative procedures. The most popular qualitative LOC estimation technique attempts to find a line that sequesters biostratigraphic FADs and LADs to opposite sides of a bivariate graphic correlation plot. The logic behind this approach to LOC estimation is based on the fundamental distinction between these two biostratigraphic datum types in terms of their ability to estimate their respective maxima. Data fulfilling these assumptions usually exhibit obvious trends, in which case LOC estimation poses no problem. For data that seriously violate these assumptions, however, or for data whose distribution within one or both of the sequences is patchy, a number of alternative LOC geometries may be possible. Adjunct qualitative criteria such as the presence of key beds, weighting of individual datum events, the 10 percent rule for doglegged LOCs, and the principle of parsimony may be of use in evaluating the likelihood of these alternative LOC geometries. In addition, extremal LOC arrangements may be used to develop "best case" (= temporally most complete with the smallest number of sediment accumulation rate changes) and "worst case" (= temporally most incomplete with largest number of sediment accumulation rate changes) LOCs. These can then be used to place limits on the true LOC geometry as well as to identify regions of commonality between both models (see MacLeod, 1991, MacLeod and Keller, 1991, MacLeod, in press, MacLeod, this volume).

The most frequently used quantitative LOC estimation procedure is linear least squares regression analysis. This technique assumes, as do all regression models, that datums will be scattered randomly about the true LOC without regard as to datum type. The least squares regression model further assumes that: (1) there is a fundamental difference between the accuracy with which datum positions in the sections or cores under consideration were measured, (2) that datum positions along the independent variable axis (the SRS) exhibit a strong deterministic influence on datum position in other sections or cores, and (3) the best way to summarize trends within the data array is to minimize the sum of squared deviations along the dependent variable (non-SRS) axis. Despite the past popularity of this regression model in graphic correlation studies, these assumptions seem highly questionable. Alternative regression models that appear to better suit the data used in a graphic correlation analysis include reduced major axis and major axis models. Both of these alternative regression models take the joint deviation along both bivariate axes into consideration when fitting the LOC. Tests of these models using Miller's (1977) dataset suggest that reduced major axis and major axis regressions provide more robust estimations of the major linear trend. Reduced major axis regressions also have the additional advantage of being extremely easy to calculate.

Simple modifications of regression models allow them to take advantage of the information provided by key beds and differentially weighted datums. Significance tests are available for the least squares and major axis models, but these only evaluate whether or not the regression slopes differ significantly from zero. Moreover, the validity of these tests is dependent on the extent to which the distribution of datums in both sections or cores is bivariate normal. Non-parametric procedures can provide a better indication as to the overall linearity of the data, while the quality of the regression is best determined through an analysis of the residual deviations.

Although linear regression models may be useful in estimating the LOC between sections that are expected to be similar (e.g., sections from the same region in a depositional basin), use of graphic correlation to provide chronostratigraphic control over extended time intervals or across broad geographic regions presupposes the ability to estimate non-linear LOCs. While qualitative LOC estimation procedures can be used to derive non-linear LOCs, quantitative counterparts to these procedures are badly needed. One promising approach to this problem involves a constrained optimization procedure in which algorithmic searches are carried out across a landscape of feasible LOC solutions. In these hypothetical landscapes, elevation is set to represent one (or a combination of several) parsimony criteria that together provide a quantitative measure of an individual LOC solution's economy of fit. The purpose of the search is to locate the lowest (or at least a very low) point on the economy of fit landscape and, in so doing, identify the LOC solution that most faithfully preserves each section's local sequence of events. Experiments conducted thus far suggest that piecewise linear LOCs with large numbers of segments, many of which are oriented parallel to one or another section axis, usually exhibit the best economy of fit characteristics. However, differential weighting of the various parsimony terms can be used to "control" the constrained optimization LOC by penalizing solutions that violate the splitting tops and bases criterion and/or that include large numbers of LOC segments. Finally, the constrained optimization approach can be extended to enable a multivariate consideration of all sequences simultaneously, thus providing more robust estimates of CSRS stability and eliminating differential weighting of the SRS.

ACKNOWLEDGEMENTS

The authors would like to thank the following for stimulating discussions on the topic of graphic correlation over the last few years that either directly or indirectly contributed to the material presented herein. These include: Gilbert Klapper, Joseph Hazel, Lucy Edwards, and W. G. Kemple. This contribution also benefitted from the comments and suggestions of Keith Mann and two anonymous reviewers. This paper is a contribution from the Natural History Museum/University College London Global Change and the Biosphere Project.

REFERENCES

BROWER, J. C., 1985, The index fossil concept and its application to quantitative biostratigraphy, *in* Gradstein F. M., Agterberg, F. P., Brower, J. C., and Schwartzacher, W. S., eds., Quantitative Biostratigraphy: Paris, D. Reidel Publishing Company, p. 43–64.

DELL, R., KEMPLE, W. G., AND TOVEY, P., 1992, Heuristically solving the stratigraphic correlation problem: Institute of Industrial Engineers, 1st Industrial Engineering Research Conference Proceedings, v. 1, p. 293–297.

DOWSETT, H. J., 1988, Diachrony of Late Neogene microfossils in the southwest Pacific Ocean: Application of the graphic correlation method: Paleoceanography, v. 3, p. 209–222.

DOWSETT, H. J., 1989, Application of graphic correlation to Pliocene marine sequences: Marine Micropaleontology, v. 14, p. 3–32.

EDWARDS, L. E., 1984, Insights on why graphic correlation (Shaw's method) works: Journal of Geology, v. 92, p. 583–597.

IMBRIE, J., 1956, Biometrical methods in the study of invertebrate fossils: Bulletin of the American Museum of Natural History, v. 108, p. 215–252.

IVANY, L. C. AND SALAWITCH, R. J., 1993, Carbon isotopic evidence for biomass burning at the K-T boundary: Comment and Reply (Reply): Geology, v. 21, p. 1150–1151.

JOLICOUER, P., 1975, Linear regressions in fishery research: Some comments: Journal of the Canadian Fisheries Research Board, v. 32, p. 1491–1494.

KEMPLE, W. G., 1991, Stratigraphic correlation as a constrained optimization problem: Unpublished Ph.D. Dissertation, University of California, Riverside, 189 p.

KEMPLE, W. G., SADLER, P. M., AND STRAUSS, D. J., 1990, A prototype constrained optimization solution to the time correlation problem, *in* Agterberg, F. P. and Bonham-Carter, G. F., eds., Statistical Applications in the Earth Sciences: Ottawa, Geological Survey of Canada Paper 89–9, p. 417–425.

KERMACK, K. A. AND HALDANE, J. B. S., 1950, Organic correlation and allometry: Biometrika, v. 37, p. 473–486.

KIRKPATRICK, S., GELATT, C. D., AND VECCHI, M. P., 1983, Optimization by simulated annealing: Science, v. 220, p. 671–680.

MACLEOD, N., 1991, Punctuated anagenesis and the importance of stratigraphy to paleobiology: Paleobiology, v. 17, p. 167–188.

MACLEOD, N., in press, Graphic correlation of high latitude Cretaceous – Tertiary boundary sequences at Nye Kløv (Denmark), ODP Site 690 (Weddell Sea), and ODP Site 738 (Kerguelen Plateau): Comparison with the El Kef (Tunisia) boundary stratotype: Modern Geology.

MACLEOD, N. AND KELLER, G., 1991, How complete are Cretaceous/Tertiary boundary sections? A chronostratigraphic estimate based on graphic correlation: Geological Society of America Bulletin, v. 103, p. 1439–1457.

MYERS, R. H., 1986, Classical and Modern Regression with Applications: Boston, Druxbury Press, 359 p.

MILLER, F. X., 1977, The graphic correlation method in biostratigraphy, *in* Kauffman, E. G. and Hazel, J. E., eds., Concepts and Methods of Biostratigraphy: Stroudsburg, Dowden, Hutchinson and Ross, p. 165–186.

OLSSON, R. K. AND LIU, C., 1993, Controversies on the placement of Cretaceous-Paleogene boundary and the K/P mass extinction of planktonic foraminifera: PALAIOS, v. 8, p. 127–139.

POLLARD, J. H., 1977, A Handbook of Numerical and Statistical Techniques with Examples mainly from the Life Sciences: Cambridge, Cambridge University Press, 349 p.

RICKER, W. E., 1973, Linear regression in fishery research: Journal of the Canadian Fishery Research Board, v. 11, p. 559–623.

SHAW, A. B., 1964, Time in Stratigraphy: New York, McGraw-Hill, 365 p.

SOKAL, R. R. AND ROHLF, J. F., 1981, Biometry: San Francisco, W. H. Freeman and Co., 859 p.

EXTENDING GRAPHIC CORRELATION TO MANY DIMENSIONS: STRATIGRAPHIC CORRELATION AS CONSTRAINED OPTIMIZATION

WILLIAM G. KEMPLE
Department of Operations Research, Naval Post Graduate School, Monterey, California 93943;
PETER M. SADLER
Department of Earth Sciences, University of California, Riverside, California 92521;
AND
DAVID J. STRAUSS
Department of Statistics, University of California, Riverside, California 92521

ABSTRACT: Stratigraphic correlation involves three distinct tasks: establishing the temporal sequence of marker events (sequencing task), determining the relative sizes of the intervals between those events (spacing task), and locating the horizons that correspond in age with each event in every section (locating task). Stratigraphic sections do not yield enough information to solve this problem exactly. Instead, stratigraphers must search for the approximation that "best" fits all local stratigraphic observations. The concept of economy of fit, as used in graphic correlation, provides a rigorous definition of "best" and implies the existence of a penalty function that can be used to rank possible solutions. Unfortunately, traditional graphic correlation requires severe simplifying assumptions about accumulation rates because it attempts to solve all three tasks at the same time, and yet it incorporates the local sections into the solution one at a time. Using a penalty function based on economy of fit, an alternative solution technique naturally emerges in the form of constrained optimization. This technique is J-dimensional, in the sense that it treats the observations in all J sections simultaneously. It can complete the sequencing task before making assumptions necessary to the spacing task.

Constrained optimization eliminates impossible solutions (constraint) and then searches for the best of all the possible ones (optimization). For realistic instances of the problem, it is not feasible to calculate the penalty function for all possible solutions. Instead, we use a probabilistic search procedure termed "simulated annealing" to find very good solutions without an exhaustive search. Simulated annealing does not maintain any memory of the search path or search exhaustively at the local scale. We reject an alternative procedure that incorporates these features because it proved difficult to tailor to produce satisfactory solutions. Our J-dimensional procedure quickly finds solutions for Palmer's (1954) classical data set ($J = 7$ sections from the Cambrian of Texas) that are comparable to the solution Shaw (1964) achieved by traditional graphic correlation. The expert judgements that the stratigrapher uses to draw the traditional lines of correlation and which give the appearance of excessive subjectivity, in fact, derive primarily from a knowledge of all the sections. By treating all sections at once, constrained optimization eliminates much of the apparent subjectivity associated with traditional graphic correlation. Constrained optimization still allows the user to explore the consequences of different and genuinely subjective judgements about the relative reliability of different taxa and sections.

THE STRATIGRAPHIC CORRELATION PROBLEM

Time correlation is a conceptually simple exercise: arrange a set of events in their correct sequence and spacing on a time scale, and locate in stratigraphic sections the horizons that correspond in age with each event. The sequencing, spacing and locating tasks may be treated separately, but the sequence must be determined in order to complete the spacing and locating tasks.

Most often the events are changes: reversals of magnetic polarity, the appearance and disappearance of fossil taxa, and shifts in isotopic composition, for example. In this paper, we emphasize the first and last appearances of fossil taxa, because these events dominate the data available for graphic correlation. The timing of these events is not known independently. It must be reconstructed from the local stratigraphic sections that we wish to correlate and which often contradict one another with respect to the sequence and spacing of events (Fig. 1). Thus, the practical problem has so many unknowns that the true solution is not determinable. Yet a solution is so fundamental to geologic enquiry that stratigraphers are willing accept simplifying assumptions that lead to approximate solutions. Unfortunately, different simplifying assumptions tend to be treated as rival methods and divert attention from the problem itself. Ideally, the approach to the problem should be flexible enough to allow many different simplifying assumptions and should yield solutions quickly enough to encourage stratigraphers to explore the consequences of changing their assumptions.

The practical difficulties of time correlation are threefold. Firstly, there are not enough reliable data. The horizons of the stratigraphically lowest and highest finds of a fossil taxon are unlikely to have the same ages from section to section. The stratigraphic range of finds underestimates the local temporal range of the taxon unless its very first and very last local representatives have been fossilized and found. Thus, we most often need to extend the locally observed taxon ranges to make them fit the true sequence and spacing of events (Edwards, 1982). Even where all other factors are uniform, the preservation and collection of fossils includes an element of chance, and the discrepancy between the true local range and the observed range varies from section to section. Furthermore, changes in geographic range during the lifespan of the taxon cause even the true local appearance and disappearance events to vary in age between sections. For each taxon, one of the local range bottoms is the earliest first observed appearance, and one of the local range tops is the latest last observed appearance. The sequence of these extreme horizons provides the best approximation of the true arrangement of evolution and extinction events in time. The search for this sequence requires expert judgements about the relative fidelity of preservation and the quality of sampling of events in different sections.

Although reliable data are insufficient to lead directly to the sequence of extreme horizons, the expert is nonetheless easily overwhelmed by the volume of information and cannot keep in mind all the varied and conflicting evidence from many localities. This is the second practical problem. Not surprisingly, stratigraphic correlations by one expert are not always readily accepted or even reproduced by others. Thirdly, the events that make correlations possible are often part of the very history that we would prefer to reconstruct after correlation by other means. It is not reasonable to force a solution by conveniently assuming, for example, that one set of index fossils at one type locality gives all the right answers. Correlation must entertain multiple working hypotheses and be able to specify the consequences of changing the expert judgements about the quality of different taxa and sections. One unaided expert can seldom manage all this.

Computer memories and the decision-making routines of statistics and operations research can render the correlation task more manageable. In fact, more quantitative techniques have been proposed for correlation than we have space to describe. Edwards (1982), Agterberg and Gradstein (1987), Tipper (1988), Kemple (1991), and Geux (1991), among others, provide reviews from which it emerges that different techniques use different assumptions to ease the computation. Regrettably, not all the assumptions are explicit or compatible with the best geological judgement. For a variety of reasons, few of the methods are widely known, and even fewer enjoy regular use (Tipper, 1988). One notable exception is Shaw's (1964) method of graphic correlation.

Several years ago we wrote of the advantages that would result from recasting Shaw's graphic method as a constrained optimization procedure (Kemple and others, 1989; Kemple, 1991). Since then we have been developing Fortran routines to implement this idea on desktop computers and work stations. Current versions of the program find the optimal sequence of first and last appearances and make the minimum range adjustments necessary to bring all local stratigraphic sections into agreement with that sequence. We are still developing routines that incorporate sedimentological information to provide more flexibility in the spacing of event horizons. The program does not replace the stratigrapher's expertise; it requires that the user understand the underlying procedures and it performs correlations according to explicit expert decisions and assumptions. Because changing the assumptions changes the solution, it is possible to explore the sensitivity of the solution to different expert judgements. The prototype programs have only rudimentary user interfaces, but readers willing to test the programs on their own data sets are encouraged to contact the authors for the latest version.

We have written this chapter for stratigraphers who wish to understand the constrained optimization procedure. We use lay terms, wherever possible, to describe specific statistical operations and offer three analogies as a means to realize the full scope of situations in which the expert must select from a range of procedures or parameters. The chapter subdivides naturally into three parts. The first part reviews the similarity between stratigraphic correlation and regression analysis, which Shaw (1964) made evident by plotting pairs of stratigraphic sections as the orthogonal axes of a two-dimensional graph. To show that stratigraphic correlation can be more complex and flexible than the familiar least-squares linear regression, we suggest a loose analogy with the evaluation of a hypothetical aqueduct project. The second part explains how the constrained optimization procedure extends the graphic approach to many more than two dimensions. Two more analogies are drawn, the first loose and the second exact. Searching for the line of correlation that best fits the stratigraphic observations resembles finding the lowest point on a landscape for which there is no map. The most efficient search strategy that we have found exactly follows the rules that govern attempts to grow a perfect crystal by annealing and cooling. The third part applies the constrained optimization procedure to the Cambrian correlation problem that Shaw used in 1964 to introduce graphic correlation.

THE GRAPHIC APPROACH TO A SOLUTION

Simple linear regression specifies that a straight line will be fit to a set of observations. The familiar method of least squares specifies how the misfit between model and data will be quantified in the search for the best line. Like regression analysis, the graphic approach to stratigraphic correlation clearly specifies the characteristics of good solutions and the criteria for choosing the best solution. In this way, graphic correlation provides a uniquely firm foundation for formal treatment of the stratigraphic correlation problem. This foundation has supported a wide range of variants in the operational details of graphic correlation. Many of them are discussed in other chapters of this volume, and it is not our purpose to evaluate them. For simplicity, we will develop the explanation of our constrained optimization procedure out of the original graphic method described by Shaw (1964), which we will call "traditional" to avoid any ambiguity. As users of current operational enhancements are well aware, some features of the traditional method are severely limiting. We discuss these limitations and review enhancements that we borrow from current graphic practice and from rival quantitative methods.

Casting the Correlation Problem in Graphic Terms

Two Dimensional Form.—

If we draw two measured sections as orthogonal axes on an x-y graph, with their bases meeting at an arbitrary origin (Fig.

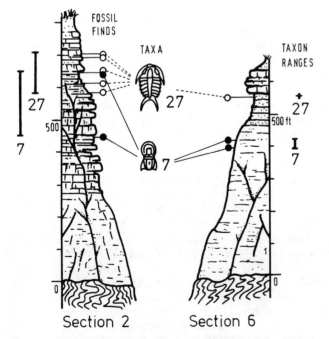

FIG. 1.—Two sections and two trilobite taxa from the Riley Formation (Palmer, 1954; Shaw, 1964) that illustrate contradictory ranges. In section six, the highest occurrence of taxon 7 is below the lowest occurrence of taxon 27; in section two, the taxon ranges overlap. Assuming that the observed coexistence in section two is real, section six must significantly under-represent the range of one or both taxa.

2A), then correlations can be shown by points on the graph. The x and y coordinates of these points represent the distances from the origin to two horizons, one in each section, that formed at the same time. Because the layers of rock are laid down younger upon older, the points for all the coeval horizon pairs must form a line whose slope is always non-negative (0 to +90°). Although in some instances the true correlation may be a straight line, the general solution of the stratigraphic correlation problem in this framework is a monotone, non-decreasing, two-dimensional, serpentine curve or "snake." We reserve the term *snake* as shorthand for the true solution to a correlation problem, leaving the term *line of correlation* (LOC) to describe any plausible approximation of the true solution, as is common practice.

Segments of the snake that are perpendicular to one axis of the plot point to hiatuses in the section plotted on that axis. Time intervals that are represented by hiatus in both sections have no part in the snake, of course. Differences in the ratio of accumulation rate between the two sections cause changes in the positive slope of the snake. Like the snake, LOCs may be curved, segmented, or straight, provided their slope is everywhere non-negative. Selection of a relatively straight LOC corresponds to a judgement that there is little variation with time in the ratio of accumulation rates in the two sections.

Events recorded in only one section may be projected into the other using the LOC to create a composite section. The composite section records on one axis an arrangement of all events that is consistent with the observations in both sections. The compositing process allows additional sections to be correlated without using more than two axes, an important consideration for manual correlation. But the LOCs remain two dimensional.

Multidimensional Form.—

The snake concept extends easily beyond two dimensions (Fig. 3). The Deep Sea Drilling Project popularized two-dimensional plots of one section against a time scale. Adding a time scale to the graph of two sections still yields a monotone non-decreasing snake, but in three dimensions. Points on the snake now have coordinates x, y, and time. A plot of J sections gives a snake in J dimensions and the addition of a time scale produces a $J + 1$-dimensional snake. So the fullest solution for a time correlation problem is a J-snake together with the ages of all points on the snake. Shaw (1964) stressed that equivalence in age can be represented without reference to an annual scale of absolute time. Thus, the essential practical problem for graphic correlation is to find LOCs that are good approxima-

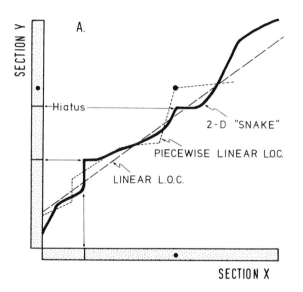

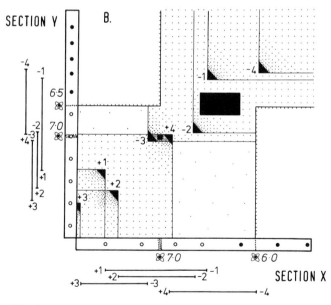

FIG. 2.—Graphic representation of pairs of hypothetical stratigraphic sections and the means of constraining the line of correlation (LOC). (A) The true solution or snake (heavy line), two LOCs of different complexity (dashed lines), and the coordinates of one range top (filled circle). (B) A variety of data types that constrain the line of correlation. Radiometric dates: atom symbols next to the section axes. Magnetostratigraphic samples: circles in the section axes, open for reversed polarity, filled for normal polarity. Biostratigraphic data: local taxon range charts for taxa 1–4 parallel to the section axes; "+" signifies the lowest observed occurrence, "-" the highest. The stipple intensity decreases away from the triangles and "good" LOCs stay close to the darkest parts of the plot. See text for explanation of the boxes.

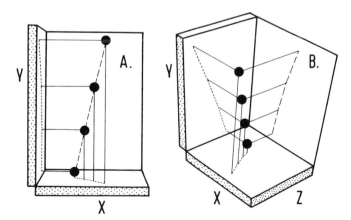

FIG. 3.—Multi-dimensional graphic correlation. Traditional graphic correlation in two dimensions (A) can be regarded as a projection of a three- or J-dimensional (B) correlation. The coordinates of events on the plots are the elevations of corresponding horizons in all sections (x,y,z, . . . n). More than three dimensions are difficult to visualize yet easy to store in coordinate or matrix form.

tions of the *J*-snake. Because the snake is unknowable, "good" LOCs cannot be defined relative to the snake. Instead, they must be recognized by their relationship to the local stratigraphic observations.

LOC now stands for any possible approximation of a *J*-snake. The *J*-dimensional LOC is the solution to the correlation problem. There is no composite section because there is no need to keep a memory of the progress of partial correlations. The *J*-dimensional LOC is completely described by a matrix in which rows are events and columns are local sections. Rows arranged in the estimated order of events complete the sequencing task. Entries in the matrix cells record the estimated level of the corresponding event horizons; they complete the locating and spacing tasks. Hypothetical composite sections might be created by using the average or maximum spacing of events across all local sections or by combining portions of the local sections that have the same sedimentary facies, but none of this is necessary.

Local Stratigraphic Observations as Error Boxes on the Graph

Figure 2B depicts the relationship between different types of local stratigraphic data, a 2-dimensional snake, and LOCs. Black rectangles represent events that can be equated with an horizon in each section. They identify boxes through which the snake and all plausible LOCs must pass. Black rectangles acknowledge better than points that the location of event horizons has variable precision in practice. The coordinates of the small black square dated at 7.0 my in Figure 2B are determined by the stratigraphic position of a thin, unique ash layer preserved in both sections. The sides of the larger black rectangle in the same figure are scaled by the sample spacing at a paleomagnetic reversal. Figure 2B also illustrates how two horizons of known but different absolute age, one in each section, define a region that the snake cannot enter. Because nothing below the 6.5-my-old horizon in section Y may map into section X above its 6.0-my-old horizon, no part of the snake can lie within the white box in the lower right corner of the diagram. The white box in the upper left corner is similarly excluded because nothing above the 6.5-my-old horizon may map into section X below the 7-my-old horizon.

In typical applications of graphic correlation, most of the available data are local lowest and highest finds of fossil taxa. As explained earlier, the lowest finds tend to occur too high in the section and the last finds too low, when compared to the horizons that correspond to either global evolution and extinction events or local immigration and emigration events. Black triangles represent these observations in Figure 2B because, assuming we can rule out reworking and borehole caving, an observed range end gives only one corner of the box (stippled) that contains the true time of first or last appearance. The triangles have two orientations because first and last occurrences fix different corners of their respective error boxes: the upper right corner for first occurrences and the lower left for last occurrences. These corner points cannot all be expected to lie on the snake and need not lie on the best LOC. We do not know where in the stippled portion of Figure 2B the snake lies.

An LOC which passes as close as possible to as many of the corner points as possible implies the best overall quality in the local stratigraphic records. It requires the minimum net range extension to make all the local range charts fit one sequence and spacing of events. Shaw (1964, p. 254–257) called this "*economy of fit*" and established that the best solution in a graphic correlation is the one that realizes the greatest economy of fit. Such a solution is desirable because it is geologically simple. An extreme example shows that this is an almost intuitive line of reasoning. We all quickly reject the hypothesis that all taxa have been in existence since the beginning and none are extinct. The argument is not that the fossil record absolutely refutes such an interpretation; it does not. The hypothesis is rejected because it implies outrageously large and systematic failures of the preservation or collection of fossils. In a way that we must later quantify precisely, the goodness of fit between LOCs and local stratigraphic observations can be measured by the sum of the stratigraphic distances by which the LOC misses the corner points. The scale on which these distances are to be measured is obviously one factor awaiting precise definition.

LOCs need not distribute the correction of local observations equally among all taxa or equally between both sections. Shaw (1964, p. 257) sought the LOC "that results in the smallest net disruption of the best established ranges." He did not weight all taxa equally when assessing the economy of fit. Some combinations of organism and sedimentary facies have higher preservation potential than others; they lead to a higher frequency of fossiliferous levels within the range and are likely to record the local taxon ranges more reliably. Such corner points may be expected to lie closer to the snake and so Shaw weighted the distance by which they misfit the LOC more heavily when assessing the economy of fit. If an LOC can be drawn that lies above all first occurrence corners and below all last occurrence corners, it forces all the range adjustments into the section on the x-axis. Thus, the position of the LOC can also be guided by judgements about the relative suitability of the sections for correlation purposes.

Because the density of stipple in Figure 2B progressively decreases away from the box corners, good LOCs must run through dark areas of the plot. (To weight taxa differently, we might have used denser stipple for better established ranges.) If most of the locally observed event horizons are in fact close to their true positions, then the darkest area of the plot will narrow as the number of box corners increases, even though the individual observations provide little constraint by themselves. There is no obvious reason why bad observations would behave in this way. Those stratigraphers who we have questioned about graphic correlation report that the choice of LOC does usually narrow as they include more taxa. This experience is one indication that economy of fit does favor LOCs close to the snake.

The task of fitting the most economical LOC to the boxes and corner points in Figure 2B is clearly an optimization problem (more precisely, it is a constrained optimization because some LOC placements are geologically impossible). But it is not an instance of optimization that is necessarily best solved by simple linear regression using the method of least squares (Miller, 1977). The following parable illustrates a range of options for ranking aqueduct routes that is comparable to the choices available to determine economy of fit.

Ranking Aqueduct Proposals: an Analog for Economy of Fit

To appreciate the full scope of the task of ranking different LOCs, imagine the task of finding the best route for an aqueduct

(LOC) that will be the nearest source of water for many villages (black boxes and corner points in Figure 2B) scattered across an uneven landscape. To model the special treatment of local range ends in Figure 2B, imagine that the villagers do not necessarily haul water from the point on the aqueduct closest to their village; instead, their haul roads are required to remain in fields (stippled boxes) that belong to the village (black corner triangles). To maintain water flow, no part of the aqueduct can be routed uphill (non-negative LOC slope). This makes it impossible to route the aqueduct through every village because some occupy hilltops and others lie in closed depressions (equivalent to the problem of contradictory local sequences of highest and lowest finds).

In the terms of this analogy, Shaw's (1964) economy of fit ranks possible aqueduct routes by the total distance that remains for villagers to haul water. Although his initial formulation kept the route straight (linear regression), Shaw gave examples in which the residual haul distance was significantly reduced by permitting an aqueduct with more than one linear segment. Shaw preferred to minimize the number of segments but did not quantify this preference. Because the optimal piecewise-linear aqueduct would be longer than the optimal straight aqueduct, we pay extra for changes in direction. We might add the length of the aqueduct to the total haul distance when ranking different plans. This strategy would select a route that strikes a balance between local hauling convenience (local stratigraphic observations) and the cost of aqueduct construction (assumptions about accumulation rate). To prepare for some later sections of this paper, notice several other options in the parable. If we weight aqueduct length more heavily than haul distance, we encourage straighter routes without completely prohibiting bends. We can regard some villages as more important, perhaps because of their size and voting habits, and weight their haul distances more heavily. And distance is not the only measurement scale; in this parable, money or votes might be used.

Limitations of the Traditional Graphic Method

Serious limitations arise because traditional graphic correlation integrates the records of local stratigraphic sections one at a time. The method proceeds by series of partial correlations, the first between two local sections, the rest between one local section and the developing composite section (Shaw, 1964; Miller, 1977; Edwards, 1989; and earlier chapters of this book for details). Many of these essentially two-dimensional steps are required to integrate all the data and each step appears deceptively simple. Unfortunately, in the early steps, the two-dimensional graphs contain only a small fraction of the total information. As a result, the early LOCs are difficult to draw and likely to be of low quality. And yet the results of the early steps have the strongest impact on the final outcome (Edwards, 1984).

The placement of LOCs depends heavily on subjective judgements, especially in the early steps. First, the partial correlations begin with the two sections judged to have the greatest abundance of reliable observations. Secondly, the LOCs are rarely fit equally to all observations. Shaw (1964, p. 235–241) explained in detail how to select a subset of presumably more reliable local observations. His selection process relies on a knowledge of observations in stratigraphic sections not yet included on the graph. Another popular tactic tries to route the LOC between the first and last occurrence observations so that range adjustments impact the section judged to be least reliable and minimize changes in the composite section (MacLeod and Sadler, this volume). This tactic extends the influence of judgements about the relative reliability of sections.

Subjective judgements that are used to constrain the early LOCs continue to influence the outcome after all sections have been integrated. In an attempt to remove the biases introduced by the early partial correlations, Shaw (1964) repeated the whole procedure several times. At the start of each reiteration, the composite section produced by the previous iteration serves as the initial section. Reiteration is a tedious exercise with progressively diminishing returns. But without it, the final composite cannot fairly reflect all local observations.

A graphic procedure in J dimensions solves the correlation problem for all J sections at once. It eliminates any need to consider the sequence of inclusion of the local sections because each occupies an axis from the start. There is no need to dedicate an axis to a composite section and there is no question of routing the LOC to separate first and last occurrences. Because all the local observed event horizons are available as objective data from the outset, none of this information is used selectively. Thus, much apparent subjectivity in the traditional method simply disappears.

In order to make partial correlations with the limited information on a two-dimensional graph, the traditional approach preferred a straight LOC (Shaw, 1964) or one with very few "doglegs" (Miller, 1977). In some instances, however, the implication that sections maintain a constant ratio of accumulation rates is too severe a simplification. Graphic correlation of cores from the Deep Sea Drilling Project with one another and against time scales (e.g., Pisias and others, 1985) has shown that straight lines of correlation can be unrealistic, even for abyssal marine accumulation. For marine sections on the shelf and continental slope, seismic stratigraphy has shown that the stratigraphic record is dominated by prograding wedges of sediment in which the site of maximum accumulation rate probably oscillates onshore and offshore with the rise and fall of sea level (Vail and others, 1977). This leads to the prediction that changes in the accumulation rates of inner shelf, outer shelf, and slope sections should be out of phase with one another and not lead to straight LOCs. Thus, sedimentologic and stratigraphic support for the assumption of a constant ratio of accumulation rates may be quite weak, compared with the paleontological evidence for the sequence of events. Modern implementations of graphic correlation routinely permit piecewise-linear LOCs (MacLeod and Sadler, this volume) and, thus, prevent sedimentologic assumptions from dominating the paleontological evidence for the sequence of events. A piecewise-linear J-dimensional LOC is desirable for the same reason.

Advantages Realized by Other Methods

The problem of arranging a set of events in their correct relative positions can be divided into two tasks: establishing the order of the events (the sequencing task) and determining the relative sizes of intervals between events (the spacing task). Traditional graphic correlation attempts the two tasks simultaneously. The RASC system (Gradstein and others, 1985; Agterberg, 1990), Edward's (1978) no-space graph method and Gradstein's (1987) extension of the latter, all show the value of isolating the sequencing task and solving it first. Sequencing requires only an ordinal scale. It can be completed with assumptions and expert judgements that concern only the recording and recognition of stratigraphic events. Broad simplifying assumptions about sediment accumulation can then be reserved for the spacing task, which is completed after the best sequence has been found.

The number of short linear segments in a piecewise-linear LOC may be allowed to range up to the total number of events, if necessary, when searching for the greatest economy of fit. This ploy effectively isolates the sequencing problem (MacLeod and Sadler, this volume). A piecewise-linear LOC can pass through the coordinates of all events that are preserved in the same sequence in all sections. These events form the inflection points or "knees" (Shaw, 1964, p. 141) in the LOC; other events are projected to the nearest knee. Because the horizons that correspond in age with each event will either be placed at the corresponding observed range end or moved to coincide with the observed level of some other event, this LOC solves the sequencing problem while avoiding the task of spacing event horizons between collecting horizons. Edwards (1978) called this approach "unscaled." Her no-space graphs are an unscaled two-dimensional graphic correlation. Currently, our computer algorithms extend her strategy to J dimensions, leaving the spacing task to be attempted after the optimal sequence has been found.

The sequencing task is complicated by events that are not everywhere recorded in the same order (Fig. 1). This difficulty is conveniently minimized by methods which apply a "democratic" approach (Hay, 1972; Edwards and Beaver, 1978; Gradstein and others, 1985; Agterberg, 1990). Like graphic correlation, these methods assume that the true local sequence of events is globally constant (spatial homogeneity); but they further assume that the true order of two events will be more frequently preserved than the wrong order (Blank, 1979). Because the true solution is unknowable, this property cannot be demonstrated. We reject this "majority rule" assumption even though it appears to lead directly to reproducible results by eliminating expert judgements.

The quality of the stratigraphic record does vary from event to event and from section to section in ways that an expert can recognize to be systematic. Just as biotas vary with environment, the preservation of fossil taxa may vary systematically with sedimentary facies. Consequently, local observed range ends are less convincing evidence of an evolutionary event horizon where they coincide with facies changes. "Majority rule" is not a convenient platform for incorporating this type of expert knowledge. Furthermore, majority rule does not necessarily respect local evidence for the coexistence of taxa. This contrasts strikingly with Guex's (1977, 1991) correlation procedure, which is built almost entirely upon the premise that two taxa whose ranges are proven to overlap in any one section must coexist in the global solution. An optimization procedure can easily exploit this coexistence information as a constraint on plausible solutions.

GRAPHIC CORRELATION AS A CONSTRAINED OPTIMIZATION PROBLEM

The graphic approach makes it clear that stratigraphic correlation is a constrained optimization problem. Such problems are solved by eliminating impossible solutions (constraint) and then searching for the best of all the possible solutions (optimization). For the stratigraphic correlation problem, Shaw's (1964) economy of fit defines what is meant by "best" and leads to a quantitative measure, called the penalty or objective function, that can be used to rank different solutions relative to the qualities of the best solution. Because the penalty function measures misfit, the solution with the best economy of fit carries the lowest penalty.

The best solution (or supposed true history of events) requires the smallest net disruption when all the local range charts are adjusted to match the supposed truth. Accordingly, the simplest quantitative penalty is the net range adjustment, measured in stratigraphic thickness and totalled across all taxon range ends in all local sections. Exactly how the penalty increments are computed and totalled is partly a programming question. The complexity of the penalty function influences the ease with which an optimization procedure can be adapted. We will derive formulae for penalty functions and discuss different measurement scales. But first we explain how constrained optimization uses penalty values. This is independent of precisely how each local range extension is converted to an increment of the total penalty.

Two Levels of Optimization

If there are I events, each possible solution of the correlation problem is both an arrangement of I events on a time scale and a placement of the corresponding $I \times J$ event horizons in the J local stratigraphic sections. For this reason, the optimization must minimize the penalty at two nested levels. The outer level searches for the best global arrangement of the I origination and extinction events (sequencing task). The inner level searches for the best local placements of event horizons in each section for each global arrangement (locating task).

The outer minimization routine permutes the supposed order of events and remembers the arrangement for which the inner minimization returns the lowest total penalty. It generates a series of arrangements, eliminates any that fail to meet all known constraints about the true sequence, and passes the others in turn to the inner minimization routine, which returns the lowest penalty for that solution. The inner minimization works with a local measurement scale, such as stratigraphic thickness, to determine penalties. It adjusts the position of expected event horizons — horizons that will be assumed to correspond to the "true" age of each event. The purpose of these adjustments is to find the placements that: (1) remain consistent with the arrangement of events that is currently selected by the outer minimization, (2) honor any constraints against range contractions, and (3) yield the lowest total penalty.

The following subsections work through the steps necessary to build a formal constrained optimization procedure that will yield reproducible solutions: (1) list constraints that identify impossible solutions, (2) decide what form the LOC will be allowed to assume, (3) formulate a penalty function that rewards economy of fit, and (4) find a search procedure that efficiently locates the best solution or an extremely good solution.

Constraints

Some constraints identify solutions that can be eliminated without recourse to the inner minimization. They reduce the number of feasible solutions and accelerate our form of the optimization. In effect, a sequence of events that violates these constraints is immediately recognized to carry an infinitely large penalty without calculating any part of the penalty function.

Obviously, solutions that place the first occurrence of any species after its last occurrence are impossible. This is one constraint. Species that can be shown to have co-existed in any one section must have overlapping ranges in the true sequence of events (Guex, 1977; Rubel, 1978). The outer minimization compares solutions with a table of coexistences and eliminates any that violate these constraints. In Figure 1, for example, section two proves that taxon "27" evolved before the extinction of taxon "7." Feasible sequences must also include all dated event horizons in the order of their absolute ages. If strong confidence intervals on the ages of two events do not overlap, this establishes the sequence that solutions will be constrained to honor. The constraint table simply specifies the correct order for these horizons. The dated events need not be recognized in more than one section to provide useful constraint, as was shown by the white areas in Figure 2B.

Constraints can also operate within the inner optimization. Obviously, the sequence of supposed event horizons must be the same in all local sections and match that chosen by the outer minimization (spatial homogeneity). This corresponds to the constraint that, like the snake, no part of the LOC may have a negative gradient. Additionally, it is common to disallow the placement of an expected event horizon within the observed local range. Such placements make "range contractions" in the sense that they imply a true range that is shorter than the observed range. Constraints against range contractions forbid the interpretation that fossils have been reworked into younger sediments or carried to anomalously deep levels by drilling operations (Jones, 1958; Wilson, 1964; Foster, 1966).

An alternative treatment recognizes that reworking is unlikely but not impossible. It charges range contractions a large penalty but not the infinite one implied by a constraint. Local ranges are so easily smeared downward in well cuttings that some stratigraphers simply delete range bottoms from the data set. If a majority of sections are derived from outcrop or continuous cores, however, penalties for upward adjustment of first appearances in cuttings might be set very low. Penalties have the advantage over constraints that they can be adjusted for individual sections and events. Constraints are universal because they cause a solution to be rejected before a penalty is calculated.

Although the table of fossil coexistences represents a set of constraints that apply to the sequences generated by the outer minimization, the evidence for coexistence is local. It may disappear during the correlation procedure if range contractions are permitted. Consequently, the coexistence table must be checked and updated as necessary whenever the inner minimization is allowed to propose a true range that is shorter than a locally observed range.

Form of the Line of Correlation

How closely the LOC can be fit to the coordinates of all the local stratigraphic observations (Fig. 2B) depends on several factors: the precision of the stratigraphic observations (size of the error boxes), the extent to which local sections record conflicting sequences of events (the scatter of the corner points), and the frequency with which the line of correlation is permitted to change slope (subject always to the constraint that its slope is non-negative). In order to appreciate the significance of the complexity of the LOC, it is useful to examine the properties of two extreme end-member forms, even though they are unlikely to be the best approximations of the snake.

The simplest LOC remains perfectly straight (Fig. 4A). It usually causes many local range adjustments but offers a solution to both the sequencing and the spacing problems. In the trivial example of Figures 2 and 4, only one pair of events (+1 and +2) is recorded in contradictory sequences, yet the straight line of correlation forces five range extensions (see Edwards, 1984, for further discussion). At the opposite extreme, a piecewise linear LOC can pass through the coordinates of all observed event horizons that have the same sequence in all sections. It accrues penalties only when forced to resolve local contradictions in the sequence of events (Figs. 4B, C). Even then, the greatest economy of fit is always achieved by extending one observed range so that its estimated end coincides with the horizon of the conflicting event.

Notice first that simplicity in the LOC can be forced by assumptions about accumulation rate in any instance of the problem. But economy of fit to the local stratigraphic observations will not force a simple LOC unless the preponderance of those data constrain a simple path. By freeing the LOC to assume complex piecewise-linear forms, we complete only the sequencing task. It is useful to know what emerges as the best sequence when the analysis is free of all assumptions about accumulation rate. The piecewise linear LOCs found by the current versions of our program successfully achieve this. We are now working on a new inner minimization that makes use of local observations of facies changes, cyclothem thicknesses, and surfaces of hiatus to estimate where the accumulation rates most likely increase or decrease. One experimental concept penalizes range extensions at a different rate in different facies. Another penalizes LOCs for changing slope where there are no sedimentological changes. Where the route of possible LOCs is already tightly constrained by many range end coordinates, the correct slope is driven by paleontological observations alone. The idea is to extract from the best constrained segments a model of relative accumulation rate as a function of sedimentary facies and then apply it to parts of the LOC where a paucity of range ends allows erratic slope changes from segment to segment.

An easier solution to the spacing problem simply applies some general mathematical smoothing function to the LOC found by the current procedure. Smoothing makes solutions appear more realistic by effectively breaking up clusters of events and moving event horizons to positions intermediate between collecting horizons. Yet it is not stratigraphically satisfying. Notice that the piecewise linear LOC with the lowest penalty resolves contradictions with segments that are perpendicular to one axis of the graph. These segments have the most extreme slopes that the LOC can assume, and they imply hiatuses. Any general smoothing algorithm will work most efficiently to realign these segments and eliminate the implied hiatuses. This exposes our biggest unresolved difficulty with the spacing task; simple economy of fit gives rise to complex piecewise-linear LOCs that place hiatuses in some manifestly ar-

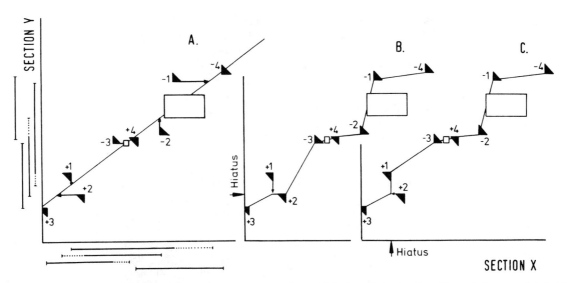

FIG. 4.—One possible straight LOC (A) and two (B,C) alternative piecewise linear LOCs through the data from Figure 2. Arrows and dotted lines indicate range extensions. Large open rectangle: magnetostratigraphic reversal. Open square: correlative ash bed.

bitrary places, yet smoothed LOCs tend arbitrarily to eliminate hiatuses altogether.

Hiatal surfaces are real. Some are marked by shell concentrations, and observed range ends tend to concentrate at these horizons because collecting is easy. If the time span of the hiatus captures several events, however, those local range ends are correctly located together even though they are not reliable evidence for the true global sequence. A low reliability rating usually means that we are relatively free to extend the observed ranges up or down section to match the global sequence, but not in this case. Moving a short distance away from a surface of hiatus may be translate to a very long range extension in time. Independent evidence for the location of hiatuses should be incorporated if possible. One interpretation of sequence stratigraphic models is that hiatuses correlate from section to section. We caution against applying this notion to the LOC or testing it with graphic correlation; the correlative parts of hiatuses have no expression in the J-snake.

A Penalty Function Based on Economy of Fit

We have explained the role of penalties; now we formulate a penalty function based upon the observed ends of taxon ranges in local sections. Kemple and others (1989) set out a formal notation, which is summarized in Figure 5. For a given taxon (i) in a given section (j), the observed local range extends from the lowest find (a_{ij}) to the highest find (b_{ij}). The unknown parameters we must estimate are coordinates of the J-dimensional snake: the elevations of all the local event horizons (α_{ij} and β_{ij}) that are presumed to correspond with the true global limits of each taxon range.

Because the penalty function sums the total discrepancy between observed range ends and estimated event horizons for all taxa and sections in the data set, the incremental discrepancies ($|a_{ij} - \alpha_{ij}|$) and ($|\beta_{ij} - b_{ij}|$) are the only essential increments to be summed. The simplest penalty function can be written:

$$\sum_i \sum_j [|a_{ij} - \alpha_{ij}| + |\beta_{ij} - b_{ij}|] \qquad (1)$$

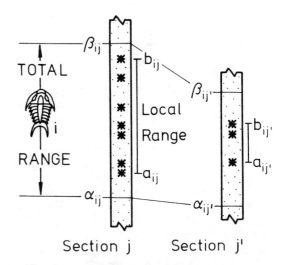

FIG. 5.—Notation for total and preserved local taxon ranges in two sections, j and j' (after Kemple and others, 1989). Asterisks mark fossil finds.

If the inner minimization is not permitted to make range contractions, expression (1) is minimized subject to the constraint that incremental discrepancies may not be negative, i.e., subject to $(a_{ij} - \alpha_{ij}) \geq 0$ and $(\beta_{ij} - b_{ij}) \geq 0$, where stratigraphic distance is measured from the base of the section. It is preferable to make range contractions unlikely rather than impossible. This can be achieved by assigning an arbitrarily large penalty increment wherever the incremental discrepancy is negative. The size of this increment may be adjusted downward to admit a greater possibility of reworking. In this way, a series of runs can search systematically for local observations that might economically be considered as candidates for reworking, in the sense that forbidding the local range to contract has forced large net range extensions elsewhere.

Weighted Observations.—

Formulation (1) is rather crude; it penalizes for stratigraphic distance between observed range ends and estimated event horizons without regard to the section or taxon involved. In practice, the expected utility of local taxon ranges for time correlation varies with the section completeness, the richness of the local fossil record, the ecology and taphonomy of the taxon, the sampling strategy, and the quality of the taxonomy. If these differences can be assessed, they might be incorporated into a set of weights (w_{ij}), which adjust the size of the penalty increment. More reliable local range ends get larger weights so that the penalty grows faster for lines of correlation that fail to pass close to them. The weights for range beginnings and range ends are kept separate (w_{1ij} and w_{2ij}) because they are not equally susceptible to degradation by processes such as contamination of borehole cuttings or reworking. Thus, (1) becomes:

$$\sum_i \sum_j [w_{1ij}(|a_{ij} - \alpha_{ij}|) + w_{2ij}(|\beta_{ij} - b_{ij}|)] \quad (2)$$

Shaw (1964) used local range charts that recorded not only the highest and lowest finds but also the total number of horizons (n_{ij}) at which each taxon had been found. He recommended (p. 235) considering both the number of finds (n_{ij}) and the range length ($b_{ij} - a_{ij}$) when judging which ranges were best established. The reasoning is nearly intuitive. Few biostratigraphers would object to a big discrepancy between the LOC and the observed range ends for a rarely fossilized taxon with a long range, such as the platypus (Marshall, 1990). Most would be outraged if the same discrepancy were suggested for a taxon that is abundantly preserved through a short range within uniform facies, such as a nannoplankton species in calcareous ooze. Our expectation is that the range extension will be of the same order of magnitude as the average gap between finds within the range (Strauss and Sadler, 1989).

The average gap length, given by $(b_{ij} - a_{ij})/(n_{ij} - 1)$, evidently includes both criteria recommended by Shaw and can contribute a part of the weighting scheme that is objective in the sense that it is derived directly from the local data. Strauss and Sadler (1989, p. 417, Eqn. 8) have shown that unbiased point estimation places the true range ends one average gap length beyond the observed range ends. Their formulae for statistical confidence limits on local taxon ranges provide a natural, but non-linear, weighting function based on average gap length. Use of a non-linear relationship between the range adjustment and the penalty considerably complicates the optimization process. A simple linear implementation of the same idea measures stratigraphic distance in units of mean gap length. This leads to a much more manageable penalty function in which the range adjustments are divided by the average gap size. Thus, decreasing the average gap increases the penalty increment for extending the range, making it less likely that a large adjustment will be accepted in the best solution:

$$\sum_i \sum_j [\{w_{1ij}(|a_{ij} - \alpha_{ij}|) + w_{2ij}(|\beta_{ij} - b_{ij}|)\}(n_{ij} - 1)/(b_{ij} - a_{ij})] \quad (3)$$

The distribution of the length of gaps observed in a taxon range is often highly skewed (Strauss and Sadler, 1989), with many small gaps and a few large ones. From his familiarity with very non-random fossil distributions, Charles Marshall (pers. commun., 1994) has suggested that the median gap length, which is usually smaller and more conservative than the average, might be a preferable basis for weights. If all the gap lengths are known, the median is computationally easy to use although it does not produce a tidy equation.

We deliberately leave w_{1ij} and w_{2ij} in the equation to accommodate subjective expert judgements about the reliability of different taxa and sections. It is not our purpose to eliminate expert judgement altogether. Rather, we insist that expert judgements take the form of an explicit table of the values assigned to every w_{1ij} and w_{2ij}. Then it is a simple matter to alter the weights and rerun the optimization to determine the consequences, if any, of changing opinions about the relative merits of different observations. This form of experimentation should educate the expert about the sensitivity of the results to the local observations. It was simply too time consuming and not easily reproducible with traditional graphic correlation.

It may be easy to decide upon the relative sizes of subjective weights for different taxa and sections: assign larger w values for taxa assumed to be more reliable. But there is no established practice to follow when choosing absolute values. Fortunately, formulation (3) helps assign some concrete meaning to the values of w_{1ij} and w_{2ij}. Because increasing the w values and increasing $(n_{ij} - 1)/(b_{ij} - a_{ij})$ have comparable effects, an increase in subjective weight may be regarded as ranking a taxon equal to another that has a smaller average gap within its observed range. Setting $w > 1.0$ is equivalent to ranking a taxon equal to one with a proportionately smaller average gap size.

The range of values assigned to w_{1ij} and w_{2ij} will vary from problem to problem but should be guided by the range of average gap lengths. The Cambrian data set (Palmer, 1954) used by Shaw (1964) illustrates this. Shaw's tally sheets record 239 collection gaps (i.e., barren rock intervals separating two finds of the same species). The average gap size is 18 ft and the range of average gap sizes from the 5th to the 95th percentile is from 5 to 45 ft. With all w_{1ij}'s and w_{2ij}'s equal to 1.0, equation (3) would weight the best constrained 5% of the ranges heavier than the worst constrained 5% by a factor of 9 or more. In equation (2), w values from 1/9 to 1 could reproduce this spread. In equation (3), w values from 1/9 to 9 would be sufficient to reverse most of the weight differences introduced by average gap length alone.

Adding the Spacing Task.—

We have already mentioned the possibility of solving the spacing and sequencing tasks simultaneously by placing a premium on smooth LOCs. The penalty function would need to be modified as follows (Kemple and others, 1989):

$$\sum_i \sum_j [\{w_{1ij}(|a_{ij} - \alpha_{ij}|) + w_{2ij}(|\beta_{ij} - b_{ij}|)\}(n_{ij} - 1)/(b_{ij} - a_{ij})]$$

$$+ K[\text{smoothing penalty}] \quad (4)$$

The value of K adjusts the relative size of the sequencing and spacing components of the penalties.

Units of Measurement.—

Stratigraphic position, the length of range extensions, and the size of penalties may be measured on any one of the five different scales illustrated in Figure 6: time, thickness, event horizons, fossiliferous horizons, or sampled horizons. Ordinal scales, which are quite sufficient for the sequencing task, mea-

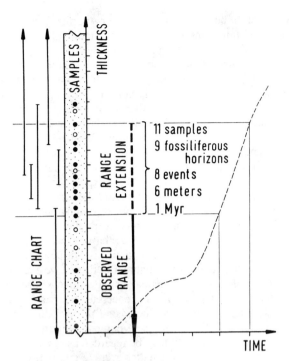

FIG. 6.—Five possible units of measurement for the length of range extensions. Circles mark the positions of sampled horizons; open circles are unfossiliferous samples.

sure stratigraphic distance by counting the number of sampled horizons, fossiliferous horizons ("successful" samples only), or event horizons (range ends). Scales based on numbers of horizons support apparent solutions to the spacing problem that appear rather imprecise relative to the results of traditional graphic correlation. But precision finer than the spacing of sampled horizons is potentially quite spurious.

Of course, the sequencing task can be solved on a continuous scale of distance or time, and the complete solution to the spacing problem certainly requires such scales. We prefer penalties based on rock thickness. Time scales based on interpolation between dated paleomagnetic or isotopic event horizons are an attractive option for very young sections. But these independent estimates of age are too rarely available, and the next paragraph explains that there is a hidden pitfall if ages of first and last appearance datums are used to make a time scale.

Absolute ages have been estimated for many Neogene FADs and LADs from the evidence of numerous different sections. Local ages for events are estimated by linear interpolation in sections where the observed range ends are bracketed between dated horizons. These local age estimates are then averaged to remove the unsystematic errors of interpolation. For some purposes the average dates can be applied to all local sections, with or without independent age control. But graphic correlation specifically seeks the oldest of the local first occurrences and the youngest of the local last appearances. Graphic correlation works with the different ages of the locally observed event horizons, and these are rarely well-constrained by independent dating techniques in all sections. The act of averaging the local estimates of the age of a first or last appearance has already ignored the basic principle of Shaw's (1964) approach: that locally observed range ends do not correlate. Evidently, it is better to win the powerful insights of graphic correlation first, using the more straightforward thickness scales to make range adjustments. The solution can be converted to a time scale after optimization.

Search Procedures

Having formulated a penalty function, we now turn to procedures by which the outer minimization may search through the possible solutions. Because the number of feasible sequences of I events is finite, we might expect to be able to evaluate the penalty function for every sequence and choose the best. Unfortunately, even relatively small data sets are likely to allow too many feasible arrangements for such a search. The outer minimization takes longer as the number of feasible sequences of events increases. The time taken by the inner minimization to find the lowest penalty for each sequence grows with both the number of sections and the number of events ($J \times I$). Most worthwhile instances of the correlation problem include enough local sections and fossil taxa that there is not enough time to calculate the penalty for every possible solution. A simple calculation shows how the number of feasible sequences easily grows beyond control.

I events can be arranged in $I!$ ways. For $I = 100$ there are about 10^{158} sequences. Because some of the sequences are likely to be impossible, this is an upper bound on the size of the search. It can be reached if all I events are taxon beginnings. The I events usually include both taxon beginnings and endings, however, and arrangements that place a beginning after the ending event for the same taxon will be unacceptable, as will arrangements that do not respect observed coexistences of taxa. Any solution that places all the beginning events in the first half of the sequence and all the ending events in the second half is clearly feasible in the sense that no taxon ending now falls before its beginning and all taxa coexist. If there are $I/2$ beginning events and $I/2$ ending events, each permutation of the beginning events can be matched with $(I/2)!$ ending events. So there are $(I/2)! \times (I/2)!$ feasible sequences of this type alone. For $I = 100$ there are approximately 10^{129} such sequences. This number establishes only a lower bound on the size of the search. We have not included any sequences in which some ranges begin after others have ended. It is not necessary that both beginning and ending events are included for all taxa in the data set.

Stratigraphic correlation belongs to a class of problems termed "strongly Non-Polynomial Complete" (Dell and others, 1992). This means that the computation time increases exponentially (or faster!) with the number of sections and taxa in the problem. A simple example, given in the bottom row of Table 1, shows how geologic time might be too short to search all solutions to a reasonably large correlation problem. Other optimization problems that can be NP-complete include searching for the shortest cladogram, locating airline hubs, and routing travelling salesmen. Fortunately, research into these problems has led to a number of general-purpose algorithms that, properly adapted, often find the best solutions for small instances and very good solutions for large ones, without calculating the penalty for every feasible solution. They start with one possible solution and attempt to apply a series of modifications that progress toward better solutions.

TABLE 1.—GROWTH OF COMPUTATION TIME WITH PROBLEM SIZE

Size of Problem	N	10	20	50	100
Number of Possible Solutions and (Solution Time)	2N	20 (2 msec)	40 (4 msec)	100 (10 msec)	200 (20 msec)
	N^2	100 (10 msec)	400 (40 msec)	2500 (250 msec)	10^4 (1 sec)
	2^N	1.0×10^3 (100 msec)	1.0×10^6 (1.7 min)	1.1×10^{15} (3.6 Kyr)	1.3×10^{30} (4.0×10^{12} Myr)

The size of the problem (N) is a function of the number of events and the number of sections. Solution time assumes it takes 1 msec to search 10 solutions. $2N$, N^2 and 2^N are simplistic hypothetical relationships between N and the number of solutions.

A Topographic Analog for Search Algorithms.—

Consider how you might walk to the lowest point on an unmapped topographic surface if you could see only your immediate neighborhood. Imagine that the visible neighborhood extends only one small step from your current position in any direction. Figure 7A shows a landscape that represents the penalties for a straight LOC and a trivial instance of the correlation problem like the one in Figure 4A. Because a straight LOC is specified by its slope and intercept, the penalties corresponding to all solutions can be contoured on a two-dimensional map. Latitude and longitude on the map are slope and intercept. Step size is fixed by the smallest permitted changes in slope and intercept. Every point on the map represents a different straight line. The elevation of the landscape at any point is the penalty associated with the corresponding LOC. A downhill step moves to a LOC with a lower penalty; uphill steps increase the penalty. The lowest point on the landscape is the best solution.

Because the landscape in Figure 7A is a basin with only one minimum, instructions for walking to the lowest point are very simple. Take any step that leads downhill from the current standpoint and repeat until all possible steps lead uphill. The descent to the minimum is more direct if we evaluate the whole neighborhood and take the steepest downhill step. With steepest descent, the search path mimics the movement of a water droplet. This "greedy" search omits two time-consuming strategies that prove useful when the search must find the deepest of many local minima in the landscape. It does not allow uphill steps, which are needed to lead the search path out of local minima and it retains no memory of past steps.

The landscape concept of Figure 7A extends easily to multidimensional piecewise linear solutions (Fig. 7B). Coordinates of each point on the landscape now give a feasible sequence of all events. Elevation is the total penalty found by the inner minimization after adjusting ranges in all sections. Steps from one point to a neighboring point represent small changes in the sequence of events. The outer minimization makes these small changes to generate a new trial solution from the current one after the inner routine has returned its penalty. For example, after evaluating the current sequence, the outer minimization might generate the next one by moving one event at random to a new position in the sequence. Another strategy reverses the order of one pair of adjacent events. Whatever the nature of the permitted changes, the neighborhood of a sequence consists of every new sequence that the outer minimization might generate from it in one step. Do not worry that it is hard to build a mental image of this new landscape and neighborhood structure. The

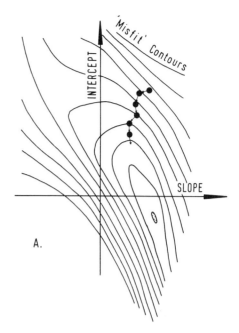

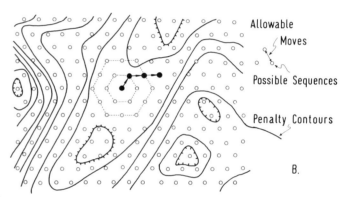

FIG. 7.—Contoured penalty landscapes for (A) a straight line of correlation such as in Figure 4A and for (B) a piecewise linear correlation in J dimensions. Hatched contours indicate the unique minimum in A and multiple local minima and maxima in B. Each small circle in B is a possible sequence of events. Each sequence differs from its immediate neighbors by one permitted move (e.g., switching a pair of adjacent events).

point is that our search procedure must anticipate a complex landscape that includes many local minima.

Every search procedure requires a starting point, which is a feasible sequence of events (perhaps based on a local section), and a neighborhood structure, which specifies the type of change that the outer minimization will use to generate new sequences. Search procedures differ in the way they decide which changes to accept. We describe three search strategies that were designed to cope with local minima. We explain the third (simulated annealing) in most detail because it treats the correlation problem more effectively than the other two.

A Greedy Algorithm with Random Restarts.—

Greedy searches are easy to program, but rarely successful. They evaluate the whole neighborhood, always accept the steepest downhill move, and readily get trapped in local minima. The chance of finding the true minimum can be increased by restarting the greedy search several times using different starting sequences. For some problems, multiple greedy searches perform better than more complex single search algorithms.

Tabu Search.—

If a greedy algorithm could remember the path into local minima, it might learn to climb out and seek deeper minima. Tabu search (Glover, 1990) attempts to give the greedy algorithm this kind of intelligent memory. The search keeps "tabu lists" of past moves and updates them as it moves about the landscape. It still evaluates the entire neighborhood at each step but selects its best move after eliminating steps that are on the current tabu list. The lists have fixed length; each update adds a new tabu move to the top of the list and drops one from the bottom.

A tabu search program must specify the general character of sequence changes that can become tabu. Obvious tabu rules attempt to prevent repeated visits to the same location. An event that has recently been moved higher to generate a new sequence might be prohibited from moving back down as long as it is listed. A pair of events that have switched places might not be allowed to switch back until dropped from the list. At run time, the tabu lists keep track of specific events to which the rules will apply each time the outer minimization generates a new sequence. Multiple lists with different lengths are permitted. Long lists resemble long-term memory because past events remain longer on the list. Obviously tabu search is not a rigidly specified procedure. It gives memory and whole-neighborhood searching their place in a flexible template for tailoring an intelligent search. In addition to fixing parameters like the number and length of tabu lists, it is necessary to discover rules that identify potentially bad sequence changes.

Simulated Annealing.—

It takes time to evaluate the entire neighborhood and to update memory at each step in the search. Simulated annealing is the name given (Kirkpatrick and others, 1983) to a search strategy that abandons both of these "intelligent" features, without losing the ability to climb out of local minima. The name arose because the procedure is modelled on the physical law that describes the probability of growing a perfect crystal by annealing. In this section we explain the procedure qualitatively in terms of the landscape analogy. The succeeding section summarizes the relationship with Boltzmann's distribution law. This law provides a much more exact quantitative analogy, but the section may be skipped by readers not interested in the computational strategy.

Simulated annealing picks possible steps at random in the current neighborhood. Downhill steps are always accepted. If an uphill move is selected, a carefully adjusted probability structure determines whether to accept the move or try another random selection in the same neighborhood. The probability of accepting a large uphill move is always set smaller than the probability of accepting a small uphill move. The probability of accepting an uphill move of a given size is made to decrease progressively throughout the search. In effect, the searcher progressively tires. Early in the search, it is possible to climb large-scale barriers in the landscape because the searcher has strength for large uphill steps and long runs of successive uphill steps. Thus, simulated annealing begins by sounding the large-scale topography and tends to home in on the slopes around the global minimum. Toward the end of the search, an exhausted searcher makes mostly downhill steps and can descend precisely to the lowest point.

For human searchers on a landscape, we would need to specify the initial vigor and a tiring schedule. The initial vigor determines the chances of moving uphill away from the starting point; a high value is more desirable if the starting point is separated from the lowest point by significant relief on the landscape (e.g., a random starting sequence that is far from the optimal sequence). In practice, a stepped schedule for tiring the searcher is easier to manage than a smooth curve. After a fixed number of moves, vigor is reduced by a constant tiring ratio. The fixed tiring ratio (less than one) ensures that the likelihood of uphill moves decreases less rapidly toward the end of the search and never reaches zero. The slowest tiring schedules give the searcher a high initial vigor, a high tiring ratio (close to 1.0), and a large number of trials between each reduction in vigor. Because slow tiring schedules allow significant uphill movement late into the search, they work best from bad starting points on very rough landscapes. They waste too many moves if the starting point is good or the landscape simple.

All that remains to be specified is the formula that converts vigor into a probability distribution for accepting uphill moves. Kirkpatrick and others (1983) derived their formula by analogy with Boltzmann's distribution law from statistical mechanics. Readers who prefer to skip the details given below should simply note that it is usual practice to speak in terms of initial "temperature" and a "cooling" schedule, rather than the initial vigor and tiring schedule of our qualitative analog.

The Mineralogical Analogy for Simulated Annealing.—

For crystals growing from a cooling melt, Boltzmann's distribution law explains why slower cooling increases the probability of growing perfect crystals. Perfect crystals are the minimum energy state. Although the system prefers low energy states at any temperature, higher energy states do arise but with a small probability that decreases as temperature falls. Slow cooling maximizes the likelihood that the final state will achieve the minimum energy. For a fixed temperature (T) greater than zero, Boltzmann's distribution gives the probability $P_T(\pi)$ that a state π will arise:

$$P_T(\pi) = K_T e^{-\{E(\pi)/k_b T\}} \qquad (5)$$

where $E(\pi)$ is the energy of state π, K_T is a normalizing constant which brings the area under the probability distribution to 1, and k_b is Boltzmann's constant. Kirkpatrick and others (1983) recognized that the energy of a state is exactly analogous to the penalty for a solution to an NP-complete problem. They proposed a search that allowed "bad" moves with the same probability that a cooling physical system assumed non-minimal energy states. They reasoned that such a search would reliably find the best solution if the probabilities were reduced according to the right schedule, just as efficient annealing schedules can be found that yield perfect crystals.

Using a computer to simulate sampling from the distribution given above is difficult because K_T is unknown. In fact, enumeration of all possible states for a given $T > 0$ would be required to evaluate K_T. Metropolis and others (1953) developed a method of simulation that samples the distribution without evaluating K_T. Their algorithm starts at one state, generates a new "neighboring" one, calculates the change in energy (δE) that will occur if the new state (π) is adopted, and accepts the new state according to the probability distribution:

$$P_T(\pi) = \begin{cases} 1, & \delta E \leq 0 \\ e^{-\{\delta E/k_b T\}}, & \delta E > 0 \end{cases} \quad (6)$$

The analogy with the outer optimization and the neighborhood structure of our searches is obvious. Hammersley and Handscomb (1964, p. 117–121) give a fairly readable proof that this algorithm simulates sampling from Boltzmann's distribution. Simulated annealing exploits the Metropolis algorithm to allow the outer minimization a comparable probability of accepting a change to a new sequence (π):

$$P_T(\pi) = \begin{cases} 1, & \delta F \leq 0 \\ e^{-\{\delta F/T\}}, & \delta F > 0 \end{cases} \quad (7)$$

where δF is the change in penalty (elevation) associated with the move to π from the current sequence, and T decreases during the search. The constant k_b is not needed because we are no longer using real temperature and energy scales. The equation says that, although the search prefers to move "downhill" to a sequence with a lower penalty, uphill moves do occur and such moves have a small probability, which decreases with T as the search progresses.

The appearance of T in the denominator of the exponent $-(\delta F/T)$ is the key to the behaviour of the search. The range of values assumed by dF depends on the data set and the distance scale used. If the search begins at a value of T that is very large compared with typical values of δF, $P_T(\pi)$ approximates a uniform distribution on the set of all feasible sequences; all moves are approximately equally likely to be chosen regardless of δF. If the search continues until T becomes very close to zero, the probability of accepting any move that increases the penalty now tends to zero regardless of δF. At intermediate values of T, the search is more sensitive to δF; moves that increase the penalty are more likely to be accepted if the penalty increment is small.

To simulate the annealing process, the search starts with a high value of T (analogous to a molten state) and lowers it slowly while testing feasible sequences. If T is reduced slowly enough, the search path converges to a sequence with the global minimum penalty (analogous to a perfect crystal). The components of the best cooling schedule are all problem specific and may need to be altered for different sets of data. To get a starting temperature T, first specify an approximate initial value for accepting an uphill move (0.5–0.8 worked for us). Next generate a series of neighbors for one sequence and determine the distribution of δF. Finally solve backward for the initial value of T that gives the desired probability. We tried logarithmic cooling as suggested by Geman and Geman (1984). After 100,000 trials, searches on our stratigraphic data sets had not stabilized, so we returned to the stepwise cooling schedule of Kirkpatrick and others (1983). We chose cooling ratios between 0.90 and 0.98 by trial and error and adjusted the number of trials at each temperature to prevent $P_T(\pi)$ from falling too fast. The next section describes the performance of different cooling schedules on one stratigraphic data set.

REPLICATION OF SHAW'S CASE STUDY

To ensure that the automated constrained optimization procedure leads to reasonable results, we examined its performance with an instance of the problem for which a very good solution has already been found manually. Shaw (1964, Appendix A, p. 225) illustrated his graphic method by correlating seven sections with 62 taxa from the Cambrian Riley Formation of the Llano Uplift in Texas (Palmer, 1954). We have repeated the example, using constrained optimization to treat all seven sections simultaneously. We describe the results of runs for which the subjective weights (w_{1ij} and w_{2ij}) were all set to the value of 1.0, and no limits were placed on the number of segments in the piecewise linear LOC. In other words, only the sequencing task is completed, and the penalty function is given by formulation (1), constrained to avoid range contractions.

Measured (in ft) against penalty function (1), Shaw's (1964) solution scores 4,139. This is presumably a very good score because the local observations in the two best sections constrain the LOC to be very nearly linear. It is not necessarily the lowest score because Shaw used some subjective expertise in fitting LOCs. After examining the results of dozens of runs on this data set using different search procedures, we are confident that the minimum penalty is 3,546. After some initial trial-and-error, we have found a wide range of search protocols that lead to a best solution with this penalty. It represents an economy-of-fit just a little better than Shaw's, as would be expected if our procedures are stratigraphically reasonable because they searched for the greatest economy of fit without adjusting to any subjective judgements or weights. The penalty for the best solution increases if some taxa or sections are weighted more heavily than 1.0. This is effectively what Shaw did by fitting his LOCs to what he judged to be the best established ranges. We briefly explain the results of runs that illustrate how we selected our search procedure, initial sequence, and annealing schedule.

Search Procedure

Four different search procedures have been tried for the outer optimization. Runs using a greedy algorithm and multiple starting sequences fail to reach good scores (Dell and others, 1992). Perhaps the stratigraphic correlation problem generates very rough penalty landscapes with too many minima even at a very local scale. The difficulty might be overcome by increasing the neighborhood size. This means allowing larger sequence changes between penalty calculations and could be counterproductive when close to the best sequence. Simulated annealing reduces the scale of the search as it progresses.

Simulated annealing has been applied in two versions, one with a large neighborhood structure and the other with a small neighborhood structure. The neighborhood size results from the way the outer optimization generates new sequences. The small neighborhood version allows the penalty to be calculated much faster as a reoptimization. Both versions find solutions with the lowest known penalty. With the large neighborhood structure (Kemple, 1991), the outer optimization generates new sequences by selecting one event at random; if a beginning event, it is moved to a random position anywhere before the corresponding ending event; if an ending event, it is moved to a random position anywhere after the corresponding beginning event. In the small neighborhood structure, moves are much more conservative (Dell and others, 1992); an adjacent pair of events is selected at random and their sequence reversed.

The large neighborhood structure allows relatively large-scale changes, produces a rougher landscape, and requires the inner minimization to recalculate the penalty completely with every change. The small neighborhood structure restricts changes in sequence to such a local scale that the inner minimization can calculate directly the penalty change caused by the move, rather than recalculate the whole penalty and compare it with the last. This very fast reoptimization procedure exploits the fact that only a very local change has been made to a sequence for which the placement of the α_{ij}'s and β_{ij}'s had already been optimized. Reoptimization calculates the whole penalty only once—for the initial sequence. When both optimization and reoptimization used the same random starting sequences, the same hardware, and the small neighborhood structure, reoptimization was more than 5.7 times faster than optimization for Palmer's (1954) data set. Larger data sets should yield larger performance gains under reoptimization.

Reoptimization has been used to measure the performance of tabu search against simulated annealing (Dell and others, 1992). Although some tabu searches lead to the best known solution for Palmer's (1954) data, we have not found search rules that make it as reliable as simulated annealing. Tabu search often proceeded quickly to a very good solution and then failed to move to the best. Perhaps we could discover better tabu rules and list lengths for the time correlation problem. After learning that tabu search produces only poor solutions to some other stratigraphic data sets, however, we abandoned it in favor of simulated annealing.

Speed of the Search

All runs were considerably more efficient than the traditional manual method. We present our results in terms of the number of sequences tried (Fig. 8) and offer some specific examples of the time taken. The time taken depends upon the efficiency of the computer code and the hardware. The precise number of trial sequences examined before simulated annealing finds the best solution is a matter of chance, but the range of values depends upon the data set and the annealing schedule. Critical factors under our control are the sequence of events from which the search starts and the cooling schedule. With a good initial sequence, the large neighborhood structure, and a well-adjusted annealing schedule, optimization can routinely find the best solution in as few as 25,000 to 40,000 trials. With a very poor initial sequence and a small neighborhood structure, reoptimization might try more than a million sequences.

Initial Sequence.—

Shaw (1964) started his graphic compositing with section 1 (Morgan Creek). We experimented with several different initial sequences. A few were random, one was Shaw's solution, and most were based upon the sequence of observed event horizons in a local section. Unlike an initial section in traditional graphic correlation, our initial sequence must include all events and respect all constraints. To generate an initial sequence from an observed local section, all missing taxon bases are placed in random order before the lowest event horizon, and missing tops

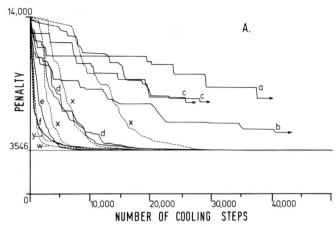

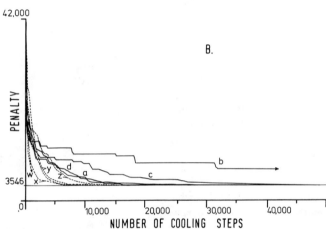

FIG. 8.—Rate of approach to lowest penalty during simulated annealing with the large neighborhood structure. (A) annealing from an initial solution based on the best section (Morgan Creek; section 1 of Shaw, 1964) with different cooling parameters. Parameters are starting temperature, cooling ratio, and number of sequences tried at each temperature; a: 200,0.98,1000; b: 100,0.98,500; c: 200,0.98,500; d: 200,0.98,80; e: 200,0.98,20; f: 200,0.98,4; w: 500,0.9025,5; x: 500,0.9025,100; y: 100,0.9025,20. The paths of approach curves with the same cooling parameters differ because of the chance selection of uphill moves. (B) annealing from an initial solution based on a relatively poor section (Pontotoc; section 6 of Shaw, 1964); a: 200,0.98,80; b: 100,0.98,40; c: 200,0.98,200; d: 100,0.98,80; w: 500,0.9025,10; x: 500,0.9025,5; y: 500,0.9025,100; z: 500,0.9025,200. Aa, Ab, and Ac cool too slowly; they allow too many early bad moves early. Bb cools too fast; it needs more bad moves late in the annealing process. Ax and By have the same cooling schedule; it provides an efficient search from a poor initial section but not from the best section.

are placed at random above the highest event horizon. Each taxon pair that is constrained to coexist is tested to ensure that both beginning events precede both ending events. Violations may arise in pairs of locally observed ranges but not from the placement of missing taxa. For pairs in violation, the positions of the first ending event and the second beginning event are reversed. The resulting sequence honors all coexistences.

Table 2 summarizes the results of several runs that used simulated annealing and the large neighborhood structure with different initial permutations, but same "cooling" schedule. From an initial value of 500, "temperature" was lowered every 1,000 trials to 0.9025 of its previous value. We wanted an initial acceptance probability of about 0.5 for uphill moves and had an

TABLE 2.—PENALTIES FOR SOLUTIONS TO THE RILEY FORMATION

	Initial Permutation				
	Random	Section 1	Section 4	Section 6	Shaw's Solution
Initial Penalty	>60,000	12,457	28,675	42,057	4,139
After 40,000 trials	<4,139	3,904	3,935	3,898	3,546*
After 80,000 trials	3,546*	3,546*	3,555	3,546*	3,546*

*best known economy of fit

average δF of around 350. Solving backward led to the initial "temperature" of 500.

The advantage of starting with a good solution is obvious in terms of the number of trials needed. The penalty function may be used to chose the best section. An initial sequence based upon the Morgan Creek section scores between 12,450 and 13,990, depending upon the random placement of missing events. Initial sequences based on the White Creek and James River sections score 20,000–25,000. The Little Llano River, Lion Mountain and Streeter sections give rise to initial penalties in the range 29,000–33,000. The Pontotoc section leads to initial penalties above 40,000, still much better than most random starting sequences for which penalties exceed 60,000. Higher initial penalties result where larger numbers of taxa are missing from the local section and must be placed at random in the initial sequence. Missing taxa do not contribute to the penalty for the best solution. Consequently, the section that produces the highest starting penalty does not necessarily make the largest contribution to the final penalty. The latter depends upon the under-representation of the ranges of taxa that are found in the section

"Cooling" Schedule.—

Simulated annealing typically makes rapid initial progress in reducing the penalty function and then suffers diminishing returns in the final approach to the best solution. Figures 8A and 8B show examples of approach curves from a series of runs that start from the same initial section but experiment with different "cooling" schedules. From initial "temperatures" of 100, 200, and 500, "cooling" was applied at intervals of 40 to 1,000 trials, with "cooling" ratios from 0.90 to 0.98. High initial "temperatures" and a slow "cooling" schedule allow the search to move rapidly away from the initial sequence. This is undesirable when starting from a good sequence based upon the best section; it is more effective from a poor initial section or a random starting sequence. The "cooling" ratio and the number of trials between "cooling" events must be high enough that the search can make occasional uphill steps near the end of the search. If the search "cools" too fast, it rapidly moves to a good solution but is unable to step away and find the best solution. If the search "cools" too slowly, many steps are wasted calculating penalties for moves away from good solutions.

Running Time.—

We offer a few examples of the running times on different hardware platforms in order that the reader may gauge the practicality of the procedure. Of course, the running time will vary widely for data sets of different size. Early versions of our programs had relatively inefficient data structures and did not take advantage of reoptimization. The early runs required 5–6 CPU hours to reach the best solution for Palmer's data on an IBM 3033AP mainframe computer. The latest versions run on IBM personal computers using 32-bit Fortran and much more efficient data structures. With well-tuned annealing schedules they solve Palmer's problem within 40,000 trials. Typical running times are 13 minutes with an i486DX processor at 33MHz, 7 minutes with an i486DX processor at 66 MHz, and less than 4 minutes with a Pentium processor at 66 MHz. (including simultaneous charting of progress on the screen). These runs do not take advantage of reoptimization. If the number of sections or taxa increase, more time will be required to calculate the penalties. Reoptimization requires more trials; but the calculation time increases only as the number of sections increases. Regardless of the total number of taxa, reoptimization determines the new penalty by considering only one pair of events in each section. As the number of taxa increases, therefore, reoptimization will eventually become the more efficient procedure.

Comparison with Shaw's Solution

Solutions found by constrained optimization and traditional graphic correlation are very similar. The results can be compared qualitatively on range charts (Fig. 9) or with graphic correlations of the two solutions (Fig. 10). Differences are small

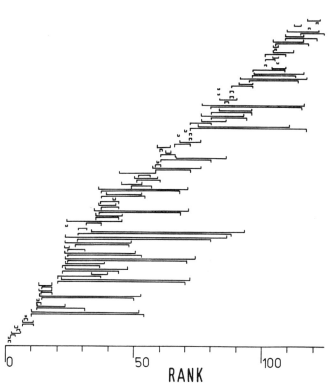

FIG. 9.—A range chart for the Riley Formation that compares the best solution given by constrained optimization (range ends turned down) with Shaw's (1964) solution (range ends turned up). Ranges for corresponding taxa are placed back to back. In order to compare the two solutions, range lengths and positions are plotted against an ordinal scale of the 124 beginning and ending events. Tied ranks are additive (i.e., if two range ends tied for the 4th rank, the next assigned rank is sixth). This ploy brings the two solutions to a common scale despite the fact that Shaw's solution has fewer range ends with tied ranks.

80 WILLIAM G. KEMPLE, PETER M. SADLER, AND DAVID J. STRAUSS

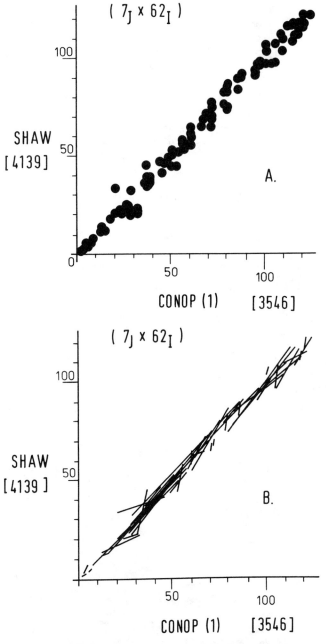

FIG. 10.—Graphic correlation of the best solution given by constrained optimization (CONOP) starting from section 1 with Shaw's (1964) solution. Penalties are shown in square brackets. Axes have rank scales as in Figure 9. (A) circles show coordinates of range ends. (B) lines connect range ends for single taxa; line slopes steeper than +1 mean that Shaw's ranges are longer; slopes gentler than +1 arise when Shaw's ranges are shorter.

TABLE 3.—KENDALL'S RANK CORRELATION OF SOLUTIONS TO THE RILEY FORMATION

	Simulated Annealing Section 4	(Initial Permutations) Section 6	Shaw's Solution	Traditional Graphic Correlation (Shaw's Solution)
Section 1	0.9831	0.9864	0.9899	0.9509
	0.9754*	0.9733*	0.9770*	0.9361*
	0.9814**	0.9900**	0.9938**	0.9505**
Section 4		0.9826	0.9881	0.9525
		0.9708*	0.9843*	0.9374*
	0.9827**	0.9843**	0.9566**	
Section 6			0.9899	0.9484
			0.9768*	0.9321*
			0.9973**	0.9507**

* beginning events only
** ending events only

and non-systematic, except that constrained optimization produces slightly better economy of fit. Kendall's (1975, p. 94) coefficient of concordance, W, quantifies the similarity of the solutions derived by annealing from different initial sections (Table 3). Perfect concordance results in a value of 1.00. For pairs of solutions derived from different local sections, W is always greater than 0.98; comparison of these solutions with the one derived from Shaw's solution produces W's of 0.95. For correlation purposes, the differences are probably not worth the effort of making multiple searches from different starting points.

Shaw's solution required many pages of *ad hoc* justification for LOCs. Mostly he was selecting the better established local ranges, and the exercise had an air of subjectivity. Our comparable solution is particularly noteworthy because we used no subjective weights for the taxa or the sections. We conclude that the complete set of local observations serves to identify the better established ranges. A reasonable solution in J dimensions can arise without elaborate weighting schemes. But constrained optimization is fast enough that it is possible to experiment with truly subjective opinions about the quality of different taxa and sections. Differences in the final solution that result from different subjective weights will indicate where subjective opinion really matters and which parts of the data set have little influence on the outcome. Similar insights might be gained from studying the contributions of different taxa and sections to the penalty for the best solution. Such sensitivity analyses indicate where additional sampling would be most profitable.

Although the discrete scale of event horizons is entirely sufficient for the sequencing task, the solution obtained by constrained optimization appears crude by comparison with Shaw's (1964) results when both are plotted as a fence diagram (Fig. 11). Constrained optimization bunches the lines of correlation because it places expected event horizons (a_{ij}'s and b_{ij}'s) only at horizons where fossils have been collected or sought. The more even spacing of events in Shaw's (1964) solution results solely from the assumption that the ratios of accumulation rates between sections do not change.

SUMMARY

The principle of economy of fit, which is the basis of Shaw's (1964) method, defines the best solution for a stratigraphic correlation problem as that which results in the smallest net disruption of the best established local taxon ranges. Economy of fit is easily quantified and allows the correlation problem to be described in terms precise enough to admit a solution by constrained optimization. The constrained optimization may be performed for all sections simultaneously. The simulated annealing procedure described by Kirkpatrick and others (1983) completes the search in acceptably short times for moderately large stratigraphic data sets.

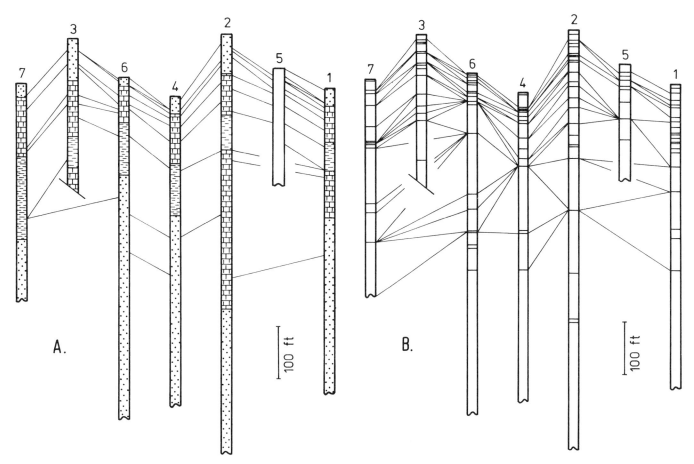

Fig. 11.—Fence diagrams of correlation between the seven sections of the Riley Formation that result from Shaw's (1964) solution (A) and the best solution from constrained optimization (B). The seven sections have been projected onto an east-west line. Shaw's section names are: 1- Morgan Creek, 2- White Creek, 3- James River, 4- Little Llano River, 5- Lion Mountain, 6- Pontotoc, and 7- Streeter. Columns in 11A have been ornamented according to lithology as reported by Shaw (1964): dots for sandstone, dashes for siltstone, and brickwork for limestone; no information for section 5. The columns in 11B indicate the distribution of horizons at which local range ends were recorded.

We have experimented most extensively with the instance of 62 taxa and 7 measured sections in the Cambrian Riley Formation of Texas. Shaw (1964) used the same data to introduce graphic correlation. Without using any subjective weighting of different taxa and sections, constrained optimization finds a solution that has marginally better economy of fit than Shaw's composite sequence of events. The traditional method builds up its composite range chart by integrating the observed ranges one section at a time. Some steps in the compositing process include so few local observations that the LOC cannot be usefully constrained without *ad hoc* expert selection of the best established observations. Our successful replication suggests that most of the expertise is not as subjective as it might appear. Rather, the expertise derives from knowledge of the other sections not currently on the graph. Thus, by working in J dimensions simultaneously (J = number of sections), the constrained optimization allows better separation of objective and subjective aspects of the problem. The procedure is sufficiently fast that it becomes feasible to examine the sensitivity of the solution to different subjective judgements about the relative quality of different parts of the stratigraphic data. It also allows better separation of the paleontological information from assumptions about relative accumulation rates.

ACKNOWLEDGMENTS

Michael Murphy and Lucy Edwards introduced us to Shaw's method. Keith Mann, Charles Marshall and an anonymous reviewer suggested improvements to an earlier draft of this paper. Parts of this work were supported by NSF grant EAR8721192 to Sadler and EAR9219731 to Sadler and Kemple.

REFERENCES

AGTERBERG, F. P., 1990, Automated Stratigraphic Correlation: Amsterdam, Elsevier, Developments in Paleontology and Stratigraphy, v. 13, 424 p.
AGTERBERG, F. P. AND GRADSTEIN, F. M., 1988, Recent developments in quantitative stratigraphy: Earth Science Reviews, v. 25, p. 1–73.
BLANK, R. G., 1979, Applications of probabilistic biostratigraphy to chronostratigraphy: Journal of Geology, v. 87, p. 647–670.
DELL, R., KEMPLE, W. G., AND TOVEY, P., 1992, Heuristically solving the stratigraphic correlation problem: Institute of Industrial Engineers, First Industrial Engineering Research Conference Proceedings, p. 293–297.
EDWARDS, L. E., 1978, Range charts and no-space graphs: Computers and Geosciences, v. 4, p. 247–255.
EDWARDS, L. E., 1982, Quantitative Biostratigraphy: the methods should suit the data, *in* Cubitt, J. M. and Reyment, R. A., eds., Quantitative Stratigraphic Correlation: Chichester, Wiley, p. 45–60.
EDWARDS, L. E., 1984, Insight on why graphic correlation (Shaw's method) works: Journal of Geology, v. 92, p. 583–587.

EDWARDS, L. E., 1989, Supplemented graphic correlation: a powerful tool for paleontologists and non-paleontologists: Palaios, v. 4, p. 127–143.

EDWARDS, L. E. AND BEAVER, R. J., 1978, The use of a paired comparison model in ordering stratigraphic events: Mathematical Geology, v. 10, p. 261–272.

FOSTER, N. H., 1966, Stratigraphic leak: American Association of Petroleum Geologists Bulletin, v. 50, p. 2604–2606.

GEMAN, S. AND GEMAN, D., 1984, Stochastic relaxation, Gibbs distributions, and the Bayesian restoration of images: Institute of Electrical and Electronics Engineers Transactions on Pattern Analysis and Machine Intelligence, v. 6, p. 721–741.

GLOVER, F., 1990, Tabu search: a tutorial: Boulder, Author's Course Notes, University of Colorado, 47 p.

GRADSTEIN, F. M., AGTERBERG, F. P., BROWER, J. C., AND SCHWARZACHER, W. S., 1985, Quantitative Stratigraphy: Dordrecht, UNESCO, Reidel Publishing Company, 598 p.

GRADSTEIN, F. M., 1987, Four quantitative stratigraphic methods: Ouro Preto, Proceedings of the 4th Cogeodata Symposium, p. 67–82.

GUEX, J., 1977, Une nouvelle méthod d'analyse biochronologique; note preliminaire: Laboratoire de Géologie, Mineralogie, Geophysique et Mussée Géologique de l'Université de Lausanne Bullétin, v. 224, p. 309–321.

GUEX, J., 1991, Biochronological Correlations: Berlin, Springer Verlag, 252 p.

HAMMERSLEY, J. M. AND HANDSCOMB, D. C., 1964, Monte Carlo Methods: London, Methuen, 169 p.

HAY, W. W., 1972, Probabilistic stratigraphy: Eclogae Geologica Helvetica, v. 65, p. 255–266.

JONES, D. J., 1958, Displacement of microfossils: Journal of Sedimentary Petrology, v. 24, p. 453–467.

KENDALL, M. G., 1975, Rank Correlation Methods: London, Griffin, 202 p.

KEMPLE, W. G., 1991, Stratigraphic correlation as a constrained optimization problem: Unpublished Ph.D. Dissertation, University of California, Riverside, 189 p.

KEMPLE, W. G., SADLER, P. M., AND STRAUSS, D. J., 1989, A prototype constrained optimization solution to the time correlation problem, *in* Agterberg, F. P. and Bonham-Carter, G. F., eds., Statistical Applications in the Earth Sciences: Ottawa, Geological Survey of Canada Paper 89-9, p. 417–425.

KIRKPATRICK, S., GELATT, C. D., AND VECCHI, M. P., 1983, Optimization by simulated annealing: Science, v. 220, p. 671–680.

MARSHALL, C. R., 1990, Confidence intervals on stratigraphic ranges: Paleobiology, v. 16, p. 1–10.

METROPOLIS, N., ROSENBLUTH, A. W., ROSENBLUTH, M. N., TELLER, A. H., AND TELLER, E., 1953, Equation of state calculations by fast computing machines: Journal of Chemical Physics, v. 21, p. 1087–1092.

MILLER, F. X., 1977, The graphic correlation method in biostratigraphy, *in* Kauffman, E. G. and Hazel, J. E., eds., Concepts and Methods of Biostratigraphy: Stroudsberg, Dowden, Hutchinson and Ross, p. 165–186.

PALMER, A. R., 1954, The faunas of the Riley Formation in Central Texas: Journal of Paleontology, v. 28, p. 709–786.

PISIAS, N. G., BARRON, J. A., NIGRINI, C. A., AND DUNN, D. A., 1985, Stratigraphic resolution of leg 85 drill sites: an initial analysis: Initial Reports of the Deep Sea Drilling Project, v. 85, p. 695–708.

RUBEL, M., 1978, Principles of construction and use of biostratigraphic scales for correlation: Computers and Geoscience, v. 10, p. 97–105.

SHAW, A. B., 1964, Time in Stratigraphy: New York, McGraw Hill, 365 p.

STRAUSS, D. J. AND SADLER, P. M., 1989, Classical confidence intervals and the Bayesian probability estimates for ends of local taxon ranges: Mathematical Geology, v. 21, p. 411–427.

TIPPER, J. C., 1988, Techniques for quantitative stratigraphic correlation: a review and annotated bibliography: Geological Magazine, v. 125, p. 475–494.

VAIL, P. R., MITCHUM, R. M. Jr., AND THOMPSON, S., 1977, Global cycles of relative changes in sea level, in Payton, C. E., ed., Seismic Stratigraphy—Applications to Hydrocarbon Exploration: Tulsa, American Association of Petroleum Geologists Memoir 26, p. 83–98.

WILSON, L. R., 1964, Recycling, stratigraphic leakage, and faulty techniques in palynology: Grana Palynologica, v. 5, p. 427–436.

EVALUATING THE USE OF AVERAGE COMPOSITE SECTIONS AND DERIVED CORRELATIONS IN THE GRAPHIC CORRELATION TECHNIQUE

KENNETH C. HOOD

Exxon Exploration Company, 440 Benmar, Houston, TX 77060

ABSTRACT: Two enhancements to the graphic correlation technique, average composite sections and derived correlations between comparison sections, help refine correlations and increase the robustness of geological interpretations. Average composite sections match the observed ranges of most taxa more closely than maximum composite sections and therefore better constrain the respective lines of correlation. Comparison of the two approaches using a hypothetical taxon first-occurrence surface and a published dataset comprising 10 DSDP sites from the Atlantic and Pacific ocean basins documents the refinements possible using average composite sections. Analysis of the hypothetical taxon first-occurrence surface demonstrates that the expected error of correlating with a maximum composite section increases rapidly as additional comparison sections are included; whereas, the expected error of correlating with an average composite section remains relatively stable. Analysis of Pliocene planktonic foraminifers and calcareous nannofossils from 10 DSDP sites documents that correlations based on average composite sections match the independent paleomagnetic constraints more closely than correlations based on maximum composite sections alone. Average composite sections should be based on potential maximum events which excludes those ranges that are truncated by a substantial hiatus or by the limits of sampling in a comparison section.

Derived correlations use the implied correlation between comparison sections to help evaluate and refine the original interpretations of how each comparison section correlates with the composite section. Analysis of three DSDP sites from the eastern and central equatorial Pacific document that substantial refinements are possible. Using derived correlations, a hiatus missed on an original interpretation was identified, and several smaller hiatuses were adjusted to better match the additional constraints. Derived correlations take advantage of the multidimensional nature of biostratigraphic datasets that has traditionally been under utilized, and they provide a powerful mechanism for evaluating alternative interpretations.

INTRODUCTION

Alan Shaw first introduced the graphic correlation technique in 1964, and many recent papers that utilize the technique attest to its subsequent increase in popularity (e.g., Shaw, 1964; Sweet, 1984, 1992; Prell and others, 1986; Edwards, 1989; Hazel, 1989; Martin and others, 1990; MacLeod and Keller, 1991; Melnyk and others, 1992). One reason for this popularity is that graphic correlation is based on relatively simple concepts which makes the results easy to understand. Graphic correlation has many other strengths that contribute to its success as a biostratigraphic tool. For example, large amounts of data can be incorporated into the correlation process without requiring *a priori* assumptions about the biostratigraphic utility of any particular event. Instead, each event can be evaluated in the context of all other available data for a well or stratigraphic section. Additionally, graphic correlation simplifies multidimensional datasets by reducing correlations into multiple 2-dimensional (pairwise) comparisons, which in turn are used to combine all available biostratigraphic information into a single composite reference framework. Some of the methods used to achieve the advantages of graphic correlation, however, also serve as limitations of the technique. This paper addresses the limitations of the technique relating to the method of constructing composite sections and the simplification of correlations into multiple 2-dimensional comparisons.

Development of a composite section is a critical aspect of graphic correlation, because the composite section provides the foundation both for subsequent correlations and for evaluating the chronostratigraphic ranges of taxa. The approach used to construct the composite section can have a major impact on the results. Generally, the composite section is constructed by combining data from many comparison sections to produce the total composite range for each taxon (Shaw, 1964; Miller, 1977; Edwards, 1984). This process results in a maximum composite section, because it contains the maximum observed range for each taxon (event). As the number of comparison sections incorporated into the composite section increases, the composite ranges of taxa become progressively longer than the ranges observed in most individual comparison sections. Shaw (1964) referred to the pattern resulting from this progressive divergence as channeling, because most tops plot to one side of the line of correlation (LOC) and most bases plot to the other. Channeling decreases the level of constraint the data provide on correlations because it reduces the number of events that plot near the true LOC. For graphs exhibiting pronounced channeling, the LOC generally is interpreted to plot within the interval separating the bases from the tops. With this approach, however, the potential for error increases as a function of the width of the channel. Moreover, in some instances the true LOC does not follow the channel (see below). Despite these limitations, maximum composite sections are used routinely to obtain the advantages of having all available taxa integrated into a single composite section.

Graphic correlation reduces correlations into multiple 2-dimensional comparisons. By using the same reference section in each comparison, all correlations can be expressed in terms of the same (composite) units. However, this approach under-utilizes potentially useful information from direct correlation between individual comparison sections. Commonly two or more comparison sections will be quite similar in terms of facies, sedimentation rate, taxa (events) present, and so forth. Correlation between such comparison sections provides an opportunity to identify and refine subtle hiatuses and relative changes in sedimentation rate that might not otherwise be apparent. Much of this information is lost when correlating with a composite section, particularly a global composite section that contains information from many sections encompassing widely dissimilar areas.

Despite frequent use of the graphic correlation technique, few improvements have been made to the original methodology. A notable exception is Miller (1977) who recommended that LOCs be positioned subjectively by operator judgment rather than by least-squares regression to accommodate greater utilization of all available geological constraints. This paper introduces two new enhancements to the graphic correlation technique: (1) average composite sections, and (2) derived correlations between comparison sections (see discussion below). Combined use of these two enhancements reduces several limitations of the original methodology and can produce more reliable and geologically robust interpretations for many biostratigraphic datasets. This paper uses a hypothetical taxon first-occurrence surface and a published dataset comprising 10 DSDP wells to document the refinements possible using average composite sections, and it uses an additional dataset comprising 3 DSDP wells to document the refinements possible using derived correlations between comparison sections. In the following discussion on these enhancements, only those aspects of the methodology which differ from the traditional approach are considered in detail. Excellent discussions of the basic concepts of graphic correlation can be found in Shaw (1964), Miller (1977), and Edwards (1984).

AVERAGE COMPOSITE SECTIONS

Average composite sections, which are based on the average ranges of taxa, provide an alternative method of establishing a composite which helps to overcome several limitations of maximum composite sections. An average composite section is constructed as follows: (1) convert all events from each comparison section into composite units by projecting them to the LOC, (2) identify potential maximum events (using the methodology presented below), and (3) calculate the average value for each event (base or top) using only potential maximum events. Median bases and tops can also be used and provide results similar to those based on average ranges for many datasets. In contrast to a maximum composite section, as more information is incorporated into an average composite section, the ranges of taxa become more similar to ranges observed in many comparison sections. Average events will therefore plot closer to the true LOC and better constrain its position.

Implementation of an average composite section requires bookkeeping and computations beyond those required for a maximum composite section. For either method of constructing a composite, the composite value of each event must be recorded for every comparison section. These values are required for rounds of correlation, when a comparison section that has previously been incorporated into the composite section is replotted (e.g., Shaw, 1964; Edwards, 1984). Any events based on the section being recorrelated must be removed from the composite section and replaced with the next most appropriate value. Failure to remove events based on the current comparison section results in those events being plotted against themselves, thus biasing the correlation. Average ranges are less sensitive to this limitation because most events in the composite section incorporate information from multiple comparison sections.

Average ranges should be based on *potential maximum events*. Potential maximum events are events from comparison sections which contain the interval of strata necessary for a taxon (event) to potentially be present at its maximum range. This prevents the average ranges in the composite section from being unduly shortened by ranges in a comparison section that are truncated by a substantial hiatus (time not represented in the rock record due to erosion or nondeposition) or by the limits of sampling. Evaluation of potential maximum events utilizes a maximum composite section to determine the potential range of each taxon. The base and the top of each taxon range (event) must be evaluated separately in every comparison section to determine its status as a potential maximum event. In practice, potential maximum events represent those events that project to the LOC perpendicular to both axes (Fig. 1). Horizontal and vertical segments in the LOC should be excluded from the analysis of potential maximum events, because they violate the concept of unique point by point equivalency between sections.

Each event in a maximum composite section is derived from only one of the multiple sections used, whereas most events in an average composite section incorporate information from many comparison sections. Compared to an average composite section, therefore, a maximum composite section is more subject to errors resulting from erroneous taxon occurrences or incorrectly positioned segments in the LOC. As noted by Shaw (1964, p. 194), these errors will not impact the (maximum) composite section if they fall within the established range of a taxon, but a single occurrence outside of the established range for a taxon can result in an event being excluded from contributing to all subsequent correlations. In contrast, an average composite section will tend to mitigate the effects of erroneous taxon occurrences or incorrectly positioned segments in the LOC. Note, however, that average composite sections are sensitive to unusually short as well as unusually long ranges.

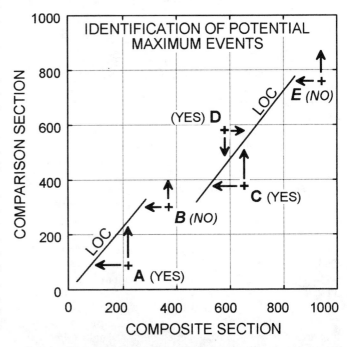

FIG. 1.—Method used to determine whether an event represents a potential maximum event. By including only potential maximum events, apparent bases and tops that actually represent the limits of sampling or ranges that are truncated by a substantial hiatus are excluded from the calculation of average composite values. Potential maximum events are those events which project to the LOC from both the comparison section and the maximum composite section. Events A, C, and D are potential maximum events and would be included in the calculation of their respective average composite values. Event B represents a range that is truncated by a substantial hiatus and would be excluded. Event E represents a range that is truncated by the limits of sampling and would be excluded.

The use of only maximum composite sections in the correlation process will, to some extent, reduce the advantages of graphic correlation. Graphic correlation enables large amounts of data to be evaluated, but maximum composite sections greatly limit the number events that can be utilized in the correlation process (e.g., Shaw, 1964). Edwards (1984) stated that correlations should be based on events that are synchronous within the resolution of the available data. Although synchronous events are certainly preferred for making biostratigraphic interpretations, the availability and distribution of truly synchronous events is usually limited. Moreover, the synchroneity of many events cannot be determined until the correlation between sections is established, and the correlation between sections cannot be established without some interpretation of the synchroneity of various events. Average composite sections reduce the impact of this limitation because they increase the number of nearly synchronous events. Because correlations based on average composite sections generally use data from a larger number of sections, they more fully utilize the entire dataset.

Evaluation of a Hypothetical Taxon First-Occurrence Surface

To evaluate differences between maximum and average composite sections, the two methods are applied to a hypothetical taxon first-occurrence surface based on a 20 × 20 grid (Fig. 2). The X and Y dimensions of the grid represent the geographic distribution of the surface, and the Z dimension represents the temporal occurrence of the surface at any location within its extent. For this example, each of the 400 locations (grid nodes) would represent a well or stratigraphic section in which the surface (event) is identified. Using a hypothetical surface removes any ambiguities relating to correlations among sections, sampling technique, sample size or spacing, taxon identification, and so forth. The surface in Figure 2 is diachronous, and could represent an allopatric speciation event with delayed migration into a wide range of environments.

Results of the maximum and average methods of constructing a composite event are quantified using an *error of correlation* which is calculated as the unsigned difference between the temporal (Z) value at a site and the associated maximum or average composite value. The error of correlation measures how closely an event would constrain the correlation for a given location. Error values of 0 indicate that an event would plot exactly on the true LOC, whereas larger values indicate that an event would plot further from the true LOC and provide less constraint. Each of the 400 locations in this example was compared to a composite event constructed from the remaining 399 locations. The mean error of correlation using the maximum composite approach (Fig. 3A) is 69% of the total temporal range (Z dimension) of the surface, compared to only 6% using the average composite approach (Fig. 3B). Although the magnitude of differences between the two methods will vary with the shape of the surface, the overall error of correlation will always be less (or the same in a few special cases) using the average composite approach than using the maximum composite approach. As a result, most events in a comparison section generally will correlate better with an average composite section than with a maximum composite section. Nevertheless, in any given comparison section a few taxa (events) may be at or near their maximum observed range and will correlate better with a maximum composite section. Alternating between the maximum and average methods of constructing composite sections enables the strengths of each approach to be used to full advantage.

Effect of Sample Size on Average and Maximum Composite Sections

Maximum and average composite events differ markedly in how they change with increasing sample size. To evaluate this difference, the hypothetical taxon first occurrence surface was analyzed using each method of constructing a composite event for 10 sample sizes ranging from 1 to 399. Each trial was performed by selecting a location at random and comparing it to a composite event constructed from the given number of other locations, each also selected at random. Figure 4 shows the mean error of correlation based on 500 trials for each of the 10 sample sizes. For a composite event based on a single location, the results of the average and maximum methods are identical. As the number of locations increases, the mean error of correlation for the maximum composite event increases rapidly. In contrast, the mean error of correlation for the average composite event decreases slightly and stabilizes at relatively small sample sizes. This difference illustrates why channeling increases as more comparison sections are incorporated into a maximum composite section but does not develop relative to an average composite section.

Example Using Pliocene Planktonic Foraminifers and Calcareous Nannofossils

Pliocene planktonic foraminifers and calcareous nannofossils from 10 DSDP sites within the Atlantic and Pacific ocean basins

MODEL OF TAXON FIRST OCCURRENCE

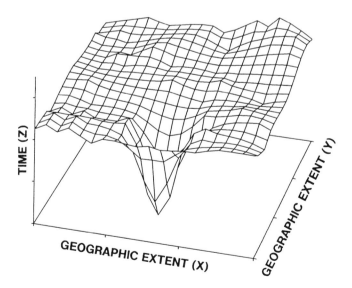

Fig. 2.—Hypothetical taxon first-occurrence surface (with modification from Dowsett, 1988) depicting a localized speciation event with delayed migration into different environments. The shape of this surface emphasizes differences between the maximum and average methods of constructing composite sections. No vertical or lateral scale is implied.

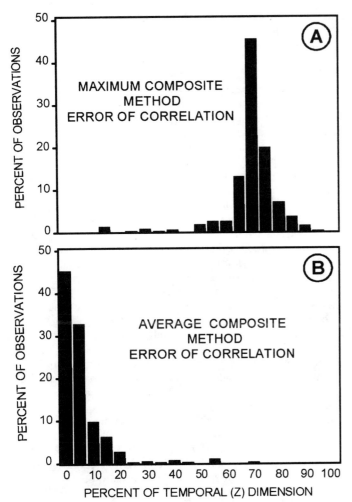

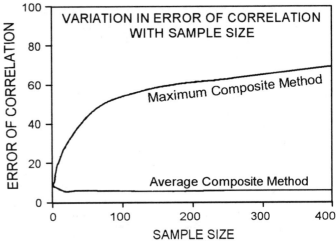

Fig. 3.—Analysis of the error of correlation for each of the 400 locations compared in turn to a composite event constructed from the remaining 399 locations. The error of correlation measures how closely an event would constrain the respective LOC for that site, with a value of 0 indicating the event would plot exactly on the true LOC. Larger values indicate that the event would plot further from the true LOC. Because no vertical scale is implied for the hypothetical surface, the error value is expressed as a percent of the temporal range (Z dimension) of the surface. (A) Histogram of the error from correlating to a maximum composite event ($\bar{x}$ = 69%). (B) Histogram of the error from correlating to an average composite event ($\bar{x}$ = 6%).

Fig. 4.—Analysis of the error of correlation with sample size for maximum and average composite events. The lines represent the mean error of correlation for the two methods. Each line is based on 500 trials for each of 10 sample sizes (1, 2, 4, 8, 16, 32, 64, 128, 256, 399). For a composite based on a single location, the two methods give identical results. As additional locations are incorporated, the mean error of correlation for a maximum composite event increases rapidly while the mean error of correlation for an average composite event remains relatively stable.

(Dowsett, 1988, 1989) provide an excellent geologic dataset for contrasting results based on average and maximum composite sections. Nine of the 10 sites have paleomagnetic data to independently constrain the LOC. Dowsett (1988, 1989) previously analyzed these data using graphic correlation, and because of the high confidence in correlations provided by the paleomagnetic data, he was able to document substantial diachroneity of several fossil datums. Most biostratigraphic datasets, however, do not have detailed paleomagnetic data available to constrain the interpretations. This study analyzes the combined Pliocene planktonic foraminifers and calcareous nannofossils using both a maximum and an average composite section to determine which approach would best match the interpretation constrained by paleomagnetic data if the paleomagnetic events were absent.

DSDP site 502 (Caribbean Sea) is used as the Standard Reference Section (SRS), following Dowsett (1988), because it contains a large number of biostratigraphic and paleomagnetic events, it records a high overall sedimentation rate, and it is interpreted to be relatively complete. Each of the remaining nine DSDP sites is correlated directly with the SRS to eliminate any possible effects on the correlations relating to the method of constructing the composite section (maximum versus average). The LOCs used here are similar to those of Dowsett (1988, 1989); each is subjectively positioned to closely honor the paleomagnetic data with a minimum number of segments (except site 586, which lacked paleomagnetic data) and to obtain a parsimonious fit to the available biostratigraphic data.

DSDP site 606 (north Atlantic) is closely constrained by paleomagnetic data and provides a good comparison of the maximum and average methods of constructing a composite section. Figure 5A shows site 606 graphed against the SRS, including the paleomagnetic control and the LOC. Note how closely the LOC matches both the paleomagnetic data and all available biostratigraphic data in the SRS. That same LOC is included on the graph of site 606 plotted against the maximum composite section (Fig. 5B) and the graph of site 606 plotted against the average composite section (Fig. 5C). Each composite section was constructed from the remaining nine DSDP sites. Note the pronounced channeling on the graph of site 606 plotted against the maximum composite section; only a small number of composite biostratigraphic events plot near the LOC (Fig. 5B). Moreover, the LOC indicated by paleomagnetic data does not split the bases and tops of the channel. In contrast, most of the events on the graph of site 606 plotted against the average composite section closely constrain the LOC (Fig. 5C). In the absence of paleomagnetic data, an LOC based on the average composite section would match the independent constraints

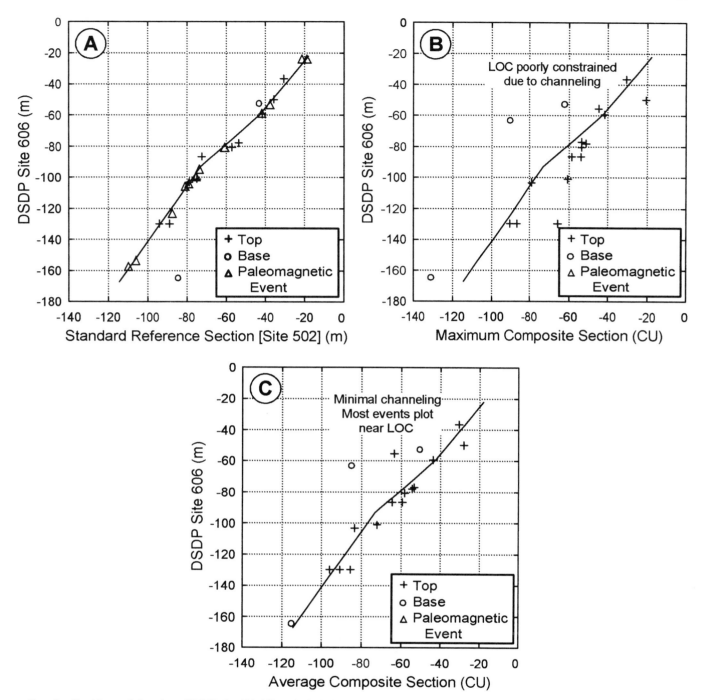

FIG. 5.—Graphic correlation plots of DSDP site 606. All graphs include the same LOC. (A) Site 606 plotted relative to the original SRS (site 502) showing the biostratigraphic and paleomagnetic control used to constrain the LOC. Eight of the biostratigraphic events are within 5 m of the LOC. (B) Site 606 plotted relative to a maximum composite section based on the remaining nine sites. Note the pronounced channeling and that the LOC indicated by paleomagnetic data does not follow the center of the channel. Only five of the biostratigraphic events are within 5 m of the LOC. (C) Site 606 plotted relative to the average composite section based on the remaining nine sites. The biostratigraphic data are more compatible with the LOC independently constrained by paleomagnetic data. Ten of the biostratigraphic events are within 5 m of the LOC.

more closely than an LOC based on the maximum composite section.

Effective use of channeling to constrain the LOC requires both first- and last-occurrence events, preferably with each being distributed uniformly through the interval of interest. Only 3 of the 17 biostratigraphic events for site 606 represent first occurrences (bases), and some datasets (particularly those based on well cuttings) may lack reliable first-occurrence data altogether. If only last occurrences (tops) are available, a maximum composite section may result in all events plotting to one side of the true LOC with no data available to constrain the other side. In contrast, an average composite section is more effective for correlating with datasets comprising only tops or only bases, because channeling is not expected. Those bases and tops which do not plot directly on the LOC are equally likely to plot on either side of true LOC. The practical effect of this difference is that the LOC is positioned along the greatest density of high-quality data for an average composite section, rather than within the gap separating bases from tops as for a maximum composite section.

DSDP site 586 (southwest Pacific) does not have paleomagnetic data available and makes a good test case for comparing the two methods of constructing a composite section under circumstances involving a higher level of uncertainty. Figure 6A shows site 586 graphed against the SRS, including the LOC. Note how closely the LOC matches the available biostratigraphic data. That same LOC is included on the graph of site 586 against the maximum composite section (Fig. 6B) and the graph of site 586 against the average composite section (Fig. 6C). As with site 606, the average composite section constrains the LOC more closely than does the maximum composite section. Furthermore, an LOC interpreted to split bases and tops within the channel relative to the maximum composite section would not provide a reasonable fit to the original data; such a line could deviate from the original interpretation by as much as 15 composite units. Using Dowsett's (1988) estimate of 0.0336 my/composite unit, this error could potentially represent more than 0.5 my, or almost 20% of the range of data on the graph.

The pronounced differences between maximum and average composite sections for sites 606 and 586 were developed from a dataset consisting of only 10 wells. As additional wells or stratigraphic sections are incorporated into the analysis, the differences relating to the method of constructing the composite section would increase. In this example, the revised method did not improve upon the correlations proposed by Dowsett (1988, 1989) because the original interpretations were constrained by paleomagnetic data. However, these data clearly demonstrate that in the absence of paleomagnetic data the average composite section would constrain the position of the LOC better than would the maximum composite section. Although maximum composite sections are quite useful for evaluating the chronostratigraphic significance of taxa, average composite sections are generally better for establishing the correlations. Once reasonable LOCs are established, switching between the two methods of constructing a composite section enables the advantages of each to be used to full advantage.

Comparison of Average Composite Sections to Probabilistic Stratigraphy

Probabilistic stratigraphy was originally presented by Hay (1972) as a method for determining an average sequence of events. Graphic correlation based on average composite sections differs fundamentally from probabilistic techniques in being a *sequence of average events*, rather than an *average sequence of events*. Probabilistic techniques order events using binomial or trinomial analysis of crossover frequency without regard to the relative stratigraphic dispersion (e.g., Edwards and Beaver, 1978; Blank, 1979; Agterberg and Nel, 1982a, 1982b; Gradstein and Agterberg, 1982). In contrast, average composite sections analyze the relative stratigraphic dispersion of each event without regard to its crossover frequency with any other event. While probabilistic techniques are more amenable to automation and are not subject to error introduced by incorrect correlations among sections, they are subject to a number of limitations. For example, because stratigraphic separation is not incorporated into the analysis, a crossover between closely spaced samples is not differentiated from a crossover between widely spaced samples. When considered in geological context, however, events from closely spaced samples can crossover and still be nearly isochronous, whereas events from widely spaced samples can be in the same order yet be diachronous. Moreover, determining that two events have changed in sequence does not necessarily indicate which event moved. Because stratigraphic separation is distorted in an average sequence of events, graphing against a comparison section (e.g., Blank, 1979) could produce terraces that represent intervals with many closely spaced events rather than hiatuses. In addition, probabilistic techniques may not correctly order rare events or events that do not overlap geographically (e.g., Hay, 1972). In contrast, the average composite methodology can analyze the relative isochroneity of each event and correctly model rare or geographically restricted events, but it is subject to error introduced by the correlation process. The choice between graphic correlation and probabilistic stratigraphy therefore depends on the objectives of the analysis and the quality of the data available, but graphic correlation is generally better for establishing correlations among stratigraphic sections containing high quality data.

DERIVED CORRELATIONS BETWEEN COMPARISON SECTIONS

Rigorous correlation of a group of wells or stratigraphic sections is a multidimensional process, with each stratigraphic section representing one dimension in data space. A common objective of this correlation process is to obtain the best possible understanding of the regional stratigraphic framework, although other outcomes such as evaluating the distribution of taxa in space or time are also possible. Several metric ordination techniques, including principal components analysis (Hohn, 1978), principal coordinates analysis (Hazel, 1977), and detrended correspondence analysis (Hill and Gauch, 1990), have been used to analyze biostratigraphic data in multidimensional space. The use of multivariate techniques, however, has limitations. For example, stratigraphic hiatuses or relative changes of sedimentation rate among sections are difficult to accommodate. These features would cause offsets or changes in the trend of the correlation between sections and might have to be analyzed as separate segments. Many events are not present in all sections, and methods of estimating missing values increase the uncertainty of the results (e.g., Hohn, 1978). In addition, supplemental data and geologic experience, such as the relative biostratigraphic utility of different events and known correlation points, are difficult to incorporate into the analysis.

Graphic correlation approaches the multidimensionality of stratigraphic datasets by subdividing the correlations into multiple 2-dimensional (pairwise) comparisons. By using the same composite section for each comparison, the results can be expressed in terms of common (composite) units. While this approach circumvents many limitations of performing the analysis in multidimensional data space, it under-utilizes potentially useful information available from direct correlations among com-

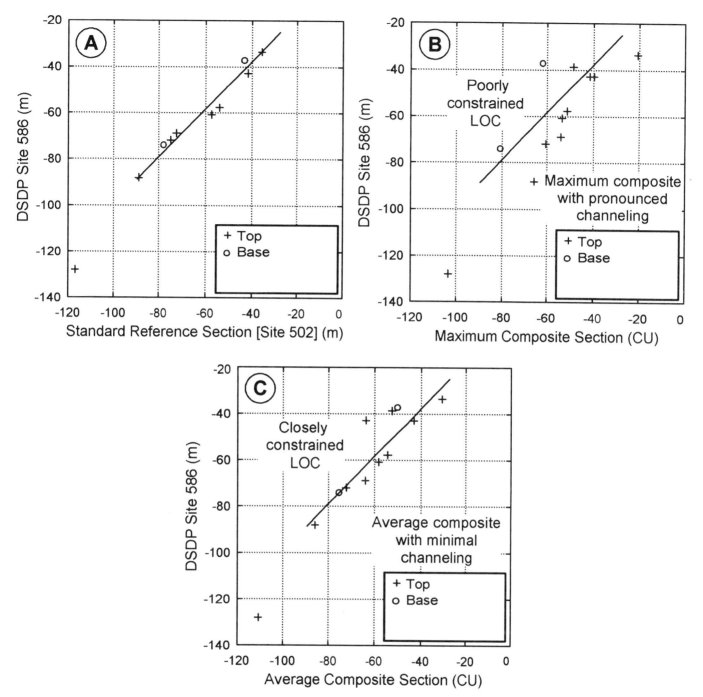

Fig. 6.—Graphic correlation plots of DSDP site 586. All graphs include the same LOC. (A) Site 586 plotted relative to the original SRS (site 502) showing the biostratigraphic control used to constrain the LOC. Seven of the biostratigraphic events are within 5 m of the LOC. (B) Site 586 plotted relative to the maximum composite section based on the remaining nine sites. Note the pronounced channeling and that the LOC does not follow the center of the channel. Only one biostratigraphic event is within 5 m of the LOC. Because the composite incorporates data from all nine additional sites, channeling is more pronounced than for Dowsett's (1988) Figure 5a which is based on only three sites. (C) Site 586 plotted relative to the average composite section based on the remaining nine sites. The biostratigraphic data constrain the LOC much closer than with the maximum composite section. Seven of the biostratigraphic events are within 5 m of the LOC.

parison sections. Any two sections that have been correlated with the same composite section are, by implication, correlated with one another through the interval in which they overlap in the composite section. This implied correlation between two comparison sections is here referred to as a derived correlation (Fig. 7). As implemented here, derived correlations do not enable the original correlations or the ranges of taxa to be adjusted directly, but rather provide a method to evaluate the original interpretations against possible alternatives. For example, if one comparison section shows a hiatus or change in sedimentation

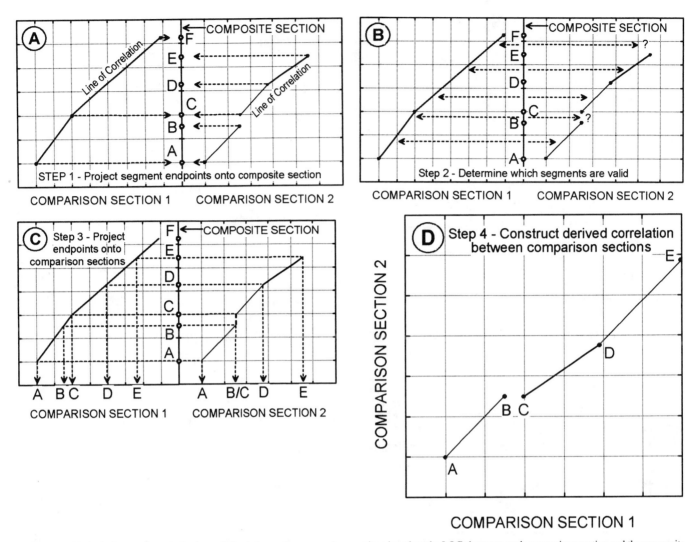

FIG. 7.—Method of extracting a derived correlation between two comparison sections based on the LOCs between each comparison section and the composite section. (A) Compile a list of unique endpoints for all segments in both LOCs (in composite units). (B) Identify valid segments by attempting to project a horizon between each pair of endpoints into both comparison sections. In this example, segment EF is not valid. (C) For each valid segment, project the endpoints into both comparison sections. Note the method of accommodating stratigraphic gaps. (D) Combine the valid segments to construct the LOC between comparison sections.

rate relative to the composite section and in the same interval a second does not, then that feature should be present on a graph between the two comparison sections. Furthermore, the magnitude of the feature should be in accord with that defined by the correlation of each comparison section to the composite section. By incorporating information from derived correlations, the LOC between the composite section and each comparison section can be adjusted until a parsimonious fit is obtained for all three sections. The process can then be repeated for additional pairs of comparison sections.

Radiolarian and diatom data from DSDP sites 573, 574, and 575 (Pisias and others, 1985) in the eastern and central equatorial Pacific provide a dataset well suited to illustrate the potential benefit of using derived correlations. DSDP site 573 was selected as the SRS because it is interpreted to be the most complete. To avoid introducing uncertainty by the method used to construct a composite section, both comparison sections (DSDP sites 574 and 575) are correlated directly with the SRS (DSDP site 573). Although site 573 was also included in the dataset from Dowsett (1988, 1989), which was used to evaluate average composite sections, the Pisias and others' (1985) example includes a greater stratigraphic interval and does not contain any of the same taxa.

The LOCs used here are positioned based on parsimonious fit to the combined radiolarian and diatom data using a minimum number of segments. They differ slightly from the LOCs of Pisias and others (1985), who correlated the diatom and radiolarian data separately in order to analyze the level of stratigraphic resolution available using this type of data. Although the initial LOCs for site 574 (Fig. 8A) and site 575 (Fig. 8B) are closely constrained by the available biostratigraphic data, the derived correlation between sites 574 and 575 (Fig. 8C) suggests several areas where the original interpretations can be refined. Most notably, the trend of biostratigraphic events re-

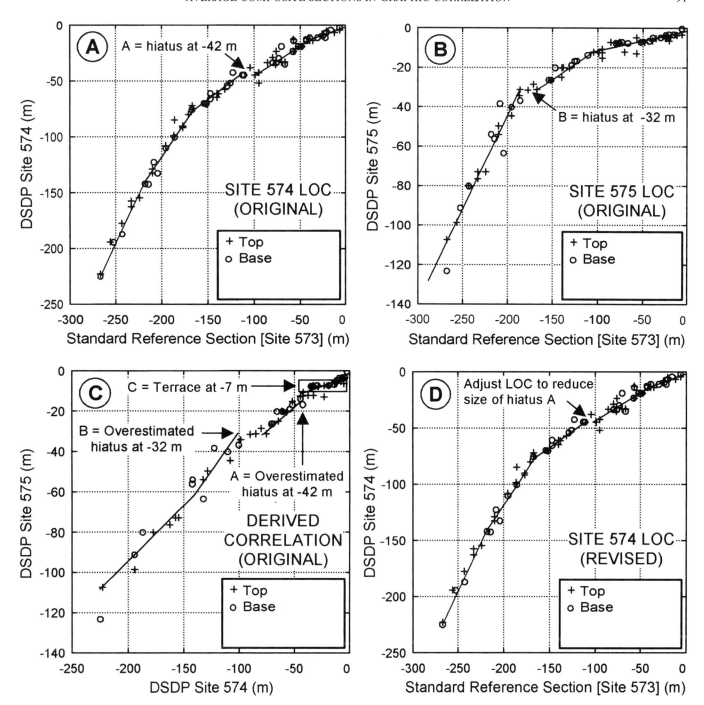

FIG. 8.—Illustration of how derived correlations between comparison sections can be used to refine interpretations. (A) Site 574 correlated with the SRS (site 573). (B) Site 575 correlated with the SRS (site 573). (C) Graph of comparison sections 574 and 575, including the derived LOC (see text for discussion). (D) Revised correlation between site 574 and the SRS (site 573). (E) Revised correlation between site 575 and the SRS (site 573). The LOCs for these two graphs were revised based on the derived correlation. (F) Graph of comparison sections 574 and 575 including the new derived LOC based on the revised correlation of each to the composite section. Note that this derived correlation provides a more parsimonious fit to the original data.

veals a probable hiatus at -7 m in site 575. This hiatus is marked by a terrace on Figure 8C and was probably overlooked on the original interpretation presented here because of the low sedimentation rate in this interval at site 575 compared to site 573. The hiatus is more apparent when correlated with site 574 because the relative sedimentation rate is more similar between the two comparison sections. In addition, offsets in the LOC at -32 m in site 575 and at -42 m in site 574 appear to overestimate the magnitude of the respective hiatuses. The LOCs between site 574 and the SRS (Fig. 8D) and site 575 and the SRS (Fig. 8E) were subjectively adjusted to reduce the observed discrepancies. The new derived correlation between sites 574

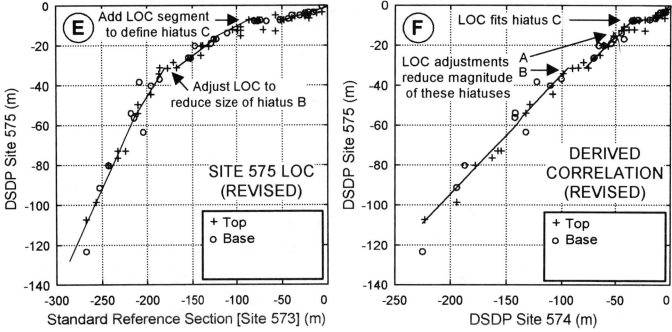

FIG. 8.—Continued

and 575 (Fig. 8F), based on these revisions, provides a more parsimonious fit to the original data which I interpret to more accurately reflect the true correlation.

This example illustrates that even with relatively well constrained data, derived correlations between comparison sections can substantially refine correlations. A hiatus that had been overlooked on the original interpretation was identified and two small offsets in the LOC were adjusted to better match all available data. Channeling, which was not a factor in this example because both comparison sections were correlated directly with the SRS, would have further obscured these features on the original interpretations and increased the benefit of the derived correlation. For large datasets, the task of evaluating derived correlations between all possible combinations of comparison sections would be overwhelming. As a starting point, a comparison section can be graphed against one or two of the other more similar comparison sections. Additional sections can be evaluated as necessary to achieve the desired level of refinement. Derived correlations between comparison sections which include only a small number of events in common may not be particularly instructive for refining the respective LOCs and can be skipped to save time unless the correlation of those sections is of interest for other reasons.

SUMMARY

Development of a composite section is a critical aspect of graphic correlation, because the composite section provides the foundation both for subsequent correlations and for evaluating the chronostratigraphic ranges of taxa. The approach used to construct the composite section, however, can have a major impact on the results. Maximum composite sections are useful for evaluating the chronostratigraphic significance of taxa, but average composite sections are generally better for establishing the correlations between sections. Once reasonable LOCs are established, switching between the two methods of constructing composite sections enables the characteristics of each to be used to full advantage.

Analysis of Pliocene planktonic foraminifer and calcareous nannofossil data from 10 DSDP sites in the Atlantic and Pacific ocean basins demonstrates that substantial refinements are possible using average composite sections. As compared to the maximum composite sections, events in the average composite sections more closely match the LOC for these wells as independently constrained by paleomagnetic data. For example, without the paleomagnetic constraints the potential error of correlating DSDP site 586 with a maximum composite section is 0.5 my, or about 20% of the range of the data.

Derived correlations between comparison sections are designed to take advantage of the multidimensional nature of most biostratigraphic data within the graphic correlation framework. Analysis of three DSDP sites from the central and equatorial Pacific illustrates the potential of derived correlations to identify and refine hiatuses or changes in sedimentation rate. In combination, average composite sections and derived correlations between comparison sections provide a way to refine correlations beyond what is possible using the traditional graphic correlation approach and can provide more reliable and geologically robust interpretations.

ACKNOWLEDGMENTS

I thank Exxon Exploration Company for permission to publish this paper. P. McLaughlin, L. Edwards, D. Advocate, W. Burroughs, K. Mann, and an anonymous reviewer made helpful comments on the manuscript.

REFERENCES

AGTERBERG, F. P. AND NEL, L. D., 1982a, Algorithms for the ranking of stratigraphic events: Computers and Geosciences, v. 8, p. 69–90.

AGTERBERG, F. P. AND NEL, L. D., 1982b, Algorithms for the scaling of stratigraphic events: Computers and Geosciences, v. 8, p. 163–189.

BLANK, R. G., 1979, Applications of probabilistic biostratigraphy to chronostratigraphy: Journal of Geology, v. 87, p. 647–670.

DOWSETT, H. J., 1988, Diachroneity of late Neogene microfossils in the southwest Pacific Ocean: application of the graphic correlation method: Paleoceanography, v. 3, p. 209–222.

DOWSETT, H. J., 1989, Application of the graphic correlation method to Pliocene marine sequences: Marine Micropaleontology, v. 14, p. 3–32.

EDWARDS, L. E., 1984, Insights on why graphic correlation (Shaw's method) works: Journal of Geology, v. 92, p. 583–597.

EDWARDS, L. E., 1989, Supplemented graphic correlation: a powerful tool for paleontologists and nonpaleontologists: Palaios, v. 4, 127–143.

EDWARDS, L. E. AND BEAVER, R. J., 1978, The use of paired comparison model in ordering stratigraphic events: Mathematical Geology, v. 10, p. 261–272.

GRADSTEIN, F. M. AND AGTERBERG, F. P., 1982, Models of Cenozoic foraminiferal stratigraphy—northwestern Atlantic margin, in Cubitt, J. M. and Reyment, R. A., eds., Quantitative Stratigraphic Correlation: Chichester, Wiley and Sons, p. 119–170.

HAY, W. W., 1972, Probabilistic stratigraphy: Eclogae Geologica Helvetiae, v. 65, p. 255–266.

HAZEL, J. E. 1977, Use of certain multivariate and other techniques in assemblage zonal biostratigraphy: Examples utilizing Cambrian, Cretaceous, and Tertiary benthic invertebrates, in Kauffman, E. G. and Hazel, J. E., eds., Concepts and Methods of Stratigraphy: Stroudsburg, Dowden, Hutchinson, and Ross, p. 187–212.

HAZEL, J. E., 1989, Chronostratigraphy of Upper Eocene microspherules: Palaios, v. 4, p. 318–329.

HILL, M. O. AND GAUCH, H. G., 1980, Detrended correspondence analysis: an improved ordination technique: Vegetation, v. 42, p. 47–58.

HOHN, M. E., 1978, Stratigraphic correlation by principal components: effects of missing data: Journal of Geology, v. 86, p. 524–532.

MACLEOD, N. AND KELLER, G., 1991, How complete are Cretaceous/Tertiary boundary sections? A chronostratigraphic estimate based on graphic correlation: Geological Society of America Bulletin, v. 103, p. 1439–1457.

MARTIN, R. E., JOHNSON, G. W., NEFF, E. D., AND KRANTZ, D. W., 1990, Quaternary planktonic foraminiferal assemblage zones of the northeast Gulf of Mexico, and tropical Atlantic Ocean: Graphic correlation of microfossil and oxygen isotope datums: Paleoceanography, v. 5, p. 531–555.

MELNYK, D. H., ATHERSUCH, J., AND SMITH, D. G., 1992, Estimating the dispersion of biostratigraphic events in the subsurface by graphic correlation: an example from the Late Jurassic of the Wessex Basin, UK: Marine and Petroleum Geology, v. 9, p. 602–607.

MILLER, F. X., 1977, The graphic correlation method in biostratigraphy, in Kauffman, E. G. and Hazel, J. E., eds., Concepts and Methods of Stratigraphy: Stroudsburg, Dowden, Hutchinson, and Ross, p. 165–186.

PISIAS, N. G., BARRON, J. A., NIGRINI, C. A., AND DUNN, D. A., 1985, Stratigraphic resolution of leg 85 drill sites: an initial analysis, in Mayer, L., Theyer, F., and others, Initial Reports, DSDP 85: Washington, D. C., United States Goverment Printing Office, p. 695–708.

PRELL, W. L., IMBRIE, J., MARTINSON, D. G., MORLEY, J. J., PISIAS, N. G., SHACKLETON, N. J., AND STREETER, H. F., 1986, Graphic correlation of oxygen isotope stratigraphy application to the late Quaternary: Paleoceanography, v. 1, p. 137–162.

SHAW, A. B., 1964, Time in Stratigraphy: New York, McGraw-Hill, 365 p.

SWEET, W. C., 1984, Graphic correlation of upper Middle and Upper Ordovician rocks, North America midcontinent province, USA, in Bruton, D. L., ed., Aspects of the Ordovician System: Oslo, Paleontological Contributions from the University of Oslo 295, Universitetsforlaget, p. 23–35.

SWEET, W. C., 1992, Middle and Late Ordovician Conodonts from southwestern Kansas and their biostratigraphic significance: Oklahoma Geological Survey Bulletin 145, p. 181–191.

INTEGRATION OF THE GRAPHIC CORRELATION METHODOLOGY IN A SEQUENCE STRATIGRAPHIC STUDY: EXAMPLES FROM NORTH SEA PALEOGENE SECTIONS

JACK E. NEAL[1]

Department of Geology and Geophysics, Rice University, PO Box 1892, Houston, Texas 77251

AND

JEFF A. STEIN AND JAMES H. GAMBER

Amoco Production Company, PO Box 3092, Houston, Texas 77253

ABSTRACT: The composite standard method of biostratigraphy, described by Shaw (1964), provides a consistent temporal framework for stratigraphic analysis of a basin. The method enables geologists to graphically correlate deposition in individual sections to an ideal composite standard reference section, quantifying deposition and stratigraphic lacuna. Sequence stratigraphy (Posamentier and Vail, 1988; Van Wagoner and others, 1990) is a stratigraphic interpretation method that genetically relates deposits in a dip profile to relative changes in sea level, based on physical surfaces in the rock record. Weaknesses in graphic correlation (underuse and static application) match well with strengths in sequence stratigraphy (widespread use and dynamic application). Weaknesses in sequence stratigraphy (documentation and consistency) can be equally well matched with the strengths of graphic correlation. Sequence stratigraphic key bounding surfaces cause predictable patterns in the graphic correlation of biostratigraphic data. Integration of graphic correlation and sequence stratigraphy enhances the utility of both stratigraphic tools and provides a powerful basin analysis technique.

INTRODUCTION

Since its development in the early 1960's (Shaw, 1964), the practical utility of the graphic correlation methodology has been largely overlooked by the geologic community. Meanwhile, chronostratigraphic representation of the rock record (Wheeler, 1958) and its division into cratonic sequences of transgression and regression (Sloss, 1963) gained popularity. Seismic stratigraphy (Payton, 1977), then sequence stratigraphy (Wilgus and others, 1988), developed from the work of Sloss and Wheeler and captured the imagination of geologists and geophysicists worldwide. This paper seeks to demonstrate the logic and utility of integrating graphic correlation with sequence stratigraphy to produce a documented and consistent chronostratigraphic depositional framework.

Sequence stratigraphy subdivides the rock record into depositional sequences, defined by Mitchum and others (1977, p. 53) as "a stratigraphic unit composed of a relatively conformable succession of genetically related strata bounded at its top and base by unconformities and their correlative conformities." Relative changes in sea level ("apparent rise or fall of sea level with respect to the land surface"—Vail and others, 1977a, p. 63) create or destroy accommodation space for sediment along a depositional profile (Posamentier and others, 1988). In marine settings, relative changes in sea level, depositional profile, and the delivery of sediment to the sea control the development of depositional sequences (Vail and others, 1991). One of the first principles of sequence stratigraphy is that relative changes of sea level are a function of changing tectonic subsidence and eustasy (Vail, 1987; Posamentier and others, 1988). Eustasy is defined as global changes of sea level (Seuss, 1906; Fairbridge, 1961), therefore, eustatic events should have some expression in deposits of a given age worldwide. A controversial aspect of sequence stratigraphy remains the demonstration of time equivalence of events (Miall, 1992), which requires documentation not published with the eustatic curves of Exxon (Haq and others, 1988). The possibility of a miscorrelation has great significance in a sequence stratigraphic analysis because of its heavy theoretical reliance on eustasy.

A common criticism of Haq and others (1988) was the lack of documentation of sequence boundary ties around the world. Aubry (1991) examined this problem for lower-middle Eocene sediments around the world. With high-resolution biostratigraphy, she documented the complexity of sedimentation for this interval to be greater than that predicted by the simple, two-event model proposed by Haq and others (1988), although the global signal of two events ("49.5" and "48.5"—Haq and others, 1988) could be recognized.

The challenge in this type of documentation rests in precise age-equivalent correlation of events. Figure 1 (modified from Aubry, 1991) demonstrates the need for caution and precision when correlating stratigraphic gaps. Section A (in depth) shows an unconformity, separating younger from older sediments. This unconformity can be thought of as two time surfaces, the lower representing the time of subaerial exposure down to the oldest eroded bed and the upper surface corresponds to the time of resumed deposition. These surfaces bracket an interval of geologic time not represented by rock in Section A. Sections C and D contain unconformities of similar age to the unconformity of Section A, however, the unconformity in Section C erodes rocks that are younger than rock directly overlying the unconformity in Section D. The unconformities of C and D are of different ages, yet both overlap in time with the unconformity in Section A. A correlative stratigraphic gap is not shared by all three sections, therefore, we are left with two choices: (1) the unconformities are caused by local conditions (tectonics, current scour, dissolution, etc.) and no eustatic signal is observed or, (2) the eustatic signal is expressed in some way other than an unconformity in one of the sections (e.g., sharp change in sedimentation rate). For each case, documentation of a eustatic signal is dependent upon demonstration of chronostratigraphic equivalence for a chosen stratigraphic interval.

Graphic correlation has the potential to test eustasy. Prell and others (1986) demonstrated this potential by correcting $d^{18}O$ records from 13 piston cores around the world using graphic correlation. The composite standard methodology can include all types of chronological data (biostratigraphic, magnetostratigraphic, isotopic, etc.) producing a consistent temporal framework that can be used to test the stratigraphy of different well or outcrop sections. Much of the previously published work

---

[1] Present address: Exxon Production Research Company, P.O. Box 2189, Houston, Texas 77252

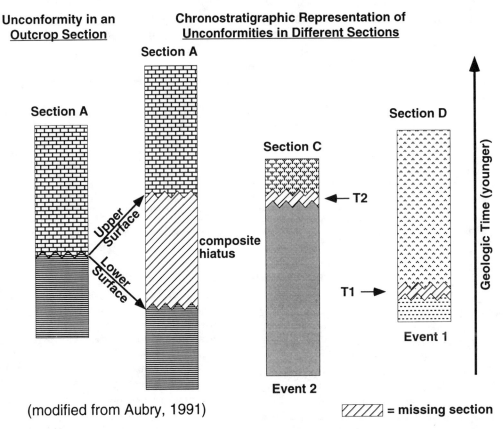

Fig. 1.—Chronostratigraphic expression of an unconformity (Section A — in depth and time), compared to the chronostratigraphic expression of unconformities in two different sections (Sections C and D). The hiatus in Section A could have been formed by either Event 1 (at time T1), Event 2 (at time T2), or a combination of both. The unconformities in Sections C and D could not have been caused by one (eustatic) event, since they do not overlap in time (see text for discussion).

using graphic correlation has utilized relatively continuous lines of correlation (LOC), with few "terraces" (Edwards, 1984, 1989; Prell and others, 1986; Hazel and others, 1984). Terraces in the LOC represent section missing from the well or outcrop being graphed against the composite standard section (Shaw, 1964; Miller, 1977). Shaw and Miller point out that LOC terraces can result from erosion, nondeposition, or section omission by faulting. There is much evidence and theory to suggest a discontinuous stratigraphic record in the marine as well as nonmarine setting (Pomerol, 1989; Aubry, 1991; Sadler, 1981; Plotnick, 1986); gaps that graphic correlation has the potential to identify and equate (Harper and Crowley, 1985). Demonstration of chronostratigraphically-overlapping stratigraphic gaps in widely separated basins is a condition required, but insufficient to prove, eustasy (Aubry, 1991).

APPLICATION AND THEORY

When developing a composite standard stratigraphy for a basin, one must search the data set and available literature for a well or outcrop section with the most complete biostratigraphy. This becomes the standard reference section (SRS). Graphic correlation is best used in areas where outcrop or continuous core data allow the identification of faunal first appearance datums (FAD) as well as last appearances datums (LAD) or first downhole occurrences (top or FDO). This control adds an extra level of confidence in the interpreted LOC by capturing the full range of any particular taxa as originally proposed by Shaw (1964). The composite standard is built by supplementing the SRS with information from other wells in many paleogeographic locations throughout the basin. This step insures the capture of the most possible depocenters and, therefore, results in a more comprehensive composite standard database. As additional wells are added, the chronostratigraphic range of each fossil will become more fully expressed within the composite standard (Shaw, 1964; Edwards, 1984). Interpretation of wells using a stabilized or mature database begins to provide real insight into depositional trends and basinwide events.

Before discussing applications, we will briefly examine our graphic correlation method for a basin where the primary data source is ditch cutting samples. Economic conditions often preclude the drilling of continuously cored wells, dictating some adjustments to the classic composite standard methodology and graphic correlation. In many subsurface databases, the geologist is fortunate to have a series of reported faunal tops, usually at 10-m or 30-ft cutting sample intervals. Constructing a chronostratigraphic framework using cutting samples alone places doubt on half the potential biostratigraphic data points since FADs are likely to be inconsistent markers. Caving younger fossils into older samples downhole causes a reported base (FAD) to be older than its correct stratigraphic position. For consistency, tops (LAD or FDO) and acme (most abundant occurrences) are the best markers in cuttings. The problem of reworked occurrences is lessened by making multiple passes through the data to stabilize composite ranges. A stable composite standard is essentially a fossil range chart (Fig. 2) that assigns each fossil datum a chronostratigraphic value in composite standard units (CSU), which indicates the marker's position relative to all other fossil datums in the composite standard. With a stable composite standard, new wells can be graphed to identify any missing or condensed intervals (Fig. 2).

Graphic correlation interpretations start by plotting reported fossil tops by depth occurrence (vertical axis) against time in CSU's (horizontal axis). The scatter of points is then used to define a depth versus time relationship in LOCs, drawn with respect to an evaluation of individual data points (Fig. 2). Fossils that do not fall on the LOC may not achieve their full range in sediments from the comparison well and plot below the LOC. Additionally, fossils plotting above or to the left of the LOC could be reworked occurrences or they may indicate that the chronological position of the particular datum needs to be updated in the composite standard (Fig. 2). These occurrences can distort chronostratigraphic frameworks that are based exclusively on key markers. A composite standard biostratigraphic data set develops from a pre-existing knowledge of the expected (knowledgeable weighting of important index fossils), and interpretation of the observed (capture of unexpected or overlooked markers and recognition of out-of-sequence occurrences) fossil succession.

The LOC approximates relative sediment accumulation rates (not accounting for differential compaction) as resolvable by fossils (Shaw, 1964). The biostratigraphy in this example (Fig. 2) can be interpreted with a "doglegged" LOC. Two horizontal terraces represent missing or condensed section separating sloping line segments characteristic of higher depositional rates (sediment pulse). Also plotted are fossil tops occurring off the LOC due to local environmental conditions. For example, in Figure 2, fossil G plots above and to the left of the LOC, indicating a probable reworked occurrence. This observation has important implications for correlation of stratigraphic breaks. The FDO of fossil G is normally found above the FDO of F and below the FDO of H, according to the mature composite standard. In the example well, the FDO's of fossils F and H define a stratigraphic break that would normally contain the top of fossil G. If fossil G is regionally used as the key marker for this break, a significant amount of section falls within an incorrect chronostratigraphic unit.

GRAPHIC CORRELATION TERRACES IN A SEQUENCE STRATIGRAPHIC MODEL

Physical surface-based stratigraphy, correlated with biostratigraphy and interpreted in terms of relative sea-level fluctuations, is the essence of sequence stratigraphy. The sequence stratigraphic model (Vail, 1987; Posamentier and others, 1988; Posamentier and Vail, 1988) proposes a way to subdivide a sedimentary section based on seismic stratal patterns and well-log interpretation as they relate to relative changes in sea level. Deposition has been modeled as a function of changing accommodation space (both on the shelf and onshore) and the response of depositional systems to fill that space (Jervey, 1988). Van Wagoner and others (1990) applied this model at outcrop scale, employing parasequence stacking patterns as a substitute for seismic scale geometries.

On a marine siliciclastic margin with a shelf-slope break, changes in accommodation space on the shelf control deposition in a dip direction (Fig. 3). Changes in shelfal accommodation space are caused by relative changes of sea level (relation

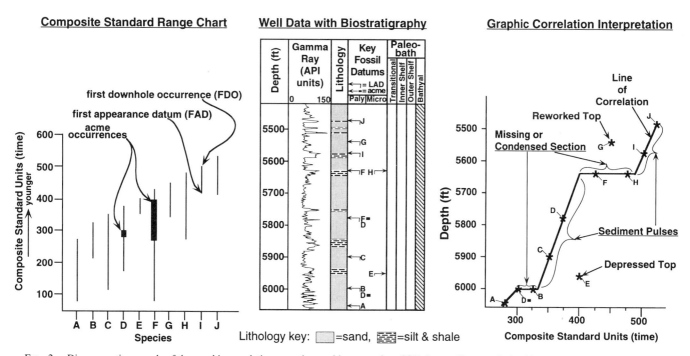

FIG. 2.—Diagrammatic example of the graphic correlation procedure and interpretation. CSUs have a llinear scale for this example only (see text for further explanation).

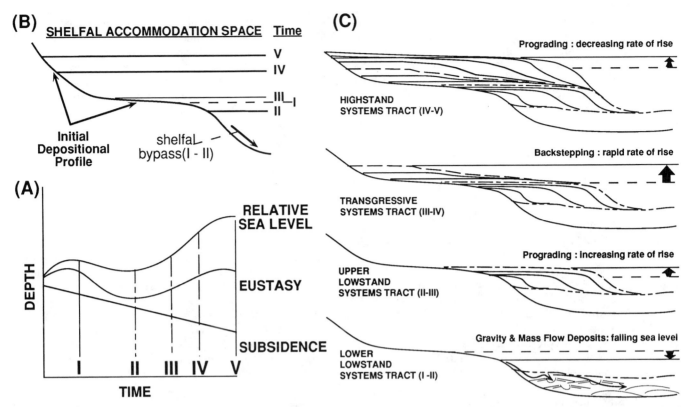

FIG. 3.—Brief overview of sequence stratigraphic concepts, showing how relative sea level is a function of subsidence (A), how changes in relative sea level create shelfal accommodation space (B), and how changes in shelfal accommodation space create stratal patterns (systems tracts) as sediment flux interacts with the rate of relative sea level rise through time (C).

of the sea surface to the inherited depositional profile-usually the previous sequence boundary). Relative sea-level change is a function of subsidence (or uplift) and eustasy. The rate of sediment delivered to the basin (flux) determines how much shelfal accommodation space is filled. If sediment flux is greater than the rate accommodation space is created, the depositional system will prograde. Sediment bypasses to the deep basin if no shelfal accommodation space exists (and the profile is too steep for normal progradation) or if shelfal marine processes sweep sediment to the slope (Posamentier and others, 1991). Unfilled accommodation space is equal to water depth in marine settings. At any given time, a depositional margin will possess some morphology that can range from steeply dipping escarpments to gentle ramps. This morphology is referred to as the depositional profile. The rate of change in shelfal accommodation space is directly linked to the depositional profile, controlling the styles of sediment fill (systems tracts) as relative sea-level changes.

This model is so robust that it is common for sequence stratigraphic studies to all but ignore biostratigraphy (e.g., Van Wagoner and others, 1990) due to the power of the method and the ability to correlate log markers and outcrop surfaces over significant distances. While this approach may be applicable for continuous outcrop sections and field studies with very dense well control, it is a questionable practice when attempting regional studies with scattered subsurface or discontinuous outcrop data. Regional studies need a rigorous chronostratigraphic framework to establish correlation. Graphic correlation can provide this framework and be updated as more detailed stratigraphic data becomes available. Graphic correlation improves the consistency of age correlations and provides a means of detecting and measuring stratigraphic hiatuses. The method will aid a stratigrapher in the interpretation of a section, but it will not identify key surfaces in the well, nor can it distinguish unconformities (subaqueous or subaerial) from section removed by faulting. These interpretations must be made by the geologist, using graphic correlation as one of many tools.

The geologic interpretation of a LOC terrace has great importance in determining its position within a sequence stratigraphic framework. Paleobathymetry estimates are a key piece of information when considering the cause of an terrace. The well in Figure 2 is given a paleobathymetry estimate of bathyal depths for the entire section. In real wells, this information is obtained from various sources (i.e., benthic foraminifera, certain dinoflagellate cysts, etc.). The section is mostly sand, with some silt and shale interbeds, and an interpretation of stacked sandy turbidites in a submarine fan is reasonable. In this setting, LOC terraces result from sediment starvation. Shallow water sections develop LOC terraces from subaerial exposure and erosion (Shaw, 1964; Miller, 1977). By incorporating graphic correlation with sequence stratigraphy, a geologist can compare and correlate the two environments and construct precise paleogeographic maps.

We propose a model for the theoretical relationship of graphic correlation terraces within a sequence stratigraphic framework (Fig. 4). This model has been modified to reflect the

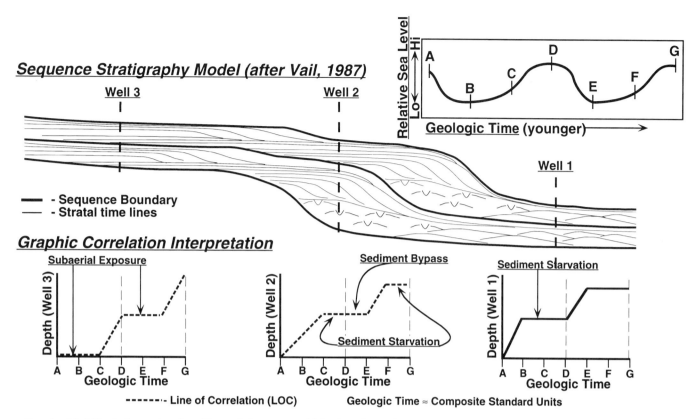

Fig. 4.—Model for the integration of graphic correlation terraces with key bounding surfaces in sequence stratigraphy. In the graphic correlation interpretation, sequence boundaries are marked as thin dashed lines at the time where they begin to form (i.e., with the first relative fall of sea level). At the sequence boundary (time D), terraces from all three wells overlap for an instant of geologic time. With erosion in the updip well (Well 3) and a lag time for delayed deposition in the downdip well (Well 1), this instant will lengthen. Chronostratigraphic units are documented from overlapping graphic correlation terraces.

separation of the highstand and transgressive shelf from lowstand basinal deposits (as is the case in the North Sea Paleogene, Armentrout and others, 1993) for two depositional sequences. This simplified 2-D model assumes deposition to occur in the plane of section, responding to shelfal accommodation space created by relative changes in sea level, coupled to the depositional profile. For discussion purposes, biostratigraphic resolution in this idealized model permits the recognition of each depositional systems tract. A well through submarine fan deposits (Well 1) will have short periods of rapid sedimentation, separated by longer intervals of sediment starvation (see also Fig. 2). A sequence boundary occurs at the base of each depositional package in Well 1, as major fan deposition results from falling sea level (Kolla and Macurda, 1988; Posamentier and others, 1991, Mutti, 1992). Graphic correlation terraces between depositional pulses contain the chronostratigraphic-equivalent distal deposits of the lowstand prograding, transgressive and highstand systems tracts. During this time interval (time B–D), most sediment is captured within coastal systems (Wells 2 & 3) and only minor suspension settling deposition occurs on the basin floor. If deposition can be demonstrated to be continuous, although at a greatly reduced rate (<1 cm/ka) across this interval, then the package is correctly termed a condensed section. However, if sampling is insufficient to document continuous sedimentation, or hiatuses can be demonstrated, additional complications arise that will be discussed below.

A well near the lowstand shelf edge (Well 2) should show a pattern of accumulation and nondeposition similar to the fan setting. However, some differences exist in the position and interpretation of LOC terraces. The first difference is the time of deposition below the first terrace. In Well 1, accumulation took place between time A and B (Fig. 4). Well 2 accumulation extends to time C, as this well encounters lowstand shelf deposits that are distally starved in Well 1. The sequence boundary (time D) occurs within the LOC terrace, dividing this lack of accumulation into periods of sediment starvation and sediment bypass. The time of bypass (time D to E) is represented by deposition above the sequence boundary in Well 1. For time E to F, Well 2 records only the deposition of lowstand shelf sediments.

Shelfal wells have potentially the most complex pattern of sediment accumulation as coastal processes of wavebase and subaerial erosion compete with point or line source sediment supply and relative change of sea level. Some of the possible complications will be discussed below. The LOC for Well 3 shows a single terrace, commencing with the sequence boundary (time D) and ending with renewed deposition at the next transgression (time F). This terrace reflects nondeposition due to subaerial exposure. Should any subaerial erosion occur, the terrace from D to F will be extended back in time to capture the time of deposition removed by this erosion.

Theoretically, there should exist an instant of time that is contained within a LOC terrace for wells in each setting (Fig.

4, time D). This instant corresponds to eroded section in Well 3, the point where starvation becomes bypass in Well 2, and the time just before sedimentation resumes in Well 1. If a time of overlapping terraces can be demonstrated, then a chronostratigraphic unit (possibly a depositional sequence) is documented. The usefulness of this model to a real data set will be demonstrated below, but first we will discuss some possible correlation pitfalls.

COMPLICATIONS CAUSED BY NESTED STRATIGRAPHIC CYCLES

Recent papers have discussed the ramifications of high- and very high-frequency sea-level signals within the sequence stratigraphic model (e.g., Mitchum and Van Wagoner, 1991; Posamentier and others, 1992). Accordingly, basinwide correlations must consider the effect of nested high- and low-frequency sea-level changes. During low-frequency sea-level falls, high-frequency falls will have greater relative magnitude (Mitchum and Van Wagoner, 1991) and regressive pulses will occur into deeper parts of the basin. Updip sections may only encounter a single exposure surface spanning the time of a multi-sequence regression in the basin. Conversely, low-frequency rises in sea level may subdue expression of high-frequency falls in the deep basin while stacking thick sedimentary packages at the basin rim (Vail and others, 1991). These packages may record a relative sea-level signal that is unresolvable in the deep basin (Fig. 5).

The graphic correlation expression of these backstepped packages depends again on biostratigraphic resolution. If resolution is fine enough to separate high-frequency systems tracts, then the ideal LOC of a well penetrating the updip section (Well 3) will contain high-frequency terraces. The relative sea-level curve (Fig. 5) shows a low-frequency rise with superimposed high-frequency events. These high-frequency cycles have less amplitude than the low-frequency trend. The sediment response to high-frequency cycles also has a smaller magnitude when compared to the low-frequency regressive wedge penetrated by Well 2 (Figs. 4, 5).

Although smaller in magnitude, deposition during the high-frequency sea-level cycle has all the characteristics necessary for a depositional sequence and should be identified as such (Mitchum and Van Wagoner, 1991). The smaller events cause a period of subaerial exposure similar to larger falls. The occurrence of high-frequency sea-level falls during a long-term sea-level rise may cause high-frequency lowstand deposition to remain on the inherited shelf. Little, if any, sediment reaches the shelf-slope break due to an excess of shelfal accommodation space that the sediment flux from sea-level falls at C' and C" cannot fill (Fig. 5). Given complete preservation and sufficient biostratigraphic resolution, the LOC for a well through this section (Well 3) will record small terraces, caused by the brief period of subaerial exposure (Fig. 5, times following C' and C"). A more distal well (Well 2) may only recognize the low-frequency signal, illustrating the need for careful correlation of events to avoid mistying high-frequency surfaces to low-frequency deposits.

Graphic correlation can provide a solution to this problem by maintaining a consistent chrono-biostratigraphic framework, regardless of resolution. Graphic correlation allows a geologist to be flexible, but accurate, in the construction of a chronostratigraphic framework. Although biostratigraphic resolution limits and data sample error bars control the accuracy and precision of a correlation, graphic correlation can help the biostratigrapher establish a baseline for resolution. Graphic correlation can result in a framework of chronostratigraphic units, verified by demonstrating LOC terraces that overlap in time for multiple sections. The chronostratigraphic resolution of these terrace-bounded units becomes a baseline, with correlation of finer resolution chronostratigraphic units possible, but less confident. Sequence stratigraphy will almost always produce a higher resolution stratigraphy than graphic correlation, but graphic correlation can help determine when a sequence stratigraphic correlation has pushed resolution limits too far. A graphic correlation approach to stratigraphy can quantify deposition within a section, but still remain flexible to enhanced correlations with improved data resolution. This flexibility and accuracy is an advantage graphic correlation has over the condensed section method of Loutit and others (1988).

Figure 6 highlights some possible stratigraphic complexities near a "condensed section" with alternative ways to interpret the same interval. This diagram focuses on the section around the LOC terrace at 5630 ft in Figure 2. The idealized log section could represent a downdip well, basinal to Figure 5, where the LADs of fossils F and H correspond to times C" and D, respectively. Deposition occurring between the F and H markers could represent the lowstand of the high-frequency sea-level fall at C." If condensed sections are "... thin marine stratigraphic units consisting of pelagic to hemipelagic sediments characterized by very low-sedimentation rates (Loutit and others, 1988, p. 186)," associated with apparent marine hiatuses, then what becomes of a condensed section if true marine hiatuses are suspected? Deposition between the LADs of F and H could have been episodic, with long periods of nondeposition or even erosion at the sequence boundaries (creating true marine hiatuses). A condensed section interpretation overlooks this possible complexity, whereas graphic correlation interpretation remains open to other interpretations with better data collection (coring). The term, "hiatal interval," provides a conservative means of relating LOC terraces to real geology. Hiatal intervals are, informally, well sections associated with a graphic correlation terrace, where data sampling limitations prohibit the recognition of possible multiple stratigraphic lacuna. In a study that relies on ditch cuttings for data collection, hiatal interval correlation seems the most conservative and flexible way to build a meaningful chronostratigraphic framework.

GRAPHIC CORRELATION AND SEQUENCE STRATIGRAPHY
INTEGRATION EXAMPLES

Graphic correlation will almost certainly not resolve all key stratigraphic surfaces in the rock. It will also not tie exactly with key surfaces because biostratigraphic precision is limited by sampling interval and available markers. Graphic correlation shares this problem with zonal biostratigraphy. If, for example, calcareous nannofossils zones are used for correlation, it is likely that more than one important surface will fall within each zone. The result is an aliasing problem where the resolution of a tool (nanno zones) is coarser than the resolution of what is being measured (stratigraphic position of key surfaces). An advantage of graphic correlation is not only that a chronostrati-

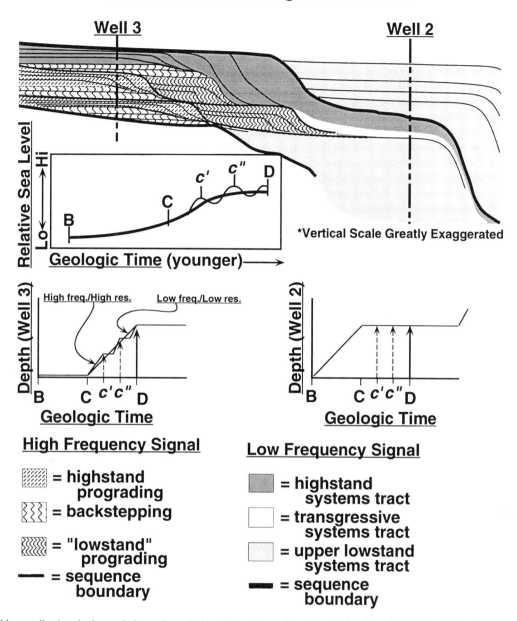

FIG. 5.—Possible complications in the resolution and correlation of deposits resulting from the interference of high- and low-frequency sea-level ssignals. Times C' and C'' mark high-frequency sea-level falls that result in complete depositional sequences, which are restricted to the shelf due to long-term sea-level rise from time B to D. High-frequency events are correlated within low-frequency cycles except where sediment supply or data resolution allows for their separation.

graphic framework of overlapping terraces demonstrates chronological correlations, but also that the method uses the maximum biostratigraphic resolution available in all taxa group within the composite standard.

The following examples of graphic correlation integration with sequence stratigraphy come from a Paleogene regional study of the central North Sea subsurface, correlated to Paleogene outcrops in NW Europe using graphic correlation. Sequence names come from lithologic units in the area and are not important for this discussion. Sequences were picked using wireline well logs (mostly the gamma-ray curve) and tied to regional seismic lines. The resulting sequence stratigraphic framework is based on an interpretation of the relative sea-level signal for each particular location as expressed in the rocks and correlated with biostratigraphy.

In a multi-disciplinary study such as this, it is often difficult to achieve total agreement among the various information sources. We generally tried to obtain agreement on interpretation with at least two of the three disciplines used in this study (log interpretation, paleontology, and seismic interpretation). Often, sequence boundaries that were clearly recognized on the gamma-ray log and in the graphic correlation of a well could not be distinguished on seismic data due to resolution limits. Conversely, some excellent surfaces on seismic data could be noted in gamma ray curves, but could not be resolved with graphic correlation. An additional source of error might come

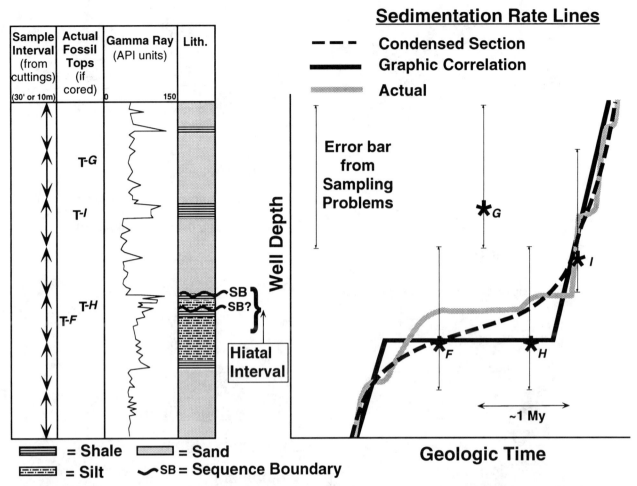

FIG. 6.—Diagramatic illustration of possible stratigraphic complexities (Actual line) around a hiatal interval and different interpretations of rock accummulation using the same fossil LADs with graphic correlation and a condensed section approach. Data points are plotted at the base of each sample interval as they appear in standard well reports. The "Actual" line is based on the exact position of each marker as they would appear in a continuously cored well. The slope of this line comes from an interpretation of relative sedimentation rates of turbidites (rapid rate = inclined LOC) and suspension drape or marine hiatus (slow rate = flat LOC). The vertical error bar accounts for only error associated with sample recovery (Armentrout and others, 1993) and does not address timing errors (horizontal axis). Timing errors are assumed to be minimal in this idealized application of graphic correlation. Significant differences exist between condensed section methodology and graphic correlation that impact regional correlations.

from the reported depth of a cutting sample. Lags in ditch cutting sample collection are common, therefore, a 30-m 'error bar' between the reported fossil depth occurrence and the well-log pick was proposed by Armentrout and others (1993), to account for operator consistency and seismic resolution. The affect of this error on a correlation is discussed above (Fig. 6). Using a 30-m error bar in the correlation of a cutting sample with its conjugate gamma log marker allowed some freedom to make a sequence boundary correlate between well-log, biostratigraphic, and seismic data. When sequences are below seismic resolution or occur within a seismically chaotic zone, well-log interpretation and fossils are the only tools that allow correlation with any confidence.

Basinal Setting

We built and utilized our composite standard in the central North Sea where most deposition occurred in the deep marine environment (Den Hartog Jager and others, 1993; Armentrout and others, 1993) and correlation of turbidite packages relies on good biostratigraphy. A sequence stratigraphic interpretation involves a series of steps that begins with interpretation of the biostratigraphic data using the composite standard. Depositional sequences are then identified on the well logs using the methodology described by Vail and Wornardt (1990) and correlated to seismic data with synthetic seismograms. Graphic correlation helps identify important depositional gaps in each well log and aids the correlation with seismic data between well control points. Figure 7 is a typical basinal well for the North Sea Paleogene interval showing the gamma-ray curve, lithology and biostratigraphic LAD or acme datums from a well report. Biostratigraphic datums are plotted to the right of the well at their reported depth and relative age within the composite standard for the central North Sea (see Fig. 2 for procedure). CSUs are non-linear in absolute time, but they show the relative position of each fossil to others in the composite standard. This means that a fossil with its LAD at 750 CSU is younger than one with its LAD at 600 CSU. One can convert CSUs to absolute time by calibrating the composite standard to a chrono-

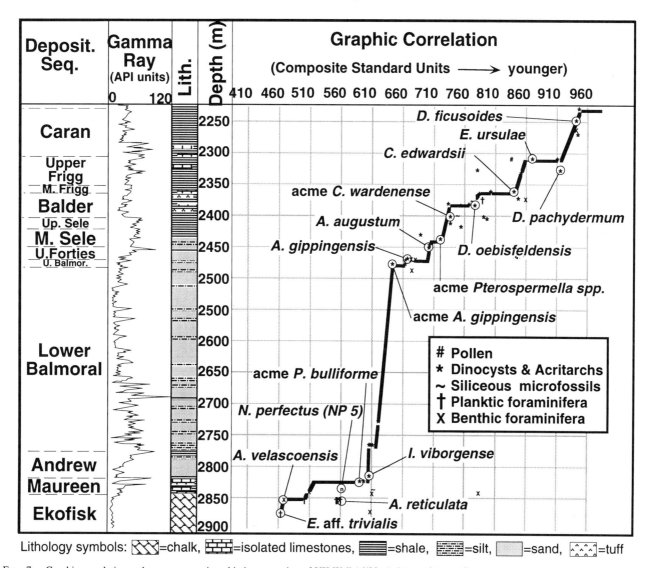

Fig. 7.—Graphic correlation and sequence stratigraphic interpretation of UK Well 16/28–1. Data points are first downhole occurrences of individual fossils and various acme events, distinguished by symbol. Paleobathymetric estimates are in the bathyal range for the entire section (courtesy of S.P.T., Robertson Research).

biostratigraphic chart (such as Haq and others, 1988, or Harland and others, 1990). The scatter of data points is interpreted to produce a LOC, which can be weighted for index fossils or simply connect the greatest number of data points, using straight line segments that always have a positive slope (unless the succession is overturned, Shaw, 1964). Since the composite standard for the North Sea was built using North Sea wells, individual wells are plotted against the standard in later iterations minus the well itself to avoid a circular interpretation.

Where graphic correlation identifies a terrace, we look for a well-log marker that becomes a candidate for a sequence boundary at or near the sample depth. Often, high gamma-ray shales are used to pick genetic sequences (Galloway, 1989; Stewart, 1987). These high gamma intervals, which are interpreted to represent very slow sedimentation, divide the section stratigraphically into sediment pulses. In basinal wells, we locate sequence boundaries at the base of each sediment pulse (see Fig. 4), since turbidite sedimentation increases with falling sea level (Posamentier and Vail, 1988). In the deep marine environment, this approach is possible because highstand and transgressive deposits are not well-represented and result in merged sequence boundaries and maximum flooding surfaces (Vail, 1987). However, in a sand-rich turbidite environment such as the Paleogene central North Sea, submarine scouring may produce a sequence boundary in wells that is locally a sand-on-sand contact. Choosing a log marker to designate as the sequence boundary must rely on the biostratigraphy. Unfortunately, not all depositional sequences identified from well-log and seismic data in this study were resolvable with the graphic correlation tool. Sequences that were unresolvable with separate graphic correlation terraces were instead correlated within a framework of units that could be resolved and documented with time overlaps in many wells.

A basinal well 15 km to the east of Figure 7 is interpreted using the same procedure to correlate sequences and demonstrate how sediment distribution within sequences changes (Fig.

8). The Lower Balmoral sequence (2475 and 2775 m) in Figure 7 is a thick stack of sandy turbidites that becomes a thinner, silty interval in Figure 8. The Maureen sequence sands of Figure 8 between 2704 m (8870 ft) and 2784 m (9130 ft) are calcareous claystone in Figure 7 (2850–2825 m). Additionally, the LOC terrace above the U. Forties sequence in Figure 7 merges with an terrace above the Upper Balmoral in Figure 8, demonstrated by a longer stratigraphic break in the LOC of Figure 8 at this position. Individual biomarkers defining each sequence may vary slightly. Fossils that define a sediment pulse in one well may be found within a terrace above or below the sequence in another well, but the markers should not define a sediment pulse for a different sequence (e.g., the acme of *Palaeocystodinium bulliforme*) is within the sediment pulse of the Andrew sequence in Fig. 8, but occurs in the terrace below the sequence in Fig. 7). Individual markers are found in different sequences if they occur out of normal succession (i.e., the LOC does not pass through them, see Fig. 2 for example). The independence graphic correlation provides from strict key marker stratigraphy is an important advantage over stratigraphic frameworks based solely on individual markers or zonal biostratigraphy.

Integrating Biostratigraphic Data With Seismic Data

Tying the well data to a seismic line with synthetic seismograms is a critical step in supporting correlations since seismic reflectors follow geologic time lines (Vail and others, 1977b). Plotting the LOC of a well next to the time-converted gamma-ray curve directly ties relative geologic time, as resolved by the biostratigraphy, to a seismic line (Fig. 9). This plot provides information concerning the variable duration of geologic time represented within the seismic section. Hiatal intervals correlate, within resolution limits, to regionally mappable seismic

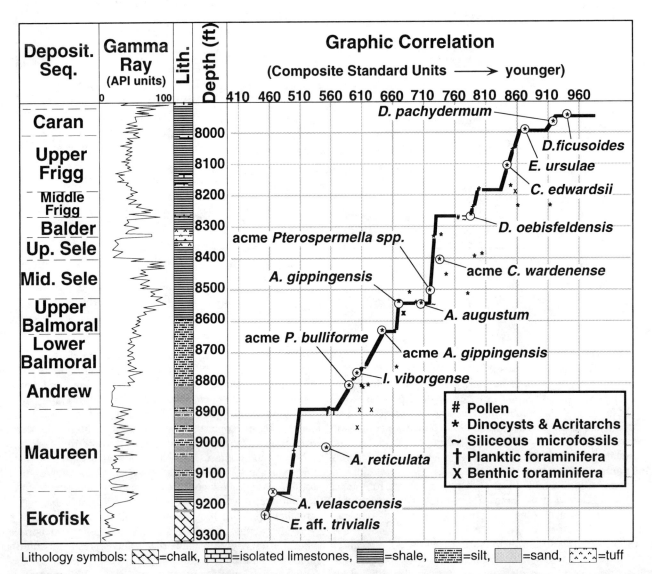

FIG. 8.—Graphic correlation and sequence stratigraphic interpretation of UK Well 16/29-4. Data points are first downhole occurrences of individual fossils and various acme events, distinguished by symbol. Paleobathymetric estimates are in the bathyal range for the entire section (courtesy of S.P.T., Robertson Research).

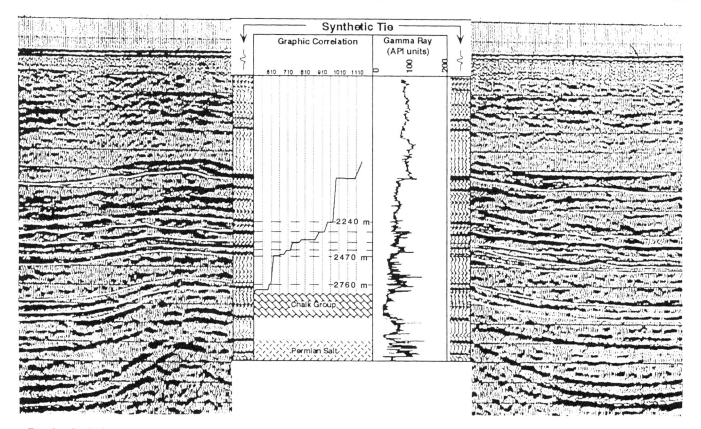

Fig. 9.—Synthetic seismogram created from the sonic and density logs of the UK Well 16/28–1 (Fig. 7) spliced into a regional seismic line at the well location. Graphic correlation plot and gamma-ray log are included with the synthetic tie to demonstrate how relative geologic time can be directly tied to seismic time. See Figure 7 for correlation of sequences as some sequences are below seismic resolution.

reflectors. Figure 9 illustrates that some sequences are too thin for the resolution of this seismic data and are represented by a single wavelet peak or trough. These units are correlated on wells until a thicker section permits seismic resolution.

Plotting both graphs together with the same horizontal (time) scale highlights time-correlative stratigraphic breaks (Fig. 10). For example, datasets from wells 16/28-1 and 16/29-4 both show LOC terraces from 860 to 910 CSU. Both wells experience slow sedimentation and/or erosion during that period, which demonstrates a correlative break in rock accumulation. Major breaks may contain more than one terrace as shown at location A (Fig. 10). Even time intervals of small correlative breaks (i.e., location B, Fig. 10) identify depositional sequences, regardless of lateral facies changes. Figure 10 also highlights the variable nature of rock accumulation between the correlative terraces in time and space. In a submarine fan setting, sedimentation is complex and varies greatly in depositional rate along strike and dip (Walker, 1978). Where both wells share a time interval of no rock accumulation, the overlapping data terraces are interpreted to represent correlative events bracketing sediment pulses. Regional correlation is necessary to verify that the stratigraphic breaks are basinwide events. Deposits between terraces do not necessarily have to overlap in time due to the episodic nature of submarine fan sedimentation. The different lengths of terraces between wells is explained by considering fan lobe avulsion and its effects on a LOC.

Shelfal Setting

Shelfal depositional settings are very important to sequence stratigraphic interpretations because the effect of relative changes in sea level are clearly evident from changes in sedimentary facies. Figure 11 shows the gamma-ray curve, lithology, graphic correlation, paleobathymetry and sequence stratigraphic interpretation of a well through the Upper Paleocene shelf of the central North Sea. The paleobathymetry column estimating the depositional environments of these sediments is critical to the interpretation of this section. We identify sequence and systems tract boundaries based on interpretation of depositional environment and basinward shifts in facies (Vail, 1987; Posamentier and Vail, 1988; Vail and Wornardt, 1990). These interpretations come from the gamma-ray log pattern, however, log patterns alone are not diagnostic. For example, the sand at a depth of 1067 m (3500 ft) has a log pattern similar to other sands below 1402 m (4600 ft). Paleobathymetry estimates place the upper sand in the transitional marine to inner shelf environment (Robertson Research, 1987, unpubl. data), while the lower sands were deposited in an upper bathyal water depth. The lower sands represent fan deposition (lower lowstand, Fig. 3) during falling sea level and the upper sand is interpreted as incised valley fill during rising sea level. Paleobathymetry estimates help the interpreter to predict updip, downdip, and lateral facies variations within a sequence stratigraphic framework. Paleobathymetry can also help track

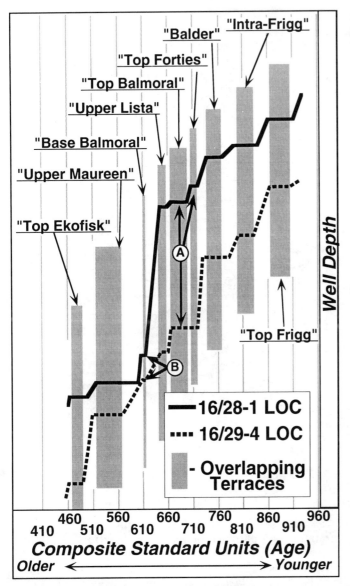

FIG. 10.—Composite graph lines of correlation for two central North Sea wells with overlapping time gaps indicated. Location A shows the collapse of two terraces into one and location B shows correlation of even short events.

low-frequency sea-level cycles, as is highlighted with peak floodings around 1372 m (4500 ft) and 823 m (2700 ft) in Figure 11.

The LOC for this well has few terraces, but these terraces provide some interesting information. The first observation that can be made is that more sequences are identified on the well log than are reflected in the LOC. Sequences are correlated within LOC terrace-bounded units. This problem was discussed above and requires regional work be done to document sequences picked on physical surface criteria (wells and seismic-see below). Another point is made by the terrace at approximately 1128 m (3700 ft) depth, which corresponds to a high gamma marker interpreted to be a maximum flooding surface. In this case, the terrace may be caused by sediment starvation or wavebase erosion (e.g., Well A, Fig. 3). Another possibility is that depositional environment conditions caused a terrace to form here because paleowater depths for overlying deposits are very shallow. Paleobathymetry estimates are inner shelf to transitional marine for deposition in well 15/6-1 between 1128 m (3700 ft) and 884 m (2900 ft), with most samples containing only terrestrial pollen (Fig. 11). With this real data set, the level of biostratigraphic resolution described in our theoretical models (Figs. 4, 5) is not achieved.

The Lower Eocene section of Figure 11 highlights a limitation in data sampling. The LOC connecting each data point between 884–823 m (2900 ft–2700 ft) has a gradual slope. Paleobathymetry estimates note a rapid deepening through this interval (Fig. 11). The sand between 884 – 823 m is interpreted as a lower lowstand deposit, due to the upper bathyal paleowater depth estimate. A possible explanation for the scatter of data points is that rapid deepening occurred above the section deposited at 884 m (2900 ft). As discussed above, turbidite deposition occurs much faster than suspension settling, and sandy turbidites have been observed to scour out and rework thin drape deposits. A great deal of time may have been represented in thin drape deposits below the Lower Eocene sand, but the sand has since eroded the thin underlying mudstones and reworked their constituent fossils.

The sequence stratigraphic and graphic correlation interpretation of the Lower Eocene section in Figure 11 highlights why "hiatal interval" terminology is preferred over "condensed section." The Lower Eocene sand was probably deposited rapidly, despite the gradual slope of the LOC. Most of the time spanned by the gently sloping LOC should be accounted for within the shales above the sand, or at the likely hiatus below the sand. Cutting sample biostratigraphic resolution is unable to verify this. The Lower Eocene interval spans 5 Ma according to Haq and others (1988). If the sand is assumed to be deposited by turbidite events, then most of that 5 Ma must be accounted for in the very thin high gamma-ray shales above and below this unit. Sedimentation rates necessary to accommodate this amount of time in such thin units are well under 1 cm/ka, unrealistically low for an intracratonic basin such as the North Sea without accounting for periods of erosion (Sadler, 1981). Since hiatuses are difficult to recognize in electric logs, the conservative interpretation would be to recognize that hiatuses occur in this section and suggest where they might be found. If "error bars" for each datum through the Lower Eocene section are used, one can interpret a possible LOC that makes geological sense (Fig. 11). Forced interpretations using error bars are not easily supported by the raw data and are not recommended. The conservative LOC is preferred to highlight the biostratigraphic correlation and suggest that further interpretation of the section involve well-log (or core) analysis. Error bars are best used to place individual cutting samples with their appropriate log marker, rather than force a biostratigraphic interpretation of a terrace when no individual sample supports this interpretation.

Graphic correlation terraces in the shelf well (Fig. 11) overlap in time with similar terraces found in distal wells (Figs. 7, 8 at ~610, ~650, and ~700 CSUs). Clearly the mechanism for the formation of a terrace in the shelf well will differ from basinal ones; however, if terraces in each location contain similar fossils, they will necessarily occupy the same time. The theoretical model should be considered in basinwide correlations, but actual data has its limitations. The interpreter must

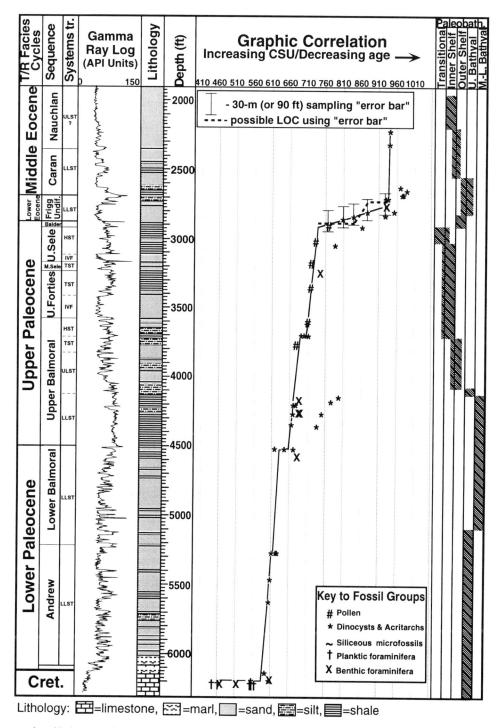

FIG. 11.—Sequence stratigraphic interpretation of UK 15/6-1, a well with shallow water section. Systems tract interpretations are based on log character and paleobathymetry: LLST = lower lowstand (bypass sediments), ULST = upper lowstand (lowstand prograding and slumps), TST = transgressive systems tract, IVF = incised-valley fill, and HST = highstand systems tract. Paleobathymetric estimates are in the bathyal range for the entire section (courtesy of S.P.T., Robertson Research).

be aware of possible pitfalls and communicate any data limitations.

Enhanced Correlation on the Shelf Using Seismic Data

Shelfal sections often have problematic LOCs when using a composite standard built mainly from basinal wells. The advantage that shelfal wells have over their distal counterparts is that deposits resulting from sea-level changes are more easily distinguished from autocyclic deposits. In the basin, sea-level changes are determined indirectly by recognizing sediment pulses that result from sea-level fall and linking drape deposits to periods of high sea level. Without good chronostratigraphic

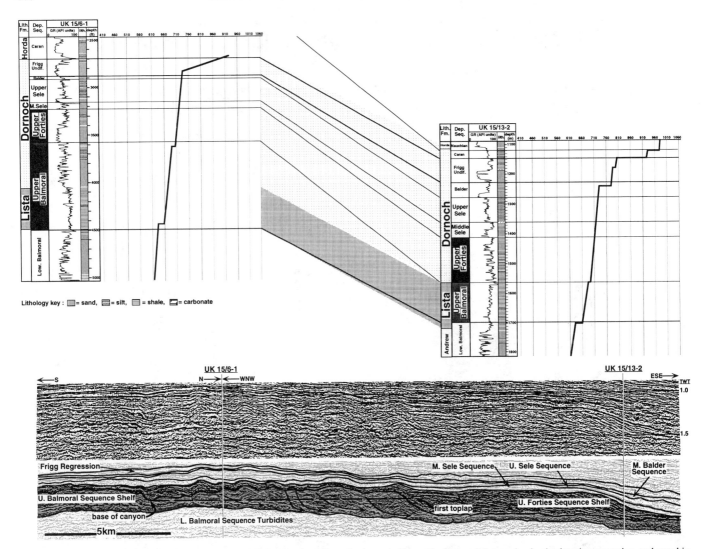

FIG. 12.—Correlation of depositional sequences and lithostratigraphic units in a shelfal setting by well-log and seismic data interpretation and graphic correlation. The base of the canyon incising into the Upper Balmoral Sequence shelf occurs at the same depth as the first onlapping deltaic toplap of the Upper Forties Sequence shelf. This relationship demostrates a 150-m relative fall of sea level and suggests that major deltaic progradation in the Paleogene central North Sea occurred during lowstands. This figure also shows how the sloping LOC for the Lower Eocene strata in Well 15/6-1 can be separated into at least two distinct units in Well 15/13-2 (see FIG. 13 and text for discussion). Lithostratigraphic units are from Robertson Research well reports. Seismic data is courtesy of S.P.T. (Horizon Geophysical).

control, any rapid increase in sedimentation rate might be mistaken for a separate depositional sequence when, in fact, it could result from fan lobe avulsion. In shelfal sections, basinward shifts of facies are the clearest indicator of sea-level change. Shelfal sections, if thick enough, also display seismic stratal patterns that relate to relative changes in sea level (see Fig. 3). The correlation of wells to seismic in a shelfal setting can provide additional support for sequences that may be biostratigraphically unresolvable.

Seismic data interpretation can strengthen correlation of shelfal sequences especially when well-log and biostratigraphic data are inconclusive. The correlation of Middle and Upper Sele sequences benefits from seismic data interpretation between wells 15/6-1 and 15/13-2 (Fig. 12). These two wells have been interpreted for depositional sequences and tied to seismic data with synthetic seismograms (see Fig. 9 for example). The Sele sequence boundaries are identified at physical surfaces in the wells and at reflector discontinuities in seismic data (Fig. 12). The Sele sequences fall between two overlapping graphic correlation terraces, which means we must rely on other tools to verify the sequence correlation. In the 15/6-1 well, both sequence boundaries are picked at the base of transgressive sands, indicated by upward increasing gamma-ray values. In well 15/13-2, a slope sand rests on the base Middle Sele sequence boundary. This sand is interpreted as a lowstand deposit with updip transgressive and highstand deposits represented as capping mudstones (1360–1380 m, Fig. 12). A thin carbonate layer marks the base Upper Sele sequence in 15/13-2 with the overlying silty section representing a lowstand deposit. The Upper Sele sequence has a highstand progradational wedge (see Fig. 11 at ~884–945 m {2900–3100 ft}), seen on seismic data, that thins rapidly toward the shelf break (Fig. 12) and is thin or

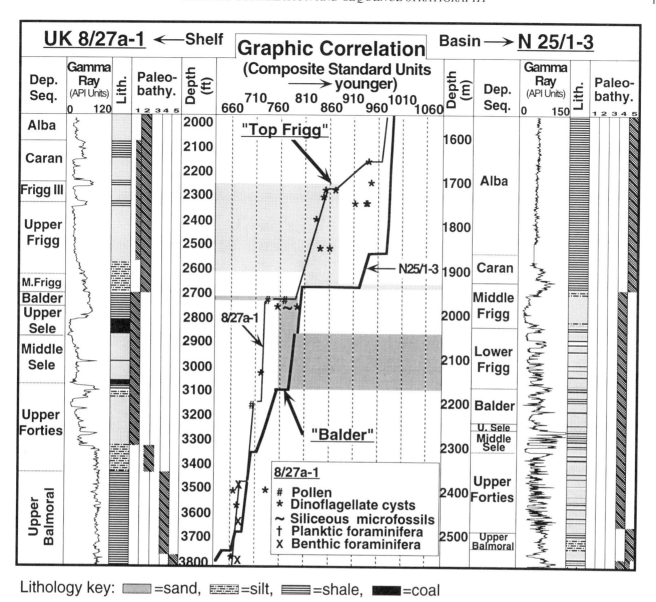

FIG. 13.—Composite plot of graphic correlation comparing the shelfal UK 8/27a-1 and basinal (Frigg Field) N 25/1–3 wells. Data points for the UK 8/27a-1 are plotted. The LOC for the N (25/1–3 is well constrained by data (not shown here). This correlation reveals sediment partitioning between shelf and basinal areas, as predicated in Figure 4, and here resolved with graphic correlation. The Lower Frigg basinal sequence correlates within the Balder terrace on the shelf, and the Upper Frigg sequence correlates within the Top Frigg terrace in the basin. Paleobathymetry zones: 1- Transitional/marginal marine, 2- Inner shelf, 3- Outer shelf, 4- Upper bathyal, 5- Middle to lower bathyal (courtesy of S.P.T., Robertson Research).

absent in well 15/13-2. Seismic data interpretation verifies and enhances well-log correlations.

Coupling correlation with LOC terraces and seismic data interpretations can help document chronostratigraphic equivalence or diachronaiety of lithostratigraphic units. The "Dornoch" deltaic lithostratigraphic unit is demonstrated to be diachronous with graphic correlation and seismic data (Fig. 12). The lower prograding unit of the Dornoch in well 15/6-1 (~1220–1098 m {4000–3600 ft}) falls within a different chronostratigraphic unit the lower prograding of well 15/13-2 (~1560 m–1410 m). Correlation to seismic data confirms two distinct prograding units, separated by a deeply incisive sequence boundary.

Shelf to Basin Correlations

If correlation of two or more wells from the same depositional environment in the same basin is a fairly straightforward exercise with good biostratigraphic and seismic control, can the same statement apply to correlation of wells from the basin to the shelf? The theoretical model we present relates LOC terraces to depositional systems tracts (Fig. 4), but acknowledges possible complications in shelf-to-basin correlations with accompanying data limitations (Fig. 5). All aspects of the theoretical model will rarely be observed; however, certain aspects of the model can explain some minor miscorrelations of units.

Correlation of two wells having very different depositional environments is attempted in Figure 13. Well 8/27a-1 is situated

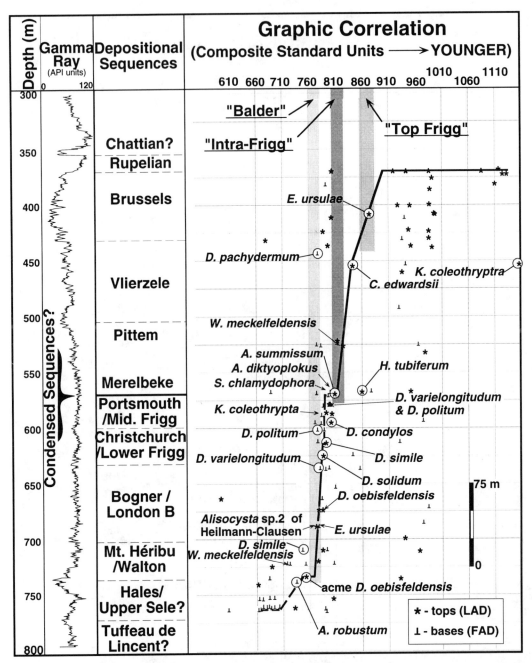

Fig. 14.—Graphic correlation of dinoflagellates in the Northern German Wursterheide borehole from Heilmann-Clausen and Costa (1989). In this section, the "Balder" terrace observed in all central North Sea wells corresponds to an interval of deposition. Interpretation of the gamma-ray log identifies depositional sequences that are observed elsewhere in Northwest Europe.

on the East Shetland Platform in the North Sea and penetrates a thick shallow-water section as evidenced by well-developed coals and verified with paleobathymetry estimates that are inner shelf to transitional (Robertson Research, 1987, unpubl. data) for much of the section. Conversely, paleobathymetries in well N25/1-3 on the Norwegian side of the South Viking Graben never gets shallower than upper bathyal (Robertson Research, 1987, unpubl. data). Well N25/1-3 penetrates thick sequences of submarine fan sands that should be genetically linked to shelfal deposits according to the sequence stratigraphic model.

Data are sparse through the shallow section of well 8/27a-1, but the resulting LOC does resolves several data terraces, and a correlation of this stratigraphy to the LOC from well N25/1-3 highlights some key points. Both wells record a "Top Forties" terrace (~710 CSU). Using the error bar to relate biostratigraphic data to well depth, this terrace can be placed near the base of a thick coal in well 8/27a-1. Thick coals often occur above sequence boundaries and represent the early transgressive systems tract of the overlying sequence (Cross, 1988). The "Top Forties" terrace occurs in well N25/1-3 at the base of a turbidite sand that is interpreted as a lower lowstand systems tract (Fig. 3). The "Top Forties" terrace in N25/1-3 starts and ends earlier than in well 8/27a-1 as would be suggested from the model (Wells 1 and 3, Fig. 4). In well N25/1-3, the upper package of sandy fans is divided into two sequences, Lower and Middle Frigg. Most of the time represented by these deposits falls within a LOC terrace in well 8/27a-1 suggesting a long period of bypass at the well 8/27a-1 site while significant accumulation occurred at the well N25/1-3 site. At the top of the Frigg package in well N25/1-3 (~1950 m), a long terrace appears in the LOC. During this time interval, a thick progradational package is deposited on the shelf (~808–692 m {2650–2270 ft} in well 8/27a-1), and it is even possible to recognize more than one sequence. This is an example of a possible correlation problem as suggested in Figure 5.

Recognition of stratigraphic incompleteness is very important to sequence stratigraphic basin analysis. Sequences will not be obvious in all parts of the basin, but with a composite standard biostratigraphy, depocenters from all parts of the basin may be fitted into a complete chronostratigraphic framework. Resolution of events depends on sedimentation rate and the resolution of the age dating tool. Even time intervals of overlapping terraces in one part of the basin may be expanded into thick sections for another area. Figure 14 demonstrates how the time of overlapping terraces below the Frigg package ("Balder" terrace) expands into 100 m of section in a cored borehole from onshore Germany. Sequences are identified by increasing and decreasing value trends on the gamma-ray log and related to sea-level change as a function of silt content. An ideal composite will incorporate all depocenters in a basin then determine in later graphing rounds what intervals are missing for any particular section.

The most important product of a basin-scale sequence stratigraphic study is a chronostratigraphic chart of relative sea-level events and their effect. In Northwest Europe, the construction of a chronostratigraphic chart revealed that different numbers of sequences could be found in each sub-basin. In the central North Sea, we find lowstand pulses of submarine fans that correlate onshore to a single exposure surface. From outcrops of shallow marine deposits, we observe multiple sequences that correlate within a single graphic correlation terrace in the central North Sea (Fig. 14). Graphic correlation provides the chronostratigraphic framework to document and correlate the complex sedimentation packages from shelf-to-basin, recognized with sequence stratigraphy.

DISCUSSION, PITFALLS, AND IMPLICATIONS FOR STRATIGRAPHY

Clearly, the integration of sequence stratigraphic and graphic correlation methodologies has potential to be a powerful tool in basin analysis. Graphic correlation also has the potential to bring a systematic documentation to questions about the role of eustasy in sequence stratigraphy. The key to this integration procedure is for the user to be aware of all the possible complications in its application. We present an idealized relationship between LOC terraces and sequence stratigraphic bounding surfaces, but the link depends on correct application of both procedures and is then limited by the data resolution.

The examples presented here come from an area with exceptionally rich biostratigraphic data due to continuous study for many years. In frontier areas, the technique is no less valid. Any stratigraphic study is limited by resolution of the available data, be they discontinuous outcrops, sparse well data or coarse biostratigraphic resolution. Workers at outcrop scale have different resolutions from those using multichannel seismic data. The scale-dependency of geology ensures that no stratigraphic study will be able to record and correlate all events, but by integrating the available stratigraphic techniques, it is hoped that some key bounding events can be confidently recognized and correlated. By compositing whatever data are available, a chronostratigraphic framework is constructed. As more data become available, the composite standard becomes more complete. Observations made on the rocks being studied do not change. The interpretation of those observations and their position within a chronostratigraphic framework may change as new information becomes available.

Sequence stratigraphic theory has developed as the resolution of seismic data improved. The coastal onlap curve of Vail and others (1977a) was not invalidated by revisions in Haq and others (1988), rather the resolution of events improved and more cycles were added to the original curve. As sequences have been observed at increasingly brief time intervals, a need to distinguish between magnitudes of cycles develops. A sequence stratigraphic study needs to be regional, if it hopes to recognize the different magnitudes of cycles. Recognition of a global signal (eustasy) will subsequently require study on a global basis. Graphic correlation can help in this endeavor. A composite in its early stages should show major stratigraphic breaks (continental floodings, basin forming events, or plate tectonic reorganizations). Within these major stratigraphic units, smaller magnitude events can be correlated as stratigraphic detail is infilled. The confidence of correlation for smaller duration events is directly related to the resolution of the correlation tool being used (marker beds, biostratigraphy, magnetostratigraphy, isotopes, etc.). All types of chronological information may be carried in a composite at any time scale.

CONCLUSIONS

We propose here a theoretical model that relates graphic correlation data terraces to key sequence stratigraphic bounding surfaces. This model shows the tremendous potential that composite standard biostratigraphy, integrated with sequence stratigraphy, has for documentation and correlation of basin- and global-scale chronostratigraphic frameworks. Sequence stratigraphy and graphic correlation should be done simultaneously, creating a system of checks and balances that significantly improves the quality of each discipline. We have found this integration invaluable in northwestern Europe, documenting a chronostratigraphic framework containing sections that have been studied for up to 150 years without reaching a consensus correlation.

Data resolution limits are the most important consideration and source of potential errors in the integration of graphic correlation and sequence stratigraphy. The model proposed here presents ideal relationships for conceptual purposes. Interpretation of real data, even from an area of excellent biostratigraphy, will not match the model. This model shows what could occur with ideal resolution, but its concepts hold true with less than perfect resolution. The concept of chronostratigraphically-overlapping graphic correlation terraces defining sequence boundaries (or composite sequence boundaries) can be applied at any time scale. Multi-million year stratigraphic cycles can be documented as chronostratigraphic equivalents within the composite standard, which can have significance for tectono-eustasy (sea-level change resulting from changes in ocean basin size) and basin-formation events. A chronostratigraphic correlation framework does not have to resolve every depositional sequence to be effective. Graphic correlation provides a complimentary and controlling temporal framework for sequence stratigraphic units that are observed at a higher resolution than the biostratigraphy.

REFERENCES CITED

ARMENTROUT, J. M., MALACEK, S. J., FEARN, L. B., SHEPPARD, C. E., NAYLOR, P. H., MILES, A. W., DESMARAIS, R. J., AND DUNAY, R. E., 1993, Log-motif analysis of Paleogene depositional systems tracts, Central and Northern North Sea: defined by sequence stratigraphic analysis, *in* Parker, J. R., ed., Petroleum Geology of Northwest Europe: Proceedings of the 4th Conference: London, Geological Society, p. 45–58.

AUBRY, M.-P., 1991, Sequence Stratigraphy: Eustasy or Tectonic Imprint: Journal of Geophysical Research, v. 96, no. B4, p. 6641–6679.

CROSS, T. A., 1988, Controls on coal distribution in transgressive-regressive cycles, Upper Cretaceous, Western Interior, U.S.A., *in* Wilgus, C. K., Hastings, B. S., Kendall, G. C. St. C., Posamentier, H. W., Ross, C. A., and Van Wagoner, J. C., eds., Sea-Level Changes: An Integrated Approach: Tulsa, Society of Economic Paleontologists and Mineralogists Special Publication 42, p. 371–380.

DEN HARTOG JAGER, D., GILES, M. R., AND GRIFFITHS, G. R., 1993, Evolution of Paleogene submarine fans of the North Sea in space and time, *in* Parker, J. R., ed., Petroleum Geology of Northwest Europe: Proceedings of the 4th Conference: London, Geological Society, p. 59–72.

EDWARDS, L. E., 1984, Insights on why graphic correlation (Shaw's method) works: Journal of Geology, v. 92, p. 583–597.

EDWARDS, L. E., 1989, Supplemented graphic correlation: a powerful tool for paleontologists and nonpaleontologists. Palaios, v. 4, p. 127–143.

FAIRBRIDGE, R. W., 1961, Eustatic changes in sea level, *in* Ahrens, L. H. and others, eds., Physics and Chemistry of the Earth: London, Pergamon Press, v. 4, p. 99–185.

GALLOWAY, W. E., 1989, Genetic stratigraphic sequences in basin analysis II: Application to northwest Gulf of Mexico Cenozoic Basin: American Association of Petroleum Geologists Bulletin, v. 73, p. 143–154.

HAQ, B. U., HARDENBOL, J., AND VAIL, P. R., 1988, Mesozoic and Cenozoic chronostratigraphy and cycles of relative sea level change, *in* Wilgus, C. K., Hastings, B. S., Kendall, G. C. St. C., Posamentier, H. W., Ross, C. A., and Van Wagoner, J. C., eds., Sea-Level Changes: An Integrated Approach: Tulsa, Society of Economic Mineralogists and Paleontologists Special Publication 42, p. 71–108.

HARLAND, W. B., ARMSTRONG, R. L., COX, A. V., CRAIG, L. E., SMITH, A. G., AND SMITH, D. G., 1990, A Geologic Time Scale 1989: Cambridge, Cambridge University Press, p. 1–263.

HARPER, C. W. AND CROWLEY, K. D., 1985, Insights on why graphic correlation (Shaw's method) works: a discussion: Journal of Geology, v. 93, p. 503–506.

HAZEL, J. E., EDWARDS, L. E., AND BYBELL, L. M., 1984, Significant unconformities and the hiatuses represented by them in the Paleogene of the Atlantic and Gulf Coast Province, *in* Schlee, J., ed., Interregional Unconformities and Hydrocarbon Accumulation: Tulsa, American Association of Petroleum Geologists Memoir 36, p. 59–66.

HEILMANN-CLAUSEN, C. AND COSTA, L., 1989, Dinoflagellate zonation of the Uppermost? Paleocene to Lower Miocene in the Wursterheide Research Well, NW Germany: Geologisches Jarhrbuch, Reihe A, v. 11, p. 431–521.

KOLLA, V. AND MACURDA, D. B., 1988, Sea-level changes and timing of turbidity-current events in deep-sea fan systems, *in* Wilgus, C. K., Hastings, B. S., Kendall, G. C. St. C., Posamentier, H. W., Ross, C. A., and Van Wagoner, J. C., eds., Sea-Level Changes: An Integrated Approach: Tulsa, Society of Economic Paleontologists and Mineralogists Special Publication 42, p. 381–392.

LOUTIT, T. S., HARDENBOL, J., VAIL, P. R., AND BAUM, G. R., 1988, Condensed sections: The key to age determinations and correlation of continental margin sequences, *in* Wilgus, C. K., Hastings, B. S., Kendall, G. C. St. C., Posamentier, H. W., Ross, C. A., & Van Wagoner, J. C., eds., Sea-Level Changes: An Integrated Approach: Tulsa, Society of Economic Paleontologists and Mineralogists Special Publication 42, p. 183–213.

MIALL, A, 1992, Exxon global cycle chart: an event for evey occasion?: Geology, v. 20, p. 787–790.

MILLER, F. X., 1977, The graphic correlation method in biostratigraphy, *in* Kauffman, E. G. and Hazel, J. E., eds., Concepts and Methods in Biostratigraphy: Stroudsburg, Dowden, Hutchinson and Ross, p. 165–186.

MILTON, N. J., BERTRAM, G. T., AND VANN, I. R., 1990, Early Palaeogene tectonics and sedimentation in the Central North Sea, *in* Hardman, R. F. P. and Brooks, J., eds., Tectonic Events Responsible for Britain's Oil and Gas Reserves: London, Geological Society Special Publication 55, p. 339–351.

MITCHUM, R. M., JR., VAIL, P. R., AND THOMPSON, S. III, 1977, Seismic stratigraphy and global changes of sea level, part 2, the depositional sequence as a basic unit for stratigraphic analysis, *in* Payton, C. E., ed., Seismic Stratigraphy – Applications to Hydrocarbon Exploration: Tulsa, American Association of Petroleum Geologists Special Publication 26, p. 53–62.

MITCHUM, R. M., JR. AND VAN WAGONER, J. C., 1991, High-frequency sequences and their stacking patterns: sequence stratigraphic evidence of high-frequency eustatic cycles: Sedimentary Geology, v. 70, p. 131–160.

MUTTI, E., 1992, Turbidite Sandstones: Milan, Agip S.p.A., Amilcare Pizzi S.p.A. arti grafiche, Cinisello Balsamo, 275 p.

PAYTON, C. E., ed., 1977, Seismic Stratigraphy – Applications to Hydrocarbon Exploration: Tulsa, American Association of Petroleum Geologists Special Publication 26, 502 p.

PLOTNICK, R. E., 1986, A Fractal Model for the Distribution of Stratigraphic Hiatuses: Journal of Geology, v. 94, p. 885–890.

POMEROL, C., 1989, Stratigraphy of the Palaeogene: hiatuses and transitions: Procedings of the Geologic Association, v. 100, p. 313–324.

POSAMENTIER, H. W., ALLEN, G., AND JAMES, D. P., 1992, High-resolution sequence stratigraphy – the East Coulee delta: Journal of Sedimentary Petrology, v. 62, p. 310–317.

POSAMENTIER, H. W., ERSKINE, R. D., AND MITCHUM, R. M., JR., 1991, Models for submarine-fan deposition within a sequence-stratigraphic framework, *in* Weimer, P. and Link, M., eds., Seismic Facies and Sedimentary Processes of Submarine Fans and Turbidite Systems: New York, Springer-Verlag, Frontiers in Sedimentary Geology Series, p. 127–136.

POSAMENTIER, H. W., JERVEY, M. T., AND VAIL, P. R., 1988, Eustatic controls on clastic deposition I—conceptual framework, *in* Wilgus, C. K., Hastings, B. S., Kendall, G. C. St. C., Posamentier, H. W., Ross, C. A., and Van Wagoner, J. C., eds., Sea-Level Changes: An Integrated Approach: Tulsa, Society of Economic Paleontologists and Mineralogists Special Publication 42, p. 109–124.

POSAMENTIER, H. W. AND VAIL, P. R., 1988, Eustatic controls on clastic deposition II—Sequence and systems tract models, *in* Wilgus, C. K., Hastings, B. S., Kendall, G. C. St. C., Posamentier, H. W., Ross, C. A., and Van Wagoner, J. C., eds., Sea-Level Changes: An Integrated Approach: Tulsa, Society of Economic Paleontologists and Mineralogists Special Publication 42, p.125–154.

PRELL, W. L., IMBRIE, J., MARTINSON, D. G., MORLEY, J. J., PISIAS, N. G., SHACKLETON, N. J., AND STREETER, H. F., 1986, Graphic correlation of oxygen isotope stratigraphy application to the Late Quaternary: Paleoceanography, v. 1, p. 137–162.

Robertson Research International, 1987. Palaeocene-Eocene Stratigraphic Study of the Central North Sea and Outer Moray Firth: The Robertson Group. plc, Llandudno, Gywnedd, 78 encl., *unpublished data*.

SADLER, P. M., 1981, Sediment accumulation rates and the completeness of stratigraphic sections: Journal of Geolology, v. 89, p. 569–584.

SARG, J. F. AND SKJOLD, L. J., 1982, Stratigraphic traps in Paleocene sands in the Balder Area, North Sea, *in* Halbouty, M. T., ed., The Deliberate Search for the Stratigraphic Trap: American Association of Petroleum Geologists Memoir 32, p. 197–206.

SHAW, A. B., 1964, Time in Stratigraphy: New York, McGraw-Hill, 365 p.

SLOSS, L. L., 1963, Sequences in the cratonic interior of North America: Geologic Society of America Bulletin, v. 74, p. 93–113.

STEWART, I. J., 1987, A revised stratigraphic interpretation of the early Palaeogene of the central North Sea, *in* Brooks, J. and Glennie, K., eds., Petroleum Geology of North West Europe: London, Graham and Trotman, p. 557–576.

SUESS, E., 1906, The Face of the Earth, v. 2: Oxford, Clarendon Press, 556 p.

VAIL, P. R., 1987, Seismic stratigraphy interpretation using sequence stratigraphy. Part I: Seismic stratigraphy interpretation procedure, *in* Bally, A.W., ed., Atlas of Seismic Stratigraphy: Tulsa, American Association of Petroleum Geologists Studies in Geology 27, p. 1–10.

VAIL, P. R., MITCHUM, R. M., JR., AND THOMPSON, S., III, 1977a, Seismic stratigraphy and global changes of sea level, part 4, global cycles of relative changes of sea level, *in* Payton, C.E., ed., Seismic Stratigraphy — Applications to Hydrocarbon Exploration: Tulsa, American Association of Petroleum Geologists Special Publication 26, p. 83–97.

VAIL, P. R., TODD, R. G., AND SANGREE, J. B., 1977b, Seismic stratigraphy and global changes of sea level, part 5, chronostratigraphic significance of seismic reflectors, *in* Payton, C. E., ed., Seismic Stratigraphy — Applications to Hydrocarbon Exploration: Tulsa, American Association of Petroleum Geologists Special Publication 26, p. 99–116.

VAIL, P. R. AND WORNARDT, W. W., 1990, Well log seismic sequence stratigraphy: An exploration tool for the 90's: Eleventh Annual Research Conference: Gulf Coast Section, SEPM, Program and Abstracts, December 2–5, Austin, Earth Enterprises, Inc. 1990

VAIL, P. R., AUDEMARD, F., BOWMAN, S., EISNER, P., AND PEREZ-CRUZ, G., 1991, The stratigraphic signatures of tectonics, eustasy, and sedimentation, *in* Seilacher, D. and Eisner, G., eds., Cycles and Events in Stratigraphy: Berlin, Springer-Verlag, p. 617–659.

VAN WAGONER, J. C., MITCHUM, R. M., CAMPION, K. M., AND RAHMANIAN, V. D., 1990, Siliciclastic sequence stratigraphy in well logs, cores, and outcrop: Tulsa, American Association of Petroleum Geologists Methods in Exploration Series 7, 55 p.

WALKER, R. G., 1978, Deep-water sandstone facies and ancient submarine fans: Models for exploration for stratigraphic traps: American Association of Petroleum Geologists Bulletin, v. 62, p. 932–966.

WILGUS, C. K., HASTINGS, B. S., KENDALL, G. C. ST. C., POSAMENTIER, H. W., ROSS, C. A., AND VAN WAGONER, J. C., eds., 1988, Sea-level Changes: An Integrated Approach: Tulsa, Society of Economic Paleontologists and Mineralogists Special Publication 42, 407 p.

WHEELER, H. E., 1958, Time-stratigraphy: American Association of Petroleum Geologists Bulletin, v. 42, p. 1047–1063.

PART III
TECHNICAL APPLICATIONS OF GRAPHIC CORRELATION

WORLDWIDE AND LOCAL COMPOSITE STANDARDS: OPTIMIZING BIOSTRATIGRAPHIC DATA

RICHARD W. AURISANO, JAMES H. GAMBER, H. RICHARD LANE, EDWARD C. LOOMIS, AND JEFF A. STEIN

Amoco Exploration and Production Technology, Houston, Texas 77253

ABSTRACT: Since the publication of Shaw's *Time in Stratigraphy* in 1964, paleontologists of Amoco Corporation have concentrated on developing a Phanerozoic biostratigraphic database for use in graphic correlation applications and composite standard development. This developmental phase brought the company to a position where it can routinely utilize its composite database to solve problems of correlation related to hydrocarbon exploration and production. This experience shows that two kinds of composite standards maximize data value. These are (1) worldwide and (2) local composite standards.

A worldwide composite standard develops a comprehensive fossil occurrence database that establishes an accurate understanding of the global ranges of fossil species. Fossil datums are placed into a chrono-sequence scaled to a framework of worldwide reference sections. With database evolution through graphic correlations, traditional zonal resolution is usually exceeded due to the factoring of all fossil data, not just zonal markers and familiar accessory forms. Paleontologists in industry commonly must use multisource, multivintage data of varying qualities within the same project. Systematic comparisons of such information to a worldwide standard are particularly beneficial in overcoming the incongruities of multiple zonal schemes and varied taxonomies that make up such datasets and which tend to confound traditional biostratigraphic analysis. They also are helpful in identifying and removing spurious data.

A worldwide composite standard may distort the value of local species ranges linked to regional paleoenvironmental and geographic restrictions. These foreshortened ranges may otherwise be very valuable in establishing intrabasinal correlations. For these reasons, the concept of a "local" composite standard developed within Amoco to maximize basin evaluations and regional stratigraphic applications.

A local composite standard captures the regional biostratigraphic significance of all fossil occurrences, including those localities calibrated initially to the worldwide standard. These calibrated localities have lines of correlations that place a local fossil sequence within a worldwide context. The resulting database provides optimum refinement for regional correlations, while resolving important basinwide time-stratigraphic breaks undetected by purely local zonations.

INTRODUCTION

The idea of graphic correlation and composite standard databases was pioneered by Dr. Alan Shaw and published in his volume *Time in Stratigraphy* in 1964. In his role as biostratigrapher and Chief Paleontologist within Amoco, he instituted this new approach to making chronostratigraphic interpretations and solutions (see Shaw, 1993). During the early research development of the composite standard database, Amoco's biostratigraphy program was not centralized and the general paleontological approach was based almost exclusively on internally generated data and identification of zonal markers (Fig. 1). In fact, a rather sophisticated, but traditional, approach to paleontology had developed and held sway within the company (see Albers, 1987). In this environment, the composite standard approach was underutilized and under appreciated for three reasons: (1) skepticism existed within the paleontological ranks, especially in the operating offices, concerning the viability and reliability of the approach versus the traditional, experience-tested tops and zonal methods; (2) the inability to produce timely composite standard project results due to lengthy sample preparation, analysis time, and extensive database building requirements; and (3) the still primitive nature of computer hardware and software systems, and composite standard concepts.

As the development and application of composite standards and graphic correlation became more accepted within the company, composite standard database building had two champions based on differing needs; these were the research department which focused on the company's international exploration program, and the domestic exploration offices which at that time enjoyed a thriving hydrocarbon exploration climate. The research department was interested in a database that captured the total stratigraphic range of fossil species through a worldwide data gathering effort, while the regional paleontology offices (Denver, New Orleans, Houston, and Calgary, Alberta) followed the path which focused on local stratigraphic ranges as they exist in their basins of economic interest. As we point out in the ensuing sections, both concepts have their strong and weak points. Fortunately, it is when one begins to think "How can I take advantage of the strengths inherent in both concepts?" that a real breakthrough occurs in the scientific, as well as the exploration applications, of composite standard databases.

WORLDWIDE VS. LOCAL COMPOSITE STANDARDS

The Concept of a Worldwide Composite Standard

The purpose of a worldwide composite standard database is to capture the total chronostratigraphic range of fossils (Miller, 1977), thus including the actual first appearance and extinction of all species. Building such a database involves sampling sections and cores with the most complete record of time, facies, climates, et cetera, known. The advantages of such an assembled database are its links to absolute time and its increased potential for accurate interregional correlation. The complete and precise chronostratigraphy afforded by such a database allows one to detect hiatuses, test eustasy, identify provincialism, and understand the migration of species through time, all of which significantly impacts the interpretation of global tectonics, paleo-oceanography, and paleogeography. The creation of a viable worldwide composite standard is a major time-consuming undertaking, involving the study of many measured outcrop sections and well cores of little economic interest to industry, but which are the key to improved correlations in the basins that are being explored. However, experience has shown Amoco that a worldwide composite standard may not be useful in all exploration settings. For example, relevance of the global standard may be limited in local basins where many of the regional fossil tops are suppressed and the bases too high because of provincialism and/or facies control. These anomalous ranges show up in graphic correlation plots as many fossils with known local biostratigraphic value lying way off of the line of correlation. Understandably, the basin specialist becomes very concerned at this stage that the reliable fossil markers he uses for

MIOCENE	UPPER	ROBULUS (E) BIGENERINA (A) CRISTELLARIA (K) CYCLAMMINA (3) DISCORBIS (12) TEXTULARIA (L)
	MIDDLE	CIBICIDES CARSTENSI UVIGERINA (3) GLOBOROTALIA ROBUSTA TEXTULARIA (W) GLOBOROTALIA FOHSI BIGENERINA HUMBLEI CRISTELLARIA (I) CIBICIDES OPIMA ROBULUS (L)
	LOWER	GYROIDINA (9) SIPHONINA DAVISI PLANULINA PALMERAE LENTICULINA HANSENI CRISTELLARIA (R)

FIG. 1.—An example of a Gulf Coast regional Cenozoic zonation—foraminifera. The parenthetic letters and numbers are species designations which may or may not have a published equivalent.

- 16,428 Localities (Outcrops & Wells) Digitally Captured
- 58 Fossil Groups in Standard
- 16,399 Fossil Species in Composite Standard
- 2,436 Localities in Composite Standard

FIG. 2.—Facts and figures about Amoco's paleontology database.

local correlation appear to have little value. We will address this issue again in the next section.

The Concept of a Local Composite Standard

The purpose of constructing a local composite standard database (Lane and others, 1993) is to capture the ranges of the fossils in a particular basin of interest, in order to preserve the provincial utility of the fossils. The advantage of such a database is the optimization of local resolution by maximizing the utility of the data, and such a graphic correlation plot generally shows many of the fossils arranged linearly along the line of correlation — seemingly an excellent correlation. However, the downside of reliance upon a purely local standard is that it may be blind to the presence of local unconformities and hiatuses that can be detected only by plotting against a global chronostratigraphic scale. This is especially true if an unconformity is bounded by the same lithology above and below (for example, shale on shale, carbonate on carbonate, *et cetera*). Additionally, for those hiatuses that do show in a local standard as a terracing of data, it is difficult to truly measure the duration of the hiatus in absolute time. Furthermore, local standards limit the ability to make interregional correlations with any real confidence, and data management issues also arise because of endemic taxonomic problems and uncertainties.

Utilizing Worldwide and Local Composite Standards

Paleontologists can have the best of both worlds in applying the composite standard approach to exploration and research problems. Amoco continues to develop and maintain its global corporate database against which an exploration or exploitation well or outcrop can be graphed. Once a line of correlation from a locality in a local basin is interpreted in the context of the global database and an absolute time scale, the interpretation can be downloaded from the worldwide database to form a local database pertinent to the basin of exploration interest. When this is accomplished, all of the fossils which lie off of the line of correlation because of local provincialism and/or facies control have their tops and bases set by the interpreted line of correlation. This means that all of the tops and bases from the well have now moved to, and lie, on the line of correlation, and each fossil range is calibrated to absolute time. Now numerous local wells and outcrops can be graphed against the new local database which has been calibrated against the global standard. The fossils which truly are useful for time correlation in the basin will continue to fall on the line of correlation with each subsequent graphic plot. Those that are falling prey to local controls on their occurrence, such as salinity fluctuations, sediment influx into otherwise clear water, anoxia, andsoforth, will begin to fall off of the line of correlation. The isochroneity of basinal events such as lithologic marker beds, spikes in the abundance of arenaceous foraminifers, pollen spikes, spikes in the occurrence of charred grass remains, andsoforth can now be tested.

The Composite Standard Approach and Amoco Today

The rapid and saltative evolution of computer technology in the last ten years has challenged the composite standard approach. It has required innovative and continuous programming in order to meet new hardware capabilities and requirements and to take advantage of new time-saving concepts of computer data capture and manipulation. Times have changed Amoco dramatically. Today, the company is internationally focused, working in numerous mature to frontier basins around the world. In many of these areas, there are no Amoco historical paleontological data from which to work, and it is not practical, nor cost-effective, to generate the data. The current business climate is not one in which past practices can simply prevail.

Although Amoco is still involved in paleontological analyses, it also must take advantage of existing non-Amoco data. In many basins of interest, there is a wealth of available multi-source, multivintage consultant paleontological reports and data containing a plethora of non-consistent zonal schemes and age determinations from which to work. In this situation, database management becomes a key issue in effectively applying biostratigraphic data. For this reason, the composite standard approach becomes a major asset.

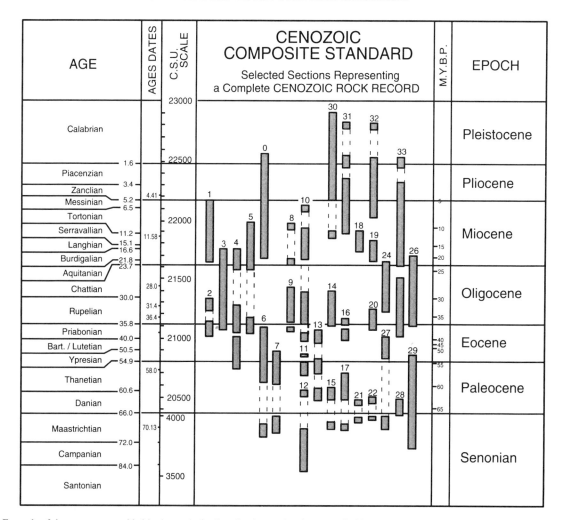

FIG. 3.—Example of the coverage provided in Amoco's database by the overlapping control of late Mesozoic and Cenozic sections.

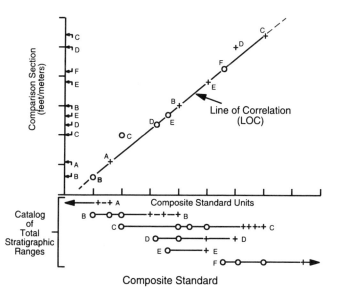

FIG. 4.—A simple graphic correlation of a comparison section against the composite standard on the horizontal axis.

THE COMPOSITE STANDARD DATABASE

Years of database building have provided Amoco with essentially complete worldwide coverage of the mid-Paleozoic and post-Triassic geological records, derived from a considerable variety of outcrop and well localities. Amoco's approach has been to divide Phanerozoic time into four different composite standard databases: Lower Paleozoic, Upper Paleozoic, Mesozoic and Cenozoic. Figure 2 illustrates some statistics about the Amoco databases. These data represent occurrences from at least 58 different fossil groups, which makes the database truly multidisciplinary in a paleontological sense. Figure 3 illustrates in graphical form, just a few of the localities that have gone into building Amoco's Cenozoic standard. Note that overlapping control gives a complete representation of Cenozoic time.

A total of over 600 wells and outcrops have gone into making the Cenozoic composite standard. The localities making up this extensive database are highly diverse in both geographic localities and paleoenvironmental coverage, and by no means restricted to wells Amoco has drilled. To the contrary, sections

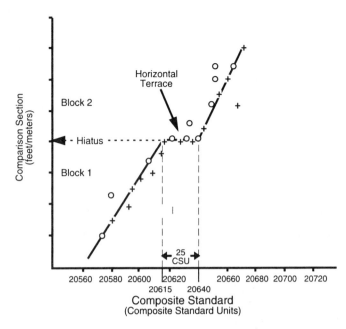

FIG. 5.—Example of a hiatus in the comparison section in graphic correlation.

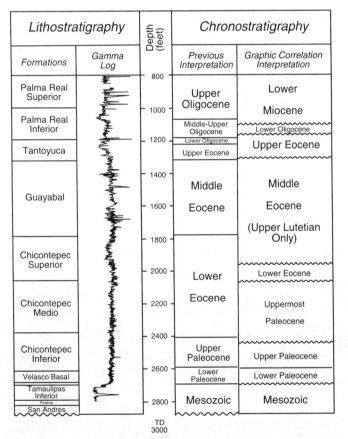

FIG. 6.—Comparison of the traditional biostratigraphic interpretation (second column from right) versus that based on graphic correlation (far right column) in a southern Gulf of Mexico Basin well. Unconformities are represented as terraces in Figure 7.

such as El Kef in Tunisia, have taken special expeditions of Amoco personnel to be sure of sample quality and stratigraphic position. Most such samples have been processed and analyzed in house. However, many of the key reference sections are industry wells, DSDP and ODP cores, and other literature data that offer critical geographic or stratigraphic coverage.

Attributes of the Composite Standard

The worldwide composite standard provides a consistent scale by which to determine the chronostratigraphy represented or not represented by strata in a particular section or area of interest. Figure 4 illustrates a simple graphic correlation of a comparison section against the composite standard on the horizontal axis. This figure emphasizes that the composite is no more than a simple catalogue of biostratigraphic ranges for continuous comparison. In this case, we use range information from many fossil groups rather than just one. Figure 5 illustrates a routine complication in drawing a line of correlation and shows a hiatus in the comparison section, which graphs as a terrace. Because Amoco's composite standards are robust databases in terms of the numbers and kinds of fossil types that have been integrated, it is not unusual for Amoco to be able to rapidly establish correlations in new areas where it has not had previous analytical experience.

EXAMPLE APPLICATIONS

Gulf of Mexico

In some cases, the initial correlations to the corporate standard provide a refined view of the stratigraphic relationships, particularly in the identification of stratigraphic breaks or sequence boundaries. Figure 6 illustrates a traditional biostratigraphic interpretation of a well from a southern Gulf of Mexico Basin. Note that in the center and left columns, no breaks in sedimentation are shown throughout the Cenozoic. However, when the data are compared with Amoco's worldwide Cenozoic composite standard (Fig. 7), there are eight major and minor breaks reflected by terraces, all of which could be of important exploration significance. This refined interpretation is shown on the right column of Figure 6.

North Sea

In other areas, such as the North Sea, consultants have developed very detailed zonations for numerous fossil groups based on abundant data and many years of experience. Figure 8 illustrates some of the wells in the North Sea in which data has been generated by various consultants and purchased by Amoco. In the past, vendor data has been underutilized as a source of information in an exploration area for input into a composite standard database because of variation in its quality and terminology. However, in some areas, vast amounts of vendor data have been generated for routine post-well appraisals. It is simply good business to find a way to put all of this data to good use. The current emphasis on reducing controllable operational costs does not allow us the luxury of generating our own data and still be timely for the exploration process. Figure 9 illustrates an example of zonations for foraminifera and dinoflagellates in the North Sea. There are at least five to six other North Sea zonations used by consultants in the industry, each

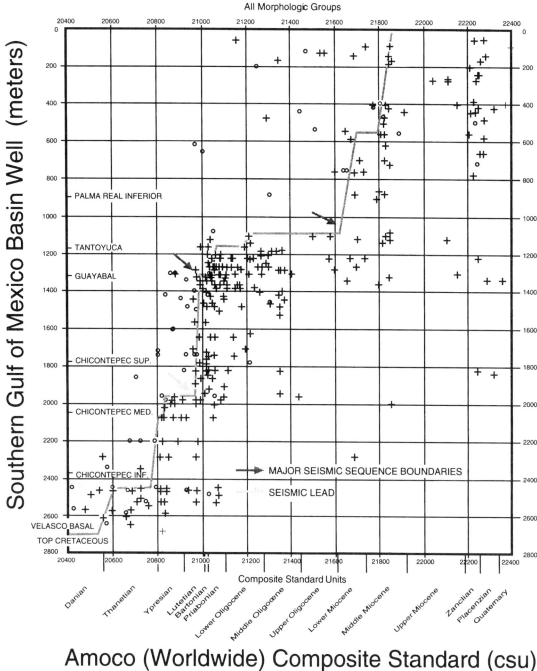

FIG. 7.—A southern Gulf of Mexico Basin well graphically correlated to the composite standard.

based on their own taxonomy, zonal markers, and stratigraphic interpretations. This creates a maze of specialized information to work with, befuddling geologists and geophysicists, and tending to isolate the paleontologist from the rest of the geoscientific world.

Figure 10 shows some key aspects of North Sea zonations. To achieve high resolution, North Sea Paleogene zonations utilize not only first downhole occurrences but abundance acmes and other biostratigraphic events too. In addition, many of the consultants have established informal taxonomies for unpublished "in-house" species. This makes it very difficult for the customer (industry paleontologists) to interpret and quality control vendor results.

Many of the correlation markers used in the North Sea have extended ranges at the global scale when they are compared with the worldwide composite standard. Comparing these local ranges with the worldwide standard tends to diminish their value locally as markers; however, the correlation value of such regional markers may be captured, fully organized, and critically evaluated in a local composite standard. The development of a local composite standard for Paleogene biostratigraphy of the North Sea illustrates this point.

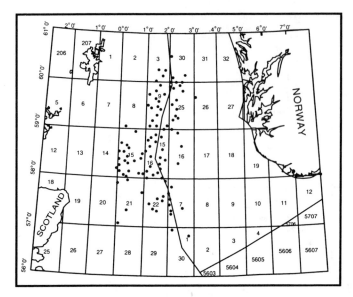

Fig. 8.—The distribution of Paleogene wells of the North Sea with vendor generated data that were used in this study (well locations are defined with numbered latitude and longitude quadrangles).

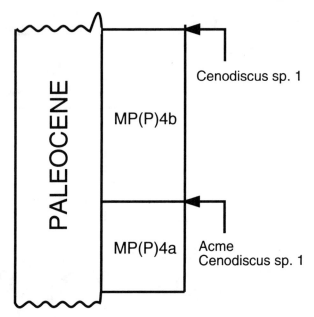

Fig. 10.—A portion of Simon Petroleum Technology's North Sea regional Paleocene siliceous microfossils zonation.

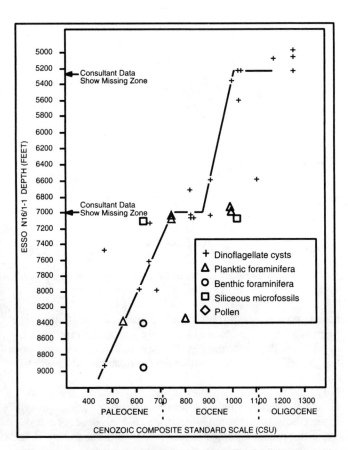

Fig. 9.—An example zonal scheme for foraminifera and dinoflagellates (after Gradstein and others, 1992.)

Fig. 11.—Initial graphic correlation of the North Sea Paleogene composite standard.

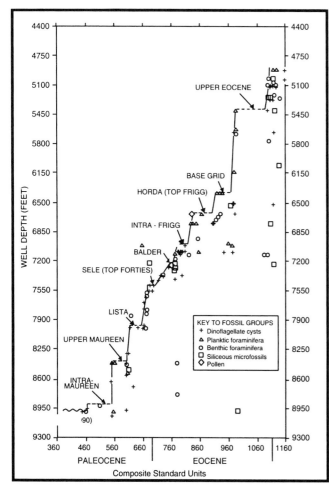

Fig. 12.—The Esso N 16/1-1 data plotted against the local Paleogene North Sea composite standard with additional control points.

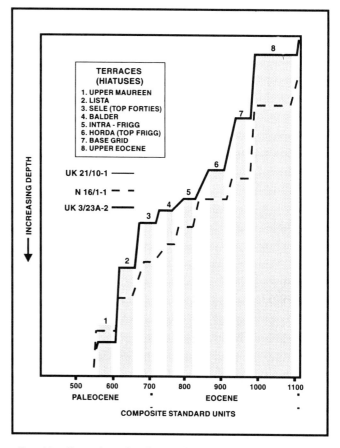

Fig. 13.—Comparison of well correlations with the Cenozoic composite standard. The gray bands show the amount of missing time in common for all terraces of the same number, representing the same event.

Over the years, Amoco has accumulated numerous North Sea Paleogene paleontologic well reports containing a wealth of raw data. However, as discussed before, these reports have been difficult to use because they did not provide a useful, coherent database. When this mass of information is integrated into the composite standard, it produces one fossil sequence, eliminating the problem of multiple zonations relating to various fossil groups and consultant-specific zonations. The composite standard also improves resolution by considering fossil information from all fossil groups and not just the marker fossils from each group. This makes age interpretations and correlations consistent across the fossil spectrum, presenting paleontology as a unified science rather than a fractionated set of competing subsciences. This method allows for the objective detection of hiatuses, rather than being based on human biases and opinions. Finally, it is a method of managing huge volumes of data that are well beyond the capabilities of a single paleontologist.

In setting up a local composite standard for the North Sea Paleogene, as is true for other such composite standards, it is important to correlate the worldwide standard to North Sea Paleogene stratigraphy. This was accomplished by graphing a key well, the Esso N16/1-1 against the Cenozoic standard and compositing the results. The Esso N16/1-1 was selected for the initial graphing because there is a high-quality, up-to-date paleontology vendor report available containing a relatively high number of taxa in common with the established worldwide standard. Also, the zonal succession is well-defined, interpreted to be largely uninterrupted, implying continuous deposition. Figure 11 shows the initial graphic correlation of the Esso N16/1-1 with the worldwide composite standard. Control in this well is limited due to the endemic nature of the Paleogene fossil sequence in the North Sea, and therefore, the paucity of fossil occurrence data on the graph. However, even with this rather open plot, graphic correlation of the data against the worldwide standard matches consultant interpretations based on traditional zonal non-sequences. That is, there are two main regional stratigraphic breaks which show up in the well at 7,000 ft and 5,280 ft, as denoted by the terraces on this graph.

The impression of mostly continuous sedimentation in the North Sea has been changed significantly with the development of a local Paleogene composite standard based on intrabasinal wells. The worldwide standard provides the gross time calibration and quantifies the major breaks. A series of North Sea reference wells is then added to the composite standard. Localities outside the North Sea Basin are then withdrawn from the database, leaving an independent local standard which has local ranges calibrated to an external standard. As shown in Figure 12, the Esso N16/1-1 plotted against the local Paleogene

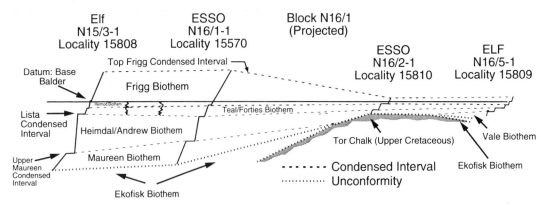

Fig. 14.—Biothemic cross section, illustrating correlations and position of Paleogene condensed intervals of the Verdi area, North Sea (not to scale).

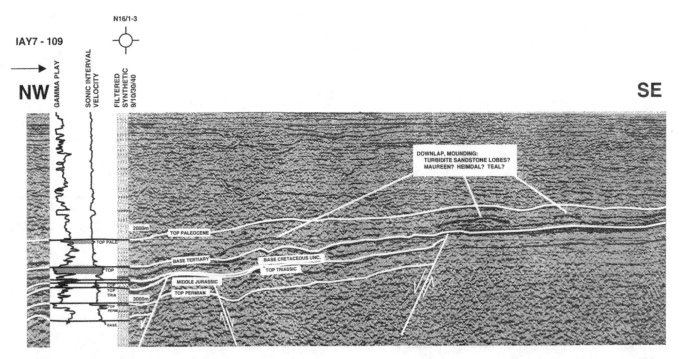

Fig. 15.—A seismic line paralleling the biothemic cross section in Figure 14.

North Sea composite standard has additional control points that provide evidence of additional hiatuses, as shown by the numerous horizontal terraces. These terraces are interpreted to represent periods of starved sedimentation based on the following criteria:

1. They appear to be time-correlative throughout the axial, deep-water part of the Paleogene North Sea Basin.
2. Deep marine conditions exist above and below the terrace intervals.
3. Distinctively deep-water fossil assemblages are associated with many of these terraces.
4. They correlate very closely with significant seismic-stratigraphic boundaries. Moreover, this highly punctuated sedimentation appears to characterize the North Sea Paleogene. Hiatuses have been found to correlate throughout the area

and as such, provide a convenient basis for defining depositional sequences (Neal and others, 1994, this volume). This is graphically demonstrated in Figure 13 where three graphs from different parts of the North Sea have been overlain to show the almost exact placement of the terraces in each within the composite standard.

The correlation framework provided by the graphic correlation terraces has been particularly useful in mapping the local and regional extents of depo-sequences and in mapping stratigraphic pinch-outs of reservoir units. The Paleogene sandstone units of the North Sea are strongly lenticular and their thickness may vary greatly over short distances. Stratigraphic pinch-outs may be easily overlooked, even in seismic sections. Figure 14 illustrates a cross section in the North Sea Paleogene which is constructed by aligning a transect of graphic correlation plots

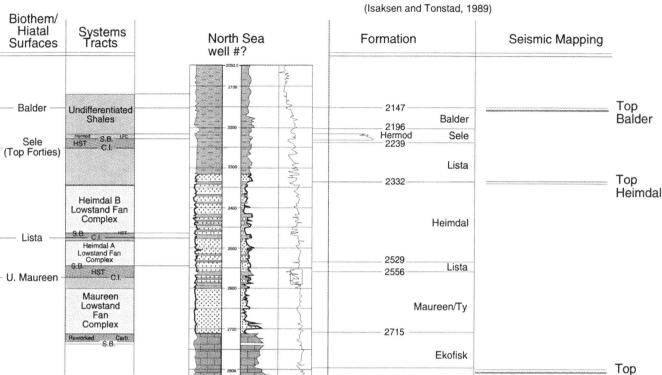

FIG. 16.—Well-log sequence stratigraphic analysis of the North Sea Paleocene. Biothem boundaries correspond to hiatal boundaries defined by graphic correlation and integrated into a sequence stratigraphic interpretation of the well logs.

and connecting time-correlative terraces along the lines of correlation. The boundaries identified define an architecture which details a regional sequence stratigraphic succession based solely on fossil distribution in time and space. Lane and others (1994) used the term biothem to denote these paleontologically defined sequence stratigraphic units. Biothems were originally identified in carbonate environments of deposition, but are also equally useful in clastic regimes. Note that in Figure 14 individual biothems thin toward the right, and up onto a high where the sand packages merged into a shale facies. This detailed, paleontologically derived architecture is not discernible either seismically or by electric logs. A seismic line paralleling the biothemic cross-section is shown in Figure 15. Note that only a couple of horizons can be detected and used seismically for defining a sequence-far fewer than those recognized biostratigraphically. However, note also that a number of seismic mounded features seemingly representing sand wedges, correspond to the biothems. Figure 16 details the chronostratigraphic or biothemic succession in the left column, the traditional formational succession as discerned by electric logs in the middle, and mappable seismic horizons in the right column. Note how different the biothem succession is versus the traditional lithic interpretations, and how only a few horizons can be carried regionally using seismic.

Figure 17 represents one long electric log cross-section using some of the same wells as in the earlier Biothem cross-section and subparalleling the seismic line. In this case, electric log interpretations have been closely integrated with the biostratigraphic framework and clearly reflect the biothemic architecture shown previously. Within the Heimdal A Biothem, for example, the geologist has interpreted a lowstand fan complex resting immediately on the condensed hiatal surface reflected as a terrace on the graph. The lengths of the terraces fit the geologic interpretation because they provide the time that is interpreted to be the condensed highstand interval. In essence, the overall architecture is one of stacked lowstand deposits separated by starved intervals or the terraces on the graph. Using the terraces and moving from west to east, the geologist was able to show how the sandstones pinch-out en echelon in an up structure sense, which of course, has very important exploration implications for geographically restricting stratigraphic pinchouts of reservoir facies in the shale prone areas to the east.

Having this ability to recognize the regional architecture of genetic rock bodies allows us to map them using different physical parameters. For example in Figures 18 and 19, net sand isopachs within the Andrew and Forties Biothems are illustrated. Note in the Andrew Biothem (Fig. 18), major areas of sand accumulation were taking place in the northwest, thinning

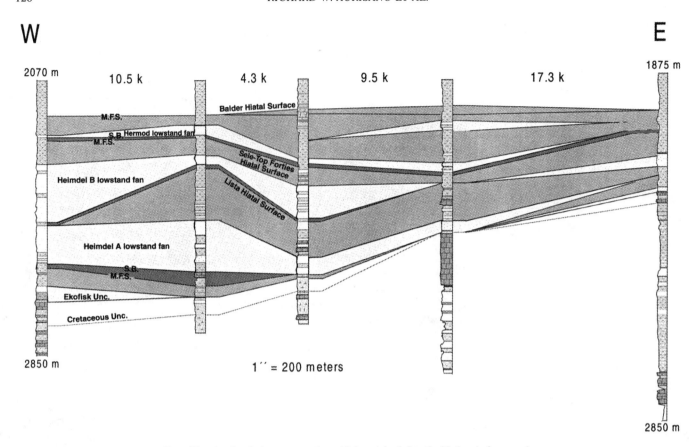

Fig. 17.—An electric-log cross section with log picks tied to the biothemic framework.

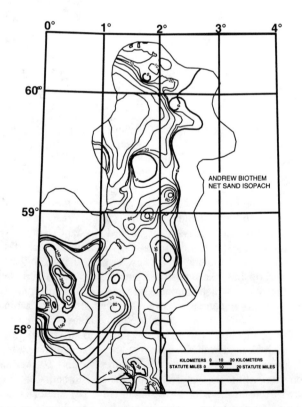

Fig. 18.—Net sand isopach within the Andrews Biothem (Contour Interval = 20 units).

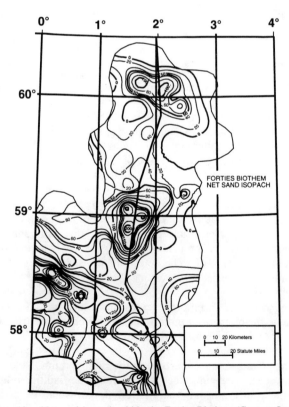

Fig. 19.—Net sand isopach within the Forties Biothem (Contour Interval = 20 units).

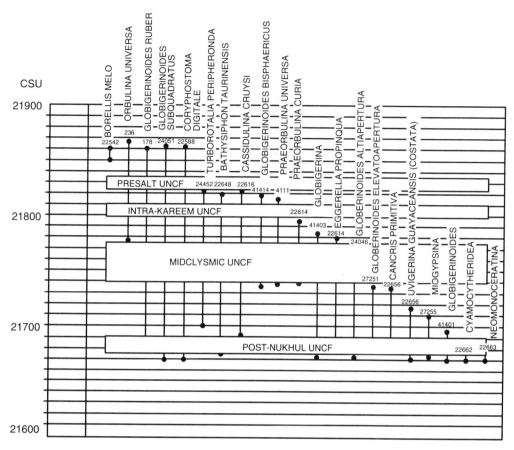

FIG. 20.—Ranges of Gulf of Suez index fossils in composite standard time showing those species that occur in more than one sequence.

to a feather-edge toward the east. A different picture is present in the Forties Biothem (Fig. 19), with major sand thicknesses being in the central and southwestern portion of the map and also thinning eastward.

Gulf of Suez, Egypt

In the following example, a methodology was developed to maximize the utility of well site or operational style paleontology in enabling chronostratigraphic correlations in a field study where production of hydrocarbons is ongoing, but the field geometry is not well understood. Operational style paleontology refers to the paleontological analyses performed at the well site to aid in picking casing points and determining when a target horizon is reached, as well as to give the geologist an idea of the stratigraphic succession penetrated.

Routine operational paleontology in the Gulf of Suez Miocene section is designed to target the identification of index or marker fossils which, for the most part, are planktonic and benthic foraminifers. A few important ostracode species are also used. Samples are examined, in detail, every 3 to 6 meters, but the marker fossils only are recorded. In this fashion, the paleontologist is able to provide timely information to the drill site personnel. Oftentimes, unless a regional study is undertaken, this is the only paleontology available to a field study team to graph against a composite standard database, and it is not the kind of data which one would consider under normal circumstances.

Because of the regional paleontological studies done at Amoco Research, in cooperation with members of the GUPCO (Gulf Of Suez Petroleum Company—a joint operating company for Amoco and the Egyptian National Oil Company) paleontology staff, there is now a wealth of data from the Miocene of the Gulf of Suez in the composite standard database. Just as in the example from the North Sea, a local composite standard was created once a correlation was made to the global standard. Comparison of the ranges of GUPCO's index fossils in composite standard unit time and by sequence, provides insight as to how these index fossils can further be used in field studies (Fig. 20). This comparison shows which fossils are restricted to a sequence and which occur in more than one sequence.

The structural complexity in the Gulf of Suez is such that some blocks are tilted greater than 20°, and the degree of tilt is variable from block-to-block. This situation makes it difficult to compare the graphic correlation plots of field wells to the plot of Well X, the correlation and control well chosen for the field study. This well is near the field being studied and is one of the wells from the joint GUPCO-Amoco study. Research quality data was generated for well X, as shown in Figure 21, with the generation of foraminifera, ostracoda, palynology, and calcareous nannoplankton data. The conversion of sample depths to true stratigraphic thickness (TST) theoretically enables us to make direct comparisons between wells. The limiting factor of course, is the accuracy of the dipmeter readings upon which the TST conversions are dependent.

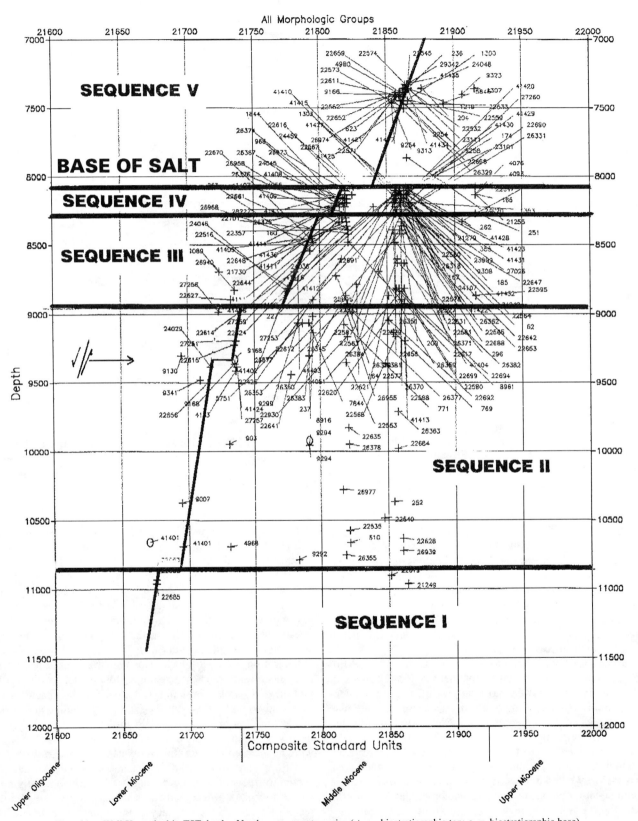

FIG. 21.—Well X graphed in TST depths. Numbers represent species (+ = biostratigraphic top; o = biostratigraphic base).

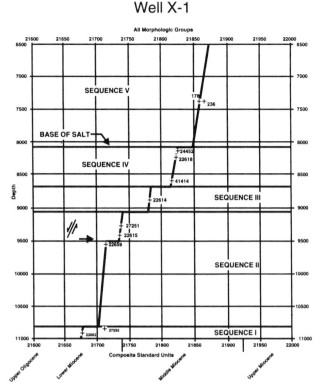

FIG. 22.—Well X-1, a sidetrack to Well X, shows the same general sequence, and so it is easy to graph and interpret the operational paleontology (+ = biostratigraphic top).

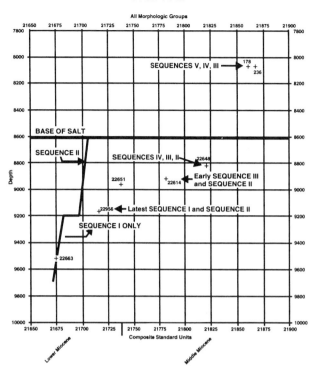

FIG. 24.—Well X-2 operational paleontology interpreted through the overlay to Well X (refer to Fig. 21; + = biostratigraphic top).

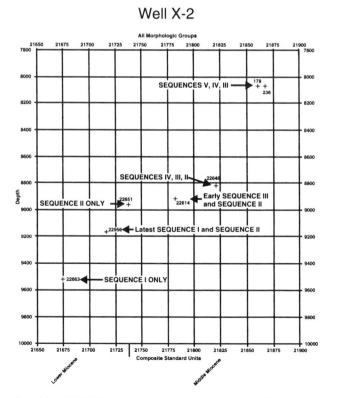

FIG. 23.—Well X-2 operational paleontology shows a different pattern. How does one interpret it? (+ = biostratigraphic top).

Well X and all of the field wells with operational paleontology were converted to TST depths. Graphic correlation plots in the same depth and composite standard unit time (CSU) scales (200 ft/in and 10 CSU/in) were generated. This made it possible to overlay the plots of operational paleontology on top of the plot of the control well. The basic assumption is that the slopes of the LOCs would remain fairly constant within depositional sequence numbers (that is, genetically related sedimentary packages over a small area, such as a producing field.). Any anomalous character of the graph, in combination with known geologic aspects, would indicate that this assumption was indeed unacceptable and unrealistic. No such indications were observed, and this basic technique dramatically put each one of the stratigraphic sections penetrated in a much different perspective than previously witnessed (Figs. 22–24).

The Miocene section consists of five sequences recognizable by graphic correlation of checklisted paleontological data from Well X. The operational paleontology reports are usable for sequence stratigraphic work within certain limitations. Some of the sequences are decipherable using a combination of fossils and logs. The overlay technique developed for this study, helped to pick the sequence that gives more replicable results than that derived from operational paleontology and logs alone. The original interpretation was that every field well had penetrated the Kareem Formation (Sequence IV) below the Belayim Salt because the tops of either some or all of the following species were identified: Turborotalia peripheronda, Bathysiphon taurinensis, Cassidulina cruysi, Globigerinoides bisphaericus, and Praeorbulina universa. The overlay technique demonstrates that this is not true. For example, in Well X-2, the Belayim Salt (Sequence V) rests atop Sequence II (Fig. 22).

CONCLUSIONS

In every facet of the hydrocarbon exploration and production industry, technical specialists are being called upon to continually improve their efficiency of operation. They are being asked to do more with less. The old saying "Time is money" has never been more true. The present and future application of biostratigraphy will require faster turn around of results and greater chronostratigraphic resolution. Quantitative approaches, such as the composite standard, address this need.

Paleontology is crippled by the continued development and use of multiple zonations for different fossil groups of the same age. These multiple zonations require an additional level of interpretation because they are once removed from the real data. They fractionate our science and befuddle other geoscientists. The composite standard approach circumvents this issue by using all fossil data available and by eliminating the zonal approach. Biostratigraphy should play a leading role in developing sequence stratigraphic models. Sequences should just not be checked by biostratigraphy, but they should also be defined by biostratigraphy. Perhaps if less emphasis is place on the zonal concept, and paleontologists seek to unify the approach, other geoscientists would find paleontology less confusing and biostratigraphy would take its rightful place in the development of sequence-stratigraphic models.

REFERENCES

ALBERS, C. C., 1987, Forty years behind the scope: Recollections of a recalcitrant bug-picker or reflections on the evolution of Gulf Coast foraminiferal biostratigraphy: Houston, Eighth Annual Research Conference, Gulf Coast Section, Society of Economic Paleontologists and Mineralogists, Selected Papers and Illustrated Abstracts, p. 1–5.

DEEGAN, C. E. AND SCULL, B. J., 1977, A standard lithostratigraphic nomenclature for central and northern North Sea: Institute Geological Sciences Report 77/25, Norwegian Petroleum Directorate, Bulletin No. 1, p. 1–36.

GRADSTEIN, F. M., KRISTIANSEN, T. L., RISTIANSES, I. L., LOEMO, L., AND KAMINSKI, M. A., 1992, Cenozoic foraminiferal and dinoflagellate cyst biostratigraphy of the central North Sea: Micropaleontology, v. 38, p. 101–137.

ISAKSEN, D. AND TONSTAD, K., 1989, A revised Cretaceous and Tertiary lithostratigraphic nomenclaure for the Norwegian North Sea: Norwegian Petroleum Directurate Bulletin No. 5, p. 1–59.

LANE, H. R., AURISANO, R. W., AND STEIN, J. A., 1993, Worldwide and local composite standards: Having your cake and eating it too: Northeastern Section, Geological Society of America, Abstracts with Programs, v. 25, p. 32.

LANE, H. R., FRYE, M. W., AND COUPLES, G. D., 1994, The biothem approach: a useful biostratigraphically-based sequence-stratigraphic procedure: Willi Ziegler-Festschrift I, Sonderdruk aus CFS, Bd.168, p. 281–297.

MILLER, F. X., 1977, The graphic correlation method in biostratigraphy, in Kauffman, E. G. and Hazel, J. E., eds., Concepts and Methods in Biostratigraphy: Stroudsburg, Dowden, Hutchinson and Ross, p. 165–186.

NEAL, J., STEIN, J. A., AND GAMBER, J. H., 1994, Graphic correlation and sequence stratigraphy in the Paleogene of NW Europe: Journal of Micropalaeontology, v. 13, p. 55–80.

SHAW, A. B., 1964, Time in Stratigraphy: New York, McGraw-Hill, 365 p.

SHAW, A. B., 1993, The Origin of Graphic Correlation: Northeastern Section, Geological Society of America, Abstracts with Programs, v. 25, p. 78.

HIGH-RESOLUTION BIOSTRATIGRAPHY IN THE UPPER CAMBRIAN ORE HILL MEMBER OF THE GATESBURG FORMATION, SOUTH-CENTRAL PENNSYLVANIA

JAMES D. LOCH
Department of Earth Sciences, Central Missouri State University, Warrensburg, Missouri 64093
AND
JOHN F. TAYLOR
Geoscience Department, Indiana University of Pennsylvania, Indiana, Pennsylvania, 15705

ABSTRACT: Microstratigraphic sampling of four measured sections in the Upper Cambrian Ore Hill Member of the Gatesburg Formation in south-central Pennsylvania produced a data set adequate for construction of a composite standard superior to the composite range chart previously erected by Wilson (1951) for trilobite faunas of that unit. This new data set revealed the presence of two thin, widespread units not previously reported from the central Appalachians: the *Irvingella major* Subzone at the top of the *Elvina* Zone and the *Parabolinoides* Subzone at the base of the *Taenicephalus* Zone. A new subzone, the *Cliffia lataegenae* Subzone, is proposed to include that portion of the *Elvina* Zone below the base of the *Irvingella major* Subzone. Restriction of several species to the lower third of that subzone and others to the upper two-thirds on the composite standard suggests the possibility of further subdivision of the *Elvinia* Zone in the Ore Hill Member, but additional sampling is needed to confirm the mutually exclusive occurrence of those faunas. Some species of *Pseudosaratogia* and *Conaspis* display preference for specific lithofacies at the base and top of the Ore Hill, respectively, warranting caution in the use of these genera for chronocorrelation.

INTRODUCTION

The Gatesburg Formation in south-central Pennsylvania is a lithologically variable unit deposited near the boundary between nearshore clastics and more offshore platform carbonates in the Late Cambrian (Wilson, 1952). Butts (1918, 1945) subdivided the Gatesburg into four members which record an alternation of primarily clastic and dominantly carbonate deposition. The higher of the two carbonate members, which he designated the Ore Hill Member, contains a diverse Franconian (medial Upper Cambrian) trilobite fauna where it is preserved primarily as limestone; elsewhere, dolomitization of the member has obliterated all calcareous macrofossils. The Ore Hill was deposited at the inner margin of the carbonate platform. The sandstones and sandy dolomites of the Upper Sandy and Lower Sandy Members, which occur immediately above and below the Ore Hill, accumulated at the outer edge of the nearshore clastic zone (Wilson, 1952; Read, 1989).

Wilson (1951) described the Franconian trilobite faunas of the Ore Hill Member in exceptional detail. Although he established correlations with portions of the Conococheague Limestone to the east, Wilson used the biozonal nomenclature developed by Howell (1944) for the Upper Cambrian in nearshore clastic sequences of the Upper Mississippi Valley region to the west. Over the last four decades, additional work in the upper Mississippi Valley region (Berg, 1953, 1954; Bell and others, 1956) and detailed studies of coeval assemblages in carbonate sequences in central and western North America (Grant, 1965; Palmer, 1965; Longacre, 1970; Stitt, 1971, 1977; Kurtz, 1975; Westrop, 1986) greatly refined the trilobite-based biostratigraphy available for temporal correlation (chronocorrelation) of Franconian strata in North America.

Wilson (1951, 1952) detailed the distribution of lithologies and trilobite species through 13 measured sections in south-central Pennsylvania. No individual section exposes the entire Ore Hill Member, and all sections include significant covered intervals. Additionally, the thinly bedded, moderately bioturbated lime mudstones and wackestones that dominate the member generally yield few fossils. Isolated and laterally discontinuous lime packstones and grainstones, which compose a much smaller fraction of the member, provided most of the trilobite collections. As a result, the vertical distribution of a species within a single section typically is only a local range (teilzone of Hedberg, 1976) that does not adequately represent the maximum vertical distribution of the species within the member.

Recognizing this deficiency, Wilson (1951) constructed a composite range chart that illustrated the potential (maximum) range for each species through the Ore Hill Member (Fig. 1). Wilson selected a lithostratigraphic boundary, the base of the Ore Hill Member, as the datum for correlation of his sections, allowing range data from various sections to be combined in the composite section. Unfortunately, this boundary is often difficult to place precisely in the field, and the composite range chart includes species associations that have never been documented in any measured section and are most likely artifacts of minor miscorrelations.

METHODOLOGY

To construct a composite standard that accurately represents the potential ranges of Ore Hill species, we resampled four of Wilson's best exposed and productive sections. Figure 2 shows the locations of the resampled sections. Figures 3–6 show the ranges of species recovered from the four individual sections and the biostratigraphic units recognized in the present study. In each section, bed-by-bed sampling increased the number of productive horizons, expanding the local range zones for most species. Nonetheless, owing to poor exposure and sporadic distribution of productive lithologies, most teilzones remain poorly representative of the full ranges for those species.

These limitations, along with the fine scale resolution sought in constructing our composite standard, strain the limits of the graphic correlation method. Our initial attempt to construct a composite standard through graphic correlation utilized all available tops and bases of species ranges. Transferrance of the range data from the Drab-Beavertown section onto a Potter Creek Crossroads composite section, using a line of correlation (LOC) derived from linear regression, extended the range of species known to occur only in the *Irvingella major* Subzone downward to overlap with the ranges of species restricted to the lower part of the *Elvinia* Zone (herein named the *Cliffia lataegenae* Subzone, see discussion). The suggested association of diagnostic species of the *I. major* Subzone with species of the

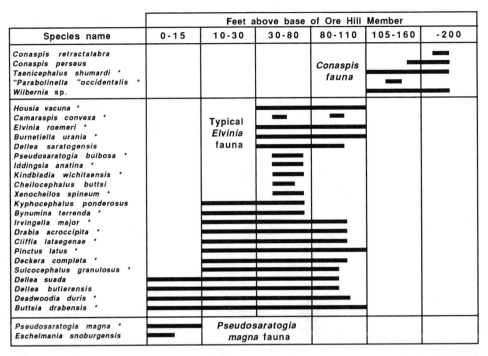

Fig. 1.—Preliminary stratigraphic range data from Wilson (1951, with taxonomic revision) for trilobite species of the Ore Hill Limestone Member of the Gatesburg Formation. Species identified in this study are indicated by an asterisk.

C. lataegenae Subzone: (1) conflicts with well established, mutually exclusive occurrences of those faunas throughout North America (Westrop, 1986; Hohensee and Stitt, 1989), and (2) has not been documented in any individual section of the Ore Hill Member. These results, which we rejected as artifacts of the technique, clearly indicated the need for a different approach.

An alternative approach that yielded good results began with conventional posting of the stratigraphic columns (including species ranges) side-by-side on a bulletin board and selecting a few horizons as plausible "time lines." Most of the lines are based on highest or lowest occurrences of individual species with emphasis upon those that are abundant, easily identifiable, and tightly constrained by productive horizons above and below. Additionally, a one- to three-ft cryptalgal laminite that occurs two to three ft below the lowest occurrence of parabolinoidid trilobites in many sections appears to be a reliable datum within the study area. This interval of supratidal lithofacies is unique within the Ore Hill (Taylor and Loch, 1989; Loch and Taylor, 1993); all other strata within the member display features indicative of intertidal to subtidal deposition. We interpret this laminite as an "event horizon" that represents a relative fall in sea level that affected the area, immediately following (or accompanying) the extinction of the *Elvinia* Zone fauna and preceeding (or accompanying) the appearance of characteristic *Taenicephalus* Zone taxa. Other horizons (e.g., those based on the lowest occurrence of *Taenicephalus shumardi* and the highest occurrence of *Bynumina terrenda*) paralleled the laminite datum, supporting the contention that it is an essentially synchronous surface. The parallelism of these horizons also indicates that sediment accumulation rates were roughly uniform for these three sections. Consequently, we constructed a composite section that combined range data from the Potter Creek, Crossroads, and Drab-Beavertown sections without employing graphic correlation techniques. This composite section served as the standard reference section (SRS) that accommodated data from the Eschelman Quarry section to create a composite standard for trilobite species in the Ore Hill Member (Figs. 7, 8).

The Eschelman Quarry section (Fig. 6), which includes only the lowest 64 ft of the Ore Hill Member, does not include any strata younger than the *Cliffia lataegenae* Subzone. It lacks the cryptalgal laminite, lowest occurrence of parabolinoidids, and base of the range of *Taenicephalus shumardi*. Only the tops and bases of ranges for species of the *C. lataegenae* Subzone were available to establish a line of correlation (LOC) that would allow range data from the Eschelman Quarry section to be incorporated in a composite standard. The first attempt at graphic correlation utilizing 40 data points for 20 species (Fig. 7, Table 1) produced unsatisfactory results. The LOC produced by least squares linear regression (Fig. 7, dashed line) did not effectively divide the tops and bases of species' ranges (below and above the LOC, respectively) as should have occurred given the expansion of ranges already accomplished in assembling the SRS. Additionally, the gentle slope of the LOC (0.51) was inconsistent with fairly constant thicknesses documented for intervals within the Ore Hill Member during construction of the SRS. Reduction of the data set to 18 points (Fig. 7, Table 1, in bold), using the same criteria as mentioned previously for identifying reliable bases and tops (abundant and easily identified species, well-constructed by closely spaced collections), produced much better results. The LOC based on the reduced data set (Fig. 7, solid lines) isolates most tops below and bases above; its slope (0.96) is highly consistent with the observed uniformity of interval thicknesses in the nearby Potter Creek and Crossroads

UPPER CAMBRIAN ORE HILL MEMBER BIOSTRATIGRAPHY

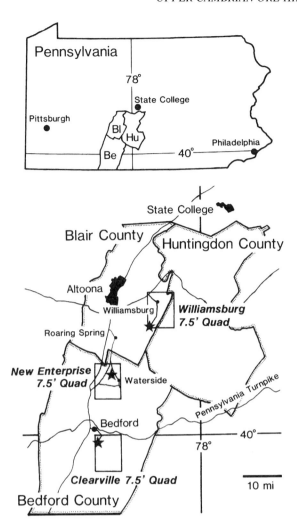

FIG. 2.—Index map for Ore Hill Limestone Member localities examined in this study (see Wilson 1951, 1952, for additional details). The Drab-Beavertown measured section is within the Williamsburg 7.5′ Quadrangle. The Potter Creek, Crossroads, and Eschelman Quarry measured sections are within the New Enterprise 7.5′ Quadrangle. The Imler Quarry measured section is within the Clearville 7.5′ Quadrangle.

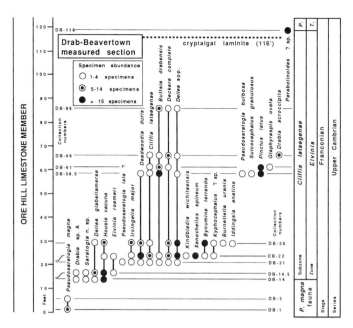

FIG. 3.—Trilobite species range chart for the Drab-Beavertown measured section (Williamsburg 7.5′ Quadrangle). Abbreviations: P.—Parabolinoides, P. magna—Pseudosaratogia magna, T.—Taenicephalus.

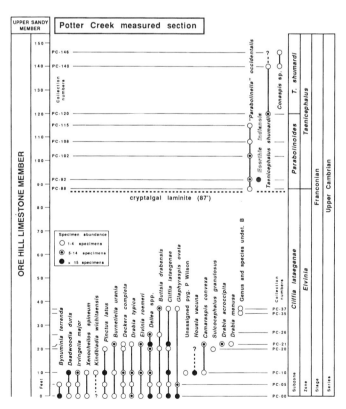

FIG. 4.—Trilobite species range chart for the Potter Creek measured section (New Enterprise 7.5′ Quadrangle). Species name in outline is a brachiopod. Abbreviation: T. shumardi—Taenicephalus shumardi.

sections. Adjustment of the ranges in the SRS to accommodate data from Eschelman Quarry produced a composite standard for trilobite species in the Ore Hill Member (Fig. 8).

DISCUSSION

Although ranges of some species in the revised composite standard are consistent with those provided in Wilson's (1951) composite range chart, most are more restricted. For some species, the shorter stratigraphic range may reflect the smaller number of sections sampled in the present study and deficiencies in those sections (barren and covered intervals). For most species, however, the reduced ranges are more accurate than those provided in Wilson's (1951) composite range chart.

The expanded data set confirms the existence of a distinct *Pseudosaratogia magna* Fauna in the basal 15 ft of the Ore Hill Member below the lowest occurrence of species characteristic of the *Elvinia* Zone. Relict textures discernible on the weathered

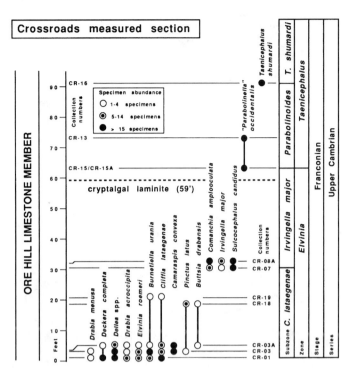

FIG. 5.—Trilobite species range chart for the Crossroads measured section (New Enterprise 7.5′ Quadrangle). Abbreviations: C. lataegenae—Cliffia lataegenae, T. shumardi—Taenicephalus shumardi.

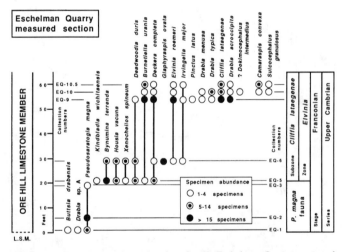

FIG. 6.—Trilobite species range chart for the Eschelman Quarry measured section (New Enterprise 7.5′ Quadrangle). Abbreviation: P. magna—Pseudosaratogia magna.

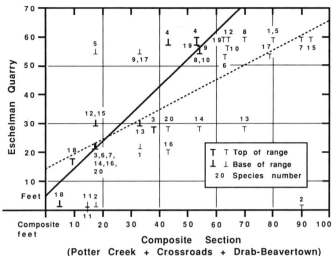

FIG. 7.—Bivariate correlation diagram for the bases and tops of trilobite species ranges from Ore Hill Limestone Member of the Gatesburg Formation. Tops and bases of ranges used in calculation of line of correlation (solid line) are in bold. Dashed line is a least-squares fit of all data points. Species numbers beside the data points refer to the species listed in Table 1.

surfaces of these basal strata, described by Wilson (1951) as "massive crystalline carbonates," revealed that they comprise recrystallized microbialitic reefs and associated inter-reef lime grainstones. We also recovered at least one additional species of *Pseudosaratogia* from non-recrystallized algal reefs discovered in the *Elvinia* Zone in Wilson's Imler Quarry section, suggesting an environmental association for that genus that warrants caution in its use for chronocorrelation. Correlation of this interval is problematic because of its limited fauna and scarcity of coeval cratonic strata elsewhere (Palmer, 1965; Kurtz 1975) attributed to regression that followed deposition of the subjacent *Aphelaspis* Zone (Lochman-Balk, 1971).

The lowest occurrence of several species characteristic of the *Elvinia* Zone, immediately above the highest occurrence of the *Pseudosaratogia magna* Fauna, defines the base of the *Cliffia lataegena* Subzone (defined below). These include *Burnetiella urania*, *Bynumina terrenda*, *Deadwoodia duris*, *Deckera completa*, and *Xenocheilus spineum*. Four very widespread species of the *Elvinia* Zone (*Cliffia lataegenae*, *Elvinia roemeri*, *Irvingella major*, and *Kindbladia wichitaensis*) appear less than a foot higher in the composite standard. The correlative interval in the Canadian Rocky Mountains encompasses Westrop's (1986) *Xenocheilus* cf. *X. spineum* and *Drumaspis occidentalis* Faunas and portions of Pratt's (1992) *Olenaspella evansi* Zone and *Proceratopyge rectispinata* Fauna in more distal settings.

Restriction of several species to the lower or upper part of the *Cliffia lataegenae* Subzone suggests the possibility of finer subdivision of the *Elvinia* Zone in the Ore Hill (dashed line in Fig. 8) *Bynumina terrenda*, *Housia vacuna*, and *Xenocheilus spineum* characterize the lower third of the *C. lataegenae* Subzone. *Camaraspis convexa*, *Drabia acroccipita*, *Drabia menusa*, and *Sulcocephalus granulosus* occur only in the upper two-thirds of the subzone. The stratigraphic distribution of trilobite species through coeval clastic-carbonate transition facies in Texas (Wilson, 1949) and Oklahoma (Stitt, 1971) display a similar pattern of separation, allowing recognition of distinct lower and upper faunas within this subzone. However, any finer subdivision of the *Cliffia lataegenae* Subzone at this time would be premature. Additional sampling particularly in other sections is needed to confirm the restricted distribution indicated for the aforementioned species in the composite standard.

The microstratigraphic sampling conducted in this study also revealed the presence in the crossroads section (Fig. 5) of the *Irvingella major* Subzone of the *Elvinia* Zone, a thin but widespread unit only recently documented in the central Appala-

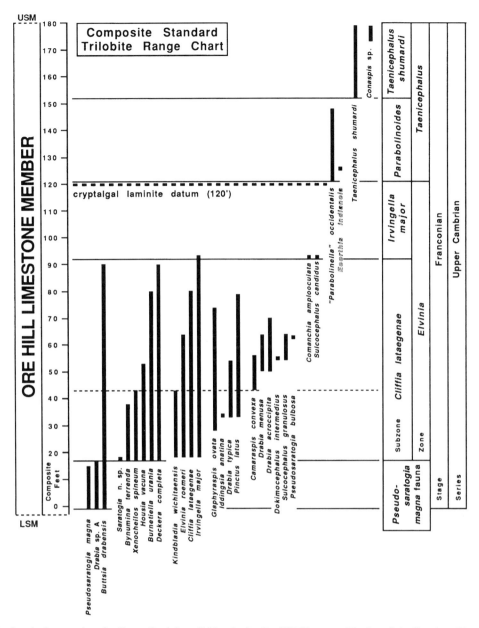

FIG. 8.—Composite Standard range chart for Upper Cambrian trilobites in the Ore Hill Limestone Member of the Gatesburg Formation. Species name in outline is a brachiopod. Horizontal dashed line within the Cliffia lataegenae Subzone separates lower and upper subzone faunas discussed in text.

chians (Taylor and Loch, 1989; Loch and Taylor, 1993). Where present, the base of the *I. major* Subzone defines the top of the *Cliffia lataegenae* Subzone. Two collections from the upper 1.5 ft of an algal reef interval yielded diagnostic species of the I. *major* Subzone (*Comanchia amplooculata* and *Sulcocephalus candidus*) and the highest occurrence of *I. major,* which ranges downward into the *Cliffia lataegenae* Subzone. The thickness of the *I. major* Subzone in the Crossroads section (Fig. 5) and the composite standard (Fig. 8) is greatly exaggerated by assignment of a 25- to 30-ft barren interval (including the cryptalgal laminite datum) to the top of the subzone. In all other known occurrences of the *I. major* Subzone (for summary, see Hohensee and Stitt, 1989), the confirmed thickness of the subzone (excluding barren intervals above the highest diagnostic collections) is six feet or less; similarly, the thickness confirmed for this subzone by productive horizons in the Ore Hill Member is only 1.5 ft. Much or all of the 25- to 30-ft non-productive interval in the Ore Hill Member may be equivalent in age to the lower part of the *Taenicephalus* Zone in areas outside the Appalachians, but the lack of data for the interval precludes assignment to the higher zone. For the same reason (stratigraphic convention of assigning barren intervals to the underlying unit), the *C. lataegenae* Subzone is extended upward to the base of the *Taenicephalus* Zone in sections lacking any evidence of the *I. major* Subzone (Figs. 2–4).

The composite standard (Fig. 8) differs strikingly from that of Wilson (1951) above the *Elvinia* Zone. Wilson assigned the upper half of the Ore Hill Member to the *Conaspis* Zone (Fig.

TABLE 1.—SPECIES RANGE DATA FOR CONSTRUCTION OF COMPOSITE STANDARD TRILOBITE RANGE CHART OF THE ORE HILL LIMESTONE MEMBER OF THE GATESBURG FORMATION.

Species	Potter Creek + Crossroads + Drab-Beavertown Composite section (composite ft)		Eschelman Quarry measured section (feet)	
	base	top	base	top
1 Burnetiella urania (Walcott, 1890)	33	80	21	60
2 Buttsia drabensis Wilson, 1951	17.5	90	1	1
3 Bynumina terrenda Wilson, 1951	**17.5**	**38**	**21**	**29**
4 Camaraspis convexa (Whitfield, 1879)	**43**	**54.5**	**57**	**60**
5 Cliffia lataegenae (Wilson, 1949)	17.5	80	54	60
6 Deadwoodia duris (Walcott, 1916)	17.5	63	21	54
7 Deckera completa Wilson, 1951	17.5	90	21	60
8 Drabia acroccipita Wilson, 1951	54	70	54	60
9 Drabia typica (Hu, 1979)	**33**	**54**	**54**	**57**
10 Drabia menusa Wilson, 1951	54	64	54	57
11 Drabia sp. A	14.5	14.5	1	1
12 Elvinia roemeri (Shumard, 1861)	17.5	64.5	29	60
13 Glaphyraspis ovata Rasetti, 1961	33	70	29	29
14 Housia vacuna (Walcott, 1890)	17	54	21	29
15 Irvingella major Ulrich and Resser, in Walcott, 1924	17.5	93	29	60
16 Kindbladia wichitaensis (Resser, 1942)	17.5	43	21	21
17 Pinctus latus Wilson, 1951	33	79	54	54
18 Pseudosaratogia magna Wilson, 1951	5	**9.5**	**1**	**18**
19 Sulcocephalus granulosus (Wilson, 1951)	53	62.5	57	60
20 Xenocheilos spineum Wilson, 1951	7.5	43	29	29

Data in **bold** type used in formation of line of correlation.

1), resulting in a correlation with faunas in nearshore sandstones of the Upper Mississippi Valley region. Little information on Franconian faunas in carbonate platform facies was available at that time. Subsequent studies in the western United States (Deland and Shaw, 1956; Grant, 1965; Longacre, 1970; Stitt, 1971; among others) replaced the *Conaspis* Zone with the *Taenicephalus* Zone and subdivided the latter, delineating a basal *Parabolinoides* Subzone. Westrop (1986) provided a good summary of the occurrences of the *Parabolinoides* Subzone and assigned the name *Taenicephalus shumardi* Subzone to the remaining, higher portion of the *Taenicephalus* Zone.

In Wilson's (1951) composite range chart (Fig. 1), the range of "*Parabolinella*" *occidentalis* (which requires reassignment to a different, probably parabolinoidid genus) overlaps completely with that of *Taenicephalus shumardi*. This association has not been documented in any Ore Hill section. To the contrary, in all sections that yielded parabolinoidids that are not assigned to *Taenicephalus*, those species occur exclusively below the range of *T. shumardi* (Figs. 3–5) but above the highest occurrences of *Elvinia* Zone species. We assign these strata that contain "*Parabolinella*" *occidentalis* (and other parabolinoidid trilobites) below the lowest occurrence of *T. shumardi*, to the *Parabolinoides* Subzone. The occurrence of brachiopods of the genus *Eoorthis* with "*P.*" *occidentalis* (Fig. 4) supports this assignment. In many areas outside the Appalachians, species of *Eoorthis* are abundant in the *Parabolinoides* Subzone, locally forming coquinas (Berg, 1953, 1954; Longacre, 1970; Kurtz, 1975). The lowest occurrence of *T. shumardi* marks the base of the overlying *Taenicephalus shumardi* Subzone.

Species of *Conaspis* have their lowest occurrence in the highest beds of the Ore Hill Member, some distance above the base of the *Taenicephalus shumardi* Subzone. These highest beds of the member represent a transition to the sandy dolomites and dolomitic sandstones of the overlying Upper Sandy Member of the Gatesburg Formation. Given the abundance of this genus in nearshore clastic facies (Berg, 1953, 1954), the appearance of *Conaspis* at that level may be environmentally controlled and of questionable value for chronocorrelation.

CONCLUSION

Because of numerous barren and covered intervals, faunal data from the Ore Hill Member strain the limits of the graphic correlation method. Nonetheless, with rigorous evaluation of the potential data points to select those that were best constrained by the collection horizons, the method proved adequate even at this level of resolution for construction of a composite standard superior to the composite range chart constructed by Wilson (1951) for trilobite species through the Ore Hill Member. Greater precision in the data from four resampled sections produced finer resolution in the composite standard. Species ranges in the new composite standard clearly define the thin subzones established for the *Elvinia* and *Taeniciphalus* Zones elsewhere in North America, significantly improving the precision possible in correlating Ore Hill sections with one another and with coeval strata outside the Appalachians. New data from additional sampling in other Ore Hill sections is needed, however, to further refine the composite standard and determine the geographic range of its applicability.

ACKNOWLEDGMENTS

We thank C. A. Roebuck for assistance in collecting and processing samples. J. F. T. acknowledges the donors of the Petroleum Research Fund, administered by the American Chemical Society, for partial support of this research. Support was also provided through an IUP Senate Fellowship Grant. We also wish to acknowledge J. L. Wilson for the exceptional quality and detail of his work on the Ore Hill, which laid the foundation of our study.

REFERENCES

BELL, W. C., BERG, R. R., AND NELSON, C. A., 1956, Croixan type area- Upper Mississippi Valley, in Rodgers, J., ed., El sistema Cambrico, Su paleograpfia y el problema de su base, Part 2: Australia, America: Mexico City, 20th International Geological Congress, p. 415–446.

BERG, R. R., 1953, Franconian trilobites from Minnesota and Wisconsin: Journal of Paleontology, v. 27, p. 553–569.

BERG, R. R., 1954, Franconian formation of Minnesota and Wisconsin: Geological Society of America Bulletin, v. 65, p. 857–882.

BUTTS, C., 1918, Geologic section of Blair and Huntingdon counties, central Pennsylvania: American Journal of Science, 4th series, v. 46, p. 523–537.

BUTTS, C., 1945, Holidaysburg-Huntingdon quadrangles: Washington, D.C., United States Geological Survey Atlas Folio 227, 20 p.

DELAND, C. R. AND SHAW, A. B., 1956, Upper Cambrian trilobites from western Wyoming: Journal of Paleontology, v. 30, p. 542–562.

GRANT, R. E., 1965, Faunas and Stratigraphy of the Snowy Range Formation (Upper Cambrian) in southwestern Montana and northwestern Wyoming: New York, Geological Society of America Memoir 96, 171 p.

HEDBERG, H. H., ed., 1976, International Stratigraphic Guide: New York, Wiley, 200 p.

HOHENSEE, S. R. AND STITT, J. H., 1989, Redeposited Elvinia Zone (Upper Cambrian) trilobites from the Collier Shale, Ouachita Mountains, west-central Arkansas: Journal of Paleontology, v. 63, p. 857–879.

HOWELL, B. F., 1944, Correlation of the Cambrian formations of North America: Geological Society of America Bulletin, v. 55, p. 993–1004.

Hu, C. H., 1979, Ontogenetic studies of a few Upper Cambrian trilobites from the Deadwood Formation, South Dakota: Transactions of the Proceedings of the Paleontogical Society of Japan, v. 114, p. 49–63.

Kurtz, V. E., 1975, Franconian (Upper Cambrian) trilobite faunas from the Elvins Group of southeast Missouri: Journal of Paleontology, v. 49, p. 1009–1043.

Loch, J. D. and Taylor, J. F., 1993, The Irvingella major and Parabolinoides Subzones: Ideal time slices for refined paleogeographic and biogeographic reconstructions for medial Upper Cambrian (Franconian) strata in North America: State College, 1993 Society of Economic Paleontologist and Mineralogists Meeting Abstracts with Program, p. 46.

Lochman-Balk, C., 1971, The Cambrian of the craton of the United States, in Holland, C.H., ed., Cambrian of the New World: New York, Wiley, p. 79–167.

Longacre, S. A., 1970, Trilobites of the Upper Cambrian Ptychaspid Biomere, Wilberns Formation, central Texas: Paleontological Society Memoir 4, Journal of Paleontology, v. 44, 70 p.

Palmer, A. R., 1965, Trilobites of the Late Cambrian Pterocephaliid Biomere in the Great Basin, United States: Washington, D.C., United States Geological Survey Professional Paper 493, 105 p.

Pratt, B. R., 1992, Trilobites of the Marjuman and Steptoean stages (Upper Cambrian), Rabbitkettle Formation, southern Mackenzie Mountains, northwest Canada: Toronto, Palaeontographica Canadiana 9, 197 p.

Rasetti, F., 1961, Dresbachian and Franconian trilobites of the Conococheague and Frederick Limestones of the central Appalachians: Journal of Paleontology, v. 35, p. 104–124.

Read, J. F., 1989, Controls on evolution of Cambrian-Ordovician passive margin, U.S. Appalachians, in Crevello, P. D, Wilson, J. L, Sarg, J. F., and Read, J.F., eds., Controls on Carbonate Platform and Basin Development: Tulsa, Society of Economic Paleontologists and Mineralogists Special Publication 44, p. 147–165.

Resser, C. E., 1942, New Upper Cambrian trilobites: Smithsonian Miscellaneous Collections, v. 103, 136 p.

Taylor, J.F. and Loch, J.D., 1989, Upper Cambrian (Franconian) faunas and lithologies across the Pterocephaliid-Ptychaspid Biomere boundary in the Ore Hill Member of the Gatesburg Formation, south-central Pennsylvania: Geological Society of America Abstracts with Programs, v. 21, p. A168.

Shumard, B. F., 1861, The primordial zone of Texas with descriptions of new fossils: American Scientist (2nd series), v. 32, p. 213–221.

Stitt, J. H., 1971, Late Cambrian and earliest Ordovician trilobites: Timbered Hills and Lower Arbuckle Groups, western Arbuckle Mountains, Murray County, Oklahoma: Oklahoma Geological Survey Bulletin 110, 83 p.

Stitt, J. H., 1977, Late Cambrian and earliest Ordovician trilobites, Wichita Mountains area, Oklahoma: Oklahoma Geological Survey Bulletin 124, 79 p.

Walcott, C. D., 1890, Descriptions of new forms of Upper Cambrian fossils: Proceedings of United States National Museum, v. 13, p. 269–279.

Walcott, C. D., 1916, Cambrian trilobites: Smithsonian Miscellaneous Collections Publication 2370, v. 64, p. 157–258.

Walcott, C. D., 1924, Cambrian and Lower Ozarkian trilobites, no. 2 of Cambrian geology and paleontology V: Smithsonian Miscellaneous Collections Publication 2788, v. 75, p. 53–60.

Westrop, S. R., 1986, Trilobites of the Upper Cambrian Sunwaptan Stage, southern Canadian Rocky Mountains, Alberta: Toronto, Palaeontographica Canadiana 3, 179 p.

Whitfield, R. P., 1880, Description of new species of fossils from the Paleozoic Formations of Wisconsin: Madison, Wisconsin Geological Survey, 1879 Annual Report, p. 44–71.

Wilson, J. L., 1949, The trilobite fauna of the Elvinia Zone in the basal Wilberns Limestone of Texas: Journal of Paleontology, v. 23, p. 25–44.

Wilson, J. L., 1951, Franconian trilobites of the central Appalachians: Journal of Paleontology, v. 25, p. 617–654.

Wilson, J. L., 1952, Upper Cambrian stratigraphy in the central Appalachians: Geological Society of America Bulletin, v. 63, p. 275–322.

GRAPHIC ASSEMBLY OF A CONODONT-BASED COMPOSITE STANDARD FOR THE ORDOVICIAN SYSTEM OF NORTH AMERICA

WALTER C. SWEET
Department of Geological Sciences, The Ohio State University, Columbus, Ohio 43210

ABSTRACT: A composite standard (CS), assembled graphically through consideration of the ranges of more than 300 conodont species in measured sections at 127 localities, will apparently be adequate as the backbone for a conodont-based chronostratigraphic framework for the Ordovician System of North America. The SRS is a 374-m core through Middle and Upper Ordovician rocks drilled at a site in north-central Kentucky. Relations between the SRS and additional Ordovician sections have been determined graphically following a compilation strategy that involves extension of the network of correlated sections into older and younger rocks by use of overlapping control sections. The weakest link is currently between Ibexian and lower Whiterockian rocks and the well-controlled upper Whiterockian-Mohawkian-Cincinnatian part of the CS. An undescribed composite section through Whiterockian and lower Mohawkian rocks in east-central Nevada is cited as a promising bridge between these two parts of the Ordovician System. It is not certain if the CS extends to the top of the system, because it is not yet possible to add described North American sections through the Ordovician-Silurian boundary to the network of correlated sections anchored by the SRS described here.

INTRODUCTION

Ordovician rocks are the most widely distributed Paleozoic strata in North America. In the aggregate, these rocks are thick, developed in many different sedimentary facies, and have yielded fossils that represent thousands of species. Trilobites, graptolites, and conodonts have been used to work out zonal biostratigraphies that are useful in parts of the system, in certain regions of North America, or in limited facies belts. However, there is at present no single biostratigraphic framework that applies to all the Ordovician rocks of North America. It is an objective of this report to show how sections related to one another graphically through use of conodont ranges might ultimately provide the framework for a pan-Ordovician chronostratigraphic framework.

For several reasons, conodonts are especially suitable for use as the primary frame-builders for an Ordovician biostratigraphy compiled graphically. They are abundant fossils in most of the marine Ordovician strata that have been sampled in North America, and many of the species recognized have both a limited stratigraphic range and a wide geographic distribution. In addition, conodonts are heavy, acid-resistant microfossils that may be collected in some abundance from almost any type of marine sedimentary rock that can be disaggregated in the laboratory. Because conodonts can not be seen in the field, those who seek them commonly collect rock samples at regularly spaced intervals through stratigraphic sections. If those sections have been measured competently and the positions of samples recorded accurately, it is possible to obtain information on the stratigraphic range of conodonts in those sections that is far more precise than is attainable for most megafossil groups. For these reasons, conodonts are exceptionally well suited for use in graphic correlation, which requires scaled range data for all the species represented in all the sections compared and ultimately assembled into a composite standard.

ORDOVICIAN CHRONOSTRATIGRAPHIC UNITS

Ross and others (1982) divided the Ordovician System in North America into four series (Fig. 1). The Ibexian Series, which includes the oldest Ordovician rocks in North America, is based on a thick succession of fossiliferous rocks in the Ibex area of western Millard County, Utah (Hintze, 1951, 1973; Ross and others, 1982; Ross and others, 1993). Conodonts have been described from strata now included in the typical Ibexian Series by Miller (1969), Ethington and Clark (1982), and Hintze and others (1988). The base of the Ibexian Series is established formally by Ross and others (1993). The top is defined by the base of the superjacent Whiterockian Series.

The Whiterockian Series has only a basal-boundary stratotype (Ross and Ethington, 1991), which is pegged at the contact between the Ninemile Shale and the Antelope Valley Limestone in a section in Whiterock Canyon, northern Monitor Range, Nevada. In the stratotype section, the base of the Whiterockian Series is also the base of the *Tripodus laevis* conodont zone and the base of the *Paralenorthis-Orthidiella* brachiopod zone (= Zone L of Ross, 1951). The top of the Whiterockian Series is defined, formally, as the base of the overlying Mohawkian Series. Ross and Ethington (1991) chart ranges of conodonts in the section in Whiterock Canyon that includes the basal-boundary stratotype and in several other sections established as secondary references.

The Mohawkian and Cincinnatian Series have their typical development and their type sections in eastern North America. Like the Whiterockian Series below, the Mohawkian Series has no stratotype, but Ross and others (1982) suggest informally that its base be drawn at the level of first occurrence of the conodont *Baltoniodus gerdae* at a locality in the Hogskin Valley, Tennessee.

The Cincinnatian Series has its area of typical development in southwestern Ohio and adjacent parts of Kentucky and Indiana (Fig. 2). The three major divisions of the typical Cincinnatian Series, now the Edenian, Maysvillian, and Richmondian Stages, were originally described as lithic units (Orton, 1873), and their boundaries have never been defined biostratigraphically. Conodont-based correlations summarized by Sweet and Bergström (1971), however, served to fix the base of the Edenian Stage, and thus the base of the Cincinnatian Series, relative to the Trenton Group of northern New York, which includes the youngest rocks of the Mohawkian Series.

For the purposes at hand, it is convenient to divide the Ordovician System of North America into just three parts: Ibexian, Whiterockian, and Mohawkian-Cincinnatian Series. The latter part is an internally coherent package that includes most of the Ordovician component of Sloss's (1963) Tippecanoe Sequence and thus is the part of the Ordovician System that is most widespread in North America. Were it not for tradition, there would be little justification for dividing this part of the Ordovician System into two series.

CINCINNATIAN Series	Gamachian Richmondian Maysvillian Edenian
MOHAWKIAN Series	Shermanian Kirkfieldian Rocklandian Blackriveran
WHITEROCKIAN Series	Upper (Chazyan?) Lower
IBEXIAN Series	4 stages, not yet formally defined

FIG. 1.—Series and stages of the Ordovician System in North America.

STANDARD REFERENCE SECTION

Nature and Function of a Standard Reference Section

In building a network of sections correlated with one another graphically, one section of the network is selected as the *standard reference section* (SRS), and data from other sections of the network are translated into terms of the SRS through use of the equation for the *line of correlation* (LOC) (Shaw, 1964). The SRS is ideally the longest section of the network, the one that has been most thoroughly studied, and the one that has yielded the greatest amount of information on the ranges of fossil species represented in it. In short, the SRS is the backbone of a composite standard (CS) developed from the sections correlated with the SRS graphically.

The long-term aim of a program begun a decade ago (Sweet, 1984) and outlined in this report is to develop on a single framework a network of closely correlated sections spanning the entire Ordovician System that might ultimately be useful as the basis for a high-resolution chronostratigraphy. Because I have chosen to use the ranges of conodont species in compiling components of the network graphically, the SRS must also be a section that yields fossils of that group in closely spaced samples. Unfortunately, I am not aware of a single, long, uniformly conodont-bearing Ordovician section that meets these requirements. Hence I report here on the possibility of extending the CS into parts of the Ordovician System both older and younger than the SRS through consideration of three major sections, which overlap the SRS and are readily related to it. These sections form the backbone of a pan-Ordovician chronostratigraphic system developed graphically.

A SRS in the Cincinnati Region

A decade ago (Sweet, 1984) I summarized an attempt to develop graphically a high-resolution biostratigraphic framework

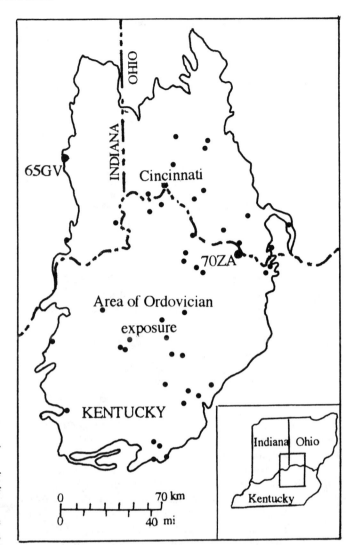

FIG. 2.—The Cincinnati Region of Kentucky, Ohio, and Indiana, showing location of section 70ZA, the SRS, and section 65GV, whose relationship to 70ZA is shown graphically in Figure 3. Unlabeled dots mark the locations of other Cincinnati Region sections that are now parts of the Ordovician CS.

for the North American Mohawkian and Cincinnatian Series. That framework was established by assembling a composite standard (CS) from data on the scaled ranges of more than 100 conodont species in stratigraphic sections in 18 sectors of the North American midcontinent and stated in CSUs based on a SRS in the Cincinnati Region of Ohio, Kentucky, and Indiana (Fig. 2). Explaining how the 1984 CS might be extended to form a pan-Ordovician biostratigraphic framework is the principal function of the report at hand.

No single one of the 40 Cincinnati Region sections for which I have information (Fig. 2) spans the entire Mohawkian and Cincinnatian Series. However, one of them, identified as section 70ZA in Figure 2 and in the Locality Register in Sweet (1979a, p. G22-G25), is a continuous core that reaches upward from a level in the upper Whiterockian Series to one in the Maysvillian Stage (Cincinnatian Series) that is about 104 m (about 340 ft) below the projected top of the Ordovician System. This core was drilled by Cominco American, Inc. in July 1970 on the T.

TABLE 1

Cat. #	Species	70ZA	65GV	70ZA of 65GV	CS1
2	*Amorphognathus superbus*	223-334	51-165	222-329	222-334
3	*A. tvaerensis*	183-220	9.0-48	182-219	182-220
5	*Aphelognathus kimmswickensis*	173	5	179	173-179
7	*A. politus*	246-369	188-301	351-459	246-459
9	*Belodina compressa*	173-181	6.0-8.0	180-182	173-182
10	*B. confluens*	267-329	124-151	291-316	267-329
11	*Yaoxianognathus abruptus*	175-314	20-24	193-197	175-314
12	*Bryantodina staufferi*	183-256	9-113	182-280	182-280
13	*Drepanoistodus suberectus*	27-369	7-301	181-457	27-457
15	*Icriodella superba*	183-398	9-185	182-348	182-398
18	*Oulodus subundulatus*	206-296	26	198	198-296
20	*O. velicuspis*	298-369	136-182	302-345	298-369
22	*Panderodus gracilis*	36-303	0-276	174-433	36-433
25	*Dapsilodus mutatus*	173-337	8-122	182-289	173-337
26	*Periodon grandis*	184-291	11-125	184-292	184-292
27	*Phragmodus undatus*	173-369	6-290	180-447	173-447
28	*Plectodina aculeata*	7-179	6.0-11	181-184	7-184
29	*P. tenuis*	181-369	15-299	188-455	181-455
30	*Polyplacognathus ramosus*	178-224	8.0-9.0	182-183	178-224
32	*Protopanderodus liripipus*	200-335	15-226	188-387	188-387
35	*Rhodesognathus elegans*	181-363	9-202	182-364	181-364
36	*Staufferella falcata*	200-352	156	321	200-352

W. Richardson farm about 0.35 mi east of Minerva in Mason County, Kentucky. It is stored permanently in the Micropaleontological Laboratory of the Department of Geological Sciences at The Ohio State University. The upper 203 m (665 feet) of core 70ZA, which represent the Lexington Limestone, the Point Pleasant Formation, the Kope Formation, and the Fairview Formation, were described in detail by Sweet and others (1974), who also sampled this part of the core for conodonts.

From 203 m to its base (at 370 m), core 70ZA penetrated carbonate rocks identifiable with units of the High Bridge Group. This segment was sampled and processed for conodonts by Robert B. Votaw (now of Indiana University Northwest, Gary, IN), whose identifications are tabulated sample-by-sample in his unpublished dissertation (Votaw, 1971).

BUILDING AN ORDOVICIAN CS

Mohawkian-Cincinnatian Series

Table 1 gives the ranges of conodont species common to core 70ZA, the SRS, and a second core, 65GV, which was drilled in 1964 and is about 75 miles northwest of 70ZA in southeastern Indiana (Fig. 2). In Figure 3, ranges from Table 1 are used graphically to determine the relationship between the portions of the two cores that overlap stratigraphically. Although it is clear from Figure 3 that the base of core 65GV is 174 m above the base of core 70ZA, it is not indicated as clearly that the top of core 65GV, which is the base of the Silurian System in southeast Indiana, is about 105 m above the top of core 70ZA. Thus,

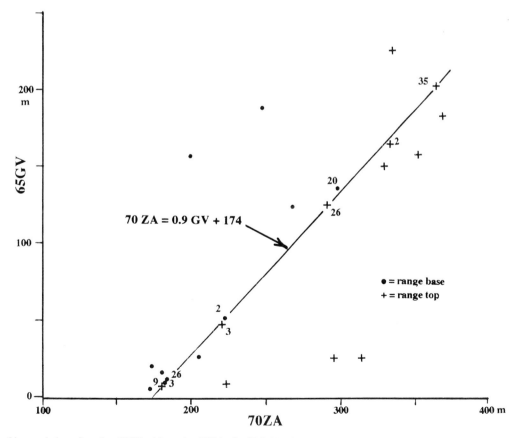

FIG. 3.—Graphic correlation of section 65GV with section 70ZA, the SRS. Numbers next to dots and crosses identify conodont species named in Table 1. Unnumbered dots and crosses indicate range limits not considered in fitting the LOC.

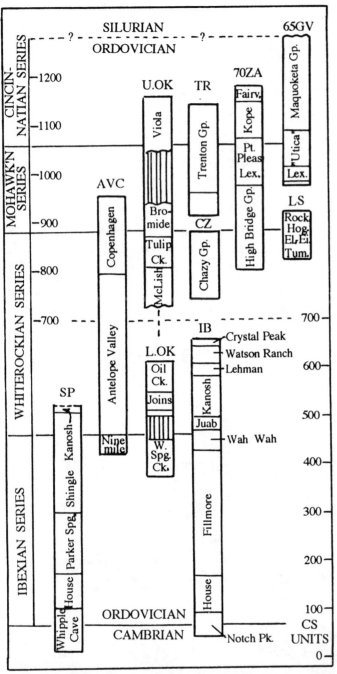

FIG. 5.—Location of the L. OK77 and U. OK77 sections, which are continuous with one another in roadcuts along U. S. Highway 77 and U. S. Interstate Highway 35 as they cross Arbuckle Anticline in south-central Oklahoma. Rows of dots indicate line of section.

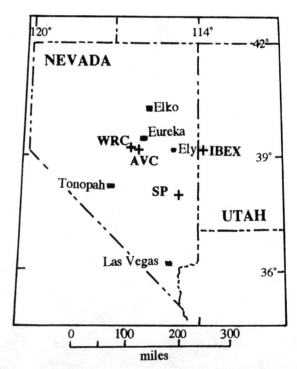

FIG. 4.—Relationship between sections that are important components of the Ordovician CS. Thicknesses of all sections are in CSUs of 70ZA, the SRS, as determined from appropriate LOC equations. **SP**—Shingle Pass section, Nevada; **AVC**—Antelope Valley composite section, Nevada; **L. OK**—pre-McLish portion of section along U.S.hwys. 35 and 77, south-central Oklahoma; **U. OK**—McLish and younger Ordovician strata in section along U.S. Hwys. 35 and 77, south-central Oklahoma; **IB**—Ibex-area composite section, Millard County, Utah; **CZ**—type section of Chazy Group, Lake Champlain Valley, New York; **TR**—composite section of Black River and Trenton Groups, northern New York (Unlabeled space beneath Trenton Group represents Pamelia and younger Black Riveran strata); **70ZA**—Minerva, Kentucky, core, the SRS (Lex. = Lexington Ls., Pt. Pleas. = Point Pleasant Fm., Fairv. = Fairview Fm.); **LS**—Lay School section, Hogskin Valley, Tennessee (Tum. = Tumbez Fm., El.-Ei. = Elway-Eidson Fm., Hog. = Hogskin Fm., Rock. = Rockdell Fm.); **65GV**—New Point core, New Point, Indiana (Lex. = Lexington Ls.).

FIG. 6.—Location of the Whiterock Canyon (WRC), Antelope Valley Composite section (AVC), and Shingle Pass (SP) sections in Nevada, and the Ibex-area composite section (IBEX) in west-central Utah.

core 65GV, related reliably by graphic means to the SRS (core 70ZA), may be used to extend the Ordovician CS to the top of the Cincinnatian Series. A combination of cores 70ZA and 65GV indicates that a CS 475 CSUs thick will be adequate for

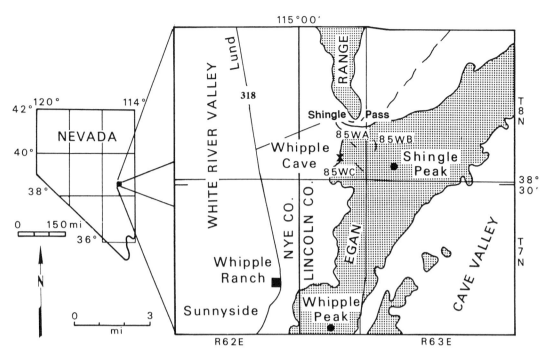

FIG. 7.—Location of the Shingle Pass section in the southern Egan Range, Lincoln County, Nevada. The section is continuous but was measured and sampled in three subsections (85WA, 85WB, and 85WC) offset from one another along strike.

relating sections within the Mohawkian-Cincinnatian segment of the Ordovician System of North America. However, some of the sections to be added have lower segments that extend well below the base of the Mohawkian Series. Thus, in order ultimately to accommodate those sections without introducing negative ranges, I have added 800 meters to the base of core 70ZA, the SRS.

After selecting core 70ZA as SRS for the Mohawkian-Cincinnatian segment of the Ordovician System in North America, and supplementing it through graphic integration of information from core 65GV, I composed the first of several composite standards (CS1 in Table 1). This synthetic section, stated in terms of core 70ZA, was then used graphically to add and evaluate distributional information from 38 additional sections in the Cincinnati Region (Fig. 2; Sweet, 1979a) and from 68 Mohawkian-Cincinnatian sections outside the Cincinnati Region. Graphs depicting the relationships of many of these sections are included in my 1984 report. Others are to be found in reports by Amsden and Sweet (1983), Bauer (1987), and Sweet (1984, 1987, 1992). The positions of five of these sections (L. OK77, U. OK77, CZ, TR, LS) are shown schematically in Figure 4, which also charts the relationship between 70ZA, the SRS, and 65GV, one of the major sections correlated with it in the Cincinnati Region.

Perhaps the most significant section integrated graphically into the Mohawkian-Cincinnatian segment of the Ordovician framework described in my 1984 report is the thick succession of Ordovician rocks that is well exposed in roadcuts along U. S. Interstate Highway 35 and parallel U. S. Highway 77 where those highways cross the Arbuckle Anticline in south-central Oklahoma (Fig. 5) (Derby and others, 1991). The upper part of this succession, from the base of the McLish Formation to the top of the Viola Group (U. OK77 of Fig. 4) has been integrated graphically with other Mohawkian-Cincinnatian sections by Dresbach (1983), Sweet (1984), and Bauer (1987). As indicated in Figure 4, this integration provides the basis for extending the Mohawkian-Cincinnatian CS downward into what is, by definition, the upper part of the Whiterockian Series. However, as is also indicated in Figure 4, McLish and younger strata of the U. OK77 section are separated from older Ordovician rocks by a major unconformity, which at this locality not only separates the Sauk and Tippecanoe Sequences of Sloss (1963), but also separates strata with profoundly different conodont faunas. Bridging this gap is a major problem in graphic development of a pan-Ordovician chronostratigraphy. Solution of the problem requires that we first consider integration of information from Ibexian rocks.

Ibexian Series

The Ibexian Series has its typical development in the Ibex area of Millard County, Utah (Hintze, 1973; Ethington and Clark, 1982; Ross and others, 1982; Ross and others, 1993). Published information on the ranges of conodonts and other fossils in typical Ibexian strata is from sections of varying length and coverage at 11 localities in the Ibex area (Fig. 6). Although carefully measured and adequately described, none of these sections spans the entire Ibexian Series, hence previous biostratigraphic discussions (e.g., Ethington and Clark, 1982; Ross and others, 1993) refer to synthetic sections compiled from some or all of the 11 principal sections in the Ibex area.

As the principal representative of the Ibexian Series in the graphically compiled network of Ordovician sections, I have

TABLE 2

Sp. #	Species	IBEX	SHINGLE PASS	IB+SP	CS
2	Acanthodus uncinatus	227	270-288	270-288	147-156
3	Acodus emanuelensis	417-866	499-923	463-923	291-465
10	Cambrooistodus cambricus	20-80	69-123	69-126	49-76
14	Clavohamulus bulbousus	110-138	180-187	156-187	91-106
19	Clavohamulus n. sp.	452-458	499	498-504	258-261
20	Cordylodus angulatus	185-266	220-312	220-312	122-167
25	Cordylodus proavus	81-150	123-208	123-208	75-117
29	Diaphorodus deltatus	335-822	364-873	364-873	193-441
32	Drepanodus arcuatus	385-859	434-999	430-999	225-503
41	Fryxellodontus inornatus	84-112	140-158	130-158	78-92
43	Glyptoconus quadraplicatus	293-823	336-853	336-869	179-439
47	Histiodella altifrons	960-990	1006-1082	1006-1082	505-543
53	Juanognathus variabilis	751-900	684-1006	684-1006	349-506
54	Jumudontus gananda	695-923	773-1006	740-1006	376-506
55	Loxodus bransoni	182-264	257-306	226-310	125-166
56	Microzarkodina flabellum	903-928	957-974	949-974	478-490
60	Oepikodus communis	614-908	672-959	659-959	337-483
62	O. sp. cf. O. minutus	835-985	896-1006	880-1030	444-518
63	Oistodus multicorrugatus	617-1156	853-1082	662-1200	338-601
66	"O." triangularis s.f.	159-276	202-306	202-322	114-172
73	Parapanderodus emarginatus	571-820	618-1082	617-1082	316-543
78	Paroistodus parallelus	450-820	485-974	485-866	252-438
79	Phakelodus tenuis	20	66	66	47
87	Protopanderodus elongatus	770-823	794-891	794-891	402-450
89	Protopanderodus gradatus	689-1161	739-1082	735-1207	374-604
90	P.? leei	328-414	336-455	336-460	179-239
91	P. leonardii	689-1000	660-1059	660-1059	337-522
93	Protoprioniodus aranda	828-940	896-999	873-999	441-503
97	Pteracontiodus cryptodens	948-1092	965-1044	965-1137	486-570
98	Reutterodus andinus	700-900	773-923	746-946	379-477
99	R.? borealis	605-644	648-660	648-690	331-352
100	Rossodus manitouensis	177-276	227-312	223-322	124-172
106	Scolopodus multicostatus	547-663	593-618	593-709	304-361
107	S. sp. aff. S. rex	304-602	318-654	318-654	165-334
108	"S." sulcatus	177-501	220-556	220-556	122-286
112	Semiacontiodus lavadamensis	133-142	188	179-188	44-107
116	Tripodus laevis	887-942	928-1082	928-1082	458-543
117	Tropodus comptus	438-859	485-891	484-905	251-457
119	Ulrichodina wisconsinensis	440-694	485-709	485-740	252-376

chosen a long, continuous, readily accessible exposure of the Pogonip Group in the southern Egan Range in Lincoln County, Nevada (Fig. 7). This section, identified as the Shingle Pass section (Sweet and Greene, 1992), is exposed in and on the flanks of a broad, northwest-projecting salient of the southern Egan Range just south of the unpaved, unnumbered road that extends through Shingle Pass and connects Nevada Highway 318 with U. S. Highway 93, some 20 miles to the east.

The Shingle Pass section begins with a sequence of thick-bedded, stromatolitic lime-grainstones and wackestones that have been somewhat altered hydrothermally and represent the upper part of the Whipple Cave Formation, which is mostly Late Cambrian in age but includes 60 m or so of Early Ordovician strata at the top. The distribution of conodonts in the upper Whipple Cave and lowermost Pogonip Group in Sawmill Canyon, 15 to 18 miles north of Shingle Pass, is established in a recent report (Taylor and others, 1989), hence has not been duplicated in our work at Shingle Pass.

Pogonip Group carbonates above the Whipple Cave Formation in the Shingle Pass section are divided into three formations (Kellogg, 1963): the House, Parker Spring, and Shingle (Fig. 4). These were sampled at approximately 6-m intervals to a level low in the Kanosh Formation, which succeeds the Shingle Formation stratigraphically. All 119 samples produced conodonts, which represent more than 120 species. A majority of these species are also represented in one or several of the 11 sections that make up the Ibexian Series in its type area, and their ranges are now controlled in a single section.

Kellogg (1963) outlines Pogonip stratigraphy in the Shingle Pass area. Wilson (1988) provides more specific information on the section and includes descriptions and illustrations of some of the more significant conodont species. Celeste M. Greene and I are preparing a comprehensive description of the Shingle Pass section and the conodonts derived from it. The ranges of 39 of the 89 species common to the Shingle Pass and Ibex sections are given in Table 2.

Following descriptions and measurements given by Hintze (1951, 1973) and Ethington and Clark (1982), I compiled a composite section for the Ibex area, and in Table 2, I list the ranges of Ibex-area conodont species that are also represented in the Shingle Pass section. In Figure 8, ranges of conodont species common to the Ibex-area composite section and the Shingle Pass section are plotted in a biaxial graph, which serves visually to indicate a very close similarity between the Ibex-area and Shingle Pass sequences. Ross and others (1993) regard the Ibex-area composite section as the Ibexian stratotype, and, integrated with the Shingle Pass section, it is very useful because it extends well above the latter stratigraphically and provides a useful bridge between the Shingle Pass section and superjacent Whiterockian and younger sections. Furthermore, the basal-boundary stratotype of the Ibex Series has recently (Ross and others, 1993) been sited in the lower part of the Ibex-area composite section, and this makes it possible to show the base of the Ordovician System at the level indicated in Figures 4 and 8.

The Whiterockian Series

It is not difficult to determine relations between sections within the Mohawkian-Cincinnatian segment of the Ordovician System of North America and between major sequences of Ibexian strata. But building a bridge between these two segments from published information continues to be a problem. Part of the reason for this is that the intervening Whiterockian Series includes strata that represent not only the latest phases in the Sauk regression but also the earliest parts of the Tippecanoe transgression. Thus, throughout much of the continental interior, strata of early Whiterockian age are missing. Transgressive later phases overlie Ibexian or older rocks and are developed in facies that contain few conodonts or other fossils. Only along the former continental margins are there unbroken Whiterockian successions that are continuous with Ibexian rocks below and Mohawkian-Cincinnatian strata above.

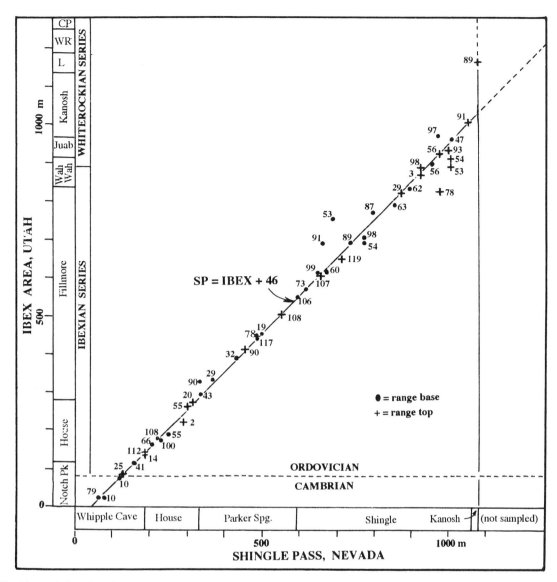

FIG. 8.—Graphic correlation of the Ibex-area composite section with the Shingle Pass section. The dashed portion of the LOC (equation, SP = IBEX + 46) is not controlled by collections in the Shingle Pass section; it is merely a continuation upward of the solid, well-controlled portion. Numbers by dots and crosses identify species whose names and ranges are given in Table 2. In Ibex-area column, L = Lehman Formation; WR = Watson Ranch Formation; CP = Crystal Peak Formation.

As indicated in Table 2 and in Figures 4 and 8, graphic integration of information from the Ibex-area and Shingle Pass sections results in a local composite section whose upper segment represents much or all of the lower half of the Whiterockian Series, and, as noted in an earlier discussion and shown in Figure 4, upper Whiterockian (Chazyan) rocks in south-central Oklahoma from the base of the McLish Formation to a level in the Bromide Formation are readily integrated into the Mohawkian-Cincinnatian CS (Sweet, 1984; Amsden and Sweet, 1983; Bauer, 1987). However, the local Ibex-area-Shingle Pass composite (IB + SP in Table 2) is overlain by the Eureka Quartzite, and the fossiliferous portion of the McLish Formation is separated from underlying strata by a thick sand. The latter represents the initial phase of the Tippecanoe transgression and unconformably overlies Oil Creek strata characterized by conodonts almost completely different than those from the McLish Formation.

To my knowledge, the most promising bridge between the Ibexian and Mohawkian-Cincinnatian segments of the Ordovician CS is an undescribed composite section in central Nevada made up of a complete section of the Whiterockian Antelope Valley Limestone and a nearly complete section of the Whiterockian-Mohawkian Copenhagen Formation. The Antelope Valley portion of this composite section, 342 m thick, is exposed in Martin Ridge in the center of the SW1/4, sec. 6, T. 15 N., R. 50 E., Eureka County, Nevada (Horse Heaven Mtn. 15-min. quadrangle). The section was measured, described, and sampled at approximately 10-m intervals in August, 1975 by Dr. Anita G. Harris of the U. S. Geological Survey, who has permitted me on several occasions to study and log the distri-

TABLE 3

Sp.#	Species	Lower AVC	Upper AVC	Lower OK77	SP+IB	MC-CS	CS
3	*Amorphognathus tvaerensis*		532			960-1025	952-1025
18	*Baltoniodus gerdae*		458-472			876-899	876-899
23	*Belodina monitorensis*		376-505			773-930	773-930
30	*Cahabagnathus friendsvillensis*		383-384			737-849	737-849
31	*Cahabagnathus sweeti*		394-411			823-885	814-885
33	*Chosonodina rigbyi*	318-376		453-520	1177-1224		576-796
41	*Dischidognathus primus*	377		520-532	1197-1209		599-797
41A	*Drepanodus arcuatus*	8.0-43		0-13	430-999		225-503
45	*Drepanoistodus angulensis*			272-557	995-1224	742	501-742
47	*Eoplacognathus elongatus*		419-438			834-893	834-893
53	*Eucharodus parallelus*	8		0-231	208-984		117-495
55	*Histiodella altifrons*	47-70		275-290	1005-1082		467-543
56	*Histiodella minutiserrata*	179-201		281-342	1029-1082		517-621
57	*Histiodella serrata*	198-376		344-557			540-796
58	*Histiodella sinuosa*	179-246		313-388	1062-1203		530-666
60A	*Jumudontus gananda*	42-74			740-1006		376-506
61	*Leptochirognathus quadratus*		318-363			727-842	727-842
66	*Neomultioistodus compressus*	186-378		272-557	984-1143	742	495-798
67	*Oepikodus communis*	0-43		0-231	659-959		337-483
68	*O. sp. cf. O. minutus*			188-256	880-1030		441-518
69	*Oistodus multicorrugatus*	7-365		0-557	662-1200		785
89	*Parapanderodus asymmetricus*	42-231		6-188	739-1082		376-651
90	*Parapanderodus striatus*	7-231		0-554	353-1209		187-651
92	*Paraprioniodus n. sp.*	272-376		465-557	1174-1231	742	580-796
95	*Phragmodus flexuosus s.l.*		370-442			749-892	749-892
97	*Phragmodus inflexus*		405-513			824-962	824-962
101A	*Plectodina joachimensis*		423-428			822-856	822-856
107	*Polyplacognathus ramosus*		532			942-1024	942-1024
112	*Protopanderodus gradatus*	0-77		188-557	735-1207		376-611
115	*Protopanderodus varicostatus*		378-438			750-859	750-859
116	*Protoprioniodus aranda*	42-74		188-231	873-999		441-503
117	*Fahraeusodus marathonensis*	0-196		275-557	618-1072		317-616
131	*Pteracontiodus cryptodens*	53-318		272-456	965-1137	742	473-742
132	*Pteracontiodus gracilis*			496-532	1181-1203		591-602
132A	*Pygodus anserinus*		410-429			828-859	828-859
132B	*Pygodus serrus*		379-384			762-828	762-828
140A	*"Scandodus" robustus*	46-81			904-968		456-499
141	*"Scandodus" sinuosus*	266-367		270-557	938-1231	742	473-789
151	*Thrincodus palaris*		378-429			805-851	798-885
152A	*Tripodus laevis*	46-109			928-1082		466-543

bution of the conodonts she collected from it. The upper part of the composite section is a very closely sampled section through the lower 120 m of the Copenhagen Formation, which overlies the Antelope Valley Limestone and is described in an unpublished Bachelor's thesis by Alice W. Spencer (1984), who identified and logged the distribution of conodonts in the section. The section is in the SW1/4, sec. 24, T. 15 N, R. 50 E, Nye County, Nevada (Horse Heaven Mtn. 15-min quadrangle). The composite section is supplemented at the base by information from Ross and Ethington (1991) on conodont distribution in the Ninemile Shale and lowermost Antelope Valley Limestone in the section in Whiterock Canyon that includes the basal-boundary stratotype of the Whiterockian Series and at the top by information published by Harris and others (1979) on conodont distribution in the uppermost Copenhagen Formation in the same section studied by Spencer. Ethington and Schumacher (1969) provide additional information on Copenhagen conodonts, but their specimens were not collected from the same section studied by Spencer (1984). Table 3 gives ranges of conodonts in the lower and upper parts of the 'Antelope Valley Composite Section' (AVC) as I have assembled it.

Neither of the two major components of the Antelope Valley composite section has been described in the literature, hence the section can not yet be established firmly as the one needed to bridge the gap between the Ibexian and Mohawkian-Cincinnatian segments of the Ordovician CS. However, the potential value of this composite section as a bridge is illustrated in Figures 9 and 10.

In Figure 9, I have plotted ranges from the upper 232 m of the Antelope Valley composite section against ranges of the same species in a CS that extends down to the base of the uppermost Whiterockian Chazyan Stage. The data I used are given in the columns headed "Upper AVC" and "MC-CS" in Table 3. At this stage in development of the Ordovician CS, there were no data against which to plot ranges in the lower 300 m of the Antelope Valley composite section.

In the graph of Figure 9, ten crosses, representing last occurrences, plot below the proposed LOC, and nine dots, representing first occurrences, plot above the LOC. Twelve dots and crosses plot on or very near the interface between first-occurrence and last-occurrence sectors of the graph, and it is to this array of well-controlled points that I have fitted the LOC, whose equation suggests a rock-accumulation rate for upper Antelope Valley and Copenhagen strata essentially the same as that in the Cincinnati Region.

Using the LOC of Figure 9 (CS = UAVC + 422), I added data from the upper part of the Antelope Valley composite section to a CS developed largely from sections through Mohawkian and Cincinnatian rocks but including data from the late Whiterockian, post-Oil Creek part of the magnificent section of Ordovician strata along Highways 35 and 77 in south-central Oklahoma (U. OK77 of Fig. 4). On the assumption that there are no major discontinuities in the Antelope Valley composite section, this procedure had the effect of adding to the Ordovician CS a segment extending downward to (and slightly below) the base of the Whiterockian Series.

Data added to the CS from graphic correlation of the upper part of the Antelope Valley composite section now enable integration into the CS of information from the lower part of the southern Oklahoma Ordovician succession (Table 3, column headed Lower OK77) and from the composite section developed graphically from range data for the Ibex area and Shingle Pass section.

Figure 10 depicts graphically the relationship between the West Spring Creek, Joins, and Oil Creek Formations of south-central Oklahoma and the CS. Figure 11 shows the relationship determined graphically between the CS and the Ibex-area and Shingle Pass local composite section (SPIB). Data plotted in these graphs are listed in Tables 2 and 3. I should point out, however, that all but two points along the LOC of Figure 10 are controlled in the CS by information from the Shingle Pass-Ibex local composite section (SPIB of Fig. 11) and that nearly all the points to which I have fit the LOC of Figure 11 are controlled in the CS by data from south-central Oklahoma. Thus, the principal function of range information from the lower 300 m of the Antelope Valley composite section (Fig. 12) is to position the Lower OK 77 and SPIB sections in the CS. As indicated in the graph of Figure 12, this has been accomplished primarily by assuming that the first occurrence of species 152A (*Tripodus laevis*) is at the same level in LAVC and SPIB (Fig. 11) and that the rate of rock accumulation remained essentially the same through the lower 200 m of the Antelope Valley composite section as in the upper 232 m. Neither of these assump-

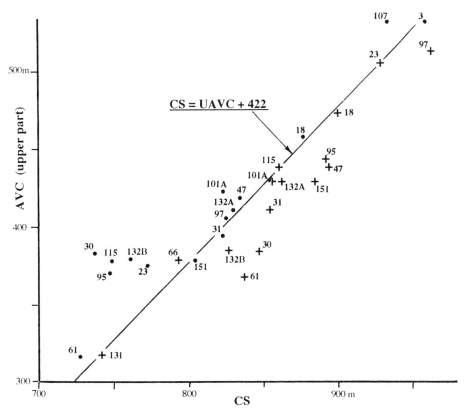

FIG. 9.—Graphic correlation of the upper part of the Antelope Valley composite section (UAVC) and the Ordovician CS. At the stage in development of the CS indicated by this correlation there were no data available below CSU 727. Addition of AVC to the CS produced a "tail" extending downward to CSU 422 and useful in determining relationships of lower Whiterockian and older rocks. Dots mark range bases; crosses mark range tops; numbers by dots and crosses identify conodont species whose names and ranges are listed in Table 3.

tions, of course, is required by the evidence; but downward projection of the LOC determined for the well-controlled upper AVC does separate most tops and bases (Fig. 12) and suggests substantial extension of the ranges of only two species (*Histiodella altifrons* and *"Scandodus" robustus*). Further, the relationship suggested in Figure 12 maintains the integrity of *Tripodus laevis* as hallmark of the basal Whiterockian Series in the section that includes its basal-boundary stratotype and also in the one that serves as the stratotype of the Ibexian Series.

THE ORDOVICIAN-SILURIAN BOUNDARY

Although it is customary to regard the Cincinnatian Series as the youngest major division of the Ordovician System in North America, it is nevertheless true that Cincinnatian strata in the type area of Kentucky, Ohio, and Indiana are separated from Silurian strata by an unconformity. Thus the Ordovician CS described in this report is likely to be incomplete at the top. Conodonts have been described from Ordovician rocks in the western midcontinent (Sweet, 1979b) that may be slightly younger than any in the Cincinnatian type area, and additional forms are known from the Ellis Bay Formation of Anticosti Island (McCracken and Barnes, 1981) and the Whittaker Formation of the Mackenzie Mountains (Nowlan and others, 1988), which may also contain Ordovician rocks younger than those typical of the Richmondian Stage. The latter are included in the Richmondian Stage by some authors but referred to a separate, presumably post-Richmondian Gamachian Stage by others (Fig. 1). Scaled ranges of conodonts in the Ellis Bay Formation and the underlying Vauréal Formation (Nowlan and Barnes, 1981), and in the Whittaker Formation (Nowlan and others, 1988) are readily available in the literature and should provide the basis for graphic integration. However, species especially distinctive of the critical Gamachian interval have not been identified in sections that are currently parts of the Ordovician CS, so integration can not be effected. Thus, until I am able to add to the CS sections known with certainty to cross the boundary between the Ordovician and Silurian systems, it will not be possible to determine the extent to which the framework described in this report is incomplete.

SUMMARY

A composite standard, stated in composite-standard units (CSUs) based on a SRS (section 70ZA) in the Cincinnati Region and some 1,260 CSUs thick, will apparently be adequate as the framework for a pan-Ordovician chronostratigraphy assembled graphically. In addition to section 70ZA and other sections through Mohawkian and Cincinnatian rocks mentioned by Sweet (1984), primary contributors to Ibexian and Whiterockian segments of the CS are the Ibex-area composite section in western Utah; the Shingle Pass section in eastern Nevada; a

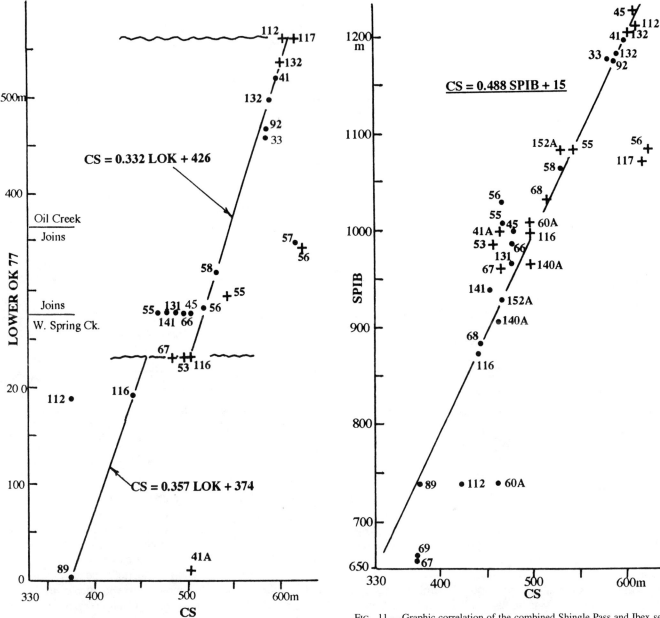

FIG. 10.—Graphic correlation of the West Spring Creek, Joins, and Oil Creek Formations (L. OK77) of south-central Oklahoma with the composite standard (CS), which now includes information from AVC (Fig. 9), and SPIB (Fig. 11; the Shingle Pass-Ibex-area local composite section). The position of most dots (first occurrences) and crosses (last occurrences) is controlled by CS-equivalent values from SPIB (Fig. 11, Table 2). Position in the CS is controlled by lower part of AVC (Fig. 12). The discontinuity shown is within the upper West Spring Creek Formation and is suggested by the distribution of dots and crosses, not by field observations. The two equations are for LOCs fit to points below and above the inferred discontinuity. Numbers near dots and crosses identify species whose names and ranges are listed in Table 3.

FIG. 11.—Graphic correlation of the combined Shingle Pass and Ibex sections (SPIB; Fig. 8; Tables 2, 3) with a CS that now includes information from L. OK77 (Fig. 10) and LAVC (Fig. 12). The Shingle Pass-Ibex composite includes at least 650 m of rock older than any previously included in the CS, hence the part of the LOC that projects below that level is not shown. Data plotted are from Table 3, which also gives the names and ranges of species identified by numbers near dots (first occurrences) and crosses (last occurrences).

composite section of Antelope Valley and Copenhagen strata in east-central Nevada herein termed the Antelope Valley composite section (or AVC); and a section through West Spring Creek, Joins, and Oil Creek strata in the Arbuckle Mountains of south-central Oklahoma. Conodonts in the AVC and Shingle Pass sections need to be described in detail and taxonomic problems clarified. Work on the Shingle Pass section is well advanced as is integration into the composite standard of information from other sections (e.g., Sweet, 1988, 1992).

ACKNOWLEDGMENTS

Data on conodont ranges at the many localities mentioned in this report have been gathered by a number of persons whose

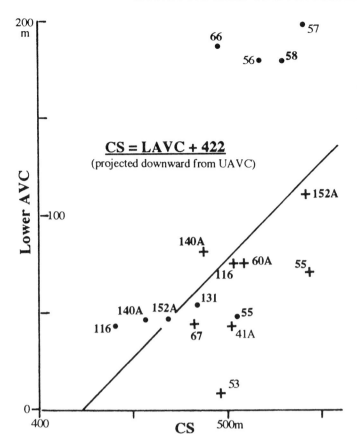

FIG. 12.—Graphic correlation of the lower 200 m of the Antelope Valley composite section (LAVC) and the Ordovician CS. LOC projected downward from UAVC (Fig. 9) includes first occurrence of *Tripodus laevis*, which marks the base of the Whiterockian Series in the stratotype section, and serves for the present to determine the position of SPIB (Fig. 11) and L.OK (Fig. 10) in the Ordovician CS. Numbers by dots (first occurrences) and crosses (last occurrences) identify species listed in Table 3.

work is acknowledged specifically in the text. I am particularly grateful, however, to Celeste M. Wilson (now Greene), who measured, sampled, and processed half the samples from the Shingle Pass section; to Dr. Anita G. Harris of the U. S. Geological Survey for allowing me access on several occasions to the collections on which the Antelope Valley Composite is based; to Dr. John Repetski, also of the U. S. Geological Survey, for help in identifying many of the conodonts Celeste Greene and I collected from Shingle Pass samples; and to Professor Raymond L. Ethington of the University of Missouri-Columbia who kindly provided access to his collections from the Ibex area, Utah and advice on many stratigraphic and taxonomic matters.

REFERENCES

AMSDEN, T. W. AND SWEET, W. C., 1983, Upper Bromide Formation and Viola Group (Middle and Upper Ordovician) in eastern Oklahoma: Norman, Oklahoma Geological Survey Bulletin 132, 76 p.

BAUER, J. A., 1987, Conodonts and conodont biostratigraphy of the McLish and Tulip Creek Formations (Middle Ordovician) of south-central Oklahoma: Norman, Oklahoma Geological Survey Bulletin 141, 58 p.

DERBY, J. R., BAUER, J. A., CREATH, W. B., DRESBACH, R. I., ETHINGTON, R. L., LOCH, J. D., STITT, J. H., MCHARGUE, T. T., MILLER, J. F., MILLER, M. A., REPETSKI, J. E., SWEET, W. C., TAYLOR, J. F., AND WILLIAMS, M., 1991, Biostratigraphy of the Timbered Hills, Arbuckle, and Simpson Groups, Cambrian and Ordovician, Oklahoma: a review of correlation tools and techniques available to the explorationist: Norman, Oklahoma Geological Survey Circular 92, p. 15–41.

DRESBACH, R. I., 1983, Conodont biostratigraphy of the Middle and Upper Ordovician Viola Group of the south-central Oklahoma Arbuckle Anticline: Unpublished M.Sc. Thesis, The Ohio State University, Columbus, 101 p.

ETHINGTON, R. L. AND CLARK, D. L., 1982 [imprint 1981], Lower and Middle Ordovician conodonts from the Ibex Area, western Millard County, Utah: Brigham Young University Geology Studies, v. 28, p. 1–160.

ETHINGTON, R. L. AND SCHUMACHER, D., 1969, Conodonts of the Copenhagen Formation (Middle Ordovician) in central Nevada: Journal of Paleontology, v. 43, p. 440–484.

HARRIS, A. G., BERGSTRÖM, S. M., ETHINGTON, R. L., AND ROSS, R. J., JR., 1979, Aspects of Middle and Upper Ordovician conodont biostratigraphy of carbonate facies in Nevada and southeast California and comparison with some Appalachian successions: Brigham Young University Geology Studies, v. 26, p. 7–43.

HINTZE, L. F., 1951, Lower Ordovician detailed stratigraphic sections for western Utah: Salt Lake City, Utah Geological and Mineralogical Survey Bulletin 39, 99 p.

HINTZE, L. F., 1973, Lower and Middle Ordovician stratigraphic sections in the Ibex area, Millard County, Utah: Brigham Young University Geology Studies, v. 20, p. 3–36.

HINTZE, L. F., TAYLOR, M. E., AND MILLER, J. F., 1988, Upper Cambrian-Lower Ordovician Notch Peak Formation in western Utah: Washington D.C., United States Geological Survey Professional Paper 1393, 30 p.

KELLOGG, H. E., 1963, Paleozoic stratigraphy of the southern Egan Range, Nevada: Geological Society of America Bulletin, v. 74, p. 685–708.

MCCRACKEN, A. D. AND BARNES, C. R., 1981, Conodont biostratigraphy and paleoecology of the Ellis Bay Formation, Anticosti Island, Quebec, with special reference to Late Ordovician-Early Silurian chronostratigraphy and the systematic boundary: Ottawa, Geological Survey of Canada Bulletin 329, part 2, p. 51–134.

MILLER, J. F., 1969, Conodont fauna of the Notch Peak Limestone (Cambro-Ordovician), House Range, Utah: Journal of Paleontology, v. 43, p. 413–439.

NOWLAN, G. S. AND BARNES, C. R., 1981, Late Ordovician conodonts from the Vauréal Formation, Anticosti Island, Quebec: Ottawa, Geological Survey of Canada Bulletin 329, part 1, p. 1–49.

NOWLAN, G. S., MCCRACKEN, A. D., AND CHATTERTON, B. D. E., 1988, Conodonts from Ordovician-Silurian boundary strata, Whittaker Formation, Mackenzie Mountains, Northwest Territories: Ottawa, Geological Survey of Canada Bulletin 373, 99 p.

ORTON, E., 1873, Report on the Third geological district — geology of the Cincinnati Group, Hamilton, Clermont, Clarke Counties: Columbus, Ohio Geological Survey Report, v. 1, pt. 1, p. 365–480.

ROSS, R. J., JR., 1951, Stratigraphy of the Garden City Formation in northeastern Utah, and its trilobite faunas: New Haven, Peabody Museum Natural History Bulletin 6, 161 p.

ROSS, R. J., JR. AND ETHINGTON, R. L., 1991, Stratotype of Ordovician Whiterock Series, with an appendix on graptolite correlation by C. E. Mitchell: Palaios, v. 6, p. 156–173.

ROSS, R. J., JR., HINTZE, L. F., ETHINGTON, R. L., MILLER, J. F., TAYLOR, M. E., AND REPETSKI, J. E., 1993, The Ibexian Series (Lower Ordovician), a replacement for "Canadian Series" in North American chronostratigraphy: Washington, D.C., United States Geological Survey Open-File Report 93–598, 75 p.

ROSS, R. J., JR., ADLER, F. J., AMSDEN, T. W., BERGSTROM, D., BERGSTRÖM, S. M., CARTER, C., CHURKIN, M., CRESSMAN, E. A., DERBY, J. R., DUTRO, J. T., ETHINGTON, R. L., FINNEY, S. C., FISHER, D. W., FISHER, J. H., HARRIS, A. G., HINTZE, L. F., KETNER, K. B., KOLATA, D. L., LANDING, E., NEUMAN, R. B., SWEET, W. C., POJETA, J., JR., POTTER, A. W., RADER, E. K., REPETSKI, J. E., SHAVER, R. H., THOMPSON, T. L., AND WEBERS, G. F., 1982, The Ordovician System in the United States, Correlation Chart and Explanatory notes: Ottawa, International Union of Geological Sciences Publication 12, 73 p.

SHAW, A. B., 1964, Time in Stratigraphy: New York, McGraw-Hill Book Company, 365 p.

SLOSS, L. L., 1963, Sequences in the cratonic interior of North America: Geological Society of America Bulletin, v. 74, p. 93–113.

SPENCER, A. W., 1984, The conodont biostratigraphy and paleoecology across the Whiterock-Mohawkian Series boundary in the Copenhagen Formation,

Antelope Range, Nye County, Nevada: Unpublished B. S. Thesis, Colorado School of Mines, Golden, 81 p.

SWEET, W. C., 1979a, Conodonts and conodont biostratigraphy of post-Tyrone Ordovician rocks of the Cincinnati Region: Washington, D. C., United States Geological Survey Professional Paper 1066-G, p. G1-G26.

SWEET, W. C., 1979b, Late Ordovician conodonts and biostratigraphy of the western Midcontinent Province: Brigham Young University Geology Studies, v. 26, p. 45-85.

SWEET, W. C., 1984, Graphic correlation of upper Middle and Upper Ordovician rocks, North American Midcontinent Province, U. S. A., *in* Bruton, D. L., ed., Aspects of the Ordovician System: Oslo, Universitetsforlaget, Paleontological Contributions from the University of Oslo 295, p. 23-35.

SWEET, W. C., 1987, Distribution and significance of conodonts in Middle and Upper Ordovician strata of the Upper Mississippi Valley region: St. Paul, Minnesota Geological Survey Report of Investigations 35, p. 167-172.

SWEET, W. C., 1988, Mohawkian and Cincinnatian chronostratigraphy: Albany, New York State Museum Bulletin 462, p. 84-90.

SWEET, W. C., 1992, Middle and Late Ordovician conodonts from southwestern Kansas and their biostratigraphic significance: Norman, Oklahoma Geological Survey Bulletin 145, p. 181-190.

SWEET, W. C. AND BERGSTRÖM, S. M., 1971, The American Upper Ordovician Standard, XIII, A revised time-stratigraphic classification of North American upper Middle and Upper Ordovician rocks: Geological Society of America Bulletin, v. 83, p. 613-628.

SWEET, W. C. AND GREENE, C. M., 1992, A standard reference section in the southern Egan Range, Nevada, for a conodont-based chronostratigraphy of the North American Ibexian and Early Whiterockian (Abstract): Geological Society of America Abstracts with Programs, v. 24, p. 64.

SWEET, W. C., HARPER, H., AND ZLATKIN, D., 1974, The American Upper Ordovician Standard. XIX. A Middle and Upper Ordovician reference standard for the eastern Cincinnati Region: Ohio Journal of Science, v. 47, p. 47-54.

TAYLOR, M. E., COOK, H. E., AND MILLER, J. F., 1989, Late Cambrian and Early Ordovician biostratigraphy and depositional environments of the Whipple Cave and House Limestone, central Egan Range, Nevada, *in* Taylor, M. E., ed., Cambrian and Early Ordovician Stratigraphy and Paleontology of the Basin and Range Province, Western United States: Washington, D.C., American Geophysical Union, p. 37-44.

VOTAW, R. B., 1971, Conodont biostratigraphy of the Black River Group (Middle Ordovician) and equivalent rocks of the eastern Midcontinent: Unpublished Ph.D. Dissertation, The Ohio State University, Columbus, 170 p.

WILSON, C. M., 1988, Lower and lower Middle Ordovician conodont biostratigraphy of the Pogonip Group in the southern Egan Range at Shingle Pass, Nevada, and decriptions of stratigraphically important species: Unpublished M. S. Thesis, The Ohio State University, Columbus, 126 p.

GRAPHIC CORRELATION OF MIDDLE ORDOVICIAN GRAPTOLITE-RICH SHALES, SOUTHERN APPALACHIANS: SUCCESSFUL APPLICATION OF THE TECHNIQUE TO APPARENTLY INADEQUATE STRATIGRAPHIC SECTIONS

BARBARA J. GRUBB
Raymond M. Alf Museum, 1175 West Base Line Road, Claremont, California 91711
AND
STANLEY C. FINNEY
Department of Geological Sciences, California State University, Long Beach, California 90840

ABSTRACT: Middle Ordovician graptolite-rich shales exposed in the southern Appalachians provide an opportunity to apply graphic correlation to what appear to be inadequate data. The fact that the resulting composite standard (CS) is used successfully to address significant geologic problems demonstrates the effectiveness and versatility of the technique. The shales are so strongly diachronous that no single stratigraphic section spans the entire biostratigraphic interval represented by the shales. The Standard Reference Section (SRS) included only a part of this interval. Therefore, many partially overlapping sections had to be used in the construction of the CS with each section extending the CS upwards or downwards. The completed CS contains the data from 22 stratigraphic sections, is composed of 89 composite standard units (CSUs), and ranges from the lower *G. teretiusculus* Zone to the *C. bicornis* Zone.

Once we established the CS, we used it as a basis for correlating the basal shale contact. This contact records the subsidence and migration of the foreland basin in which the shales were deposited, and its diachroneity can be expressed in terms of CSUs. With the biostratigraphic correlation of radiometric dates into the CS, the duration of the CSUs could be calibrated in terms of years, and so, too, could the age differences between sections of the diachronous basal shale contact. Because this contact recorded the migration of the basin axis, we used its age difference and the palinspastically restored distance between sections to calculate migration rates. Calculated rates indicate that the basin migrated 50 km cratonward at an average of 13 mm/yr. We were also able to demonstrate that the rate decreased from 40 mm/yr to 9 mm/yr before the migration completely ceased. The foreland basin migration and its subsequent deceleration and halt were produced by arc-continent convergence and collision. Our calculated rate of 40 mm/yr is comparable to modern rates of convergence.

We also calculated sediment accumulation rates for selected sections. To do so, we used the slope of the line of correlation (LOC) to determine the number of CSUs and their duration in years corresponding to the stratigraphic thickness of graptolite shales in the section compared to the CS. The rates of 1.8 to 2.8 cm/1000 yrs calculated for intervals of 1.3 to 2.8 Ma are within the ranges of those reported in other studies of pelagic environments.

Our rates of basin migration and sediment accumulation are virtually identical to rates determined from a variety of other methods, demonstrating the validity of a CS constructed from what appear to be inadequate biostratigraphic data.

INTRODUCTION

A critical step in graphic correlation is selection of a Standard Reference Section (SRS) to which all other sections are compared. Ideally, the SRS should be continuous and unfaulted, contain a large and varied fauna, and include the entire biostratigraphic interval of interest. Many continuous, unfaulted, and fossiliferous sections were available for our studies of Middle Ordovician graptolitic shales in the southern Appalachians (Fig. 1), but not one spanned the entire biostratigraphic interval of the shales (Finney, 1986; Grubb, 1991). The shales are so strongly diachronous that there is very little overlap between sections separated by modest geographic distance, even though the full thickness of the shales was exposed and collected at most of the sections (Fig. 2). Because each section represents only part of the full biostratigraphic interval spanned by the shales, selection of a SRS and compilation of the composite standard section (CS) was unusual. It meant that the CS had to be assembled from many partially overlapping sections, including the section selected as the SRS. Although in most studies the CS is compiled from overlapping sections, these sections are typically substantial in that they span several zones, and the section selected as the SRS spans most of the relevant biostratigraphic interval (e.g., Sweet, 1979). In contrast, our sections, each of which spanned only a partial zone, appeared to be too limited to effect a graphic correlation.

In spite of these apparently inadequate stratigraphic sections, Grubb (1991) was able to establish a CS for the graptolite-rich shales of the southern Appalachians, and used the CS to correlate the basal contact of the shales between 22 stratigraphic sections in Alabama, Tennessee, and Virginia (Fig. 1). This contact is significant because it records the subsidence and migration of the Sevier foreland basin. Its correlation expressed in terms of composite standard units clearly demonstrated its diachroneity. In addition, by calibrating the CS with radiometric dates, Grubb and others (1991) expressed the diachroneity of the basal shale contact in terms of numeric ages, which allowed for calculation of rates of foreland-basin migration and sediment accumulation.

The purpose of this paper is to demonstrate the application of graphic correlation in circumstances that required the CS to be constructed from several partially overlapping sections and to report the innovative application of graphic correlation in calibrating the rate of migration of the Sevier foreland basin and the rate of accumulation of pelagic sediment within it. Notwithstanding the nature of the apparently inadequate stratigraphic sections, we obtained valid results and successfully used them to address important geologic problems.

GEOLOGIC SETTING

Middle Ordovician, graptolite-rich shales extend the length of the southern Appalachians from central Alabama to northern Virginia where they are exposed in fold and thrust belts of the Valley and Ridge province (Fig. 1). These shales, the underlying shallow-water carbonates, and the overlying thick sequence of clastic turbidites (Fig. 3) record the sudden, dramatic subsidence of the Sevier foreland basin and its subsequent filling (Drake and others, 1989). The foreland basin subsided in response to deformational loading of the lithosphere (Shanmugam and Lash, 1982) during the Blountian phase of the Taconic orogeny (Rodgers, 1971) and was produced by arc-continent

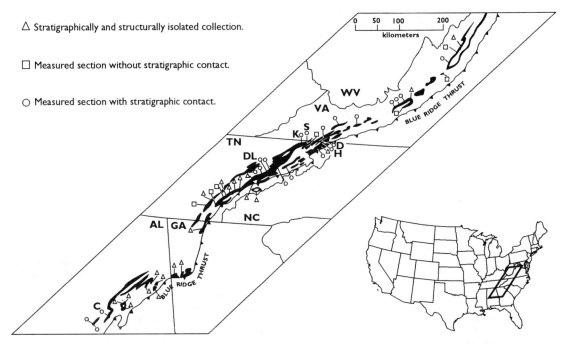

FIG 1.—Map showing outcrops of Middle Ordovician shales in southern Appalachians, collection sites, and their nature shown by squares, circles, and triangles. Localities mentioned in text and figures include Calera (C), Denton Valley (D), Douglas Lake (DL), Holston Lake (H), Kingsport (K), and Steele Creek East (S).

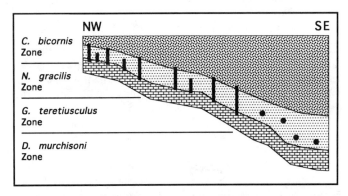

FIG 2.—Schematic cross section through the southern Appalachians illustrating the diachronous nature of the Middle Ordovician shales. Cross section is oriented perpendicular to depositional strike. Note that no single stratigraphic section encompasses the entire interval from *G. teretiusculus* Zone to *C. bicornis* Zone. Brick pattern = carbonate units; dashed = graptolite shales; stippled = turbiditic sandstone; vertical bars represent stratigraphic sections; dots represent collections from outcrops too limited to provide measured sections.

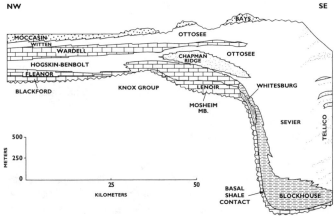

FIG 3.—Palinspastically restored, stratigraphic cross section of Middle Ordovician strata of eastern Tennessee, modified from Walker (1977). Contact between Lenoir Limestone and onlapping Blockhouse Shale (i.e., the basal shale contact) records foreland basin subsidence.

convergence and collision. As the deformational load advanced onto the craton, the foreland basin subsided in front of it, migrating cratonward (Ettensohn, 1991; Hatcher, 1989). Accordingly, the basal shale contact, which records the initial basin subsidence, is diachronous becoming younger to the northwest (Walker, 1977; Read, 1980; Fig. 2). The upper contact of the shales with the overlying clastic turbidites is also younger to the northwest. The turbidites derived from orogenic highlands to the southeast accumulated in submarine fans that prograded to the northwest across the basin and covered the shales deposited in front of their advance (Shanmugam and Lash, 1982).

Because the graptolitic shales are highly diachronous and Alleghanian thrusting telescoped lateral facies, nearby sections, located along NW-SE transects perpendicular to depositional strike and parallel to the direction of foreland basin migration, differ substantially in age. In the southeastern belts of the Valley and Ridge Province, the shales correlate with the *Didymograptus murchisoni* Zone. In successive sections to the northwest, the shales correlate with the *Glyptograptus teretiusculus* Zone, the upper *G. teretiusculus* Zone and the lower *Nemagraptus gracilis* Zone, the lower to upper *N. gracilis* Zone, the upper *N. gracilis* Zone to lower *Climacograptus bicornis* Zone, and the *C. bicornis* Zone (Fig. 2). Although in their entirety the shales range from the *D. murchisoni* Zone to the *C. bicornis* Zone (Finney, 1986; Grubb, 1991), no single section ranges through even half of this interval. Most sections represent less than one zone, and it is fortunate that several sections range across zonal boundaries. They were used to build the CS from one biostratigraphic interval to the next.

COMPILATION OF THE COMPOSITE STANDARD

Finney (1986) collected graptolites from the Middle Ordovician shales at 50 localities in the southern Appalachians (Fig. 1). The collections provide an excellent set of biostratigraphic data for the Middle Ordovician of the southern Appalachians as summarized in Figure 4 and in Finney (1986) and Grubb (1991). At 30 of these localities, Finney collected bed-by-bed through measured sections. These measured sections range from the lower *G. teretiusculus* Zone to the lower *C. bicornis* Zone, but do not include the *D. murchisoni* Zone (Fig. 2).

Twenty-two sections were selected for graphic correlation and compiled with GraphCor, a software package for microcomputers (Hood, 1986). We recorded the lowest and highest stratigraphic occurrences of each species in each section in terms of measurements above the basal shale contact. Many range bases and tops are preservational artifacts. For example, the long-ranging biserial graptolite species such as *G. teretiusculus*, *Pseudoclimacograptus angulatus angulatus*, and *Cryptograptus marcidus* consistently appear at the base of the shales in most stratigraphic sections, regardless of the fact that the shales are highly diachronous. Additionally, many last occurrences are at the top of the shales where they are overlain by turbiditic siltstones and sandstones. However, within each section, several range bases and some range tops are clearly not facies controlled. For example the range bases of *Dicellograptus geniculatus*, *Dicellograptus gurleyi* n. ssp. A, *Nemagraptus gracilis*, *Azgograptus incurvus*, and *Dicellograptus gurleyi gurleyi* consistently occur in the same relative order in those sections that correlate with the lower part of the *N. gracilis* Zone (Figs. 5, 6). We weighed these range bases heavily in the correlation process and used them to locate the line of correlation (LOC).

Selection of the SRS was problematic in that no single stratigraphic section spans the entire Middle Ordovician graptolite zonation from the *G. teretiusculus* Zone to the *C. bicornis* Zone. After evaluating each section and attempting a few rounds of graphic correlation with various sections as the SRS, we chose Calera (Alabama) as the SRS (Fig. 5). It provides a relatively thick, fossiliferous interval, although it only ranges from the upper *G. teretiusculus* Zone to the upper *N. gracilis* Zone and does not include the lower *G. teretiusculus* Zone (Fig. 7). Therefore, it was necessary to build a CS that would encompass the interval from lower *G. teretiusculus* Zone to the *C. bicornis* Zone.

The correlation process started with those sections that overlap with Calera and extend below it. These include the Denton Valley, Holston Lake, and Steele Creek East sections (Fig. 7). These sections include the range bases of several key species that occur in the same relative order as they do at Calera and were used to position the LOC (Fig. 8). If ranges were extended, however, the range bases of species that occur in the lowest collections at Denton Valley would be adjusted downward into negative values in the CS. This is due to the fact that the key species used to position the LOC and establish correlation occur only a few meters above the basal shale contact at Calera, but they occur 20–30 m above the basal shale contact at Denton Valley, Holston Lake, and Steele Creek East sections (Figs. 5, 6). The shales at these sections are lithologically identical.

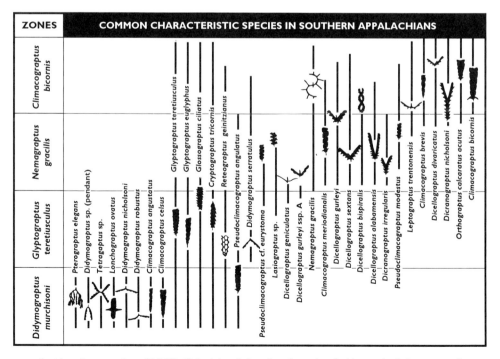

FIG 4.—Summary range chart based on zonation of Middle Ordovician shales of southern Appalachians and relative ranges of common species.

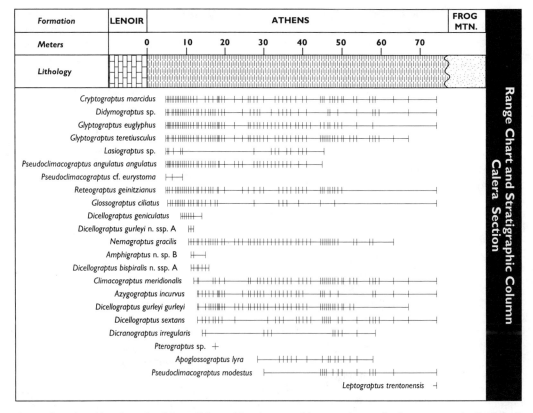

Fig 5.—Range chart and stratigraphic column for Calera, Alabama. Note that several long-ranging species first occur at a level 4.6 m above the base of the shale and approximately 4 m below those species occurring in the middle of the section. Compare with Denton Valley, Figure 6.

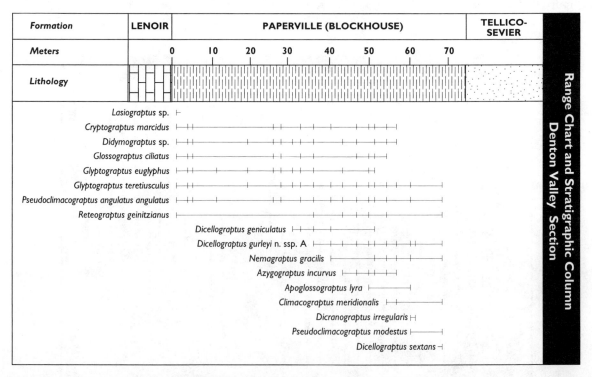

Fig 6.—Range chart and stratigraphic column for Denton Valley, Virginia. Long-ranging species (for example, *G. teretiusculus*) appearing near the base and terminating near the top of the shale are probably facies controlled. Species that occur in the middle of the section (for example, *D. geniculatus*, *D. gurleyi* n. ssp. A, and *N. gracilis*) have consistent order of occurrence in many sections and are not facies controlled.

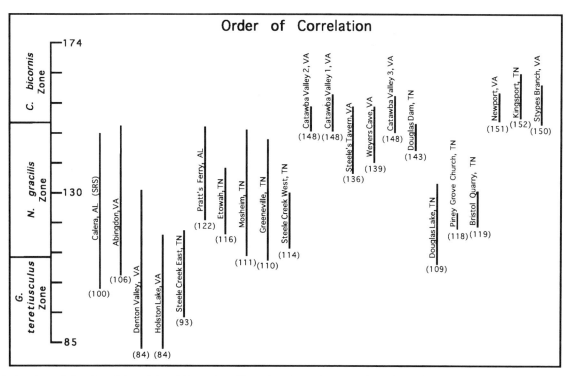

FIG 7.—Order of correlation, from left to right, of 22 stratigraphic sections. Note that no single section spans the entire biostratigraphic interval from the lower *G. teretiusculus* Zone to the *C. bicornis* Zone. To build the CS, the order was chosen such that each section extends the biostratigraphic range of the SRS (Calera). Y-intercept values (given in composite units), within parentheses, are the level at which the basal shale contact correlates into the CS.

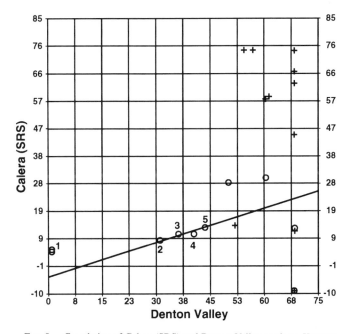

FIG 8.—Correlation of Calera (SRS) and Denton Valley sections. Key species with consistent order of first occurrences were used to locate the LOC. Range bases located above the LOC would be adjusted downward and incorporated into the CS. Note that *G. teretiusculus* (1) and several other species first appear together at a level of 1 m in the Denton Valley section (Fig. 6). These would be extended to a level of −4 CSUs (see text). Numbered species are: *G. teretiusculus* (1); *D. geniculatus* (2); *D. gurleyi* n. ssp. A (3); *N. gracilis* (4); and *A. incurvus* (5). Range bases = o, range tops = +.

Thus, one can assume that they accumulated at uniform rates. Accordingly, these occurrences of key species clearly demonstrate that the basal shale contact and the range bases that occur there are much older than at the Calera section. To avoid the incorporation of negative values in the CS, we added 100 m to the ranges of all species in the Calera (SRS) section and then restarted the correlation. Those sections extending below Calera could now be correlated into the SRS without the inclusion of negative values. The remaining sections were correlated in the following order: (1) those that overlap with Calera but extend into higher parts of the *N. gracilis* Zone; (2) those that span the upper part of the *N. gracilis* Zone and extend into the lower *C. bicornis* Zone; and (3) those that are restricted to the *C. bicornis* Zone (Fig. 7).

Upward extension of the CS into the *C. bicornis* Zone entailed the same considerations as its downward extension. Most of the range bases and tops in the comparison sections were preservational artifacts and thus provided limited constraints on the location of the LOC. Each section did, however, include a few (2–5) range bases that were given considerable weight for correlation because of their consistent position relative to other range bases and tops. We used these bases to locate the LOC and in turn used it to extend the CS upward and to add new species to it from the comparison sections. After one round of correlation, the initial SRS that spanned the upper *G. teretiusculus* Zone to upper *N. gracilis* Zone evolved into a CS that encompassed the lower *G. teretiusculus* Zone to *C. bicornis* Zone.

Correlation continued through four rounds until a stable network of data points was achieved where there was little or no adjustment of the LOC. The completed CS serves as a final range chart and includes the ranges of 37 species, reflecting the maximum stratigraphic ranges of all the species in the 22 sections (Fig. 9). The duration of the completed CS, 89 CSUs, could vary depending upon how the LOC was located during the rounds of correlation. However, the relative order and spacing of the species remain the same throughout the correlation process. Thus, we are confident that the CS is a true representation of graptolite biostratigraphy in all the sections.

APPLICATION TO GEOLOGIC PROBLEMS

The basal shale contact records the subsidence and migration of the Sevier foreland basin. This contact is the reference level from which we measured all stratigraphic sections. For this reason, the Y-intercept of the LOC represents the relative level in the CS to which the basal shale contact correlates. Assuming that the rate of sediment accumulation was constant for the entire thickness of the shales, we can even correlate the basal contact of those sections lacking graptolites in the lower few meters by projecting the LOC to the Y-axis (Fig. 10). The relative age distribution of the contact, expressed in CSUs, clearly illustrates the diachroniety between sections and the pattern of basin migration to the northwest (Fig. 11) and does so with much greater precision than zonal biostratigraphy.

By incorporating radiometric dates into the CS, numeric ages can be calculated for the basal shale contact at each section. Ordovician ash beds in Britain provide dates of 464 ± 1.8 Ma for the upper Llanvirn Series (*D. murchisoni* Zone) and 457 ± 2.2 Ma for the lower Caradoc Series (Tucker and others, 1990). These are the only reliable dates that can be correlated into the Middle Ordovician shales of the southern Appalachians. They correlate with the *D. murchisoni* and *C. bicornis* zones, respectively (Ross and others, 1982). Because the CS extends through most of this biostratigraphic interval, we estimate its duration to be approximately 5 my. The CS is composed of 89 composite units. Assuming a uniform rate of sediment accumulation for pelagic shale in our sections, each unit represents approximately 56,000 years. When applied to the Y-intercept (ex-

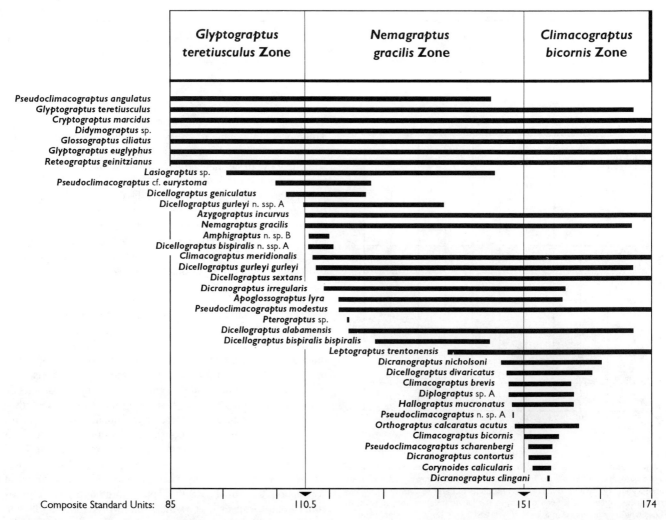

FIG 9.—Final range chart and composite standard (CS) produced from graphic correlation of 22 sections. Zonal boundaries can be defined at the bases of *N. gracilis* and *C. bicornis* and are placed at 110.5 and 151 composite units, respectively.

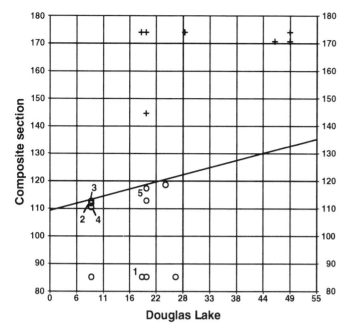

Fig. 10.—Fourth round of correlation of Douglas Lake with the CS. The LOC, projected through the barren interval near the base of the section, marks the relative level in the CS at which the basal contact correlates. Numbered species are: *G. teretiusculus* (1); *N. gracilis* (2); *D. gurleyi gurleyi* (3); *A. incurvus* (4); and *P. modestus* (5). Range bases = o, range tops = +.

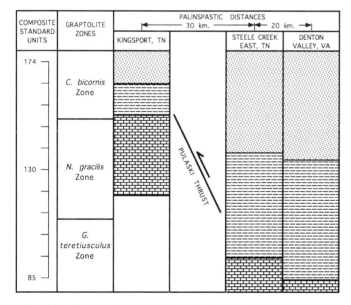

Fig. 11.—The diachronous nature of the basal shale contact expressed in terms of zonal biostratigraphy and composite standard units between three localities along a NW to SE transect perpendicular to depositional strike. Note that contact between shale and overlying turbiditic siltstone and sandstone is also diachronous. Palinspastic distances between localities based on reconstruction of Hatcher (Grubb and others, 1991). Considerable structural telescoping of facies occurs along Pulaski thrust fault. Patterns as in Figure 2.

pressed in CSUs), we used this value to determine age differences of the basal shale contact between sections.

With this temporal information, the rate of basin migration can be calculated if original distances between sections are known. Although shortened by Alleghanian thrusting, these distances can be measured from a palinspastic reconstruction produced by R. D. Hatcher, Jr. (Grubb and others, 1991). We calculated the rate of migration using three stratigraphic sections located along a transect perpendicular to the depositional axis of the basin in northeastern Tennessee and southwestern Virginia (Figs. 1, 11). These sections are, from oldest to youngest: Denton Valley, Steele Creek East, and Kingsport. Between Denton Valley and Kingsport, the relative age difference of the basal shale contact is 68 CSUs. Because each CSU may represent approximately 56,000 years, the numeric age difference is 3.8 Ma. With a palinspastic distance of 50 km between these section, the age difference represents an average migration rate of 13 mm/yr. However, the rate of migration may not have been constant across the entire 50 km. Accordingly, the rate was calculated for shorter distances using an intermediate stratigraphic section. From Denton Valley to Steele Creek East, a distance of 20 km, the rate of migration is 40 mm/yr. Similarly, from Steele Creek East to Kingsport, a distance of 30 km, the rate of migration is 9 mm/yr. This rate change from 40 mm/yr to 9 mm/yr along the southeast to northwest transect may represent the deceleration and halt of the advancing foreland basin.

The foreland basin deposits do not occur to the northwest of Kingsport, Tennessee. As the foreland basin advanced towards Kingsport, it must have been decelerating. Its migration was driven by arc-continent convergence and collision. The rate of 40 mm/yr for the more outboard sections is comparable to modern rates of convergence. With collision, the cratonward advance of the lithospheric load and migration of the foreland basin in front of it would be expected to slow and then halt. The rate of 9 mm/yr for the more inboard sections supports this expectation.

In addition to determining the rate of basin migration, graphic correlation can be used to calculate the rate of pelagic sediment accumulation, assuming that the shales accumulated at a uniform rate as recorded in the identical lithology of the shales within and between sections. We used the slope of the LOC to determine the numbers of CSUs for a selected stratigraphic thickness in the comparison section. With the conversion of CSUs to years, we could directly calculate the rate of accumulation of the selected stratigraphic interval. For example, at Denton Valley the 70 m stratigraphic interval corresponds to 50 CSUs (Figs. 6, 11), indicating that the 70 m of shale accumulated in 2.8 my. Therefore, the rate of sediment accumulation was 2.5 cm/1000 yrs. We calculated comparable rates of 2.2 and 1.8 cm/1000 yrs for the sections at Steele Creek East and Calera, respectively. These rates fall within the range of 1–3 cm/1000 yrs interpreted for the Blockhouse Shale of Tennessee by Shanmugam and Walker (1980). They also compare exactly with rates compiled by Sadler (1993) from both modern and ancient pelagic settings.

CONCLUSION

Although Middle Ordovician shales in the southern Appalachians are richly fossiliferous, provide the opportunity to collect from many measured sections, and can be readily correlated with traditional zonal biostratigraphy, they presented a challenge for the application of graphic correlation. Because of their extreme diachroneity, the best section available as the SRS represented less than half of the interval for which the CS was to be constructed (Fig. 7). This required assembly of the CS from many short and partially overlapping sections. Our subsequent use of the CS as the basis for determining rates of basin migration and sediment accumulation served to test the validity of the CS. Errors may have been introduced in the process of constructing the CS and calculating rates. Potential sources of error include: (1) precision with which the positions of the LOC and Y-intercept were set, (2) the 5 my estimate for the duration of the CS, (3) the accuracy of the palinspastic reconstruction, and (4) nonuniform rates of rock accumulation at the SRS. No doubt, these factors did introduce some degree of inaccuracy into our calculations. However, the fact that the calculated rates match so well rates determined from a variety of methods in both modern and ancient settings gives us confidence in our results.

ACKNOWLEDGMENTS

Field work, carried out in 1983–1985, was supported by a grant (#EAR-8206038) from the National Science Foundation. Complete range data, locality information, taxonomic notes, and composite range values are included in the appendices of a paper by S. C. Finney, B. J. Grubb, and R. D. Hatcher, Jr. being prepared for publication in the Geological Society of America Bulletin. We thank Lucy Edwards for her review of an early draft of this paper and Peter Sadler for advice on calculation of rates of sediment accumulation. The Raymond M. Alf Museum and the Department of Geological Sciences at California State University, Long Beach contributed to the costs of manuscript preparation.

REFERENCES

DRAKE, A. A., JR., SINHA, A. K., LAIRD, J., AND GUY, R. E., 1989, The Taconic orogen, in Hatcher, R. D., Jr., Thomas, W. A., and Viele, G. W., eds., The Appalachian-Ouachita Orogen in the United States: Boulder, Geological Society of America, The Geology of North America, v. F-2, p. 101–177.

ETTENSOHN, F. R., 1991, Flexural interpretation of relationships between Ordovician tectonism and stratigraphic sequences, central and southern Appalachians, U.S.A., in Barnes, C. R. and Williams, S. H., eds., Advances in Ordovician Geology: Ottawa, Geological Survey of Canada Paper 90-9, p. 213–224.

FINNEY, S. C., 1986, Graptolite biostratigraphy and depositional history of Middle Ordovician shales, southern Appalachians: American Association of Petroleum Geologists Bulletin, v. 70, p. 589–590.

GRUBB, B. J., 1991, Graptolite biostratigraphy and graphic correlation of Middle Ordovician shales, southern Appalachians: Unpublished M.S. Thesis, California State University, Long Beach, 241 p.

GRUBB, B. J., FINNEY, S. C., AND HATCHER, R. D., JR., 1991, Potential of graphic correlation for addressing geologic problems: Determination of migration rates of Taconic Foreland Basin, southern Appalachians: Geological Society of America Abstracts with Programs, v. 23, p. A421.

HATCHER, R. D., JR., 1989, Tectonic synthesis of the U.S. Appalachians, in Hatcher, R. D., Jr., Thomas, W. A., and Viele, G. W., eds., The Appalachian-Ouachita Orogen in the United States: Boulder, Geological Society of America, The Geology of North America, v. F-2, p. 511–535.

HOOD, K., 1986, GraphCor, Interactive Graphic Correlation for Microcomputers, version 2.2, available from K. Hood, 9707 Arrowgrass, Houston, Texas.

READ, J. F., 1980, Carbonate ramp-to-basin transitions and foreland basin evolution, Middle Ordovician, Virginia Appalachians: American Association of Petroleum Geologists Bulletin, v. 64, p. 1575–1612.

RODGERS, J., 1971, The Taconic Orogeny: Geological Society of America Bulletin, v. 82, p. 1141–1178.

ROSS, R. J., JR., ADLER, F. J., AMSDEN, T. W., BERGSTROM, D., BERGSTROM, S. M., CARTER, C., CHURKIN, M., CRESSMAN, E. A., DERBY, J. R., DUTRO, J. T., JR., ETHINGTON, R. L., FINNEY, S. C., FISHER, D. W., FISHER, J. H., HARRIS, A. G., HINTZE, L. F., KETNER, K. B., KOLATA, D. L., LANDING, E., NEUMAN, R. B., SWEET, W. C., POJETA, J., JR., POTTER, A. W., RADER, E. K., REPETSKI, J. E., SHAVER, R. H., THOMPSON, T. L., AND WEBERS, G. F., 1982, The Ordovician System in the United States. Correlation Chart and Explanatory Notes: Ottawa, International Union of Geological Sciences Publication 12: 73 p.

SADLER, P. M., 1993, Time scale dependence of the rates of unsteady geologic processes, in Armentrout, J. M., Bloch, R., Olson, H. C., and Perkins, B. F., eds., Rates of Geological Processes: Houston, Gulf Coast Section Society of Economic Paleontologists and Mineralogists Foundation 14th Annual Research Conference, p. 221–228.

SHANMUGAM, G. AND LASH, G. G., 1982, Analogous tectonic evolution of Ordovician foredeeps, Southern and Central Appalachians: Geology, v. 10, p. 562–566.

SHANMUGAM, G. AND WALKER, K. R., 1980, Sedimentation, subsidence, and evolution of a foredeep basin in the Middle Ordovician, southern Appalachians: American Journal of Science, v. 280, p. 479–496.

SWEET, W. C., 1979, Late Ordovician conodonts and biostratigraphy of the western Midcontinent province: Brigham Young University Geology Studies, v. 26, p. 45–74.

TUCKER, R. D., KROGH, T. E., ROSS, R. J., JR., AND WILLIAMS, S. H., 1990, Time-scale calibration of high-precision U-Pb zircon dating of interstratified volcanic ashes in the Ordovician and lower Silurian stratotypes of Britain: Earth and Planetary Science Letters, v. 100, p. 51–58.

WALKER, K. R., 1977, A brief introduction to the ecostratigraphy of the Middle Ordovician of Tennessee (southern Appalachians, U.S.A.), in Ruppel, S. C. and Walker, K. R., eds., The Ecostratigraphy of the Middle Ordovician of the Southern Appalachians (Kentucky, Tennessee, and Virginia) U.S.A.: Knoxville, University of Tennessee, Studies in Geology 77-1, p. 12–17.

A CONODONT- AND GRAPTOLITE-BASED SILURIAN CHRONOSTRATIGRAPHY

MARK A. KLEFFNER
Department of Geological Sciences, The Ohio State University at Lima, Lima, Ohio 45804

ABSTRACT: Graphic correlation of 12 previously uncompiled stratigraphic sections with the Silurian composite standard of Kleffner (1989) results in a revised Silurian composite standard (CS) that has worldwide applicability as a high-resolution chronostratigraphy. The additional range-data on 52 graptolite species, 39 conodont species, 10 events, and one boundary stratotype make it possible to graphically correlate virtually any stratigraphic section (which meets the data requirements of the graphic correlation method) containing representatives of diagnostic conodont and/or graptolite species with the newly revised Silurian CS. The nonannular absolute chronology based on the Silurian CS divides with confidence into 92 standard time units (STUs), a resolution that is a minimum of twice that of any previously proposed Silurian chronostratigraphy. Most sections graphically correlate with the Silurian CS by fitting a straight line of correlation, indicating that the standard reference section (Cellon, Austria) consists of rock which accumulated at a relatively constant rate. The absolute chronology based on the Silurian CS is thereby consonant (or nearly so) with an annular scale, and the 92 STUs it divides into are of equal annular length. Conodont and graptolite chronozones are defined in the Silurian CS according to international rules of stratigraphy and, if they contain at least one STU, can be recognized with confidence in any section that is a part of the Silurian CS or that can be added to it by the graphic correlation method. The conodont and graptolite chronozones defined in the Silurian CS are based on zones proposed by Walliser (1964), Barrick and Klapper (1976), Jeppsson (1988), Aldridge and Schönlaub (1989), Kleffner (1989), and Cocks and Nowlan (1993, for a proposed standard left-hand column for international Silurian correlation charts).

All post-Aeronian Silurian series and stage boundaries can be recognized with confidence in any section that is already or can become a part of the Silurian CS. Three of the series boundaries, the Llandovery/Wenlock, Ludlow/Pridoli, and Pridoli/Lochkovian (Silurian/Devonian), are recognized in the Silurian CS based on the position of the "golden spike" in their boundary stratotypes. Lower boundaries of the graptolite zones that are at the same or approximate levels as the other Silurian boundaries are used to recognize the positions of those boundaries in the Silurian CS. The Silurian chronostratic scale, based on the range-data on conodont species, graptolite species, events, and boundary stratotypes represented in the Silurian CS, is calibrated by using the Wenlock/Ludlow and Silurian/Devonian tie-points of Harland and others (1989) and STUs as chrons of equal duration to interpolate between and below those tie-points. The Silurian time scale developed by this method provides the best means at present for estimating the durations of the Wenlock, Ludlow, and Pridoli Epochs, and all of the ages that comprise them (except perhaps for the Sheinwoodian and Homerian). The Pridoli was the longest epoch with a duration of 8.4 Ma, compared to 7.1 Ma for the Ludlow and 2.6 Ma for the Wenlock.

INTRODUCTION

Kleffner (1989) developed a conodont-based Silurian composite standard (CS) for upper Llandovery-Pridoli strata using the graphic correlation method of Shaw (1964). This composite standard provides both a high-resolution chronostratigraphic framework for correlation of Silurian stratigraphic sections which can be graphically correlated with the Silurian CS and a means for recognition of all series and stage boundaries. The paucity of graptolite range-data in the Silurian CS makes it difficult or impossible to graphically correlate with the Silurian CS sections containing diagnostic graptolites but lacking diagnostic conodonts. As a result, the Silurian CS provides a high-resolution chronostratigraphic framework mainly for strata representative of shelf depositional environments. The small number of boundary-stratotype sections in the Silurian CS and the few graptolite range-data also make it difficult or impossible to accurately recognize all series and stage boundaries in those sections that can be graphically correlated with the Silurian CS.

In order for the Silurian CS of Kleffner (1989) to provide a high-resolution stratigraphic framework for sections containing diagnostic graptolites but lacking diagnostic conodonts, the number of graptolite range-data must be greatly increased in it. The central Nevada composite standard consisting of several sections comprised of part or all of the Roberts Mountains Formation provides this additional data. Eight additional sections, including two previously not possible to add to the Silurian CS because of the paucity of graptolite range-data, the Klonk, Czechoslovakia section described in Jeppsson (1988) and the Gräfenwarth, Germany section of Jaeger (1991), are also graphically correlated with the Silurian CS. The Klonk section adds another boundary stratotype and the Gräfenwarth section adds range-data from 18 more graptolite species to the Silurian CS.

The number of sections comprising the Silurian CS increased from 30 to 42. The revised Silurian CS now truly does have worldwide applicability as a high-resolution chronostratigraphy for any section containing strata representative of a marine depositional environment and containing diagnostic conodonts and/or diagnostic graptolites that meet the other data requirements of the graphic correlation method (see Section II of this volume for information on the other data requirements). The position of the "golden spike" in boundary-stratotype sections that are a part of the Silurian CS is used to recognize the Llandovery/Wenlock, Ludlow/Pridoli, and Silurian/Devonian boundaries. Graptolite chronozones developed from the graptolite range-data are used to approximately position all other boundaries between Silurian series and stages that have been designated in boundary stratotypes that are not a part of the Silurian CS.

METHODS

Development of Central Nevada Composite Standard

In order to greatly increase the amount of graptolite range-data in the Silurian CS of Kleffner (1989), a central Nevada composite standard was compiled using the graphic-correlation method and range-data for 37 graptolite species, 12 conodont species, and two events in five stratigraphic sections in the Roberts Mountains and Simpson Park Range in central Nevada. Data on 11 of the 14 sections in the Roberts Mountains and Simpson Park Range, described in detail in Berry and Murphy (1975) and Klapper and Murphy (1975), were assembled. Five of the 11 were graphically correlated to develop a central Nevada composite standard (Table 1). Locations of those sections are shown in Berry and Murphy (1975). A central Nevada CS was developed to graphically correlate with the Silurian CS, rather than simply graphically correlating each of the five sec-

TABLE 1.—SECTIONS IN CENTRAL NEVADA CS LISTED IN ORDER OF COMPILATION

Central Nevada Section	Range-bases/tops Co/Gr/Ev	Equation of LOC and Controlling Points	Figure
1. Simpson Park Range I (SRS)	0/59/0	NA	NA
2. Pete Hanson Creek IA	0/48/1	Segment 1: CS = 0.550PIA + 5.64 B5,B11,B39 Segment 2: CS = 1.755PIA - 35.43 T9,B11 Segment 3: CS = 0.980PIA + 21.27 T9,B20,B26,43	2A
3. Pete Hanson Creek IB	0/45/1	Segment 1: CS = 0.401PIB - 3.69 B1,B3,B11 Segment 2: CS = 1.734PIB - 97.15 B11,B21 Segment 3: CS = 1.072PIB - 22.93 B21,43	2B
4. Pete Hanson Creek IIC&F	26/12/2	Segment 1: CS = 1.121PIIC&F + 121.36 T4,43 Segment 2: CS = 0.685PIIC&F + 140.26 43,44 Segment 3: CS = 0.810PIIC&F + 129.52 44,T27	2C
5. Pete Hanson Creek III	0/33/1	Segment 1: CS = 0.846PIII - 22.03 B5,B11 Segment 2: CS = 1.507PIII - 58.36 B11,B20 Segment 3: CS = 0.895PIII + 8.51 B20,44	2D

Note: Co, conodont; Gr, graptolite; Ev, event; LOC, Line of Correlation; Figure, figure to see for graphic correlation of section with central Nevada CS; SRS, Standard Reference Section; NA, not applicable; CS, central Nevada CS; B, base; T, top.

tions individually with the Silurian CS, because the former method made it easier to graphically correlate the Simpson Park Range I section in Berry and Murphy (1975) with the Silurian CS. This correlation was important for that section is clearly the thickest, uninterrupted stratigraphic section in the Roberts Mountains Formation in central Nevada and contained the greatest amount of graptolite range-data.

Only the five sections listed in Table 1 were used to develop a central Nevada CS, rather than nine sections used by Murphy (1989) to develop a similar central Nevada CS. Three of the sections (Birch Creek II, Pete Hanson Creek II, Pete Hanson Creek IID) used by Murphy (1989) were already a part of the Silurian CS. They were kept in the Silurian CS so that it would be easier to graphically correlate the central Nevada CS with the Silurian CS. Murphy (1989) shows only a composite range chart for the species represented in the central Nevada sections. He did not show which section controlled the range of the species. Those data have since been lost or destroyed (M. Murphy, pers. commun., 1992). The Willow Creek I section used by Murphy (1989) was not used for the central Nevada CS because it was not possible to graphically correlate it with the central Nevada CS. The Simpson Park Range I section in Berry and Murphy (1975) was selected as the central Nevada standard reference section because it contained the greatest amount of range-data (Table 1) and is the thickest, uninterrupted Silurian stratigraphic section in central Nevada described by Berry and Murphy (1975) and Klapper and Murphy (1975). The four remaining sections were then ranked and compiled in the order shown in Table 1.

Lines of correlation were fitted by visual inspection to an array of points and drawn through what, in the best judgment of the compositor, were determined to be the best-controlled points. For example, in Figure 1A, segment 1 of the LOC is controlled by the first appearance datums (FADs) of *Monograptus flemingii* (Table 2, species 39), its subspecies *M. flemingii flemingii* (Table 2, species 5), and *Monograptus testis* (Table 2, species 11). Both species are represented in most of the central Nevada sections and the FADs of both are important in standard graptolite biozonation. Segment 2 of the LOC is controlled by the FAD of *M. testis* and the last appearance datum (LAD) of *Monoclimacis flumendosae* (Table 2, species 9), which is also represented in most of the central Nevada sections and is important in standard graptolite biozonation. Segment 3 of the LOC is controlled by the LAD of *M. flumendosae*, the FADs of *Pristiograptus ludensis* (Table 2, species 20) and *Saetograptus chimaera* (Table 2, species 26), and Roberts Mountains Formation bed A of Berry and Murphy (1975) (Table 2, event 43). *Pristiograptus ludensis* and *S. chimaera* are both represented in most of the central Nevada sections and are both important in standard graptolite biozonation. Roberts Mountains Formation bed A is a limestone conglomerate which was interpreted as a bed deposited synchronously throughout its geographic distribution (Winterer and Murphy, 1960).

The central Nevada CS reached stability at the end of two recorrelation rounds. Figure 1 shows graphs of sections and the central Nevada CS in the final recorrelation round. Table 2 lists the ranges of graptolite and conodont species and events in the central Nevada CS.

Revising the Silurian Composite Standard

The Silurian CS of Kleffner (1989) was revised by graphically correlating the newly developed central Nevada CS (section 4, Table 3) and many other sections previously not a part of the Silurian CS framework (sections 22–29, Table 3) with the Silurian CS and then recorrelating all of the sections comprising the Silurian CS with it (Table 3; Fig. 2 shows locations of sections). The next two sections graphically correlated with the Silurian CS were Klonk, Czechoslovakia (section 22, Table 3), and Gräfenwarth, Germany (section 23, Table 3). They were incorporated next because they provided another boundary stratotype and range-data on 18 additional graptolite species, respectively, thereby making it easier to correlate other sections with the Silurian CS. The remaining sections (sections 24–29,

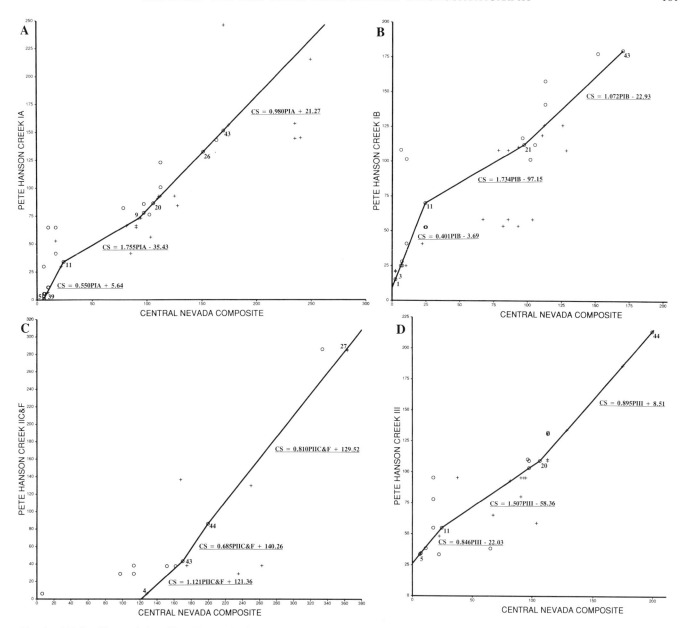

FIG 1.—(A) Graphic correlation of Pete Hanson Creek IA, Nevada section with central Nevada CS. (B) Graphic correlation of Pete Hanson Creek IB, Nevada section with central Nevada CS. (C) Graphic correlation of Pete Hanson Creek IIC & F, Nevada section with central Nevada CS. (D) Graphic correlation of Pete Hanson Creek III, Nevada section with central Nevada CS. All data from Berry and Murphy (1975), except for Pete Hanson Creek IIC & F, which also includes data from Klapper and Murphy (1975). A–D all show results of second recorrelation round.

Table 3) were correlated in the same order as they are listed in Table 3. All the sections in the Silurian CS (Table 3) were recorrelated with the Silurian CS until values for FADs and LADs stabilized (eight rounds of recorrelation).

Each LOC was fitted by visual inspection to an array of points and drawn through what, in the best judgment of the compositor, were determined to be the best-controlled points. Arrays composed of most or all of those best-controlled points were subsequently used to calculate an equation for the LOC by the method of least squares that was as close as possible to the equation for the LOC determined by its more qualitative visual fit. This calculation was done so that the resolution of each LOC could be quantified. The standard error of estimate (**S**) (Shaw, 1964, p. 175–181) was calculated for each of the LOC equations at a 95% confidence level (Table 3) and adjusted for small numbers of points using the appropriate values of t shown in Table 23–3 of Shaw (1964). The calculated value of **S** for each LOC equation measures the dispersion within the array used in calculating the LOC. **S** is also a measure of the dispersion that may be expected around the best points when projected into the Silurian CS. The maximum **S** calculated for all of the LOC equations was considered to be the limit of resolution for the Silurian CS. That is, no value smaller than S may be resolved with a high level of confidence in the Silurian CS. Figures 3, 4, 5, and 6 are graphs of sections and the Silurian CS in the eighth recorrelation round. Table 3 gives the location

TABLE 2.—CENTRAL NEVADA COMPOSITE STANDARD

Species/Event and No. of Sections in	CSS Range	Range-base*	Range-top*	Species/Event and No. of Sections in	CSS Range	Range-base*	Range-top*
1 *Cyrtograptus* cf. *C. sakmaricus* 2	2.40-4.85	1,3	3	29 *Monograptus microdon* 1	395.30-507.80	1	1
2 *Retiolites geinitzianus geinitzianus* 1	2.40	1	1	30 *Linograptus posthumus posthumus* 1	507.80	1	1
3 *Cyrtograptus* aff. *C. lapworthi* 2	2.40-4.60	1,3	3	31 *Pristiograptus* sp. ex. gr. *P. dubius* 3	161.69-362.70	2	1
4 *Pristiograptus* cf. *P. praedubius* 5	6.30-128.20	5	4,5	32 *Monograptus uniformis uniformis* 1	395.30-507.80	1	1
5 *Monograptus flemingii flemingii* 3	6.30-103.60	2,5	1	33 *Cucullograptus pazdroi* 1	164.04-185.50	2	2
6 *Monograptus flemingii primus* 2	10.10-57.03	1	2	34 *Saetograptus fritschi* 1	174.82	2	2
7 *Cyrtograptus rigidus* 3	6.33-10.10	3	1	35 *Monograptus spiralis* 1	1.80	2	2
8 *Cyrtograptus perneri* 4	10.44-22.30	5	1	36 *Monoclimacis* cf. *M. crenulatus* 1	1.80-4.85	3	3
9 *Monoclimacis flumendosae* 4	7.06-93.00	5	1,2	37 *Retiolites geinitzianus angustidens* 1	2.40-4.44	3	3
10 *Monograptus priodon* 1	7.06	5	5	38 *Retiolites geinitzianus* 2	2.40-4.85	1,3	3
11 *Monograptus testis* 4	24.40-94.63	1,2,3,5	3	39 *Monograptus flemingii* 4	6.30-103.60	2,3,5	1
12 *Cyrtograptus hamatus* 3	17.43-67.10	3	1	43 Roberts Mountains Formation bed A	170.00	2,3,4	
13 *Cyrtograptus radians* 4	17.43-85.45	3	5	44 Roberts Mountains Formation bed B	199.39	4,5	
14 *Cyrtograptus lundgreni* 4	17.43-81.83	3	5	46 *Polygnathoides siluricus* 1	205.31-213.17	4	4
15 *Monograptus retroflexus* 4	10.44-90.46	5	3	47 *Ozarkodina crispa* 1	280.06	4	4
16 *Retiolites nevadensis* 4	6.30-90.46	5	3	48 *Kockelella variabilis* 1	161.60-205.31	4	4
17 *Pristiograptus jaegeri* 4	97.40-112.41	3	3	49 *Ozarkodina excavata excavata* 1	125.85-362.73	4	4
18 *Gothograptus nassa* 3	78.85-110.54	3	2	51 *Ozarkodina confluens* 1	147.03-362.73	4	4
19 *Spinograptus spinosus* 4	96.23-125.51	2	5	52 *Polygnathoides emarginatus* s. f. 1	205.31	4	4
20 *Pristiograptus ludensis* 3	106.10-112.40	2,5	2	54 *Panderodus unicostatus* 1	147.03-364.52	4	4
21 *Pristiograptus dubius frequens* 5	97.36-235.30	3,5	1	55 *Ozarkodina confluens* alpha morphotype 1	148.71-356.09	4	4
22 *Bohemograptus bohemicus* 4	112.80-262.62	1	2	56 *Ozarkodina confluens* beta morphotype 1	148.71-349.70	4	4
23 *Neodiversograptus nilssoni* 4	112.80-240.38	1	4	57 *Ozarkodina confluens* delta morphotype 1	321.03-356.09	4	4
24 *Saetograptus colonus* 4	112.80-235.30	1	1	58 *Ozarkodina* n. sp. A Klapper & Murphy, 1975 1	284.75	4	4
25 *Monograptus uncinatus* 4	112.80-174.82	1	2	59 *Pelekysgnathus index* 1	276.17-364.52	4	4
26 *Saetograptus chimaera* 4	151.50-249.90	1,2	1	60 *Ozarkodina confluens* gamma morphotype 1	296.33-362.73	4	4
27 *Saetograptus willowensis* 2	333.80-362.70	1	1	61 *Ozarkodina douroensis* 1	205.31-213.17	4	4
28 *Monograptus birchensis* 1	382.50-507.80	1	1	62 *Icriodus woschmidti* 1	427.92	4	4

*Numbers refer to sections in Table 1.

for the graph of each section in Figures 3, 4, 5, and 6 and summarizes information about each section in the Silurian CS after eight recorrelation rounds.

Figure 3A provides an example of how LOCs were fitted to the arrays of points and how **S** was determined for each of the LOC equations. In the graphic correlation (Fig. 3A) of the Birch Creek II, Nevada section of Berry and Murphy (1975) and Klapper and Murphy (1975) with the Silurian CS, the LOC was fitted by visual inspection to the FADs of *Icriodus woschmidti* (Table 4, species 12), *Pelekysgnathus index* (Table 4, species 46), and *Monograptus uniformis* (Table 4, species 48); and to the LADs of *Ancoradella ploeckensis* (Table 4, species 6), *Pedavis latialata* (Table 4, species 8), *Ozarkodina remscheidensis eosteinhornensis* (Table 4, species 10), *Kockelella variabilis* (Table 4, species 13), and *P. index*. All of those species are represented in at least three sections in the Silurian CS, and all are important in standard conodont and graptolite biozonations. In order to get an equation for the LOC calculated by using the least squares method that was as close as possible to that determined by the equation of the LOC fitted by visual inspection to the points listed above, all points but the LADs of *O. remscheidensis eosteinhornensis* and *P. index* were used. S for the LOC calculated using those six points (0.0017) at the 95% confidence level must be adjusted by multiplying it by the value of t for a sample size of six (2.776) (Table 3, section 2, S = 0.00, or 0.0017 × 2.776 (= .0048) rounded to the nearest hundredth).

RESULTS

Definition and Resolution of the Silurian Composite Standard

The revised Silurian CS at the end of eight recorrelation rounds is given in Table 4 and Figure 7. Table 4 lists ranges of all zonal conodont and graptolite species and boundary stratotypes. Table 4 also lists ranges of all other conodont and graptolite species and events that occur in two or more sections in the Silurian CS. Figure 7 shows the Silurian chronostratic scale based on those ranges. The Silurian CS is controlled from a level in the Aeronian Stage to the Silurian/Devonian boundary. Part of the Aeronian and all of the Rhuddanian Stage are not included in the Silurian CS because that part of the Silurian System (and the rest of the Aeronian Stage as well) is missing at Cellon, Austria, the standard reference section (SRS) of the Silurian CS (Schönlaub, 1971), and because at the time of this study there were insufficient conodont and graptolite range-data in available literature on Rhuddanian to Aeronian stratigraphic sections.

The maximum **S** for the Silurian CS after eight recorrelation rounds is 0.52 units (Table 3). A division of this CS into 0.52 units is thereby the limit of resolution for recognition with a high level of confidence. The Silurian CS includes approximately 48 units; thus, the nonannular absolute chronology based on it divides into 92 Standard Time Units (STUs) (Fig. 7, column D). Equivalents of each STU can be recognized in all sections that are a part of the current Silurian CS or that can be graphically correlated with it. Most sections graphically correlated with the Silurian CS (SRS: Cellon, Austria) by fitting a straight LOC rather than several doglegged segments (Figs. 3–6). Several possibilities could account for a straight LOC between a section and the Silurian CS. (1) The control provided by FADs, LADs, and events could be insufficient to determine if the LOC between the section and the Silurian CS should be a straight line or consist of several doglegged segments. (2) The relative rock accumulation rate could be changing at the same

TABLE 3.—SECTIONS IN SILURIAN CS LISTED IN ORDER OF COMPILATION

Section	Range-bases/-tops Co/Gr/Ev	Equation of LOC and Controlling Points	S	Fig.
1. Cellon, Austria (SRS)	66/15/0	NA	NA	NA
2. Birch Creek II, Nevada	43/8/0	CS = 0.170BCII + 27.97 T6,T8,B12,T13,B46,B48	0.00	3A
3. Jacksonville, Ohio	44/0/0	Segment 1: CS = 0.030JAK + 14.48 B2,T49	0	3B
		Segment 2: CS = 0.051JAK + 14.79 B4,B18,B51	0.52	
4. Central Nevada Composite* (5 sections: Simpson Park Range 1, SRS)	29/75/2	Segment 1: CS = 0.036CNV + 15.53 B80,B113,B116	0.09	3C
		Segment 2: CS = 0.105CNV + 9.38 B72,144,145	0.01	
		Segment 3: CS = 0.128CNV + 4.91 145,T46	0	
		Segment 4: CS = 0.188CNV − 17.06 T46,B48	0	
5. Haragan Creek, Oklahoma	20/0/0	CS = 0.235HC + 17.13 T3,T26,T51,T53,B61,B63	0.05	3D
6. Pete Hanson Creek II, Nevada	24/48/2	Segment 1: CS = 0.014PII + 15.43 T49,B80	0	3E
		Segment 2: CS = 0.079PII + 10.76 B80,144	0	
		Segment 3: CS = 0.070PII + 12.83 144,145	0	
7. Pete Hanson Creek IIE, Nevada	24/4/0	CS = 0.148PIIE + 26.06 T6,T41	0	3F
8. Pete Hanson Creek IID, Nevada	25/1/0	CS = 0.148PIID + 29.01 B12,B46	0	3G
9. CT, Tennessee	23/0/0	CS = 0.376CT + 15.44 T2,B63	0	3H
10. Sinking Spring, Ohio	20/0/0	CS = 0.066SS + 16.90 B4,B51,T52	0.05	4A
11. Highway 77, Oklahoma	17/0/0	CS = 0.445H77 + 16.69 B13,T26,B63	0.01	4B
12. VU, Tennessee	20/0/0	CS = 0.055VU + 17.59 B13,T51	0	4C
13. M2, Oklahoma	21/0/0	CS = 0.200M2 + 18.22 B13,T52,T53	0.05	4D
14. Central Appalachians Composite** (5 sections: Pinto, Maryland, SRS)	35/0/0	CS = 0.088CA + 26.23 T9,B12,B89	0.05	4E
15. Niagara Gorge, New York and Ontario	36/0/0	CS = 0.109NG + 15.37 T2,B4,B51	0.02	4F
16. C1, Oklahoma	32/0/0	CS = 0.180C1 + 15.68 T2,T26,T51	0.01	4G
17. Malmøykalven, Norway	31/0/0	CS = 0.112MKN + 9.49 B1,T1,B2,B71	0.10	4H
18. Brisants Jumpers, Anticosti Island	32/0/0	CS = 0.092BJ + 12.47 B54,B71	0	4I
19. Crabbottom, Virginia	14/0/0	CS = 0.141CB + 21.65 T9,B89	0	6A
20. Bohemia Composite*** (6 sections: Marble Quarry, Czechoslovakia, SRS)	26/29/1	Segment 1: CS = 1.337BO + 27.22 B8,T13	0	5A
		Segment 2: CS = 3.126BO + 21.26 T8,B95	0	
		Segment 3: CS = 1.304BO + 30.90 B95,B12	0	
21. Leasows, Great Britain	26/0/2	CS = 1.326L + 13.92 B2,T2	0	5D
22. Klonk, Czechoslovakia	17/4/1	CS = 1.000K + 52.00 T18,B48,B164	0.00	5B
23. Gräfenwarth, Germany	0/58/0	CS = 2.676G + 16.40 B72,B80	0	5C
24. Mason Porcus, Sardinia	18/0/0	CS = 0.375MP + 54.41 B12,T18	0	6B
25. CA2, Oklahoma	15/0/0	CS = 0.422CA2 + 16.67 B13,T26,T52,B63	0.08	6G
26. Jõhve, Estonia	69/0/0	CS = 0.059J + 13.01 B1,T1,B2	0.14	6D
27. Ireviken, Gotland	0/0/8	CS = 0.078I + 15.38 205,206,207	0.02	6E
28. BGS Eastnor Park, Great Britain	10/0/3	CS = 0.023BGS + 15.44 205,206,207	0.15	6F
29. Greene County, Ohio	31/0/0	CS = 0.277GC + 14.40 T2,T26,B61	0.10	6C

*The following sections comprise the central Nevada CS: A, Simpson Park Range I; B, Pete Hanson Creek IA; C, Pete Hanson Creek IB; D, Pete Hanson Creek IIC & F; E, Pete Hanson Creek III. All data from Berry and Murphy (1975), except for Pete Hanson Creek IIC & F, which also includes data from Klapper and Murphy (1975).

**The following sections comprise the central Appalachians CS: A, Pinto, Maryland; B, Lambert Gap, West Virginia; C, Fulks Run, Virginia; D, McDowell, Virginia; E, New Creek, West Virginia. Data from Helfrich (1975) and from unpublished charts assembled by K. Denkler and A. Harris for Denkler and Harris (1985).

***The following sections in Czechoslovakia comprise the Bohemia Composite: A, Marble Quarry; B, Mramorovy; C, Muŝlovka; D, Koledník; E, Požáry; F, Na Pozarech. Data from Kříž and others (1986), Mehrtens and Barnett (1976), and Chlupáč and others (1980).

Note: Co, conodont; Gr, graptolite; Ev, event; LOC, Line of Correlation; **S**, Standard Error of Estimate; Fig., figure to see for graphic correlation of section with Silurian CS; SRS, Standard Reference Section; NA, not applicable; CS, Silurian Composite Standard; T, top; B, base; 0, no error for only two points used to determine equation of LOC.

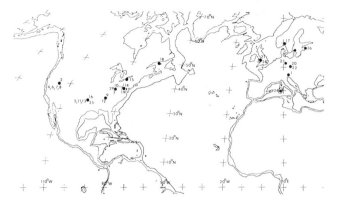

FIG 2.—General location of sections that comprise Silurian Composite Standard. Sections within a 160-km (100-mi) radius are shown by the same black circle. Sections are numbered as in Table 3.

relative rate at the same time at both the section being correlated with the Silurian CS and at the SRS of the Silurian CS. (3) The relative rock accumulation rate could be constant at the section graphically correlated with the Silurian CS and at the SRS of the Silurian CS.

Some of the straight LOCs between sections and the Silurian CS, like the one between the Brisants Jumpers, Anticosti Island section and the Silurian CS (Fig. 4I), could have resulted because of the first possibility, a very limited control afforded by FADs and LADs. However, many other straight LOCs between sections and the Silurian CS were fitted to arrays of FADs and LADs that provided good control, like the straight LOCs between the Birch Creek II, Nevada section and the Silurian CS (Fig. 3A), and the Pete Hanson Creek IID, Nevada section and the Silurian CS (Fig. 3G). The straight LOCs of the Pete Hanson Creek IID and Birch Creek II sections are also well controlled for one-third to one-half of the total thickness of the SRS of the Silurian CS. The straight LOCs of the Pete Hanson Creek IID and Birch Creek II sections with the Silurian CS most likely indicate the third possibility, relatively constant rock accumulation rates at those two sections and at the SRS of the Silurian CS. It is highly unlikely that the second possibility, relative rock accumulation rates changing at the same relative rate at the same time at both the section being correlated with the Silurian CS and at the SRS of the Silurian CS, could account for the straight LOCs of the Pete Hanson Creek IID and Birch Creek II sections, for both of those sections contain such great thicknesses of strata. There are reasonably- to well-controlled straight LOCs for most of the sections graphically correlated with the post-Sheinwoodian Wenlock through Pridoli strata of the Silurian CS (Figs. 3–6), indicating relatively constant rock accumulation rates for those strata of the SRS. A section that does not create a doglegged pattern when plotted against other sections will be that which is most nearly consonant with an annular scale (Shaw, 1964). Since most, if not all, of the SRS of the Silurian CS fits that description, the absolute chronology based on the Silurian CS is thereby consonant (or nearly so) with an annular scale. The 92 STUs it divides into are of approximately equal annular length. There are a few sections that have doglegged LOCs with the Silurian CS, the Bohemia CS (Fig. 5A), the central Nevada CS (Fig. 3C), and the Pete Hanson Creek II (Fig. 3E) sections. The doglegged LOCs of those sections with the Silurian CS are an indication that those sections did not experience relatively constant rock accumulation rates. This conclusion is further supported in the case of the central Nevada CS by the doglegged patterns created when all of the central Nevada sections were graphically correlated with the central Nevada CS (Fig. 1).

Conodont Chronozones

All but one of the Silurian conodont zones proposed by Walliser (1964) can be recognized in the Silurian CS with a high level of confidence as chronozones. Conodont chronozones are defined in the Silurian CS according to international rules of stratigraphy; each conforms to a Global Boundary Stratotype Section and Point (GSSP). The strata of the SRS (Cellon, Austria) of the Silurian CS serve as the Global Boundary Stratotype Section for those chronozones. The GSSPs for the bases of the *Pterospathodus celloni* (Fig. 7, column H, species 1), *Ozarkodina sagitta sagitta* (Fig. 7, column H, species 4), *O.? crassa* (Fig. 7, column H, species 5), *Ancoradella ploeckensis* (Fig. 7, column H, species 6), *Polygnathoides siluricus* (Fig. 7, column H, species 7), *Pedavis latialata* (Fig. 7, column H, species 8), *O. crispa* (Fig. 7, column H, species 9), *O. remscheidensis eosteinhornensis* (Fig. 7, column H, species 10), and *Icriodus woschmidti* (Fig. 7, column H, species 12) Chronozones are the FADs or projected FADs of representatives of those subspecies or species in the SRS (Fig. 7, Columns E, H). The top of each of those chronozones is the GSSP of the succeeding chronozone (FAD or projected FAD in the SRS of the zonal index of the succeeding chronozone; Fig. 7, columns E, H). The GSSPs for the base and top of the *Pterospathodus amorphognathoides* Chronozone (Fig. 7, column E) are the projected FAD and LAD of *Pt. amorphognathoides* in the SRS, respectively (Fig. 7, column H, species 2). A chronozone based on the range of *Kockelella patula* (Fig. 7, column H, species 3) would include less than 1 STU; thus, it cannot be recognized with confidence in the Silurian CS and is not shown on Figure 7. The three additional chronozones added by Kleffner (1989) to the sequence established by Walliser can also be recognized with a high level of confidence. The GSSPs for base and top of the *K. variabilis* Chronozone (Fig. 7, column E) are the projected FAD and LAD of *K. variabilis* in the SRS, respectively (Fig. 7, column H, species 13). The *Distomodus staurognathoides* Chronozone (Fig. 7, column E) is also recognized. The GSSP for the base of the *D. staurognathoides* Chronozone is the projected FAD in the SRS of *D. staurognathoides* (Fig. 7, column H, species 25); the top of that chronozone is the base of the *P. celloni* Chronozone (Fig. 7, column E). The lower boundary is dashed in Figure 7 because *D. staurognathoides* has been reported from uncompiled stratigraphic sections older than any that are currently a part of the Silurian CS. Although a *K. ranuliformis* (Fig. 7, column H, species 26)-*K. amsdeni* (Fig. 7, column H, species 63) Chronozone, as defined by Kleffner (1989), can be recognized with a high level of confidence, it is replaced by two different chronozones and not shown on Figure 7.

Four additional chronozones are added to the sequence shown in Kleffner (1989). The *Kockelella ranuliformis* Chronozone (Fig. 7, column E) is based on a Wenlock conodont zone proposed by Barrick and Klapper (1976). The base of the *K. ranuliformis* Chronozone is the top of the *Pterospathodus*

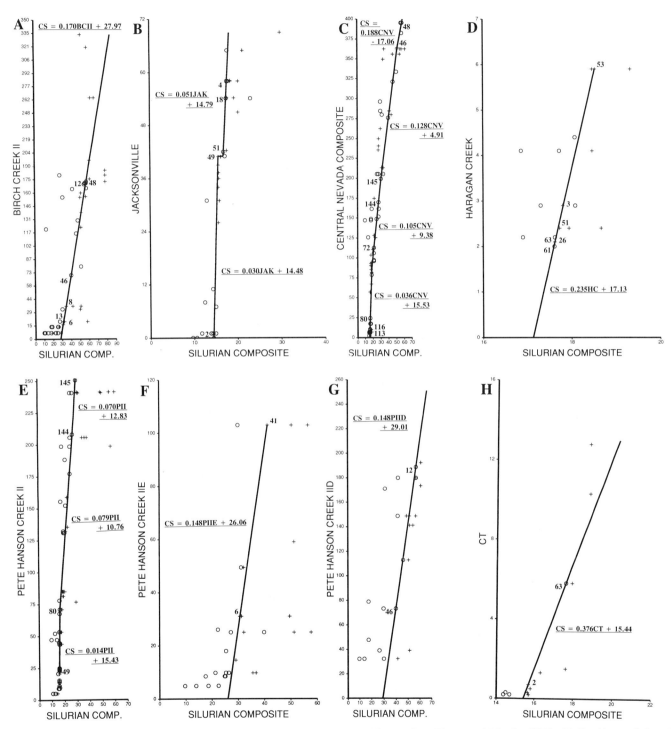

FIG 3.—(A) Graphic correlation of Birch Creek II, Nevada section with Silurian CS. Data from Klapper and Murphy (1975). (B) Graphic correlation of Jacksonville, Ohio section with Silurian CS. Data from Kleffner (1987, 1990). (C) Graphic correlation of central Nevada CS with Silurian CS. Data from Berry and Murphy (1975) and Klapper and Murphy (1975). (D) Graphic correlation of Haragan Creek, Oklahoma section with Silurian CS. Data from Barrick and Klapper (1976). (E) Graphic correlation of Pete Hanson Creek II, Nevada section with Silurian CS. Data from Berry and Murphy (1975) and Klapper and Murphy (1975). (F) Graphic correlation of Pete Hanson Creek IIE, Nevada section with Silurian CS. Data from Berry and Murphy (1975) and Klapper and Murphy (1975). (G) Graphic correlation of Pete Hanson Creek IID, Nevada section with Silurian CS. Data from Berry and Murphy (1975) and Klapper and Murphy (1975). (H) Graphic correlation of CT, Oklahoma section with Silurian CS. Data from Barrick (1983). A-H all show results of eighth recorrelation round.

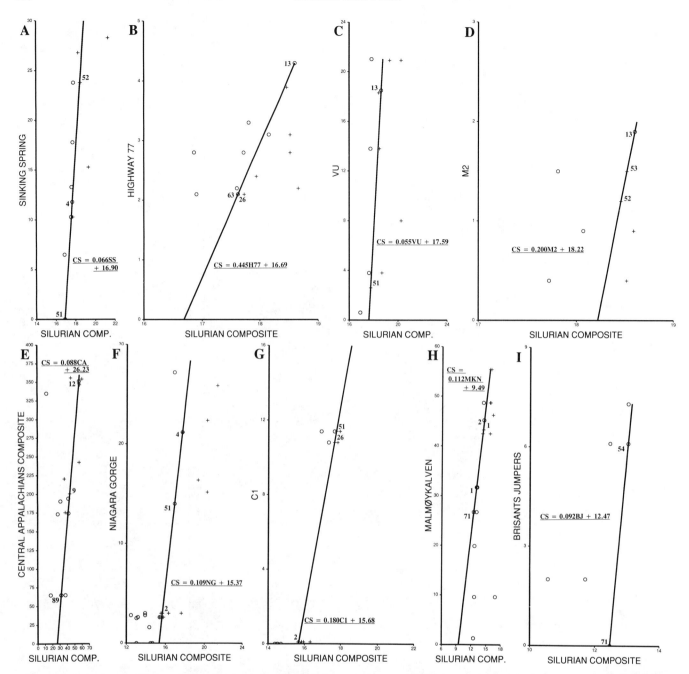

FIG 4.—(A) Graphic correlation of Sinking Spring, Ohio section with Silurian CS. Data from Kleffner (1987, 1990). (B) Graphic correlation of Highway 77, Oklahoma section with Silurian CS. Data from Barrick and Klapper (1976). (C) Graphic correlation of VU, Tennessee section with Silurian CS. Data from Barrick (1983). (D) Graphic correlation of M2, Oklahoma section with Silurian CS. Data from Barrick and Klapper (1976) and Barrick (1977). (E) Graphic correlation of central Appalachians CS with Silurian CS. Data from Helfrich (1975) and from unpublished charts assembled by K. Denkler and A. Harris for Denkler and Harris (1985). (F) Graphic correlation of Niagara Gorge, New York and Ontario section with Silurian CS. Data from Kleffner (1991). (G) Graphic correlation of C1, Oklahoma section with Silurian CS. Data from Barrick and Klapper (1976) and Barrick (1977). (H) Graphic correlation of Malmøykalven, Norway section with Silurian CS. Data from Nakrem (1986). (I) Graphic correlation of Brisants Jumpers, Anticosti Island section with Silurian CS. Data from Uyeno and Barnes (1983). A–I all show results of eighth recorrelation round.

amorphognathoides Chronozone (Fig. 7, column E), and the upper boundary is the base of the *Ozarkodina sagitta rhenana* Chronozone (Fig. 7, column E). The *O. sagitta rhenana* Chronozone is based on a Wenlock conodont zone proposed by Aldridge and Schönlaub (1989). The GSSP for the base of the *O. sagitta rhenana* Chronozone (Fig. 7, column E) is the projected FAD in the SRS of representatives of that species (Fig. 7, column H, species 51). The top of the *O. sagitta rhenana* Chronozone is the base of the *O. sagitta sagitta* Chronozone (Fig. 7, column E). The *O. remscheidensis remscheidensis* (Fig. 7, column E), *Oulodus elegans detorta* (not shown on Fig. 7), and previously defined *O. crispa* (Fig. 7, column E) Chronozones

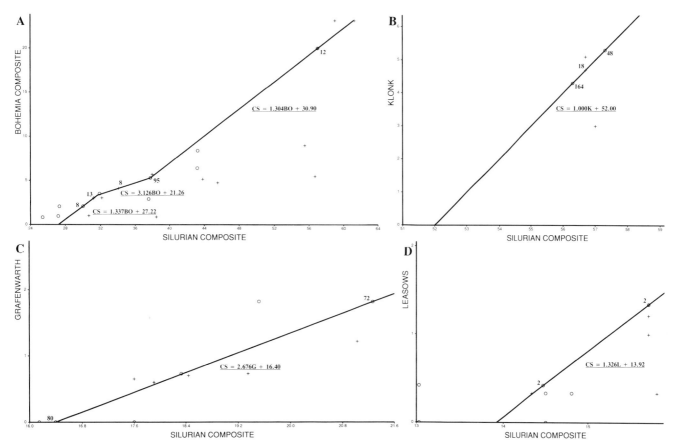

FIG 5.—(A) Graphic correlation of Bohemia Composite (includes Požáry section, the Ludlow/Pridoli boundary stratotype) with Silurian CS. Data from Kříž and others (1986), Mehrtens and Barnett (1976), and Chlupáč and others (1980). (B) Graphic correlation of Klonk, Czechoslovakia section (Silurian/Devonian boundary stratotype) with Silurian CS. Data from Jeppsson (1988). (C) Graphic correlation of Gräfenwarth, Germany section with Silurian CS. Data from Jaeger (1991). (D) Graphic correlation of Leasows (Hughley Brook), Great Britain section (Llandovery/Wenlock boundary stratotype) with Silurian CS. Data from Mabillard and Aldridge (1985). A–D all show results of eighth recorrelation round.

are recognized within the *O. remscheidensis eosteinhornensis* Chronozone. The GSSP for the base of the *O. remscheidensis remscheidensis* Chronozone is the projected FAD of representatives of *O. remscheidensis remscheidensis* in the SRS (Fig. 7, column H, species 11); the top of that chronozone is the base of the *Icriodus woschmidti* Chronozone (Fig. 7, column E). The *Ou. elegans detorta* Chronozone is based on the uppermost Pridoli conodont zone proposed by Jeppsson (1988); the GSSP of its base is the FAD of representatives of that species (Fig. 7, column H, species 164) in the SRS. The top of the *Ou. elegans detorta* Chronozone is the base of the *I. woschmidti* Chronozone.

Graptolite Chronozones

Most of the post-Llandovery graptolite zones proposed by Cocks and Nowlan (1993) for the standard left-hand column (SLHC) for international Silurian correlation charts can be recognized in the Silurian CS with a high level of confidence as chronozones. They are defined as chronozones using the same procedure used to define conodont chronozones. The GSSPs for the bases of the *rigidus/ellesae* (Fig. 7, column H, species 82/species 115?), *lundgreni* (Fig. 7, column H, species 139), *nassa/deubeli* (Fig. 7, column H, species 138/species not shown), *ludensis* (Fig. 7, column H, species 83), *formosus* (Fig. 7, column H, species 106), *parultimus/ultimus* (Fig. 7, column H, species 95/species 68), *lochkovensis* (Fig. 7, column H, species 69), *bouceki/transgrediens* (Fig. 7, column H, species 44/species 70), and *uniformis* (Fig. 7, column H, species 48) Chronozones are the projected FADs of representatives of those species (first-named species in all of the combined zones except for the *bouceki/transgrediens* Chronozone for which the second-named species is used) in the SRS (Fig. 7, column F). The top of each of those chronozones and the *bohemicus/koslowskii* (Fig. 7, column H, species 73/species not represented in Silurian CS) Chronozone is the GSSP for the succeeding chronozone (FAD or projected FAD in the SRS of the zonal index of the succeeding chronozone; first-named species in all of the combined zones except for the *bouceki/transgrediens* Chronozone for which the second-named species is used) (Fig. 7, columns F, H). The GSSP for the base of the *leintwardinensis* Chronozone (Fig. 7, column F) is the projected FAD in the SRS of *Monograptus fritschi linearis* (Fig. 7, column H, species 74). The GSSP for the top of the *leintwardinensis* Chronozone is the projected LAD in the SRS of *M. fritschi linearis*. The top of the *leintwardinensis* Chronozone also defines the base of the *bohemicus/koslowskii* Chronozone (Fig. 7, column F). Chronozones based on the *scanicus*, *centrifugus/murchisoni*, and *ric-*

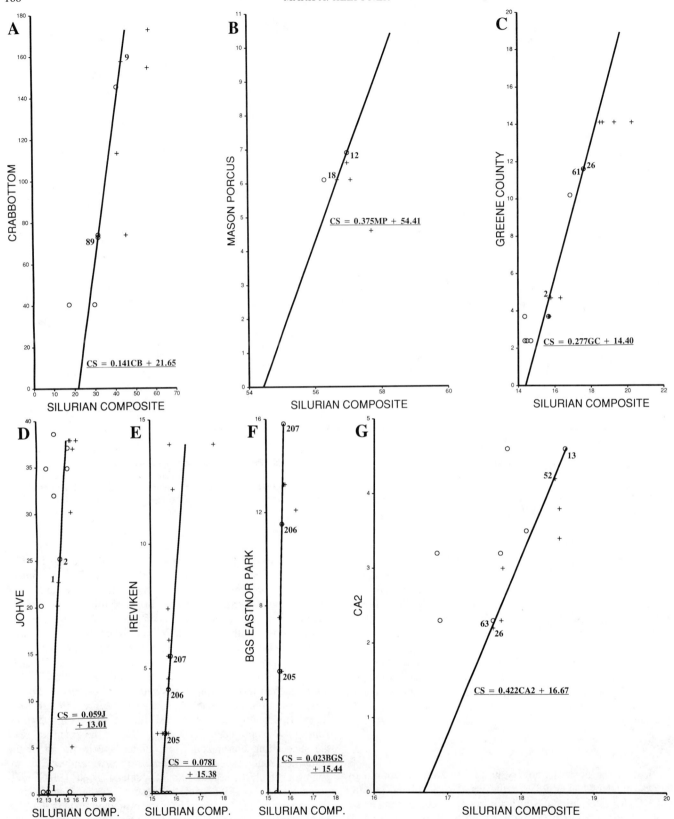

FIG 6.—(A) Graphic correlation of Crabbottom, Virginia section with Silurian CS. Data from Helfrich (1975). (B) Graphic correlation of Mason Porcus, Sardinia section with Silurian CS. Data from Gnoli and others (1988) and Olivieri and Serpagli (1990). (C) Graphic correlation of Greene County, Ohio section with Silurian CS. Data from Kleffner (1994). (D) Graphic correlation of Jõhve, Estonia section with Silurian CS. Data from P. Männik (pers. commun., 1993). (E) Graphic correlation of Ireviken, Gotland section with Silurian CS. Data from Aldridge and others (1993). (F) Graphic correlation of BGS Eastnor Park, Great Britain section with Silurian CS. Data from Aldridge and others (1993). (G) Graphic correlation of CA2, Oklahoma, section with Silurian CS. Data from Barrick and Klapper (1976). A–G all show results of eighth recorrelation round.

TABLE 4.—SILURIAN COMPOSITE STANDARD

Species/Event and No. of Sections in*	CSS Range	Range-base**	Range-top**	Species/Event and No. of Sections in*	CSS Range	Range-base**	Range-top**
1 Pterospathodus celloni 4	13.03-14.36	17,26	21	71 Aulacognathus bullatus 3	12.47-14.24	17,18	17
2 Pterospathodus amorphognathoides 11	14.48-15.70	3,21	9,15,16,21,27,28,29	72 Neodiversograptus nilssoni 6	21.27-35.64	4A,23	4D
3 Kockelella patula 3	17.30-17.81	1	5	73 Bohemograptus bohemicus 8	21.27-38.48	4A,23	4B
4 Ozarkodina sagitta sagitta 4	17.68-21.30	10,15	1	74 Monograptus fritschi linearis 2	27.22-28.82	20A	20C
5 Ozarkodina? crassa 2	18.60-22.30	13	1	75 Pristiograptus dubius 5	16.40-37.81	23	20C
6 Ancoradella ploeckensis 5	22.30-30.67	1	2,7	79 Belodella silurica 3	12.47-18.59	17	12
7 Polygnathoides siluricus 8	27.20-32.16	1	4D	80 Monograptus testis 6	16.40-19.35	4A,4B,4C,4E,6,23	4C
8 Pedavis latialata 6	30.00-34.08	1,20F	2,20A	82 Cyrtograptus rigidus 5	15.74-17.60	4C,6	1
9 Ozarkodina crispa 11	31.90-43.81	14A	14B,19	83 Pristiograptus ludensis 3	20.56-21.27	4E	6
10 Ozarkodina remscheidensis eosteinhornensis 16	31.10-57.00	20F	1,20F	84 Saetograptus chimaera 6	25.35-36.86	4A,4B	4A
11 Ozarkodina remscheidensis remscheidensis 12	41.84-61.24	20F	2	85 Ozarkodina remscheidensis trans. eosteinhornensis 5	42.91-56.64	14C	14B
12 Icriodus woschmidti 9	57.00-72.89	2,8,14D,20F,24	2	86 Ozarkodina remscheidensis trans. remscheidensis 6	41.50-57.39	14C	14D
13 Kockelella variabilis 11	18.60-31.23	11,12,13	2,20A	87 Oulodus cristigalli 2	56.70-57.32	14E	14D
14 Ozarkodina excavata 25	13.92-61.11	21	20F	88 Homeognathus penicullus 4	45.63-47.93	14B	14C
15 Ozarkodina excavata hamata 3	22.50-29.11	1	2	89 Ozarkodina highlandensis 4	31.90-41.58	14C,19	14A,14B,14C
16 Ozarkodina excavata posthamata 2	25.20-30.25	1	2	90 Distomodus dubius 3	41.39-42.19	14C	14C
18 Ozarkodina confluens 21	17.53-56.70	3	22,24	92 Belodella sp. Denkler & Harris 3	45.63-57.08	14B	14E
19 Polygnathoides emarginatus s. f. 8	25.40-31.27	1	20B	95 Monograptus parultimus 4	37.80-38.33	1,20D	20A
20 Kockelella absidula 8	17.81-29.90	5	1	106 Monograptus formosus 2	35.33-38.33	20A	20A
21 Oulodus elegans 7	31.92-57.70	14A	1	107 Monograptus fragmentalis 3	34.39-37.81	20A	20A
22 Neoprioniodus latidentatus s. f. 4	25.10-57.70	1	1	109 Cyrtograptus cf. C. sakmaricus 2	15.61-15.70	4A,4C	4C
23 Panderodus unicostatus 18	9.63-84.67	17	2	111 Cyrtograptus cf. C. lapworthi 2	15.61-15.69	4A,4C	4C
24 Carniodus carnulus 10	13.03-15.70	17,26	15,16,27	112 Pristiograptus cf. P. praedubius 6	15.75-22.89	4E	4D
25 Distomodus staurognathoides 12	10.56-16.32	17	27	113 Monograptus flemingii flemingii 4	15.75-20.30	4B,4E	4A
26 Kockelella ranuliformis 12	13.92-17.62	21	5,11,16	114 Monograptus flemingii primus 3	15.89-17.56	4A	4B
28 Pterospathodus procerus 5	14.51-15.85	3	27	115 Cyrtograptus perneri 5	15.90-16.32	4E	4A
30 Apsidognathus tuberculatus 5	13.18-15.66	26	15	116 Monoclimacis flumendosae 5	15.77-19.18	6	4A,4B
32 Panderodus sp. cf. P. recurvatus 8	11.71-15.70	17	16	117 Monograptus priodon 4	15.00-18.14	1	23
35 Pseudooneotodus beckmanni s. f. 7	12.47-82.29	17	2	118 Cyrtograptus hamatus 5	16.15-18.01	4C	23
36 Ozarkodina confluens alpha morphotype 6	17.53-50.43	3	4D	119 Cyrtograptus radians 5	16.15-18.57	4C	4E
37 Ozarkodina confluens beta morphotype 4	24.85-49.62	6	4D	120 Monograptus retroflexus 5	15.90-18.91	4E	4C
38 Ozarkodina confluens delta morphotype 3	45.74-54.37	8	2	121 Retiolites nevadensis 5	15.75-18.91	4E	4C
39 Ozarkodina confluens epsilon morphotype 2	41.48-51.07	2	8	122 Pristiograptus jaegeri 5	19.65-21.27	4C	6
40 Dapsilodus obliquicostatus 5	10.56-29.11	17	2	123 Spinograptus spinosus 6	19.52-23.38	4B	6
41 Ozarkodina n. sp. A Klapper & Murphy, 1975 5	30.00-41.32	1	4D,7	124 Pristiograptus dubius frequens 6	19.64-34.99	4E	4A
42 Ozarkodina n. sp. E Klapper & Murphy, 1975 2	55.68-72.89	8	2	125 Saetograptus colonus 6	21.27-34.99	4A,23	4D
43 Monograptus pridoliensis 2	39.50-39.90	20E	20A	129 Monograptus uncinatus 5	21.27-27.81	4A	4B
44 Monograptus bouceki 1	40.66-42.29	20E	20E	130 Saetograptus willowensis 3	47.58-55.18	4A	2
45 Aulacognathus latus 2	15.42-15.66	29	15	131 Pristiograptus sp. ex. gr. P. dubius 5	26.42-51.28	4B	4A
46 Pelekysgnathus index 3	39.87-51.51	2,8	2,4D	132 Monograptus birchensis 2	54.89-78.46	4A	4A
47 Pseudooneotodus bicornis s. f. 15	13.03-19.31	17	9	133 Monograptus microdon 2	56.06-78.46	2	4A
48 Monograptus uniformis 4	57.30-78.46	2,4A,22	4A	134 Linograptus posthumus posthumus 3	28.00-78.46	1	4A
49 Ozarkodina hadra 6	14.69-15.71	3	3,6	135 Retiolites geinitzianus 3	15.00-15.70	1	4C
50 Johnognathus huddlei s. f. 5	14.36-15.70	21	15	138 Gothograptus nassa 4	18.33-21.03	4C	4B
51 Ozarkodina sagitta rhenana 8	16.90-17.73	10,15	12,16	139 Cyrtograptus lundgreni 5	16.15-18.44	4E	4E
52 Ozarkodina bohemica bohemica 7	17.72-18.46	3	10	143 Monograptus flemingii 4	15.75-20.30	4B,4E	4A
53 Kockelella stauros 4	18.07-18.52	11	5,13	144 Roberts Mts. Fm. bed A Berry & Murphy, 1975 4	27.30	4B,4C,4D,6	
54 Ozarkodina polinclinata 8	13.03-15.70	17,18,26	15	145 Roberts Mts. Fm. bed B Berry & Murphy, 1975 3	30.40	4E,6	
55 Oulodus petila 8	14.36-15.80	21	27	164 Oulodus elegans detorta 3	56.30-57.10	1,22	1,24
56 Oulodus equirectus 8	16.86-18.52	3	5	176 Ozarkodina polinclinata polinclinata 7	14.48-15.70	3	15
57 Wallisserodus sanctíclairi 9	13.03-20.26	18,26	9	178 Apsidognathus walmsleyi 2	15.19-15.56	26	27
58 Pseudooneotodus tricornis 7	13.03-15.95	17,26	27	179 Panderodus langkawiensis 3	13.03-15.73	26	27
59 Decoriconus fragilis 11	12.47-20.26	17	9	191 Panderodus fluegeli 2	15.38-15.56	27	27
61 Kockelella walliseri 5	17.60-18.66	5	10	192 Panderodus recurvatus ssp. p Jeppsson 2	15.38-15.61	27	28
63 Kockelella amsdeni 5	17.62-17.94	9,11	25	194 Panderodus serratus 2	15.38-15.80	27	27
64 Ozarkodina confluens gamma morphotype 3	29.77-51.28	7	4D	203 Ozarkodina aff. O. confluens Jeppsson 2	15.80-16.46	27,28	27
65 Ozarkodina douroensis 2	31.16-33.38	4D	7	205 Ireviken Datum Point 2 2	15.56	27,28	
66 Ozarkodina snajdri 12	27.34-45.54	19	14D	206 Ireviken Datum Point 3 3	15.70	21,27,28	
67 Oulodus siluricus 3	27.94-38.23	20B	20A	207 Ireviken Datum Point 4 2	15.80	27,28	
68 Monograptus ultimus 4	38.22-38.73	20B	20A	212 Llandovery-Wenlock boundary 1	15.30	21	
69 Monograptus lochkovensis 3	38.87-42.31	20B	20E	213 Silurian-Devonian boundary 1	57.30	22	
70 Monograptus transgrediens 4	39.25-55.50	20A	22	214 Ludlow-Pridoli boundary 1	37.81	20E	

*Each of the sections that make up the central Nevada CS, the central Appalachians CS, and the Bohemia CS is counted as a separate section.
**Numbers refer to section numbers in Table 3.

cartonensis biozones proposed by Cocks and Nowlan (1993) for the proposed SLHC cannot be recognized on the basis of their zonal indexes. The GSSP for the base of the nilssoni/scanicus Chronozone (Fig. 7, column F) is the projected FAD in the SRS of representatives of Neodiversograptus nilssoni (Fig. 7, column H, species 72). The top of the nilssoni/scanicus Chronozone is the base of the leintwardinensis Chronozone

(Fig. 7, column F). A chronozone based on the combined centifugus/murchisoni and riccartonensis biozones proposed by Cocks and Nowlan (1993) for the SLHC, the centrifugus/murchisoni/riccartonensis Chronozone, is defined by the projected level in the SRS of the base of the Wenlock at its GSSP and by the base of the rigidus/ellesae Chronozone (Fig. 7, columns C, F).

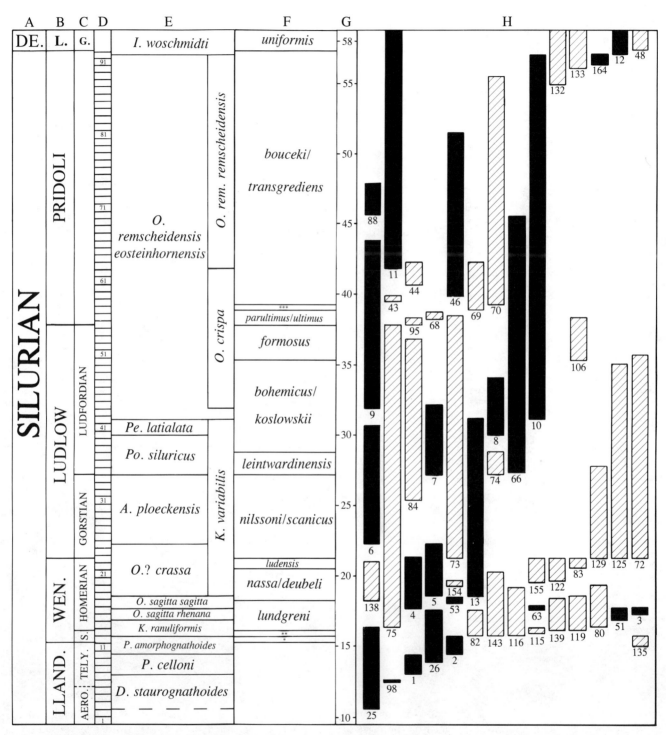

FIG 7.—Silurian Composite Standard chronostratigraphy. Column A: system; DE., Devonian. Columns B, C: series and stage, respectively; LLAND., Llandovery; WEN., Wenlock; L, Lochkovian; AERO, Aeronian; TELY., Telychian; S., Sheinwoodian; G., Gedinnian. Column D: Silurian CS divided into Standard Time Units. Column E: conodont chronozones. Column F: graptolite chronozones; *centrifugus/murchisoni/riccartonensis*; **rigidus/ellesae*; ***lochkovensis*. Column G: Silurian CS values (unscaled units) for conodont and graptolite ranges in column H. Column H: selected conodont and graptolite ranges in the Silurian CS including all zonal indices. Solid bars mark ranges of conodonts and diagonally ruled bars mark ranges of graptolites. Names of species listed by number in Table 4.

Silurian System, Series, and Stage Boundaries

Silurian system, series, and stage boundaries are recognized in the Silurian CS in one of two ways. The projected levels in the Silurian CS of boundaries fixed in stratotype sections that comprise it define those boundaries in the Silurian CS. Boundaries that are not represented by stratotype sections in the Silurian CS are recognized by using the lower boundaries of graptolite zones that are considered to approximate those chronostratigraphic boundaries. The projected positions in the Silurian CS of the "golden spikes" in the Llandovery/Wenlock, Ludlow/Pridoli, and Silurian/Devonian boundary-stratotype sections define those boundaries (Fig. 7, columns A, B). Neither the Aeronian/Telychian boundary stratotype nor any graptolites indicative of the *sedgwickii* and/or *turriculatus/crispus* Biozones are a part of the Silurian CS; thus, the Aeronian/Telychian boundary cannot be recognized with confidence. The Aeronian/Telychian boundary is tentatively recognized within the upper part of the *Distomodus staurognathoides* Chronozone (Fig. 7, column C), the approximate position in relation to that conodont zone as is shown on the proposed SLHC. The Sheinwoodian/Homerian boundary is recognized at the base of the *lundgreni* Chronozone, the Wenlock/Ludlow boundary at the base of the *nilssoni/scanicus* Chronozone, and the Gorstian/Ludfordian boundary at the base of the *leintwardinensis* Chronozone (Fig. 7, columns B, C).

The Wenlock, Ludlow, and Pridoli Series, and the stages that comprise them are delineated in the Silurian CS by their boundaries as defined above. Each series and stage (except for Aeronian and Telychian Stages as reported above) can be recognized with a high level of confidence when projected into any section that is a part of the Silurian CS or that can be correlated with it by the graphic correlation method. Durations of the epochs and ages are the number of STUs between their boundaries. The Wenlock Epoch includes approximately 11.5 STUs; the Ludlow Epoch, 31.8; the Pridoli Epoch, 37.5; the Sheinwoodian Age, 1.6; the Homerian Age, 9.8; the Gorstian Age, 11.4; and the Ludfordian Age, 20.3 (Fig. 7, columns B, C). The duration of the Llandovery Epoch cannot be determined because that epoch is incompletely represented in the Silurian CS. The portion of the Llandovery Epoch represented includes approximately 11.2 STUs (Fig. 7, column B). The duration of the Telychian Age can only be estimated at approximately 6 STUs because of the tentative recognition of its base in the Silurian CS.

DISCUSSION

Resolution of Silurian Composite Standard

The chronostratigraphy shown in Figure 7 greatly increases resolution within the Silurian System for any section that is a part of or can be correlated with the Silurian CS. The 92 STUs between the Llandovery (probably upper Aeronian) and Silurian/Devonian boundary compare with 12 conodont zones and 19–20 graptolite zones recognized for the same interval for the proposed SLHC. Even under the best conditions in localized regions only 40–50 graptolite zones can be recognized (Zalasiewicz, 1990), which represents only half the resolution attained through graphic correlation. Zalasiewicz (1990) estimated that the present state of graptolite stratigraphy of the Silurian Welsh basin subdivided the sequence into intervals of about 450 ka duration. Rickards (1989) suggested that by the next decade graptolite assemblages or arbitrary segments of evolutionary lineages will be down to around 100–200 ka. The time scale developed by Harland and others (1989) showed a duration of 21 Ma for the Wenlock through Pridoli Epochs and 15.5 Ma for the Ludlow through Pridoli Epochs. The Wenlock/Ludlow and Silurian/Devonian boundaries were defined in Harland and others (1989) by good (class B) chronograms with no interpolation; whereas, the Llandovery/Wenlock boundary was defined by interpolation for the chronogram developed for it was a class C one, with the second greatest range of error for any Silurian chronogram they constructed. Since the age of the Llandovery/Wenlock boundary is not as well constrained as the age of the Wenlock/Ludlow boundary in Harland and others (1989), the duration of the combined Wenlock, Ludlow, and Pridoli Epochs is not as reliable as the duration of the combined Ludlow and Pridoli Epochs in Harland and others (1989). Thereby, the number of STUs within the Ludlow and Pridoli in the Silurian CS is used to calculate the length of time represented in each STU. It is possible to calculate the length of time represented in each STU, for they all are of approximately equal annular length as was previously demonstrated. The Ludlow and Pridoli include a total of about 69.3 STUs in the Silurian CS, representing about 15.5 Ma. This means the Silurian chronostratic scale based on the Silurian CS subdivides into intervals of about 220 ka duration, a resolution that compares favorably to that predicted for future graptolite zonation.

Chronozones

Since the terms biozone and chronozone are used throughout this paper, it is important to recognize that they are not basically synonymous but have unique meanings. In order to show the difference, consider an example in which both a biozone and a chronozone are based on the range of the same taxon. A biozone with a lower boundary defined by the FAD of its nominative taxon and an upper boundary defined by the LAD of that same taxon can be recognized in strata only if the appropriate taxon is represented. The FAD and LAD of that taxon are both highly unlikely to be at precisely the same stratigraphic levels in all strata in which that taxon is represented. As a result, although the biozone defined by that taxon may be recognized in strata exposed at many localities, the boundaries of that biozone are not at the same stratigraphic levels at all of those localities. The boundaries of a chronozone defined by the FAD and LAD of that same taxon are at the same stratigraphic levels at all localities, for they are fixed by "golden spikes" in a Global Boundary Stratotype Section, conforming to a GSSP. However, it may be difficult or impossible to accurately recognize those chronozone boundaries in strata at other localities due to problems of resolution. The conodont and graptolite chronozones defined in this study can be recognized with a high level of confidence in strata at any section that can be graphically correlated with the Silurian CS. In fact, it is possible to recognize a conodont or graptolite chronozone in some sections even if the taxon it is based on is not represented there. The "golden spikes" in the SRS (Cellon, Austria section) of the Silurian CS and Global Boundary Stratotype Section for all of the conodont and graptolite chronozones defined in this study can be projected into any section that can be graphically correlated with the Silurian CS.

The 15 conodont and 12 graptolite chronozones recognized and defined in the Silurian CS afford far less resolution than the 92 STUs the Silurian CS is divisible into. In fact, conodont and graptolite chronozones were not defined as an aid to subdivide the Silurian CS. Graptolite and conodont biozones have typically been used as aids in correlating most Silurian strata and have been designated as the primary biozones on the proposed SLHC for international Silurian correlation charts. Graptolites and conodonts do not generally have the same distribution patterns in Silurian strata; index species of the two groups are rarely represented in the same section, even more rarely in the same bed. As a result, there have been many different correlations of Silurian graptolite biozones with Silurian conodont biozones (e.g., Berry and Boucot, 1970; Harland and others, 1989; Cocks and Nowlan, 1993). On the other hand, the succession of graptolite chronozones is easily compared to the succession of conodont chronozones in the Silurian CS. As discussed above, graptolite and conodont chronozones are not the same as graptolite and conodont biozones, even if they are based on the same taxa. Some of the graptolite and conodont chronozones defined in the Silurian CS may also differ from graptolite and conodont biozones based on the same taxa in the way that the taxa are used for definition of the zone. Nevertheless, definition of graptolite and conodont chronozones in the Silurian CS provides an excellent means (perhaps the best means currently available) for correlation of the zones most widely used in Silurian stratigraphy. Another reason graptolite and conodont chronozones are defined in the Silurian CS is that it is not possible to graphically correlate all Silurian sections with the Silurian CS. STUs recognized in the Silurian CS provide the greatest resolution available for the Silurian System, but they cannot be recognized in any section that cannot be graphically correlated with the Silurian CS. If STUs were the only way that the Silurian CS was divided, only sections which can and would be graphically correlated with the Silurian CS could use it as a means for correlation. Chronozones defined in the Silurian CS provide a means for some sections, which cannot or would not be graphically correlated with the Silurian CS, to still use the Silurian CS for correlation. Although the "golden spikes" of graptolite and conodont chronozones in the Silurian CS cannot be projected into in any section that cannot or would not be graphically correlated with the Silurian CS, the boundaries of those chronozones can be qualitatively recognized if the appropriate taxa are represented, much the same as in traditional biostratigraphy. The resolution provided would be far less than if the section was graphically correlated with the Silurian CS, but at least the section could be correlated with the Silurian CS.

Conodont Chronozones.—

The succession of conodont chronozones shown in Figure 7 differs from that shown on the proposed SLHC (Cocks and Nowlan, 1993). The differences are not solely due to the difference between a biozone and chronozone as was previously discussed, for the conodont biozones comprising the proposed SLHC have implied chronostratigraphic applicability. The succession of conodont chronozones in the Silurian CS can then be compared with those zones. The *P. amorphognathoides* and *O. sagitta rhenana* Biozones are shown as contiguous on the proposed SLHC. There is a significant gap in the Silurian CS between the LAD of *Pterospathodus amorphognathoides* and the FAD of *Ozarkodina sagitta rhenana* (Table 4; Fig. 7, column H, species 2, species 51). The *Kockelella ranuliformis* Chronozone (Fig. 7, column E) is recognized between LAD of the former and FAD of the latter in the Silurian CS. As a result, in the Silurian CS the upper part of the *P. amorphognathoides* Chronozone does not extend above the LAD of its index species. An *O. bohemica bohemica* Biozone was recognized between the *O. sagitta sagitta* and *A. ploeckensis* Biozones on the proposed SLHC, whereas the *O.? crassa* Chronozone is recognized in the Silurian CS in that position. *Ozarkodina bohemica bohemica* is more widely represented than *O.? crassa* in the Silurian CS and in uncompiled stratigraphic sections that are not currently a part of it. The range of *O. bohemica bohemica* is too restricted in the Silurian CS to be used as a basis for a chronozone between the *O. sagitta sagitta* and *A. ploeckensis* Chronozones, however, for although it is represented in upper Wenlock and Ludlow strata at several localities, none of those are currently a part of the Silurian CS. In stratigraphic sections that are a part of the Silurian CS in which *O. bohemica bohemica* is represented, its FAD is lower than was recognized in any previous conodont zonation, nearly coincident with the FAD of *O. sagitta sagitta* (Table 4). This distribution of representatives of *O. bohemica bohemica* suggests that if recognition of the *O. bohemica bohemica* Biozone is to be continued on the proposed SLHC, its lower boundary should be extended downward to the base of the *O. sagitta sagitta* Biozone. The latter zone could then become a subzone of the expanded *O. bohemica bohemica* Biozone.

The proposed SLHC recognized the *O. snajdri* and *O. crispa* Biozones between the *P. siluricus* and *O. remscheidensis eosteinhornensis* Biozones, whereas only the *Pe. latialata* Chronozone is recognized between the two chronozones based on those species in the Silurian CS. The base of the *O. remscheidensis eosteinhornensis* Biozone in the proposed SLHC is defined above the LAD of *O. crispa*, rather than by the FAD of its nominative subspecies. Aldridge and Schönlaub (1989) cited several reports which indicated that away from Cellon, Austria, the range of *O. remscheidensis eosteinhornensis* encompassed that of *O. crispa* and the upper part of the range of *O. snajdri*. The range of *O. remscheidensis eosteinhornensis* in the Silurian CS also encompasses that of *O. crispa*, extending well below the LAD of *Pe. latialata* (Table 4; Fig. 7, column H, species 10, 9, 8). Its range in the Silurian CS also overlaps with the upper part of the range of *Polygnathoides siluricus* (Table 4; Fig. 7, column H, species 7). *Ozarkodina remscheidensis eosteinhornensis* is represented in more stratigraphic sections that are a part of the Silurian CS framework than are *O. crispa* and *O. snajdri*. Recognizing a more extended zone in the Silurian CS based on *O. remscheidensis eosteinhornensis* than is recognized in the proposed SLHC facilitates qualitative correlation of stratigraphic sections with the Silurian CS that cannot be added to it by the graphic correlation method. Based on the distribution discussed above, if the lower boundary of the *O. remscheidensis eosteinhornensis* Chronozone was defined by the LAD of *O. crispa*, samples in stratigraphic sections with representatives of *O. remscheidensis eosteinhornensis* but lacking representatives of *O. crispa* and *O. snajdri* could easily be assumed to be Pridoli samples rather than Ludlow samples in which the latter two species are not represented because of ecological controls, small sample size, or other reasons. The extinction of *O. crispa* near or at the Ludlow/Pridoli boundary in the global standard area of the Barrandian and in the Cellon section of the Carnic Alps is a local extermination event only because the LAD in the Silurian CS of that species extends into the lower Pridoli (Fig. 7, column H, species 9), based on its representation in stratigraphic sections in the central Appalachians and central Nevada.

Graptolite Chronozones.—

The succession of graptolite chronozones in Figure 7 was based on the succession of graptolite biozones on the proposed SLHC (Cocks and Nowlan, 1993); thus, it is not surprising that the two are quite similar. There are some differences in the correlation of the graptolite and conodont biozones on the proposed SLHC and the correlation of the graptolite and conodont chronozones in the Silurian CS, especially in the Sheinwoodian and lower Homerian Stages. Some of those differences probably result because of the differences in the conodont chronozones and biozones just discussed. However, some of those differences are also probably due to the fundamental difference between a biozone and a chronozone and their use in stratigraphy. The successions of Sheinwoodian-lower Homerian graptolite zones in the Silurian CS and proposed SLHC are identical, but the stratigraphic position of those zones in relation to conodont zones is lower in the Silurian CS. FADs of *Cyrtograptus lundgreni* and *C. rigidus* are lower in western North American than in European stratigraphic sections that are a part of the Silurian CS. The FAD of *C. lundgreni* in Pete Hanson Creek IB, Nevada and Simpson Park Range I, Nevada is significantly lower than the FAD of *C. rigidus*, in Cellon, Austria when projected into the SRS of the Silurian CS. If the FAD of *C. rigidus* in the Silurian CS was based on its distribution in the section at Cellon, the *lundgreni* Chronozone would encompass the *rigidus/ellesae* Chronozone. However, the first appearance of *C. rigidus* at Pete Hanson Creek IB is over 11 m below the first appearance of *C. lundgreni* (Berry and Murphy, 1975). A separate *rigidus/ellesae* Chronozone can still be recognized in the Silurian CS based on the FAD of *C. rigidus* at Pete Hanson Creek IB when projected into the SRS of the Silurian CS. There are no diagnostic taxa other than graptolites represented in Sheinwoodian-lower Homerian strata in those western North American sections, and the FADs of the diagnostic graptolites occur in the same succession as in European sections. Only chronostratigraphy, not biostratigraphy, can indicate that the FADs of those graptolites are lower than the FADs of those same graptolites in European sections.

The graptolite biozonation shown on the proposed SLHC (Cocks and Nowlan, 1993) was considered to be too generalized by many Silurian workers (based on comments published in the *Silurian Times* No. 2, March 1994, p. 6–10). Most of the upper Sheinwoodian through lower Ludfordian and Pridoli zones recognized in a more detailed graptolite biozonation for a SLHC suggested by Teller (1994) can also be recognized in the Silurian CS as chronozones and in the same stratigraphic/chronologic order. The *rigidus*, *perneri*, *lundgreni* (including lower *radius* and upper *testis* divisions), *nassa*, *praedeubeli*, *deubeli*, *ludensis*, *nilssoni*, *leintwardinensis*, *parultimus*, *ultimus*, *lochkovensis*, *bouceki*, and *transgrediens* chronozones can all be recognized as unique chronozones with lower boundaries defined by the FADs in the Silurian CS of their diagnostic taxa (ranges of the index species for most of those chronozones are listed in Table 4 and shown in Fig. 7, column H) and upper boundaries defined by the bases of the suprajacent chronozones. Not all of those chronozones can be recognized with a high level of confidence in the Silurian CS, however, because of its current limit of resolution. Only the *parvus*, *scanicus*, and Pridoli *perneri* zones are not possible to recognize as unique chronozones. The *parvus* Chronozone is totally encompassed by the *nassa* Chronozone and is not possible to recognize with a high level of confidence as a subdivision of the *nassa* Chronozone because of current limits of resolution. There are no representatives of the index species of the *scanicus* and Pridoli *perneri* zones in strata that are currently a part of the Silurian CS.

Silurian Composite Standard as a Geochronologic Scale

Durations of Epochs and Ages.—

Durations of epochs and ages in the Silurian CS (Fig. 7, columns B, C), determined previously in this study based on the number of STUs between their boundaries, differ from durations shown on many previous Silurian geochronologic scales, including those developed by Snelling (1987) and Harland and others (1989). In order to compare directly the durations of the post-Llandovery Silurian epochs and ages in the Silurian CS with the durations of those epochs and ages as shown on other previously developed Silurian time scales, their durations are expressed as a fraction of the duration of the entire post-Llandovery Silurian they represent in their respective time scales. The Wenlock Epoch is the shortest in the Silurian CS (Fig. 7, column B); it is approximately 1.5 and 2 times longer in Snelling (1987) and Harland and others (1989), respectively. The Ludlow Epoch in the Silurian CS is much longer than the Wenlock Epoch and slightly shorter than the Pridoli Epoch (Fig. 7, column B). In most of the Silurian time scales compared in Harland and others (1989), separate durations were not shown for the Ludlow and Pridoli Epochs. The Ludlow Epoch is shorter in the Silurian CS than in Harland and others (1989), where it is approximately 1.6 times longer. The Pridoli Epoch is the longest in the Silurian CS (Fig. 7, column B); it was the shortest on all previous time scales compared in Harland and others (1989). The Pridoli Epoch is approximately 4.6 times longer in the Silurian CS than in Harland and others (1989). The combined durations for the Ludlow and Pridoli Epochs in the Silurian CS compare more favorably with their combined durations as shown on other time scales. The combined Ludlow and Pridoli Epochs in the Silurian CS are approximately 1.1 and 1.2 times longer than in Snelling (1987) and Harland and others (1989), respectively. The Sheinwoodian and Gorstian Ages are significantly shorter in the Silurian CS than in Harland and others (1989), where they are approximately 10 and 2.9 times longer, respectively. The Homerian Age in the Silurian CS and in Harland and others (1989) are of approximately equal duration. The Ludfordian Age in the Silurian CS is approximately 1.3 times longer than in Harland and others (1989).

Calibration of the Silurian Chronostratic Scale.—

The differences in durations of epochs and ages in the Silurian CS when compared to those in Snelling (1987) and Harland and others (1989) are not necessarily as great as they appear. There is very little difference in the duration of the combined Ludlow and Pridoli Epochs in the three time scales. Although there were good chronograms based on isotopic dates for the Silurian/Devonian and Wenlock/Ludlow boundaries, which were used for tie-points in Harland and others (1989), the chronogram based on isotopic dates for the Ludlow/Pridoli boundary was not good enough to be used as a tie-point. The age of that boundary was determined by interpolation (Harland and others, 1989). At present, separate durations for the two epochs can best be determined by using chrons, units which represent equal time durations, to interpolate between the tie-points that define the combined duration of the two epochs. Harland and others (1989) did exactly that, using as chrons seven graptolite biozones recognized for the Ludlow and Pridoli Series to equally subdivide the Ludlow and Pridoli Epochs between the Silurian/Devonian and Wenlock/Ludlow tie-points. The greater the number of chrons (graptolite biozones) within a epoch or age, the longer its duration is in Harland and others (1989). Harland and others (1989) recognized one chron (graptolite biozone) in the Pridoli Epoch and six chrons (graptolite biozones) in the Ludlow Epoch; thereby, the Ludlow Epoch was of much greater duration. The seven chrons (graptolite biozones) between the Wenlock/Ludlow and Silurian/Devonian tie-points in Harland and others (1989) each represent about 2.2 Ma, whereas the 21 chrons (graptolite biozones) between the Ordovician/Silurian and Wenlock/Ludlow tie-points in Harland and others (1989) each represent only about 0.7 Ma. That much of a difference in chron duration very convincingly demonstrates what most stratigraphers have assumed, that at present most graptolite biozones recognized represent different amounts of time. Graptolite biozones are thereby not ideal for use as chrons of equal duration, and do not provide an adequate means for determining the durations of the Ludlow and Pridoli Epochs.

The durations of the Ludlow and Pridoli Epochs in the Silurian CS are also determined based on the number of chrons included within their respective boundaries. STUs are used as chrons instead of graptolite biozones, however, since STUs were demonstrated to be of approximately equal duration in the Silurian CS. Until other, more refined chrons of demonstrably equivalent duration are recognized and/or a Ludlow/Pridoli chronogram with high-resolution is possible to construct and use as a tie-point, the method used to determine the durations of the Ludlow and Pridoli Epochs in this paper provides the only way to calculate good estimates for the durations of those two epochs and the ages recognized within them. If the Wenlock/Ludlow (424 Ma) and Silurian/Devonian (408.5 Ma) tie-points used in Harland and others (1989) are used in conjunction with those boundaries in the Silurian CS, the Ludlow/Pridoli boundary is at approximately 416.9 Ma and the Gorstian/Ludfordian boundary at 421.4 Ma. Thus, the Ludlow Epoch lasted 7.1 Ma, the Gorstian Age 2.6 Ma, the Ludfordian Age 4.5 Ma, and the Pridoli Epoch 8.4 Ma.

Differences in the durations of the Wenlock Epoch, Sheinwoodian Age, and Homerian Age in the Silurian CS compared with their durations in Snelling (1987) and Harland and others (1989) may also be a result of the different methods used to determine their durations. The present lack of good isotopic dates for strata adjacent to the Llandovery/Wenlock boundary results in the need for interpolation between the ages of the Ordovician/Silurian and Wenlock/Ludlow boundaries calculated by chronograms or other methods in order to determine an age for that boundary. As was previously discussed, interpolation is typically accomplished by using graptolite biozones as chrons of equal duration even though most stratigraphers do not consider graptolite biozones to be of equal duration, and the method itself demonstrates that they are not. There is an additional possibility for differences in durations of the Sheinwoodian and Homerian Ages, however. The FADs of several diagnostic Wenlock graptolites, including *Cyrtograptus lundgreni*, are lower in western North American than in European stratigraphic sections that are a part of the Silurian CS, as was previously discussed. The base of the Homerian at its stratotype section coincides with the base of the *lundgreni* Biozone there (Bassett, 1989). Therefore, the base of the *lundgreni* Biozone is typically used to recognize the base of the Homerian in most strata. Although the boundary stratotype for the Homerian is not a part of the Silurian CS, it is probable (but not currently verifiable) that the FAD of *C. lundgreni* in some western North American sections is lower than the officially defined Homerian base. Since the base of the Homerian Age is defined in the Silurian CS by the FAD of *C. lundgreni*, this results in a very short Sheinwoodian Age in the Silurian CS, compared to other time scales. Murphy (1989) also showed a Sheinwoodian Age with a very short duration, based on graphic correlation of nine stratigraphic sections in western North America. Murphy's (1989) results provide additional evidence that the differences in the durations of the Wenlock Epoch and its ages in the Silurian CS, compared to their durations in other time scales, is due either to shorter durations for Wenlock (especially Sheinwoodian) subdivisions based on graptolite biozones than previously recognized or a difference in distribution of graptolites during the Wenlock Epoch in European and western North American localities. The duration of the Wenlock Epoch in the Silurian CS is based on the number of chrons (STUs) between its boundaries. Thus the method used to determine duration of the Wenlock Epoch in this paper provides the only way at present to determine a good estimate of its duration and of the duration of the Sheinwoodian and Homerian Ages (if there is no difference in Wenlock graptolite distribution in western North America and Europe). If the Wenlock/Ludlow tie-point used in Harland and others (1989) is used in conjunction with the number of chrons in the Silurian CS within the Wenlock Epoch and the two ages that comprise it, the Sheinwoodian/Homerian boundary is at approximately 426.2 Ma and the Llandovery/Wenlock boundary at 426.6 Ma. Thus, the Wenlock Epoch lasted 2.6 Ma, the Sheinwoodian Age 0.4 Ma, and the Homerian Age 2.2 Ma.

CONCLUSIONS

Revision of the Silurian CS of Kleffner (1989) by graphic correlation of 12 previously uncompiled stratigraphic sections results in a Silurian CS that has worldwide applicability as a high-resolution chronostratigraphy. The Silurian CS divides with confidence into 92 Standard Time Units (STUs), a resolution that is a minimum of twice that of any previously proposed Silurian chronostratigraphy. STUs are of approximately 220 ka duration. The addition of range-data on 52 graptolite species, 39 conodont species, 10 events, and one boundary stratotype makes it possible to graphically correlate virtually any stratigraphic section which contains representatives of diagnostic conodont and/or graptolite species (and also meets the other data requirements of the graphic correlation method discussed in Section II of this volume) with the Silurian CS. Conodont and graptolite chronozones can be defined in the Silurian CS according to international rules of stratigraphy and, if they contain at least one STU, can be recognized with confidence in any section that is a part of the Silurian CS or can be graphically correlated with it. The conodont and graptolite chronozones defined in this manner may also be used qualitatively, with perhaps even greater precision of meaning than the originally proposed zones on which they are based, when correlating sections with the CS that cannot be graphically correlated with it. All Silurian series and stage boundaries that are within the boundaries of the Silurian CS (except for the Aeronian/Telychian boundary) can be recognized with a high-degree of confidence in any section that is already a part of the Silurian CS or that can be graphically correlated with it.

The Silurian chronostratic scale based on the range-data on conodont species, graptolite species, events, and boundary stratotypes represented in the Silurian CS can be calibrated using the same Wenlock/Ludlow and Silurian/Devonian tie-points calculated and used by Harland and others (1989) to develop their geologic time scale. The STUs within the Silurian CS can be used as chrons of equal duration to interpolate between and below those tie-points. The Silurian time scale developed as a result of that calibration provides the best means at present for estimating the durations of the Wenlock, Ludlow, and Pridoli Epochs, and all of the ages that comprise them (except perhaps for the Sheinwoodian and Homerian Ages).

ACKNOWLEDGMENTS

I would like to thank Walter Sweet for introducing me to the graphic correlation method and for many interesting discussions concerning graphic correlation. I am grateful to Ken Hood for developing a computerized version of the graphic correlation method, which has made it much easier to update and revise the Silurian CS, as well as to produce publication-quality graphs without drafting. I would like to acknowledge all those who shared their unpublished or in-press information with me: James Barrick, Kirk Denkler, Anita Harris, Peep Männik, Mike Murphy, and Hans Schönlaub. I would like to thank the two reviewers and Keith Mann for their many insightful comments which helped me to greatly improve the original manuscript. I would also like to thank Keith Mann again, who was responsible for organizing the symposium that eventually resulted in this volume on graphic correlation, and Keith, Rich Lane, Jeff Stein, and Dana Ulmer-Scholle for doing the work necessary to get this important volume published. Finally, I would like to thank my wife, Carol, who took on some of my family responsibilities so that I could have time to complete my revision almost by the scheduled deadline!

REFERENCES

ALDRIDGE, R. J. AND SCHÖNLAUB, H. P., 1989, Conodonts, in Holland, C. H. and Bassett, M. G., eds., A Global Standard for the Silurian: Cardiff, National Museum of Wales, p. 274–279.

ALDRIDGE R. J., JEPPSSON, L., AND DORNING, K. J., 1993, Early Silurian oceanic episodes and events: Journal of the Geological Society of London, v. 150, p. 501–513.

BARRICK, J. E., 1977, Multielement simple-cone conodonts from the Clarita Formation (Silurian), Arbuckle Mountains, Oklahoma: Geologica et Palaeontologica, v. 11, p. 47–68.

BARRICK, J. E., 1983, Wenlockian (Silurian) conodont biostratigraphy, biofacies, and carbonate lithofacies, Wayne Formation, central Tennessee: Journal of Paleontology, v. 57, p. 208–239.

BARRICK, J. E. AND KLAPPER, G., 1976, Multielement Silurian (late Llandoverian-Wenlockian) conodonts of the Clarita Formation, Arbuckle Mountains, Oklahoma, and phylogeny of Kockelella: Geologica et Palaeontologica, v. 10, p. 59–100.

BASSETT, M. G., 1989, The Wenlock Series in the Wenlock area, in Holland, C. H. and Bassett, M. G., eds., A Global Standard for the Silurian: Cardiff, National Museum of Wales, p. 51–73.

BERRY, W. B. N. AND BOUCOT, A. J., 1970, Correlation of the North American Silurian rocks: Boulder, Geological Society of America Special Paper 102, 289 p.

BERRY, W. B. N. AND MURPHY, M. A., 1975, Silurian and Devonian Graptolites of Central Nevada: Berkeley, University of California Press, v. 110, 109 p., 15 pls.

CHULPÁČ, I., KŘÍŽ, J., AND SCHÖNLAUB, H. P., 1980, Field Trip E, Silurian and Devonian conodont localities of the Barrandian, in Schönlaub, H. P., ed., Second European Conodont Symposium: Abhandlungen der Geologischen Bundesanstalt, v. 35, p. 147–180.

COCKS, L. R. M. AND NOWLAN, G. S., 1993, New left hand side for correlation diagrams: Silurian Times No. 1, p. 6–8.

DENKLER, K. E. AND HARRIS, A. G., 1985, Conodont biofacies and biostratigraphy of upper Ludlow-Pridoli shallow-water carbonates in the central Appalachians, U.S.A., in Aldridge, R. J., Austin, R. L., and Smith, M. P., eds., Fourth European Conodont Symposium (ECOS IV) Abstracts: Southampton, The University of Southampton, p. 8–9.

GNOLI, M., KŘÍŽ, J., LEONE, F., OLIVIERI, R., SERPAGLI, E., AND STORCH, P., 1988, The Mason Porcus Section as a reference for Uppermost Silurian-Lower Devonian in SW Sardinia: Bollettino della Società Paleontologica Italiana, v. 27, p. 323–324.

HARLAND W. B., ARMSTRONG, R. L., COX, A. V., CRAIG, L. E., SMITH, A. G., AND SMITH, D. G., 1989, A Geologic Time Scale 1989: Cambridge, Cambridge University Press, 263 p.

HELFRICH, C. T., 1975, Silurian conodonts from Wills Mountain Anticline, Virginia, West Virginia, and Maryland: Geological Society of America Special Paper 161, 82 p.

JAEGER, H., 1991, Neue Standard-Graptolithenzonenfolge nach der "Großen Krise" an der Wenlock/Ludlow-Grenze (Silurian): Neues Jahrbuch für Geologie und Paläontologie, Abhandlungen, v. 182, p. 303–354.

JEPPSSON, L., 1988, Conodont biostratigraphy of the Silurian-Devonian boundary stratotype at Klonk, Czechoslovakia: Geologica et Palaeontologica, v. 22, p. 21–31.

KLAPPER, G. AND MURPHY, M. A., 1975, Silurian-Lower Devonian Conodont Sequence in the Roberts Mountains Formation of Central Nevada: Berkeley, University of California Press, v. 111, 62 p., 12 pls.

KLEFFNER, M. A., 1987, Conodonts of the Estill Shale and Bisher Formation (Silurian, southern Ohio): biostratigraphy and distribution: The Ohio Journal of Science, v. 87, p. 78–89.

KLEFFNER, M. A., 1989, A conodont-based Silurian chronostratigraphy: Geological Society of America Bulletin, v. 101, p. 904–912.

KLEFFNER, M. A., 1990, Wenlockian (Silurian) conodont biostratigraphy, depositional environments, and depositional history along the eastern flank of the Cincinnati Arch in southern Ohio: Journal of Paleontology, v. 64, p. 319–328.

KLEFFNER, M. A., 1991, Conodont biostratigraphy of the upper part of the Clinton Group and the Lockport Group (Silurian) in the Niagara Gorge region, New York and Ontario: Journal of Paleontology, v. 65, p. 500–511.

KLEFFNER, M. A., 1994, Conodont biostratigraphy and depositional history of strata comprising the Niagaran sequence (Silurian) in the northern part of the Cincinnati Arch region, west-central Ohio, and evolution of Kockelella walliseri (Helfrich): Journal of Paleontology, v. 68, p. 141–153.

KŘÍŽ, J., JAEGER, H., PARIS, F., AND SCHÖNLAUB, H. P., 1986, Pridoli-the fourth subdivision of the Silurian: Jahrbuch der Geologischen Bundesanstalt, v. 129, p. 291–346.

MABILLARD, J. E. AND ALDRIDGE, R. J., 1985, Microfossil distribution across the base of the Wenlock Series in the type area: Palaeontology, v. 28, p. 89–100.

MEHRTENS, C. J. AND BARNETT, S. G., 1976, Conodont subspecies from the Upper Silurian-Lower Devonian of Czechoslovakia: Micropaleontology, v. 22, p. 491–500.

MURPHY, M. A., 1989, Central Nevada, in Holland, C. H. and Bassett, M. G., eds., A Global Standard for the Silurian: Cardiff, National Museum of Wales, p. 171–177.

NAKREM, H. A., 1986, Llandovery conodonts from the Oslo Region, Norway: Norsk Geologisk Tidsskrift, v. 66, p. 121–133.

OLIVIERI, R. AND SERPAGLI, E., 1990, Latest Silurian-early Devonian conodonts from the Mason Porcus Section near Fluminimaggiore, Southwestern Sardinia: Bollettino della Società Paleontologica Italiana, v. 29, p. 59–76.

RICKARDS, R. B., 1989, Exploitation of graptoloid cladogenesis in Silurian stratigraphy, in Holland, C. H. and Bassett, M. G., eds., A Global Standard for the Silurian: Cardiff, National Museum of Wales, p. 267–274.

SCHÖNLAUB, H. P., 1971, Zur Problematik der Conodonten-Chronologie an der Wende Ordoviz/Silur mit besonderer Berücksichtigung der Verhältnisse im Llandovery: Geologica et Palaeontologica, v. 5, p. 35–57.

SCHÖNLAUB, H. P., 1980, Field Trip A, Carnic Alps, in Schönlaub, H. P., ed., Second European Conodont Symposium: Abhandlungen der Geologischen Bundesanstalt, v. 35, p. 5–57.

SHAW, A. B., 1964, Time in Stratigraphy: New York, McGraw-Hill, 365 p.

SNELLING, N. J., 1987, Measurement of geologic time and the geologic time scale: Modern Geology, v. 11, p. 365–374.

TELLER, L. M., 1994, Comment by Lech Teller: Silurian Times No. 2, p. 9–10.

UYENO, T. T. AND BARNES, C. R., 1983, Conodonts of the Jupiter and Chicotte Formations (Lower Silurian), Anticosti Island, Quebec: Geological Survey of Canada Bulletin 355, 49 p.

WALLISER, O. H., 1964, Conodonten des Silurs: Abhandlungen des Hessischen Landesamtes für Bodenforschung, Heft 41, 106 p., 32 plates.

WINTERER, E. L. AND MURPHY, M. A., 1960, Silurian reef complex and associated facies, central Nevada: Journal of Geology, v. 68, p. 117–139.

ZALASIEWICZ, J. A., 1990, Silurian graptolite biostratigraphy in the Welsh Basin: Journal of the Geological Society, London, v. 147, p. 619–622.

GRAPHIC CORRELATION OF A FRASNIAN (UPPER DEVONIAN) COMPOSITE STANDARD

GILBERT KLAPPER
Department of Geology, University of Iowa, Iowa City, Iowa, 52242;
WILLIAM T. KIRCHGASSER
Department of Geology, State University of New York, Potsdam College, Potsdam, New York, 13676;
AND
JOHN F. BAESEMANN
Amoco Production Company, P.O. Box 3092, Houston, Texas 77253

ABSTRACT: The line of correlation (LOC) in the graphic correlation of the 27 sections in a Frasnian composite standard conforms empirically to the pattern of biostratigraphic events, the bases and tops of ranges of conodont species. The guiding principle is to position the LOC so as to avoid range overlaps of well understood species that have not been observed to overlap in an actual section. Several of the graphs used in the Frasnian composite standard display linear arrays through which the LOC effectively splits the bases and tops of species ranges. The patterns of arrays in other graphs, however, indicate a doglegged solution to the LOC, implying major changes in the accumulation rate within the section plotted against the composite axis. We do not assume that a uniform rate of accumulation at the standard reference section (from which the scale of the composite axis derives) is necessary for the empirical effectiveness of graphic correlation.

The problem of correlation between the mostly mutually exclusive *Palmatolepis* and *Polygnathus* Frasnian conodont biofacies has been insoluble by traditional zonal biostratigraphy. The graphic correlations proposed here advance an initial hypothesis toward resolving the correlation of these biofacies. Subdivision of the Frasnian into 34.5 composite standard units represents a far finer resolution than any available through zonal biostratigraphy.

INTRODUCTION AND METHODS

In graphic correlation, the line of correlation (LOC), representing the point-by-point equivalency of two sections or of a section plotted against the composite standard, "is always there; the geologist's task is to find it" (Edwards, 1984, p. 583). In a more pessimistic view, Tipper (1988, p. 480) stated that the LOC "is unique, defined in theory, and unknowable in practice." Although Shaw (1964) advocated a statistical solution to the LOC when he introduced the technique of graphic correlation, he used a modified form of least squares regression. That is, he favored unequal weighting of biostratigraphic events (bases and tops) by emphasizing some and disregarding others. The procedure is evident in the discussion titled "Selecting Reliable Points" (Shaw, 1964, p. 235–240; see also Edwards, 1991, p. 46). Most practitioners, including Shaw himself (pers. commun., 1964–1965) shortly after publication of *Time in Stratigraphy*, have agreed that statistical methods such as least squares regression are inappropriate for locating the LOC (e.g., Edwards, 1984; Hazel, 1989). As a consequence, calculation of the standard error of estimate (Shaw, 1964, p. 175–181) in order to quantify the resolution of graphic correlation is also inappropriate.

We are not advocating general rejection of statistical methods for the solution of the LOC, however, only the use of least squares regression (on this point, see the evaluation of regression methods for graphic correlation by MacLeod and Sadler, this volume). There is much promise in numerical-search algorithms applied to graphic correlation, for example the approach being developed by Kemple and others (1990). Perhaps these will ultimately represent a major step forward in methodology. For the time being, however, we prefer to use qualitative techniques for the solution of the LOC. Such methods are necessarily empirical, driven for example by the pattern of the array of biostratigraphic as well as geologic events.

The LOC for each section plotted against the composite standard is a testable hypothesis of stratigraphic correlation, as indeed is any biostratigraphic correlation. Qualitative methods for locating the LOC are diverse, involving geologic as well as biostratigraphic knowledge (Edwards, 1989), and include positioning it with respect to magnetic reversals (Dowsett, 1988, 1989), volcanic ash beds, and chemostratigraphic data (e.g., Hazel, 1989). Last but not least, in our opinion, is the well known method among practitioners of graphic correlation called "splitting" bases and tops of species. This is done in such a way so as to cause minimum disruption of known ranges ("economy of fit," Shaw, 1964, p. 257; Edwards, 1984, p. 588, assumption no. 7). At the Amoco Research Center in Tulsa in the period shortly after the publication of *Time in Stratigraphy*, Alan Shaw (pers. commun., 1964–1965) was stating this guideline more or less as follows: "Draw the LOC so as not to cause the ranges of well understood species to overlap when they have not been seen to overlap in any actual section." This was the main guideline used by Shaw in the construction of an Upper Devonian (Frasnian plus Famennian) composite standard, which derived from graphing sections with conodonts and ammonoids from the literature and which later formed the stratigraphic framework for his analysis of the species problem in paleontology (Shaw, 1969).

The scale of the composite standard units (CSUs) derives from the original thickness measurements in the section that is selected as the standard reference section (SRS) and by projection of sections that extend the SRS stratigraphically lower and higher. The SRS is the first section to go on the composite axis (usually the x-axis, Shaw 1964; Miller, 1977; Sweet, 1984, 1988, 1992; Kleffner, 1989; Edwards, 1991, p. 46; but others have used the y-axis, Murphy and Edwards, 1977; Edwards, 1984, 1989; Hazel, 1989; MacLeod and Keller, 1991a, b). The composite axis should be standardized to avoid confusion. As range values are projected onto the SRS from other sections, the first axis becomes the composite standard axis. The CSUs are not of equal time value, unless it can be demonstrated that the standard reference section accumulated at a uniform rate. On the assumption that the SRS accumulated at such a rate, some authors have inferred that the resulting CSUs are standard time units (e.g., Sweet, 1988, p. 265; 1992, p. 127–128), whereas the same assumption has been questioned by others (e.g., Fordham, 1992, p. 712–713). A test for uniform accumulation rates was suggested by Murphy (1987; see also the remarks by Kemple and others, 1990, p. 419). We do not regard the assumption of a uniform accumulation rate at the SRS as necessary for the empirical effectiveness of graphic correlation. But if the SRS accumulated at a variable rate (Fordham, 1992, p. 713), one can not infer that each CSU is of equal time value (we thank Steve Benoist, University of Iowa, for discussion on this point).

The SRS for the Frasnian composite standard (hereafter written either as Frasnian CS or simply CS) discussed in this preliminary report is a measured section in the Montagne Noire, southwestern France. Pelagic limestone sequences characterize the Montagne Noire sections, which are demonstrably condensed in thickness relative to most other sections in the CS. This should be understood by the reader in viewing the scaling of the CS axis. Retaining the first decimal place in the CSUs seems justified given the close spacing and thin sample intervals (generally 10–15 cm measured to the nearest cm) collected for conodonts in the Montagne Noire sections. However, the second and third places given in the output from Ken Hood's PC program for graphic correlation, GraphCor, are rounded off. The Frasnian extends from 96.9 to 131.4 CSUs on the CS axis; the scaling as already mentioned derives from measurements in meters at the SRS (1 m in the SRS = 1 CSU in the CS). Although the scale could be multiplied by a factor to yield numbers easier to remember, we prefer to retain the original scaling for the present purposes.

In this introduction, we do not review all aspects of the methodology of graphic correlation, many of which have been treated in the literature. Excellent summaries have been given in various papers by Edwards (e.g., 1984, 1991), but the original presentation by Shaw (1964) is indispensable and should be consulted by all investigators new to the method. At present, the Frasnian CS consists of 27 sections in the Montagne Noire, the Alberta Rockies, the Hay-Trout Rivers of the Northwest Territories, Canada, the midcontinent and western New York sequences in the U.S., the Canning Basin of Western Australia, and the Central Devonian Field and Timan-Pechora Basin of the Russian Platform. Some of the sections (Hay-Trout Rivers and New York) are themselves regional composites of many sections. Thus, there are actually data on conodonts, ammonoids, and key beds (in the case of the New York sequence) from 64 measured sections that are included in the Frasnian CS.

A range chart of 23 species of *Palmatolepis* is based on the fifth correlation round of the Frasnian CS, which included all 27 sections (Klapper and Foster, 1993, Fig. 2). This chart shows the alignment of the Frasnian CS with the Montagne Noire zonation (Klapper, 1989). The focus of the present paper, however, is not Frasnian conodont biostratigraphy, but rather uses the latter to illustrate methods of graphic correlation. Thus, we do not review the history of competing Frasnian conodont zonations or the details of Frasnian conodont and ammonoid biostratigraphy, as this can be found elsewhere (for entry into the literature, see discussions in Klapper and Foster, 1993; Becker and others, 1993). We do, however, give the ranges in terms of the Frasnian CS of all the species used to define bases of zones in the Montagne Noire conodont zonation (Appendix).

The taxonomy used in the present study has been described in a series of papers on Frasnian conodonts (Klapper, 1985, 1989, 1990; Klapper and Lane, 1985, 1989; Klapper and Foster, 1993; Kirchgasser, 1994; Kralick, 1994). Multielement taxonomy has been used wherever possible, and shape analysis has revised the understanding of Frasnian species of *Palmatolepis*. The result is a taxonomy that differs substantially from much of that previously published. Such disparity is predictable when different taxonomists approach essentially the same data starting from different initial purposes (Shaw, 1969). The point, however, for the application of a taxonomic data base to graphic correlation is *not* which is the "correct" taxonomy. Rather, for successful graphic correlation, the underlying taxonomy must be used consistently in all identifications of species. Thus, we have only used material from collections and sections that we have personally studied. This applies without exception to all the sections graphed in the Frasnian CS. However, for the purpose of graphing the post-Frasnian part of the Upper Coumiac Quarry, we have used a section from the literature (Sessacker Trench II, Ziegler, 1962) together with sections we have studied to develop a provisional CS for the lower Famennian. At present, the Frasnian CS includes 150 conodont species, 16 ammonoid genera, 23 ammonoid species, and 26 key beds in the New York regional composite. The graphic correlation is now in the fifth round and is continuing, but not all 27 sections were available during the initial rounds. In this preliminary report, we present a discussion of the correlation of ten sections, including the two regional composites, to illustrate methods used in constructing a Frasnian CS. Chief among these is Alan Shaw's guideline quoted previously.

THE FRASNIAN COMPOSITE STANDARD

La Serre Trench D (Montagne Noire, France)

In the fifth round graph of La Serre Trench D plotted against the Frasnian CS (Fig. 1), the Line of Correlation (LOC) intersects the bases of four species: *Ancyrodella gigas* form 1, *A. ioides*, *Palmatolepis ljaschenkoae*, and *P. winchelli*. The coincident bases of *Palmatolepis bogartensis* and *P. hassi* fall less than 0.1 CSU to the left of the line. Bases adjust only slightly; for example, *Polygnathus* aff. *P. dengleri* is 0.5 CSU to the right and *Palmatolepis jamieae* is 0.2 to the right of the LOC. The maximum adjustment of tops is 1.3 CSU (the unlabeled top at 103.8 x, 3.6 y). We positioned the LOC in the later correlation rounds (third through fifth) largely by the method of "splitting" the bases and tops (Shaw, 1964, p. 254–257; Edwards, 1991, p. 43–48). In GraphCor, when the plot first appears on the computer screen, the bases and tops are unlabeled. One can wait until a trial line is drawn and then, in another part of the program, identify the bases and tops while evaluating the range adjustments resulting from that particular LOC. At that point, Shaw's guideline to avoid causing artificial range overlaps can be followed by adjusting the LOC.

Compare the fifth-round plot (Fig. 1) with the first round graph of La Serre Trench D (Fig. 2), which was the first section plotted against Col du Puech de la Suque H, the standard reference section (SRS). The two sections (House and others, 1985, Fig. 1a, c) are presently 25 km apart but are in different tectonic units whose exact provenance is unknown. Both sections are condensed compared with many of the other sections that were plotted later to construct the Frasnian CS. In contrast to the graph in Figure 1, the positioning of the LOC in Figure 2 is on two key species of the Montagne Noire zonation, the

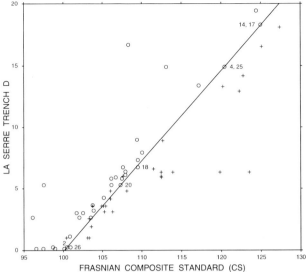

FIG. 1.—Graphic correlation of La Serre Trench D (Montagne Noire) against the Frasnian composite standard (CS). In this fifth round plot, as in Figs. 3–8, the CS is the *x*-axis. Bases of species ranges indicated by an open circle, tops by a plus, in all figures. Maximum ranges of species are used in the CS axis without exception. In all the graphs, the top of the sample interval is plotted for bases, the base of the sample interval for tops; error boxes are generally too small to be shown effectively. LOC intersects the bases of *Palmatolepis ljaschenkoae* (20), *P. winchelli* (25), *Ancyrodella gigas* form 1 (2), and *A. ioides* (4); LOC is 0.1 CSU to the left of the coincident bases of *P. bogartensis* (14) and *P. hassi* (17). Horizontal sampling terrace at 0–0.13 m in Trench D indicates that this section extends lower than the SRS (Fig. 2). See text discussion of selection of the SRS. Scale of *y*-axis in meters in all graphs.

bases of *Ancyrodella gigas* form 1 and *Palmatolepis ljaschenkoae*. In all the subsequent plots to be discussed, we tended to use the bases of key species from the zonation in early rounds but mostly split the bases and tops in later rounds. Nonetheless, the LOCs of earlier rounds had some influence on the positioning in later rounds.

If other factors were equal, we would have chosen La Serre Trench D section as the SRS. The Trench D section extends both higher and lower and has more species than Col du Puech de la Suque section H (note the sampling terrace in the lower left on the *x*-axis of Fig. 1). The lowest sample in section H is from an interval at 0 to 17 cm above the base of the measured section, which extends to 19.06 m above section base; the zero point is equated with 100 CSU to avoid negative numbers (Fig. 2). But we chose Col du Puech de la Suque section H to be the SRS because it could be extended downward to the base of the Frasnian (= base of Montagne Noire Zone 1) and, in fact, as low as the *disparilis* Zone of the subjacent upper Givetian. This downward extension results from tracing a key bed from section H into Col du Puech de la Suque section E, 90 m to the west (Klapper, 1985, Fig. 2, bed 58; 1989, p. 456), and thus accurately combining the two sections.

Pic de Bissous (Montagne Noire)

Subsequent to the correlation of La Serre Trench D with the SRS and the combining of the latter with section E, another section in the Montagne Noire was correlated with the developing Frasnian CS (Fig. 3). The LOC for Pic de Bissous intersects the bases of *Ancyrodella rotundiloba* early form and *Ozarkodina bidentatiformis*, as well as a number of unlabeled bases

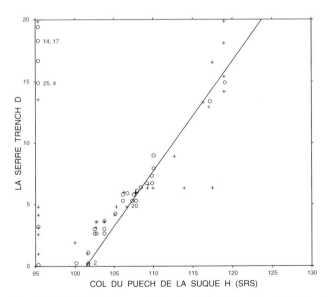

FIG. 2.—Initial plot in first round of La Serre Trench D against Col du Puech de la Suque section H (Montagne Noire, the SRS, prior to downward addition of the E section) for comparison with Figure 1. LOC positioned on the bases of *Ancyrodella gigas* form 1 (2) and *Palmatolepis ljaschenkoae* (20). Bases and tops along the *y*-axis represent species that occur at Trench D but not the SRS and later enter the Frasnian CS through projection onto the *x*-axis. Note that the bases of four species in Figure 2 (25, 4, 14, 17), which are on or close to the LOC in Figure 1, are not in the SRS. Base of SRS arbitrarily set at 100 CSU to avoid negative numbers.

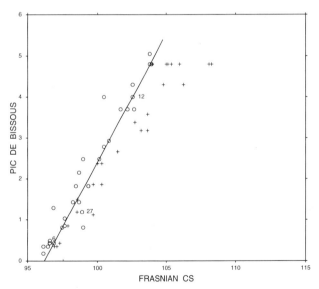

FIG. 3.—Fifth round plot of Pic de Bissous (Montagne Noire) against the Frasnian CS. LOC intersects the base of *Ancyrodella rotundiloba* early form (6) at 96.9 CSU, the lower boundary of the Frasnian, and the base of *Ozarkodina bidentatiformis* (12). Unlabeled bases intermediate between these two that are also intersected by the LOC are *A. rotundiloba* late form (at 97.5 *x*, 0.8 *y*), a transitional form between *A. africana* and *A. gigas* (100.1 *x*, 2.5 *y*, illustrated in Klapper, 1985, pl. 10, Figs. 7, 8, 11, 12), and *Palmatolepis punctata* (100.8 *x*, 2.9 *y*). *Ozarkodina* aff. *O. trepta* is the unlabeled base 0.04 CSU to the left of the line at 103.8 *x*, 4.8 *y*.

mentioned in the caption. The maximum adjustment of a base is 1.5 CSU for *Icriodus symmetricus* (unlabeled at 99.0 *x*, 0.8 *y*), a species which has not been used in the Montagne Noire zonation, and 0.8 CSU for *Polygnathus dengleri* narrow form.

Horse Spring (Canning Basin, Western Australia)

The LOC for Horse Spring (Fig. 4) consists of three segments, forming a doglegged pattern (Shaw, 1964, p. 134–136). The LOC intersects the bases of *Ancyrodella gigas* form 2, *Palmatolepis ljaschenkoae, P. winchelli*, and *P. linguiformis*. Such a doglegged pattern implies a major change in the accumulation rate at Horse Spring above 7.5 m in that section. Alternatively, one could draw a series of horizontal terraces indicating hiatuses in the Horse Spring section, connected by short diagonal segments of the same slope as the upper segment of the LOC in Figure 4 (the segment from *P. winchelli* to *P. linguiformis*). Controls for the positioning of such segments are not evident in the array of bases and tops, which rather supports the LOC drawn here. Of course, there are intervals of the LOC that are not controlled by biostratigraphic events, opening the possibility of alternative solutions.

The Montagne Noire sections discussed so far are characterized by pelagic limestones that represent condensed sedimentary accumulation by comparison with Frasnian sections in the Canning Basin of Western Australia, the Alberta Rockies, and western New York State. Nevertheless, there are exceptions because the correlation of the Horse Spring section in the Canning Basin with the Frasnian CS (Fig. 4) indicates that the lower 7.5 m at Horse Spring is comparably condensed to that of the Montagne Noire sequences. As a generalization, thin-bedded, red cephalopod limestones like those at some Montagne Noire sections, especially Coumiac (Fig. 5), dominate the lower 14 m at Horse Spring. Above this point, the Horse Spring section expands into thick-bedded, light-colored dolomitic limestones and dolomites. Thus, the doglegged pattern in Figure 4 is explicable in terms of lithofacies development. Shaw (1964, p. 141, 1993) has viewed doglegged solutions to the LOC as improbable, because of the implied major changes in the sedimentary accumulation rate within a given section. Such a view seems to disallow periods of condensed sedimentation in the formation of a section. At any rate, the proposed LOC depicted in Figure 4 derives empirically from the pattern of the biostratigraphic events, rather than from inferences about depositional environments.

Upper Coumiac Quarry (Montagne Noire)

The lower segment of the LOC for the Upper Coumiac Quarry (Fig. 5) intersects the bases of *Ancyrodella lobata* and *Palmatolepis proversa*; in this segment all bases are to the left and all tops to the right of the LOC, thus causing no changes in ranges. The short second segment, defined by a horizontal

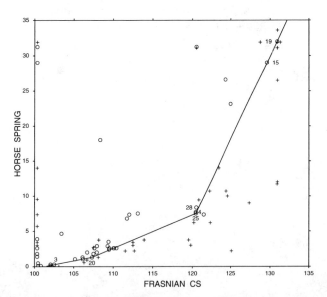

FIG. 4.—Fifth round plot of Horse Spring (Canning Basin, Western Australia) against the Frasnian CS. LOC, in three segments, intersects the bases of *Ancyrodella gigas* form 2 (3), *Palmatolepis ljaschenkoae* (20), *P. winchelli* (25), and *P. linguiformis* (19). Among unlabeled bases intersected by the LOC are *P. kireevae* (at 106.2 *x*, 1.1 *y*), and *P. semichatovae* (110.1 *x*, 2.6 *y*). The base of *Palmatolepis boogaardi* (15) is less than 0.1 CSU to the left of the line. Major change in the accumulation rate at 7.5 m at Horse Spring discussed in the text. Note that the bases of *Ancyrodella ioides* (4) and *Polygnathus samueli* (28) are close to that of *Palmatolepis winchelli* (25), as they are in Figures 5 and 7, respectively. The coincidence of tops at 130.9 CSU reflects the Frasnian-Famennian extinction. Our developing CS extends well up into the Famennian (Fig. 5), so this vertical coincidence is not explicable in terms of an unrecognized hiatus at the top of the Frasnian CS.

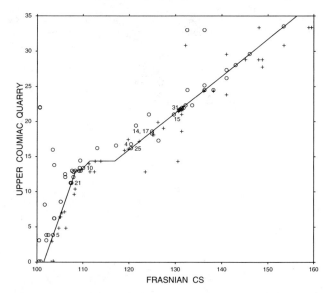

FIG. 5.—Fifth round plot of the Upper Coumiac Quarry (Montagne Noire) against the Frasnian CS, here extended into the lower Famennian. First segment of LOC intersects bases of *Ancyrodella lobata* (5) and *Palmatolepis proversa* (21); third segment, between two horizontal terraces indicating hiatuses at Coumiac (see discussion in text), intersects base of *Ancyrognathus seddoni* (10). Fifth segment positioned on bases of *Palmatolepis winchelli* (25, base of Montagne Noire Zone 12) and coincident bases of *P. bogartensis* (14, base of Zone 13) and *P. hassi* (17). Fifth segment is close to bases of *P. boogaardi* (15) and *Ancyrognathus ubiquitus* (not labeled, but in dense cluster above *P. boogaardi*). GSSP (31) for lower boundary of Famennian at 131.4 CSU, at base of Bed 32a (Becker and others, 1989, Fig. 3). Note that the fifth segment of the LOC, drawn on two upper Frasnian zonal markers, coincidentally intersects the bases of a number of lower Famennian species, some of which are also zonally defining. The fact that the LOC does not change slope between the upper Frasnian and lower Famennian indicates that there is not a geologically significant hiatus at the GSSP.

alignment of four bases at about 13 m on the y-axis, indicates a minor hiatus at Coumiac. This terrace is at the top of a 15-cm interval immediately above the base of Bed 16, a bed of 1.3 m thickness (House and others, 1985, Fig. 9b; sample 16 in Klapper, 1989, Fig. 4). The Montagne Noire Zone 10/11 boundary is somewhere within the superjacent 20 cm *within* Bed 16.

The third segment of the LOC at Coumiac (Fig. 5), just below the wider horizontal terrace at 14.3 m on the y-axis, intersects the base of *Ancyrognathus seddoni*, again resulting in no range extensions. Various attempts were tried to project the LOC from the upper end of the third segment to join with the higher, well defined linear array above about 120 in the CS axis. In the end, the only way to do this without substantially lowering the bases of *Palmatolepis winchelli* and *P. bogartensis*, among others, is to draw the fourth segment of the LOC as a second, wider horizontal terrace. This second terrace takes its position from three horizontally aligned tops (two are coincident) at 14.3 m on the y-axis and is located just above a 10-cm sample interval immediately above the base of Bed 18, a bed of 1.65 m thickness (House and others, 1985, Fig. 9b; sample 18 in Klapper, 1989, Fig. 4). The position of this second terrace is within Zone 11. There are a number of hardgrounds developed within the Frasnian of the Upper Coumiac Quarry, so the geologic implications are not improbable.

The fifth segment of the LOC at Coumiac (Fig. 5) then keys on the bases of *Palmatolepis winchelli* and *P. bogartensis*, species that define the bases of Montagne Noire upper Frasnian Zones 12 and 13, respectively (Klapper, 1989). The upper projection of this segment coincidentally intersects the bases of significant species just below the top of the Frasnian at 131.4 CSU, as well as a number of zonally defining species in the lower Famennian, none of which influenced the position of the LOC. Furthermore, the highly linear nature of the array represented by the upper segment of the LOC implies that there is not a biostratigraphically significant hiatus at the Frasnian/Famennian boundary. This is reassuring since Upper Coumiac is the boundary stratotype selected by the Subcommission on Devonian Stratigraphy; the position of the "global stratotype section and point" (GSSP, 31 on Fig. 5) at the base of Bed 32a (Becker and others, 1989, Fig. 3) projects to 131.4 on the CS axis.

New York Regional Composite (United States)

A regional composite for the lower part of the Frasnian [Montagne Noire Zones 1–7] derives from graphic correlation of 20 sections in the Genesee and Sonyea Formations in western New York. The most complete section, Beards Creek in the Genesee Valley, is the SRS for this regional composite. In applying the method of supplemented graphic correlation (Edwards, 1989), a total of 23 key beds (e.g., bases of regionally traceable black shales) were available. Because of the low density of horizons with conodonts and ammonoids, we developed a correlation hypothesis in which a total of 12 key beds provide all but two of the control points for the position of the LOC in the 19 New York graphs. In many of the graphs (nine of 19), the base of the Rhinestreet black shale of the West Falls Formation, interpreted to lie immediately above a sequence boundary, controls one point of the LOC. Two conodont species provide one controlling point each in two graphs. A detailed account of the construction of the New York regional composite will be given in a separate paper (so far only available in abstract, Kirchgasser and Klapper, 1992).

The New York regional composite, consisting of information on the ranges of 28 conodonts and 24 ammonoid genera and species, was then correlated with the Frasnian CS (Fig. 6). The LOC intersects the bases of four species: *Palmatolepis transitans*, *Ancyrodella gigas* form 1, *P. punctata*, and *Ozarkodina* aff. *O. trepta*. Correlation of the New York regional composite with the Frasnian CS serves at least two purposes. First, it provides an independent test of the assumption of geologic synchroneity of the key beds used in constructing the New York composite. If many of the ranges were extended through graphic correlation, it would call into question the validity of the assumption. Most of the New York ranges of conodonts, however, fall within the maximum ranges of bases and tops already in the Frasnian CS, a result not contradictory to the assumption. The second purpose is to add ranges of ammonoids to the Frasnian CS (part of the array of bases and tops along the y-axis in Fig. 6).

Old Bohemia Valley (Canning Basin, Western Australia)

Two sections, WCB 365 and 367, in the Old Bohemia Valley, Canning Basin, Western Australia (Becker and others, 1993, Fig. 3), span from Montagne Zone 5 into Zone 6 and contain abundant ammonoid faunas, significantly sharing some species and subspecies with the classic western New York sequence. WCB 367 has *Probeloceras lutheri lutheri* from 1.8 to 24.3–26.3 m and *Prochorites alveolatus* from 12.3–15.8 to 24.6–

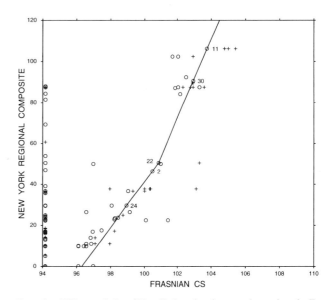

FIG. 6.—Fifth round plot of New York regional composite against the Frasnian CS. LOC intersects the bases of *Palmatolepis transitans* (24), *Ancyrodella gigas* form 1 (2), *P. punctata* (22), and *Ozarkodina* aff. *O. trepta* (11). Upper segment almost exactly (same to second decimal place) intersects the Rhinestreet Shale base (30). The latter is a key bed for controlling the LOC in many of the graphs used to construct the New York regional composite, but it was not used for positioning the LOC in Figure 6. The base of the Rhinestreet is on the Frasnian CS axis through correlation of a separate New York regional composite that extends from that horizon upward into the equivalents of Montagne Noire Zone 13. Bases and tops along the y-axis represent New York key beds and ammonoids.

24.7 m above the section base, the same order of appearance and stratigraphic overlap as that seen in the Cashaqua Shale of New York. Correlation of the two Old Bohemia Valley sections is somewhat problematical because they are in separate though nearby fault blocks and there do not appear to be any traceable key beds. Becker and others (1993, p. 304–305, Fig. 6) prefer to stack WCB 367 above 365 without overlap. Positioning the LOC strictly in terms of conodonts, 365 and 367 plot against the Frasnian CS in almost exactly the same position between 101 and 103 CSU, suggesting that the two sections overlap. Such a correlation can be approximated by zonal methods using conodonts; the graphic correlation does not alter the known stratigraphic relationship of *Probeloceras lutheri lutheri* and *Prochorites alveolatus* derived from both the Cashaqua and WCB 367 sequences. Details of this problem will be discussed in the forthcoming paper on the New York regional composite cited previously.

Luscar Mountain (Alberta Rockies, Canada)

The pattern of the array in the correlation of Luscar Mountain, Alberta Rockies (Fig. 7) with the Frasnian CS indicates a doglegged solution to the LOC, as in three previous correlations (Figs. 4–6). The LOC intersects the bases of four species: *Polygnathus timanicus*, *Ozarkodina* aff. *O. trepta*, *Palmatolepis domanicensis*, and *Polygnathus samueli*. *Palmatolepis winchelli*, the defining species for the base of Montagne Noire Zone 12, is less than 0.1 CSU to the left of the LOC and the position of *Polygnathus samueli*. Almost all of the collections on the y-axes of the graphs discussed so far represent the *Palmatolepis* conodont biofacies, including Luscar Mountain as high as 153 m (Klapper and Lane, 1989, Fig. 1). Above that level at Luscar, the sequence is representative of the *Polygnathus* biofacies.

Hay River-Trout River Regional Composite (Northwest Territories, Canada)

Ross McLean of Amoco Canada provided us a lithostratigraphic composite of 16 of the Frasnian sections in the Hay River, Trout River, and adjacent areas of the southwestern Northwest Territories, Canada. These sections are entirely dominated by faunas of the *Polygnathus* biofacies (in the sense this term is used in the analysis of Hay and Trout River faunas by Klapper and Lane, 1985). As a generalization, the *Polygnathus* biofacies occupied the inner shelf environment both in the Northwest Territories and in the Voronezh region of the Russian Platform (Aristov, 1988), while the *Palmatolepis* biofacies occurs in outer shelf to basinal environments. Correlation between these two contrasting biofacies has proven difficult because of the lack of common species. In other words, the problem is insoluble by zonal biostratigraphy, but can at least be approached by graphic correlation.

The graph of the Hay-Trout Rivers regional composite with the Frasnian CS (Fig. 8), has the least constrained LOC of all the graphs under review, because of the aforementioned biofacies problem. The doglegged LOC is positioned by three species, all of which cross the biofacies boundary: *Ancyrodella lobata*, *Palmatolepis semichatovae*, and *Polygnathus samueli*. The LOC represents an initial hypothesis to resolve the intercorrelation of the two biofacies.

SUMMARY

The Frasnian Upper Devonian composite standard consists of 27 sections, two of which are regional composites of 20 and 16 sections. The CS is divided into 34.5 CSUs, the scaling of which derives from the SRS, Col du Puech de la Suque H, a condensed section in the Montagne Noire. The SRS was extended downward to the base of the Frasnian by tracing a key

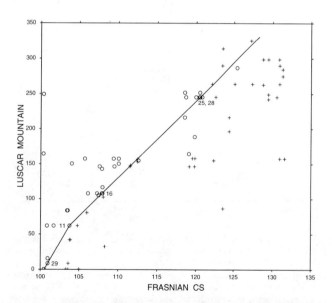

FIG. 7.—Fifth round plot of Luscar Mountain (Alberta Rockies) against the Frasnian CS. LOC intersects bases of *Polygnathus timanicus* (29), *Ozarkodina* aff. *O. trepta* (11), *Palmatolepis domanicensis* (16), and *Polygnathus samueli* (28), and nearly intersects that of *Palmatolepis winchelli* (25), shown as overlapping *P. samueli* as they are less than 0.1 CSU apart.

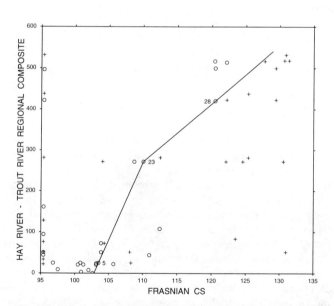

FIG. 8.—Fifth round plot of Hay River-Trout River regional composite against the Frasnian CS. LOC positioned on the bases of *Ancyrodella lobata* (5), *Palmatolepis semichatovae* (23), and *Polygnathus samueli* (28). LOC is an initial hypothesis of intercorrelation between the contrasting *Palmatolepis* and *Polygnathus* biofacies, which share few species.

bed into the nearby E section and upward to the top of the Frasnian through correlation with other Montagne Noire sections. We do not infer that each CSU is of equal time value because a uniform accumulation rate at the SRS has not been demonstrated. Nevertheless, the subdivision of the Frasnian into 34.5 composite standard units, derived primarily from biostratigraphic data, represents a far finer degree of stratigraphic resolution than any available traditional conodont or ammonoid zonation.

ACKNOWLEDGMENTS

The present paper represents a preliminary report on the Frasnian composite standard developed through our joint research, which began initially using the mainframe program for graphic correlation at the Amoco Research Center, Tulsa, in 1986. Beginning in 1989, we have carried out similar research using Kenneth C. Hood's GraphCor program for the IBM-PC. We thank Ken for providing us with a number of updates to Version 2.0, the latest being revisions to Version 2.2 in January 1992, as well as help with the program on numerous occasions. GK also thanks Joseph Hazel and Kevin Lanigan for initial help with GraphCor. Ross McLean generously provided us with a lithostratigraphic composite of the Hay-Trout Rivers sections. Discussions with Alan Shaw, who does not agree with our doglegged solutions for the LOC, were helpful in focusing us on weak points in certain of the graphs. We thank Alan Horowitz and Norman MacLeod for constructive criticism of an earlier draft of the manuscript.

Partial funding was provided by National Science Foundation, Division of Earth Sciences, grants EAR87-06567 and EAR89-03475 (GK) and a Research Opportunity Award to WTK, supplemented to the latter grant. A more detailed presentation of the Frasnian CS, ideally with all the graphs and a complete composite table of values of species ranges, publishers permitting, is planned at a later stage in the research.

REFERENCES

ARISTOV, V. A., 1988, Devonskie konodonty tsentral'nogo Devonskogo Polya (Russkaya Platforma): Akademiya Nauk SSSR, Geologicheskiy Institut, Trudy, no. 432, 119 p.

BECKER, R. T., FEIST, R., FLAJS, G., HOUSE, M. R., AND KLAPPER, G, 1989, Frasnian-Famennian extinction events in the Devonian at Coumiac, southern France: Paris, Comptes Rendus de l'Académie des Sciences, v. 309 (II), p. 259–266.

BECKER, R. T., HOUSE, M. R., AND KIRCHGASSER, W. T., 1993, Devonian goniatite biostratigraphy and timing of facies movements in the Frasnian of the Canning Basin, Western Australia, in Hailwood, E. A. and Kidd, R. B., eds., High Resolution Stratigraphy: London, Geological Society Special Publication 70, p. 293–321.

DOWSETT, H. J., 1988, Diachrony of Late Neogene microfossils in the southwest Pacific Ocean: Application of the graphic correlation method: Paleoceanography, v. 3, p. 209–222.

DOWSETT, H. J., 1989, Application of the graphic correlation method to Pliocene marine sequences: Marine Micropaleontology, v. 14, p. 3–32.

EDWARDS, L. E., 1984, Insights on why graphic correlation (Shaw's method) works: Journal of Geology, v. 92, p. 583–597.

EDWARDS, L. E., 1989, Supplemented graphic correlation: A powerful tool for paleontologists and nonpaleontologists: Palaios, v. 4, p. 127–143.

EDWARDS, L. E., 1991, Quantitative biostratigraphy, in Gilinsky, N. L. and Signor, P. W., eds., Analytical Paleontology: Knoxville, Paleontological Society Short Courses in Paleontology 4, p. 39–58.

FORDHAM. B. G., 1992, Chronometric calibration of mid-Ordovician to Tournaisian conodont zones: A compilation from recent graphic-correlation and isotope studies: Geological Magazine, v. 129, p. 709–721.

HAZEL, J. E., 1989, Chronostratigraphy of Upper Eocene microspherules: Palaios, v. 4, p. 318–329.

HOUSE, M. R., KIRCHGASSER, W. T., PRICE, J. D., AND WADE, G., 1985, Goniatites from Frasnian (Upper Devonian) and adjacent strata of the Montagne Noire: Hercynica, v. 1, p. 1–21.

KEMPLE, W. G., SADLER, P. M., AND STRAUSS, D. J., 1990, A prototype constrained optimization solution to the time correlation problem, in Agterberg, F. P. and Bonham-Carter, G. F., eds., Statistical Applications in the Earth Sciences: Ottawa, Geological Survey of Canada Paper 89–9, p. 417–425.

KIRCHGASSER, W. T., 1994, Early morphotypes of Ancyrodella rotundiloba at the Middle-Upper Devonian boundary, Genesee Formation, west-central New York, in Landing, E., ed., Studies in stratigraphy and paleontology in honor of Donald W. Fisher: New York State Museum Bulletin 481, p. 117–134.

KIRCHGASSER, W. T. AND KLAPPER, G., 1992, Graphic correlation using litho- and biostratigraphic markers in the uppermost Middle and Upper Devonian (Frasnian) of New York State: Geological Society of America Northeastern Section, Abstracts with Programs, v. 24 (3), p. 32.

KLAPPER, G., 1985, Sequence in conodont genus Ancyrodella in Lower *asymmetricus* Zone (earliest Frasnian, Upper Devonian) of the Montagne Noire, France: Palaeontographica, Abteilung A, v. 188, p. 19–34.

KLAPPER, G., 1989, The Montagne Noire Frasnian (Upper Devonian) conodont succession, in McMillan, N. J., Embry, A. F., and Glass, D. J., eds., Devonian of the World: Calgary, Canadian Society of Petroleum Geologists Memoir 14, v. 3, p. 449–468.

KLAPPER, G., 1990, Frasnian species of the Late Devonian conodont genus Ancyrognathus: Journal of Paleontology, v. 64, p. 998–1025.

KLAPPER, G. AND FOSTER, C. T., Jr., 1993, Shape analysis of Frasnian species of the Late Devonian conodont genus Palmatolepis: Paleontological Society Memoir 32 (Journal of Paleontology, v. 67, supplement to no. 4), 35 p.

KLAPPER, G. AND LANE, H. R., 1985, Upper Devonian (Frasnian) conodonts of the Polygnathus biofacies, N. W. T., Canada: Journal of Paleontology, v. 59, p. 904–951.

KLAPPER, G. AND LANE, H. R., 1989, Frasnian (Upper Devonian) conodont sequence at Luscar Mountain and Mount Haultain, Alberta Rocky Mountains, in McMillan, N. J., Embry, A. F., and Glass, D. J., eds., Devonian of the World: Calgary, Canadian Society of Petroleum Geologists Memoir 14, v. 3, p. 469–478.

KLEFFNER, M. A., 1989, A conodont-based Silurian chronostratigraphy: Geological Society of America Bulletin, v. 101, p. 904–912.

KRALICK, J. A., 1994, The conodont genus Ancyrodella in the middle Genesee Formation (Lower Upper Devonian, Frasnian), western New York: Journal of Paleontology, v. 68, p. 1384–1395.

MACLEOD, N. AND KELLER, G., 1991a, Hiatus distributions and mass extinctions at the Cretaceous/Tertiary boundary: Geology, v. 19, p. 497–501.

MACLEOD, N. AND KELLER, G., 1991b, How complete are Cretaceous/Tertiary boundary sections? A chronostratigraphic estimate based on graphic correlation: Geological Society of America Bulletin, v. 103, p. 1439–1457.

MILLER, F. X., 1977, The graphic correlation method in biostratigraphy, in Kauffman, E. G. and Hazel, J. E., Concepts and Methods of Biostratigraphy: Stroudsburg, Dowden, Hutchinson and Ross, p. 165–186.

MURPHY, M. A., 1987, The possibility of a Lower Devonian equal-increment time scale based on lineages in Lower Devonian conodonts, in Austin, R. L., ed., Conodonts: Investigative Techniques and Applications: Chichester, Ellis Horwood, p. 284–293.

MURPHY, M. A. AND EDWARDS, L. E., 1977, The Silurian-Devonian boundary in central Nevada, in Murphy, M. A., Berry, W. B. N., and Sandberg, C. A., eds., Western North America: Devonian: University of California, Riverside Campus Museum Contribution, v. 4, p. 183–189.

SHAW, A. B., 1964, Time in Stratigraphy: New York, McGraw-Hill Book Company, 365 p.

SHAW, A. B., 1969, Presidential address: Adam and Eve, paleontology and the non-objective arts: Journal of Paleontology, v. 43, p. 1085–1098.

SHAW, A. B., 1993, The origin of graphic correlation: Geological Society of America Northeastern Section, Abstracts with Programs, v. 25 (2), p. 78.

SWEET, W. C., 1984, Graphic correlation of upper Middle and Upper Ordovician rocks, North American Midcontinent Province, U. S. A., in Bruton, D. L., ed., Aspects of the Ordovician System: Palaeontological Contributions from the University of Oslo, no. 295, p. 23–35,

SWEET, W. C., 1988, A quantitative conodont biostratigraphy for the Lower Triassic: Senckenbergiana Lethaea, v. 69, p. 253–273.

SWEET, W. C., 1992, A conodont-based high-resolution biostratigraphy for the Permo-Triassic boundary interval, in Sweet, W. C., Yang Zunyi, Dickens,

J. M, and Yin Hongfu, eds., Permo-Triassic Events in the Eastern Tethys: Stratigraphy, Classification, and Relations with the Western Tethys: Cambridge, Cambridge University Press, p. 120–133.

TIPPER, J. C., 1988, Techniques for quantitative stratigraphic correlation: A review and annotated bibliography: Geological Magazine, v. 125, p. 475–494.

ZIEGLER, W., 1962, Taxionomie und Phylogenie Oberdevonischer Conodonten und ihre stratigraphische Bedeutung: Abhandlungen des Hessischen Landesamtes für Bodenforschung, no. 38, 166 p.

APPENDIX

The following values in CSUs are maximum bases and tops in the Frasnian CS (as of May 16, 1994) of species that define bases of zones in the Montagne Noire zonation (Klapper, 1989), other species labeled on Figures 1–8, and the positions of the base of the Frasnian and Famennian Stages. Numbers in parentheses correspond to those used in the figures.

1 GSSP ("global stratotype section and point") for the base of the Frasnian at Col du Puech de la Suque section E (Montagne Noire): 96.9.
2 *Ancyrodella gigas* form 1: 100.5–108.3.
3 *A. gigas* form 2: 102.5–103.5.
4 *A. ioides*: 120.4–121.3.
5 *A. lobata*: 103.5–123.5.
6 *A. rotundiloba* early form [= base of Zone 1]: 96.9–98.2.
7 *A. rotundiloba* late form [= base of Zone 2]: 97.5–99.3.
8 *A. rugosa* [= base of Zone 3]: 98.5–100.3.
9 *Ancyrognathus primus* [= base of Zone 6]: 102.1–105.4.
10 *A. seddoni*: 110.1–111.3.
11 *Ozarkodina* aff. *O. trepta* [= base of Zone 7]: 103.8–107.3.
12 *O. bidentatiformis*: 102.5–105.4.
13 *Palmatolepis* aff. *P. proversa* [= base of Zone 8]: 105.2–107.4.
14 *P. bogartensis* [= base of Zone 13]: 125.0–131.0.
15 *P. boogaardi*: 129.6–131.3.
16 *P. domanicensis* [= base of Zone 10]: 107.9–112.5.
17 *P. hassi*: 125.0–127.5.
18 *P. jamieae* [= *P.* sp. B of Klapper and Foster = base of Zone 11]: 109.3–120.2.
19 *P. linguiformis*: 130.9–131.3.
20 *P. ljaschenkoae*: 107.3–112.5.
21 *P. proversa* [= base of Zone 9]: 107.5–111.5.
22 *P. punctata* [= base of Zone 5]: 100.8–108.1.
23 *P. semichatovae*: 110.1–119.5.
24 *P. transitans* [= base of Zone 4]: 99.0–103.3.
25 *P. winchelli* [= base of Zone 12]: 120.4–131.3.
26 *Polygnathus* aff. *P. dengleri*: 100.5–105.1.
27 *P. dengleri* narrow form: 98.1–102.7.
28 *P. samueli*: 120.5–122.2.
29 *P. timanicus*: 101.0–103.8.
30 Rhinestreet Shale base (Figure 6): 102.9.
31 GSSP for the base of the Famennian Stage at Upper Coumiac Quarry (Figure 5): 131.4.

MEASURING THE DISPERSION OF OSTRACOD AND FORAMINIFER EXTINCTION EVENTS IN THE SUBSURFACE KIMMERIDGE CLAY AND PORTLAND BEDS, UPPER JURASSIC, UNITED KINGDOM

DAVID H. MELNYK
Scott Pickford plc, 256 High St., Croydon CR0 1NF, United Kingdom
JOHN ATHERSUCH
StrataData Ltd., 16 Ottershaw Rd., Ottershaw KT16 0QG, United Kingdom
NIGEL AINSWORTH
2 Millers Rise, St. Albans AL1 1QW, United Kingdom
AND
PAUL D. BRITTON
StrataData Ltd., 16 Ottershaw Rd., Ottershaw KT16 0QG, United Kingdom

ABSTRACT: Regional studies of the Upper Jurassic Wessex Basin of onshore United Kingdom have demonstrated the advantage of using digitally filtered gamma-ray logs to analyze subsequence-scale depositional packages, or cycles. This low-frequency component of the well-log trace (≥20-m wavelength) tends to be the most correlatable at a regional scale. This low-frequency component can be enhanced by (i) altering the displayed aspect ratio of the trace by squeezing the vertical scale and stretching the horizontal scale, and changing the horizontal axis to accommodate only the numerical range of log values (i.e., normalization), and (ii) digital filtering of the log values with an appropriate low-pass filter. In the Wessex Basin, gamma-ray and sonic logs treated in this manner readily reveal the longer wavelengths (>20 m) associated with major decreases and increases in log activity. These cycles can be recognized across changes in lithofacies between wells and can often be correlated over long distances (c.100 km). Missing cycles testify to hiatuses, and biostratigraphic and seismic calibration suggests that cycles are chronostratigraphic. The striking cyclicity in digitally filtered gamma-ray and sonic logs, facilitates high resolution log correlation over considerable distances, allowing the systematic calibration of biostratigraphic with lithostratigraphic data through graphic correlation. This involves the systematic cross-plotting of biostratigraphic events to produce a composite standard (CS). The scatter of biostratigraphic events is generally such that a line of correlation cannot be drawn with confidence from the biostratigraphic data alone. Independent lines of correlation are established between wells by first pattern-matching cycle boundaries on digitally filtered wireline traces and using these, in addition to the biostratigraphy, as a framework for adding biostratigraphic events to the CS. This has advantages over the conventional approach in that the effects of outliers in the data are minimized and hiatuses and subtle changes in rock accumulation rates become apparent. By adding all the biostratigraphic events from all wells to the CS a range chart is produced depicting the scatter or dispersion of all biostratigraphic events relative to the log based cycles. The end result is an integrated stratigraphy in which biostratigraphic events are calibrated against rock events giving a measure of dispersion, and hence confidence, for each biostratigraphic event.

INTRODUCTION

The aim of this paper is to demonstrate the potential of the graphic correlation technique (Shaw, 1964, Miller, 1977, Edwards, 1989) in measuring the dispersion, or scatter, of biostratigraphic events relative to major wireline log events in the subsurface, using the Upper Jurassic succession of the Wessex Basin, southern United Kingdom as an example (Fig. 1). Firstly, we illustrate the effectiveness of digitally filtering and redisplaying gamma-ray and sonic log data. Specifically, we demonstrate that the low bandpass filtering and redisplay of gamma-ray and sonic logs enhances the correlatability of low-frequency cyclic components within the log; cyclicities related to stratigraphic phenomena at the sequence and parasequence scale (≥10 m in thickness and ≥100 ka in duration). Additionally, using the filtered wireline curves, we demonstrate here a further extension of the graphic correlation method. We presuppose a situation in which a set of equivalent log events can be recognized over a large number of wells. Lines of correlation (LOC) drawn on the basis of these wireline log events can then be utilized as chronostratigraphic scales against which the dispersion, or scatter, of biostratigraphic events can be measured. This, we believe, can provide a powerful tool for measuring the utility of individual biostratigraphic events before erecting a biozonation scheme and for scaling biostratigraphic events.

STUDY AREA

The geological succession chosen for this study is the Kimmeridgian to early Portlandian interval of the Upper Jurassic System in the Wessex Basin, onshore United Kingdom. We use the terms 'Kimmeridgian' and 'Portlandian' in the sense of Cope and others (1980) and the term 'Wessex Basin' in the sense of Whittaker (1985). Our database consists of gamma-ray and sonic logs and micropaleontological data from 14 wells, most of which are concentrated in the Weald sub-basin (Fig. 1).

The Wessex Basin is a mainly onshore basin covering an area of over 250 km². The sediment fill, which is mainly Mesozoic and Cenozoic in age, is locally over 3 km thick. Sediment distribution has been explained in terms of polyphase extensional faulting and accelerated subsidence in the Early Triassic, Early Jurassic and Late Jurassic periods (Chadwick, 1986), followed by basin inversion in the mid-Tertiary (Ziegler, 1975). Previous workers have demonstrated that significant deepening of the basin occurred throughout Kimmeridgian times, while significant shallowing occurred during Portlandian times (House, 1985, 1986), implying an overall large scale cycle of relative sea-level change (Hallam and Sellwood, 1976). At a smaller scale, the only large scale outcrop of this succession on the Dorset coast shows well developed rhythms consisting of alternating kerogen-rich shales and calcareous mudstones, and there is an apparent widespread correlatability of these rhythms (Cox and Gallois, 1981; Oschmann, 1988). Small scale rhythms have been correlated in the subsurface of south-east England using unfiltered gamma-ray logs (Penn and others, 1986).

SEQUENCES AND CYCLES

Gamma-ray and sonic logs consist of discrete measurements, taken at equal intervals (usually 15.42 cm or 0.5 ft) down a well bore. As such, the data are voluminous, and ideally suited to manipulation using the standard techniques of time series analysis. Processing of digital data by bandpass filtering, with subsequent redisplay of the low-frequency components, reveals

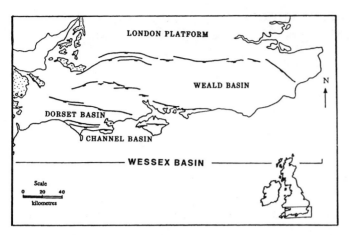

FIG. 1.—Map of the areal extent of the Wessex Basin (after Whittaker (1985).

cyclicity at an order of magnitude greater than that observed at outcrop. Figure 2 show the low-frequency bandpass component of the Standard Reference Section (Well-1); see Melnyk and others (1994) for further details of the method.

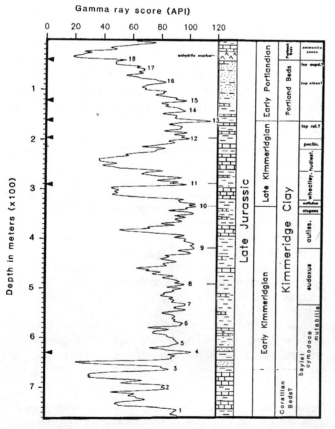

FIG. 2.—Type gamma-ray log of the Wessex Basin with chronostratigraphic, lithostratigraphic and biostratigraphic nomenclature. Correlatable log events are numbered 1 through 18; possible hiatuses are labelled as black triangles. Kerogen-rich shales intervals are labelled as horizontal dashes. Vertical divisions equal 10 m.

LOW-FREQUENCY BANDPASS FILTERING

In this paper, we utilize low-frequency bandpass filtering. Fourier theory states that the variance in any time series can be entirely accounted for by the mathematical superposition of sinusoidal waveforms of different amplitudes, frequencies, and phases. In the low bandpass filtering of wireline log data, our purpose is to attenuate the high-frequency waveforms (by reducing their amplitudes to zero) without significantly affecting the amplitudes and phases of the low-frequency components. This attenuation can be achieved in either of two ways, namely filtering in the time (or in our case depth) domain, or filtering in the frequency domain. These two methods are mathematically quite distinct, but can be shown to produce equivalent results. We prefer to achieve low bandpass filtering in the time (or depth) domain by the multiple application of moving-average filters (Melnyk and others, 1994). Figure 3 illustrates the philosophy of low bandpass filtering utilizing multiple moving averages. Figure 3A shows a gamma-ray log which has been displayed at a typical aspect ratio (as used in typical petroleum industry 1:500 scale composite logs). Simply lengthening the horizontal axis (Fig. 3B and 3C) enhances the low-frequency component, although the patterns are now obscured by high-frequency excursions. Finally, low bandpass filtering to remove all variance at wavelengths less than 2 m (Fig. 3D) and 4 m (Fig. 3E) displays low-frequency patterns (i.e., trends and anomalies at parasequence and sequence scale, van Wagoner and others, 1990) much more clearly.

Filtering in the time (or depth) domain by multiple applications of moving averages can be thought of as a process of trend-fitting (Schwarzacher, 1975). The filter is symmetrical and consists of an odd number of equal weights which sum to unity. In the specific case of the three-point moving average, data points are averaged in sets of threes to produce a new time (or depth) series, which is itself subjected to the same process. Taken as a whole, the procedure is equivalent to subjecting the original time (or depth) series to a single symmetrical filter composed of normally distributed weights (Holloway, 1958; Schwarzacher, 1975). In many algorithms, the averaging process leads to the loss of $(N/2 - 1)$ data points at both ends of the time series (where N is the length of the moving average) for each pass of the moving average filter. This can be remedied by systematically reducing the filter length to 1 at the beginning and end of the time series, so no data points are lost. The frequency response of such filters is entirely predictable (Holloway, 1958), as their transfer functions are readily calculated. For example, Figure 3F compares the frequency responses of single and multiple applications of an 11-point moving average filter by plotting the proportion of variance retained by the filter against frequency (cycles per sample distance). Note how the single application of a moving average is not an ideal filter in that it introduces spurious polarity reversals between frequencies of 0.08 and 0.18 cycles per sample distance, but that the multiple applications of the same moving average produce lowpass filters which are closer to the ideal (i.e., they pass all the low-frequency variance below the specified cut-off frequency, and remove all of the high-frequency variance without introducing systematic noise above the cut-off frequency). As a general rule, it appears that several applications of a short moving average are preferable to the single application of a long moving

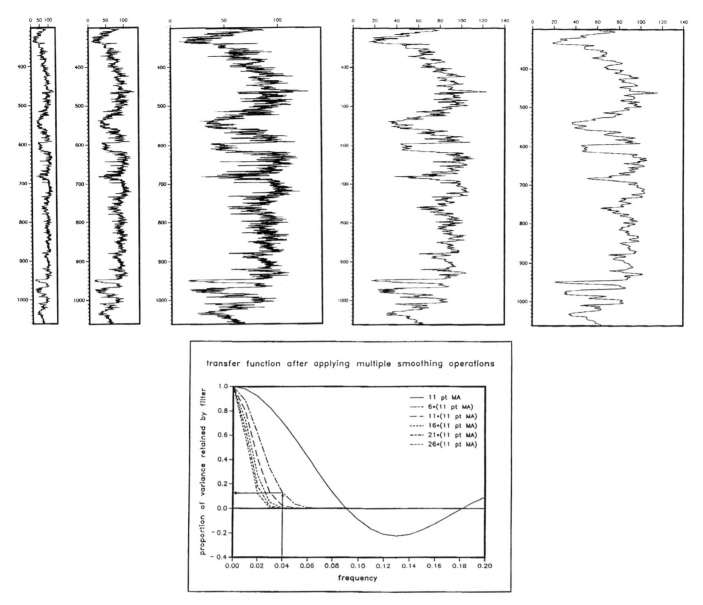

FIG. 3.—Stages in gamma-ray log enhancement. (A) A log displayed at a typical aspect ratio, with a low ratio of horizontal to vertical scales. Lengthening the horizontal axis amplifies the low-frequency component of the log signature (B, C), while low bandpass filtering removes the variance at wavelengths <2 m (D) and <4 m (E). (F) illustrates transfer functions for 11-point moving average filters. The curves represent the frequency responses (the proportion of the original variance at a particular frequency retained by the filter) of single and multiple applications of 11-point moving average filters after applying multiple smoothing operations.

average. Returning to Figure 3F, read off the frequency response (i.e., the proportion of the original variance at a particular frequency retained by the filter), project a vertical line from the abscissa to the appropriate curve, and then a horizontal line to the ordinate. For example, six iterations of an 11-point moving average will pass approximately 15% of any variance occurring at a frequency of 0.04 cycles per sample distance in the old time series into the new time series.

We interpret our filtered and amplified gamma-ray and sonic logs (Figs. 2, 4, 5, 6) as facies indicators, with the filtered gamma-ray log reflecting the continuous variation in the trend between relatively radioactive and relatively non-radioactive facies, and the filtered sonic log reflecting the continuous varia-

tion in the trend between relatively low velocity (organic-rich) and high velocity (carbonate-rich and organic-poor) facies. Similarities between filtered logs can be remarkable (Figs. 4, 5, 6). The large-scale systematic reversals in the facies trend-line appear to reflect basin-wide changes in deposition; they parallel seismic reflections (within seismic resolution) and are interpreted as time lines. These time lines can correspond to flooding surfaces, parasequence boundaries and sequence boundaries. Additionally, many of the smaller-scale patterns or cycles seen within the log traces also exhibit a high degree of correlatability despite obvious internal variations in lithology and can also be interpreted as synchronous units of probable climatic origin. It is these stratigraphic patterns that form the basis of our regional

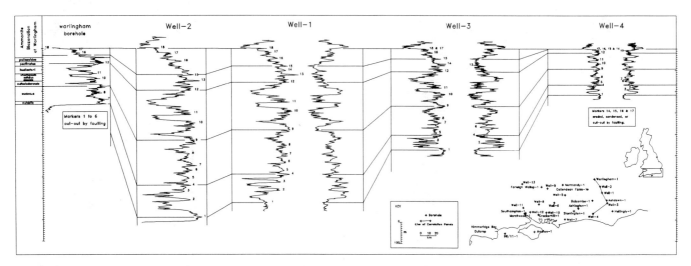

Fig. 4.—Wireline log correlation panel illustrating north to south cycle boundary correlation using filtered gamma-ray (left-hand curves) and sonic velocity (right-hand curves) logs through the Weald Basin. Correlatable log events are numbered 1 to 18 and are calibrated with the ammonite biozonation at Warlingham-1. Vertical divisions equal 10 m.

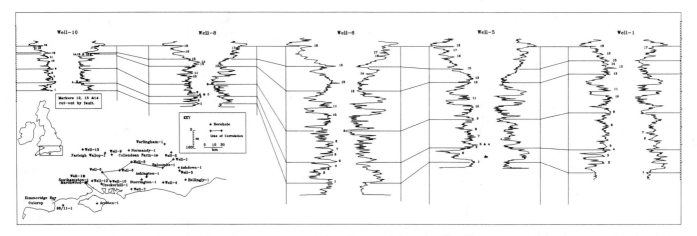

Fig. 5.—Wireline log correlation panel illustrating east to west cycle boundary correlation using filtered gamma-ray (left-hand curves) and sonic velocity (right-hand curves) logs through the Weald Basin. Correlatable log events are numbered 1 to 18. Vertical divisions equal 10 m.

and graphic correlations. How can we know that the log-events are synchronous without using the biostratigraphic data in the first place? This is, of course, a matter of interpretation. However, during most of the iterations described below, some of the log patterns between wells are so strikingly similar that the correlation is not in doubt. The correlation of these few unequivocal events is often sufficient to guide the LOC, thereby calibrating the biostratigraphy, without us having to resort to more contentious correlations. In other cases there may be considerable ambiguity, and alternative correlations should be tested and even allowed to contribute to the uncertainty or error, as discussed below.

Figures 4, 5 and 6 illustrate our regional correlations. Each well has been oriented relative to a regional marker and correlative gamma-ray highs and/or sonic velocity lows (or cycle boundaries) have been numbered 1 (oldest) through to 18 (youngest). Only the Lower Kimmeridgian through to the Lower Portlandian succession is illustrated for each well. Figure 4 shows the correlation from the northern margin (left) to the southern margin of the Weald Basin. Figures 5 and 6 illustrate lines of correlation from the northern margins of the Channel Basin (left) to the center of the Weald Basin (Fig. 1). The striking correlatability of the log cycles is immediately apparent. Generally speaking, two major trends of decreasing radioactivity (or cleaning-upwards trends) extending over hundreds of meters are apparent: 1) a lower trend (log markers 4 through to 11/12) comprises much of the Kimmeridge Clay Formation and reflects the gradual upward increase in carbonate content; and 2) an upper trend (log markers 12–18) containing the upper part of the Kimmeridge Clay Formation, the Portland Beds and the Purbeck Beds reflects an upward decrease in clay content (Fig. 2). Note that the cycles have been calibrated to the standard ammonite zonation using the Warlingham Borehole (Melnyk and others, 1992). Contained within these very long wavelength cycles or trends are cycles of shorter wavelengths (usually less than 50 m) (Figs. 4, 5, 6). The lowermost cycles

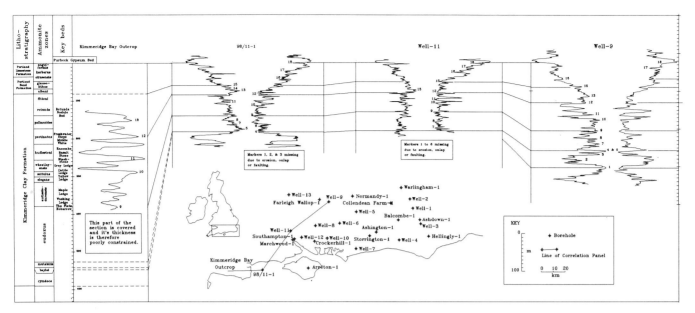

FIG. 6.—Wireline log correlation panel illustrating cycle boundary correlation using filtered gamma-ray (left-hand curves) and sonic velocity (right-hand curves) logs from the Weald Basin to the Channel Basin. Correlatable log events are calibrated with the ammonite biozones of the type-section, outcropping at Kimmeridge Bay, using a synthetic gamma-ray log. Vertical divisions equal 10 m.

containing the *cymodoce* through to the *baylei* ammonite biozones (log markers 1–4) are developed in limestones and calcareous sandstones. They are overlain by cycles of dark, occasionally bituminous, siltstones and sandstones (log markers 4–7) of the *mutabilis* ammonite biozone; these cycles show condensation over paleostructures, perhaps implying an onlap surface. Interpretation of unpublished seismic data suggests that a seismic sequence boundary coincides with the vicinity of log markers 3 and 4. Cyclicity is particularly pronounced between log markers 8 and 13, containing the *eudoxus* through to the *pallasoides* ammonite biozones (Figs. 4, 5, 6). These cycles can be correlated with confidence from Well-1, the chosen standard reference section (SRS), and can be shown to parallel the most reliable biostratigraphic indices available (Melnyk and others, 1992). It is within this interval that the kerogen-rich shales are best developed and the cyclic character of the well logs reflects long-wavelength (50–100 m) transitions from dominantly oxygenated, micritic mudstones to dominantly reducing, black shales. Lithological evidence shows the interval between log markers 12 and 13 to change character, from limestones and mudstones at Well-1, through calcareous siltstones and mudstones at Well-5, to glauconitic mudstones and sandstones at Well-6. Providing wells are reasonably close together cycle boundaries can be reliably recognized in spite of facies changes and in spite of dramatic thinning. Condensation and loss of cycles at the basin margin and on paleohighs suggests the presence of several hiatuses between log markers 1 and 15 (Figs. 4, 5, 6). A minor hiatus is present at log marker 11, within the *wheatleyensis* ammonite biozone; significant hiatuses also occur at log markers 12 and 13 at the top of the *pectinatus* and *rotunda* ammonite biozones, an observation supported by a basin-wide seismic marker at this level. Cycle loss and condensation also occurs at log marker 15, suggesting an onlap surface close to this level, and significant variation in the thickness of the cycle separating markers 17 and 18 points to a significant hiatus at this level. Generally speaking, markers 13–18 are biostratigraphically poorly constrained, but it seems that markers 16 and 18 lie close to the top of the *albani* and *anguiformis* ammonite biozones respectively. The preceding discussion, based on Figures 4, 5 and 6, indicates that low-frequency band-pass filtering of digital log data does indeed enhance the correlatability of signals at the parasequence and sequence scale, and that the log markers used are time significant.

CREATING A COMPOSITE STANDARD REFERENCE SECTION

We illustrate 14 iterations in the construction of our CS (Figs. 7–21). The microfaunas used in this present paper comprise ostracods and benthic (both agglutinating and calcareous) foraminifers. The taxa used to define events are as far as possible common to abundant in a single sample and also geographically widespread throughout the study area. Of the two groups, the ostracods provide the more useful paleontological datums due to their proliferation within many of the samples and also their shorter stratigraphic ranges. In all cases, the events are based on either their tops (extinctions) or their acme occurrences. The developed event zonation scheme used by the present authors has been based on outcrop material from the Dorset coast (Fig. 6) which has been calibrated to the standard ammonite zonation recognized throughout north-west Europe in conjunction with data from exploration wells situated within the Central Channel Basin.

As a consequence of the proprietary nature of the biostratigraphic analyses, the names of illustrated wells have been changed to Well-1, Well-2, etc . . .

Iteration 1

Well-1 was chosen as the initial SRS as it displays a thick and complete Upper Jurassic succession, contains the largest number of samples per meter and exhibits the largest number

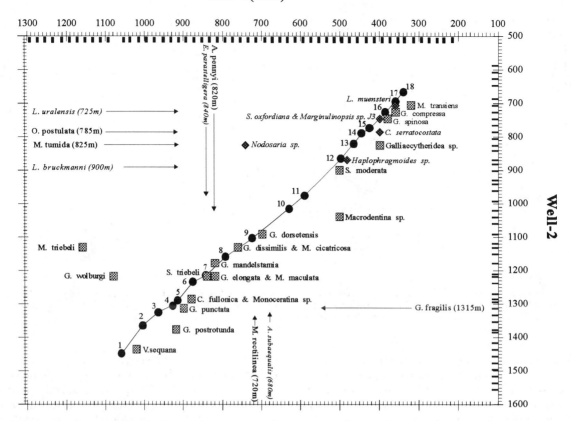

Fig. 7.—First iteration in the compilation of the CS by graphic correlation. Axes of the graph represent depths in meters in Well-1, the chosen SRS (Standard Reference Section), and Well-2. Correlatable log events are indicated by numbered black dots on the graph. Joining these points produces the hypothesized LOC. The positions of ostracod FDOs (normal type-font) and foraminifer FDOs (italicised type-font) are presented as squares and diamonds respectively. Events present in Well-2, but not in Well-1, are projected onto the SRS using the log-derived LOC. Sample positions in both wells are represented by black rectangles adjacent to the axes.

of biostratigraphic events of all the wells in our data set. A second well, geographically close to Well-1, with a thick Upper Jurassic interval, Well-2, was chosen as the ordinate.

Only gamma-ray logs were available to pattern correlate Well-1 and Well-2. However, this is not a problem as the pattern correlation of the filtered logs is extremely reliable; indeed, these logs can be compared peak for peak (Fig. 4)!

Correlative log events (1–18, Fig. 7, after Melnyk and others, 1994) are represented as black dots joined by a line. Note that the dots are plotted intentionally large. This is to imply the error associated with the log correlations (i.e., between 10 and 20 meters in the cases of these two wells). The line represents a LOC between the two wells based on log events. Changes of slope in the LOC occur at approximately 950 m and 450 m in Well-1, implying some change in relative rock accumulation rates in the vicinity of these horizons; these changes are subtle and are not readily perceived by visual comparison alone of the logs. The first downhole occurrences (FDOs) of 17 ostracod (shown in normal typeface) and 6 foraminifer species (italicised) common to both wells are plotted as squares and diamonds respectively on Figure 7. Several events lie directly on or reassuringly close to the LOC, and most of the remaining events lie to the right of the LOC. This means that these events are higher in Well-1 than in Well-2. We believe that this is an artefact of sampling, in that most of the material on which this study is based comes from cuttings samples; consequently, the recovery of fossils is influenced by the sampling interval and caving. At best, a particular taxon FDO is determined only in the highest sample below its true stratigraphic position at any one well site. Note that samples are shown as dark rectangles adjacent to the plot axes (Figs. 7–21).

Several FDOs, namely *Vernoniella sequana*, *Cytherella fullonica*, *Galliaecytheridea elongata* and *Macrodentina maculata*, *Schuleridea moderata*, *Haplophragmoides sp.*, *Galliaecytheridea spinosa* and *Galliaecytheridea compressa* are within one sample of the LOC (i.e., the interpreted true top), and the dispersion of these FDOs can be simply attributed to sampling error. Note, however, that the FDO of *Macrodentina transiens* is also found relatively lower in Well-2, but that the dispersion of this event is due to a combination of hiatus and sampling error (i.e., the FDO occurs in the last sample below the marker bed of the Purbeck Anhydrite the base of which is an hiatus). Other FDOs (e.g., *Galliaecytheridea punctata*, *Mandelstamia triebeli* and *Galliaecytheridea wolburgi*) are several samples adrift of the LOC; the dispersion of events such as these cannot be attributed to sample error alone, and a geological and/or operator mechanism should be sought. Perhaps these species are relatively rarer in Well-2 due to facies change, perhaps sam-

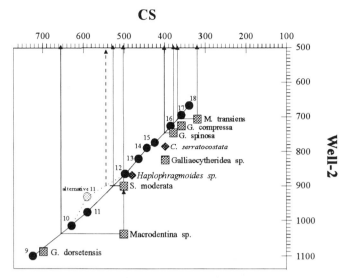

FIG. 8.—Enlarged upper portion of Figure 7 illustrating the measurement of dispersions of FDOs relative to the log derived LOC. Dispersions vary from <.30 m for *S. moderata* to ≫100 m for *Macrodentina sp.* Log events can also be subject to uncertainty. For example, if an alternative event 11 was picked at 935 m (grey circle), in addition to the preferred pick at 975 m, the dispersion for *S. moderata* (dashed line) would be larger for the second hypothesized LOC (shown as the dotted line). If both alternative events are to be added to the CS then future FDOs of *S. moderata* must be weighted twice as heavily in the calculation of its median FDO.

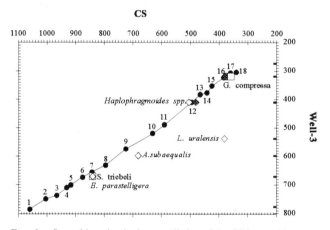

FIG. 9.—Second iteration in the compilation of the CS by graphic correlation. Axes of the graph represent depths in meters in the CS and Well-3. Correlatable log events are indicated by numbered black dots on the graph. Joining these points produces the LOC. The positions of ostracod FDOs (normal typefont) and foraminifer FDOs (italicised type-font) are presented as squares and diamonds respectively. Median, lowest and highest FDOs, from previous iterations, are presented as dark, medium and light symbols and are connected by a dashed line.

ple recovery conditions were significantly different in the two wells due to a change in drilling practice, or perhaps sample preparation was significantly different in the two wells such that fewer individuals were picked in Well-2.

In Table 1, we list the FDOs, the wells they were recognized in, the dispersions of the FDOs relative to the LOC, and the possible source of the error. Note that most of the error can be attributed to sampling.

Many ostracod and foraminiferal FDOs are not common to the two wells, and it is essential to add the FDOs present in Well-2 but absent in Well-1 to the CS. For example the ostracod *Galliaecytheridea fragilis* occurs at a depth of 1315 m in Well-2 but is absent in Well-1. *G. fragilis* can be added to the CS by projecting it horizontally to the LOC and vertically to the axis of Well-2, giving it a projected depth of 935 m on the CS. Similarly, the ostracods *Orthonotacythere pustulata* and *Mandelstamia tumida* occur at a depth of 785 m and 825 m respectively in Well-2, but are absent in Well-1. *O. pustulata* and *M. tumida* project to a depth of 425 m and 464 m respectively on the CS using the log event derived LOC. Foraminiferal FDOs *Lenticulina uralensis* and *Lenticulina brueckmanni* are also added to the CS.

To measure the dispersion, or scatter, of biostratigraphic events relative to the log picks, we project all available events to the LOC in both directions, thus comparing their relative positions in the two wells. Figure 8 is an enlargement of the upper portion of Figure 7 to illustrate this process. *S. moderata* is relatively higher in Well-1 than in Well-2. Just how much higher can be measured by projecting the event back onto the LOC and down to the abscissa; this indicates a dispersion (or uncertainty) of 18 m for the FDO of *S. moderata* relative to the gamma ray log events after the compositing of two wells. A similar exercise can be performed for *G. spinosa*, *M. transiens*, and *Macrodentina sp.*, implying dispersions of 25 m, 48 m, and 150 m respectively. It is noteworthy that these values are only meaningful within the context of the thickness of the SRS. They can be thought of as representing dispersions normalised to the depth scale of the SRS. Statistics, such as standard deviations, calculated from these values provide error estimates for any FDO.

The correlation of log events can also be subject to uncertainty. We illustrate a potential scenario in Figure 8. Suppose, for the sake of argument, we are uncertain of the correlative of log event 11 in Well-2, presently assigned a depth of 975 m, and consider 935 m to be a viable alternative. The result is two potential LOCs, shown as the dotted line and the full line in Figure 8. Fortunately, our uncertainty over log event 11 only affects the calibration of the FDO of *S. moderata* because the dispersion of this event relative to the alternative (dotted) LOC is some 20 m greater; all other FDOs remain unaffected. This complicates things somewhat with respect to the calculation of an average FDO for *S. moderata* as there are now three instead of two positions for *S. moderata* on the CS derived from one iteration. In this scenario, future iterations where only one LOC is hypothesized for *S. moderata* would have to be weighted twice as heavily in the calculation of a median to accommodate the two alternative events. Fortunately, we have not found it necessary to go to these lengths because the log events we have used have proven to be reliable in the vicinity of our biostratigraphic data

Iteration 2

The graphic correlation method facilitates the combination of data from many further wells. In Figure 9, we combine the data from a third well, Well-3, and Well-1 the CS. Well-3 is located some 20 km to the south of Well-1, but the filtered gamma-ray and sonic traces of Well-3 correlate extremely well with that of the CS (Fig. 4)

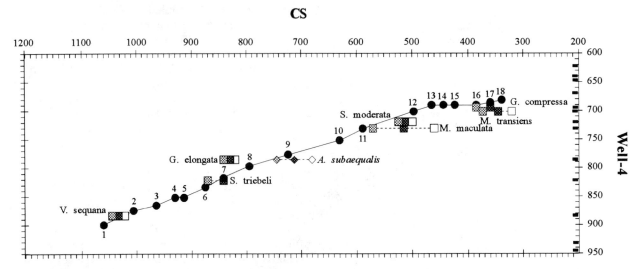

FIG. 10.—Third iteration in the compilation of the CS by graphic correlation. Axes of the graph represent depths in meters in the CS, and Well-4. The median, lowest and highest FDOs, from previous iterations, are presented as dark, medium and light coloured symbols and are connected by a dashed line.

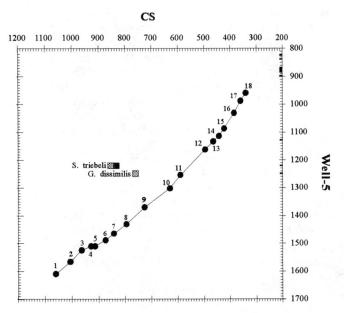

FIG. 11.—Fourth iteration in the compilation of the CS by graphic correlation. Axes of the graph represent depths in meters in the CS and Well-5.

The slope of the LOC is less than 45°, implying a lower rock accumulation rate in Well-3, which lies closer to the basin margin than Well-1 and Well-2. Again, flattening of the LOC occurs at around 950 m, 450 m and also, very significantly, around 350 m in Well-1 (CS) implying a significant decrease in rock accumulation rates at these horizons in Well-3 relative to the CS. Two ostracod FDOs, namely *S. triebeli* and *G. compressa,* and four foraminiferal FDOs, *Epistomina parastelligera, Ammobaculites subaequalis, L. uralensis,* and *Haplophragmoides sp.,* occur in common with the CS within this interval. Note that the FDOs of *Haplophragmoides sp.* and *G. compressa* are now plotted as ranges on the diagram with the darkest symbol representing the median FDO, and the lighter symbols the highest

and lowest range of FDOs; where the dispersion is small, as in the case of *G. compressa,* symbols will overlay each other. Most of the events plot within 1 sample of the LOC and the dispersions can be assigned to sampling error, with the exception of the foraminifer *L. uralensis* which is at least two samples adrift of its true top. The reason for this is unknown, but the dispersion of *L. uralensis* may be a function of its rarity. Finally, all common events are added to the CS by projecting them to the LOC; the foraminifer *Reinholdella pseudorjasanesis,* present in Well-3 but not present in the CS, is projected horizontally to the LOC and thereby added to the CS.

Iteration 3

Well-4, lies over 40 km south of the Well-1 on the margin of the Weald Basin. The section is less than half the thickness encountered in the SRS. Our favoured correlation suggests that loss of section due to hiatus or faulting has occurred between log markers 13 and 16 (Fig. 10). In spite of the high degree of thinning and loss of section, the remaining section can be readily pattern correlated with Well-3 (Fig. 4). The LOC between Well-4, and the CS is shallow, with 'doglegs' at the predicted positions (Fig. 10) and virtually all of the lower Portlandian and some of the upper Kimmeridgian section is missing in Well-4.

There are eight events, 7 ostracods and 1 foraminifer, common to Well-4 and the CS. Again we have plotted the medians and ranges of all previous occurrences (Fig. 10). Events common to both wells fall close to, and to the right of, the predicted LOC, and only *G. elongata* is relatively lower in the CS. This is very significant and represents a distance of over 5 samples on the CS. We suspect a geological or operator problem and extend the upper range of *G. elongata* by over 50 m by projecting it horizontally to the LOC (Fig. 10). Of the remaining events, the medians of *V. sequana* and *S. triebeli* lie directly on, or very close to, the LOC. *A. subaequalis, M. maculata,* and *S. moderata* occur at, or very close to, their lowest positions recorded so far, but these events are within one sample (*A. subaequalis* and *M. maculata*) or two samples (*S. moderata*) of the

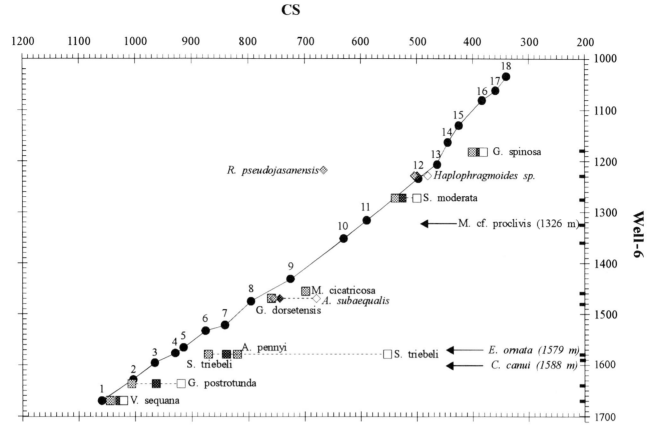

FIG. 12.—Fifth iteration in the compilation of the CS by graphic correlation. Axes of the graph represent depths in meters in the CS and Well-6.

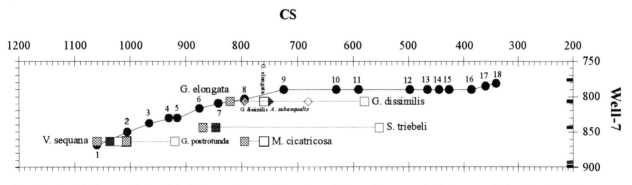

FIG. 13.—Iteration 6 in the compilation of the CS by graphic correlation. Axes of the graph represent depths in meters in the CS and Well-7.

LOC and their dispersions can be ascribed to sampling error. On the other hand, *M transiens* and *G. compressa* lie within one or two samples of the interpreted stratigraphic cut-out; as a consequence, their dispersions should be attributed to combination of a hiatus and sample spacing. The lower ranges of both these FDOs are extended by over 100 m through horizontal projection to the LOC.

Iteration 3 illustrates an interesting principle. If sampling distance remains constant and stratigraphic thickness decreases the potential error on FDOs increases proportionally.

Iteration 4

The section at Well-5 is about as thick as the Well-1 (the SRS). Pattern correlation is quite straight forward for log markers 1 to 12. Correlation of markers 13 to 18 is more difficult because this interval is slightly expanded at Well-5 compared to the SRS (Fig. 5); but a convincing correlation can be made via intermediate wells such as Collendean Farm (not illustrated) and Balcombe-1 (not illustrated); neither of which was sampled for micropaleontology. Sample cover here is very poor, and Well-5 contains only two ostracod FDOs, *S. triebeli* and *G. dissimilis*, the ranges of which are plotted on Figure 11. Note that projecting the FDOs of *S. triebeli* and *G. dissimilis* to the LOC extends the dispersions of these events by over 200m and 100m respectively (up to 10 sample intervals on the CS scale!). We attribute these dispersions to reworking and/or operator error.

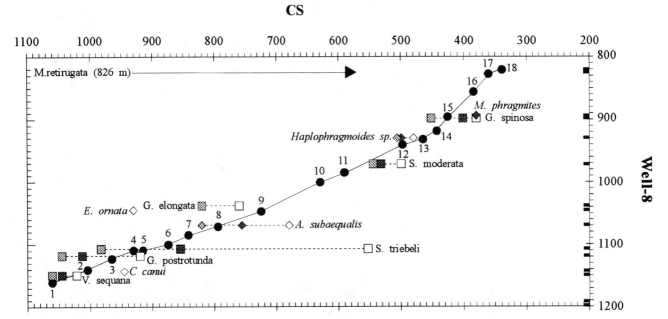

Fig. 14.—Iteration 7 in the compilation of the CS by graphic correlation. Axes of the graph represent depths in meters in the CS and Well-8.

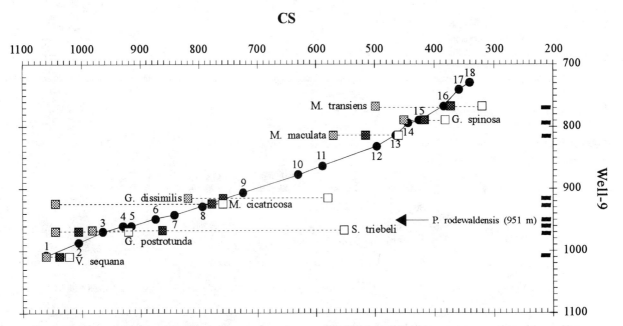

Fig. 15.—Iteration 8 in the compilation of the CS by graphic correlation. Axes of the graph represent depths in meters in the CS and Well-9.

Iteration 5

The section at Well-6 is slightly thicker than the SRS (Well-1), with much of the additional thickness attributable to expansion of the Upper Kimmeridgian and Portlandian interval containing log markers 13 to 18. Pattern correlation with the SRS is achieved via Well-5 (Fig. 5), Collendean Farm and Balcombe-1, and is believed to be straight-forward and dependable. Note the doglegs in the LOC around log markers 3 and 4, and 5, 6 and 7 (Fig. 12, implying loss of section in Well-6 relative to the CS) and 13 and 14 (implying loss of section in the CS versus Well-6).

There are 7 ostracod FDOs and 2 foraminifer FDOs in common between the CS and Well-6. The medians of two FDOs, *Haplophragmoides sp.* and *S. moderata*, lie on or close to the LOC. Of the remaining FDOs, *V. sequana, Galliaecytheridea postrotunda, S. triebeli, A. pennyi, Galliaecytheridea dorsetensis, A. subaequalis,* and *M. cicatricosa* lie within one sample interval of the LOC. We attribute these dispersions to sampling error. The FDO of *G. spinosa* occurs in the last sample; the

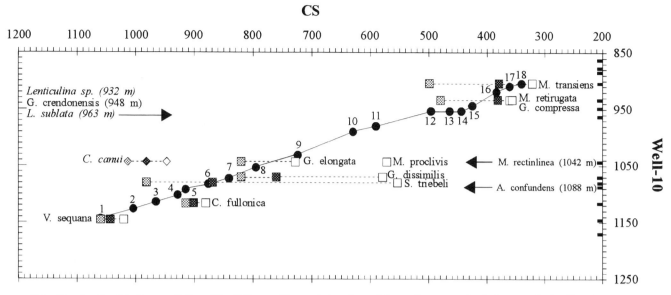

Fig. 16.—Iteration 9 in the compilation of the CS by graphic correlation. Axes of the graph represent depths in meters in the CS and Well-10.

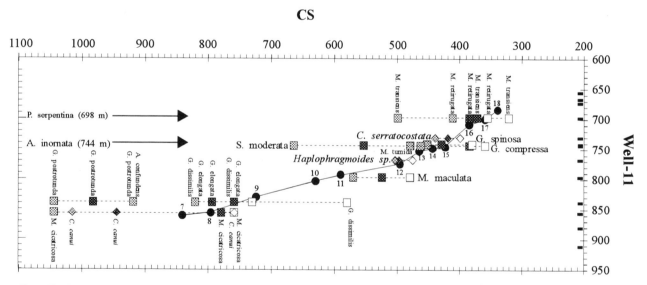

Fig. 17.—Iteration 10 in the compilation of the CS by graphic correlation. Axes of the graph represent depths in meters in the CS and Well-11.

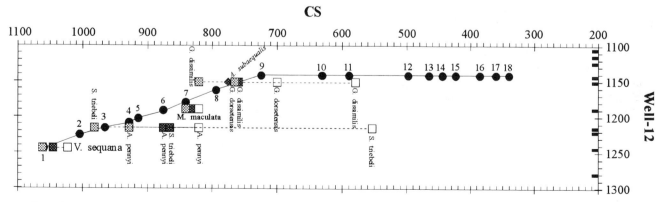

Fig. 18.—Iteration 11 in the compilation of the CS by graphic correlation. Axes of the graph represent depths in meters in the CS and Well-12.

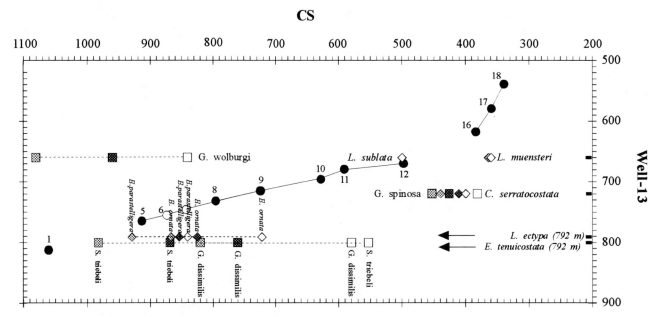

Fig. 19.—Iteration 12 in the compilation of the CS by graphic correlation. Axes of the graph represent depths in meters in the CS and Well-13.

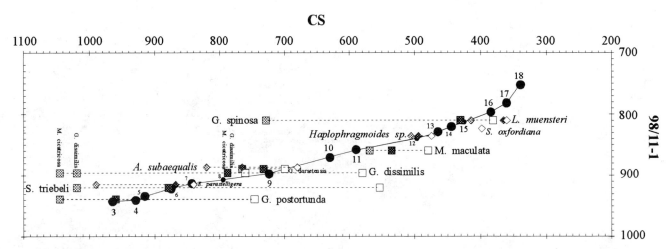

Fig. 20.—Iteration 13 in the compilation of the CS by graphic correlation. Axes of the graph represent depths in meters in the CS and offshore well 98/11–1.

large dispersion of this event is also attributable to sampling error. This is probably not true of the foraminifer FDO *R. pseudojasanensis* which appears to be circa 200 m too low in the CS. We attribute this dispersion to rarity or operator error.

Three FDOs not present in the CS, namely ostracod *Macrodentina cf. proclivis* (found at 1326 m), and foraminifers *Epistomina ornata* and *Cribrostomoides canui* (found at 1579 m and 1588 m respectively) are added to the CS using the LOC.

Iteration 6

Well-7 lies on the southern margin of the Weald Basin. Here, the study interval is very thin (less than 20% of the thickness displayed at Well-1(the SRS)) due to stratigraphic thinning and loss of section (Fig. 13). Nevertheless, correlation of log markers 1 to 8 with the CS is relatively straight forward via intermediate wells Balcombe-1, Ashington-1 and Storrington-1 (none of which is illustrated or has been sampled for micropaleontology).

The LOC forms a striking dogleg attributable to the loss of events 9 to 15 through hiatus or faulting. Sampling density is poor at Well-7 with approximately 5 samples covering the entire section of interest, and no samples taken in the upper part of the Portlandian! Nevertheless, six ostracod FDOs and one foraminiferal FDO were recognized and these are displayed in the usual way, as a median and range per FDO (Fig. 13). As a result of the poor sampling, all the medians (with the exception of *V. sequana*) lie on or well to the right of the LOC. The FDOs of *G. postrotunda, S. triebeli, G. elongata, G. dissimilis,* and *A. subaequalis* lie within one sample of the LOC, and their dis-

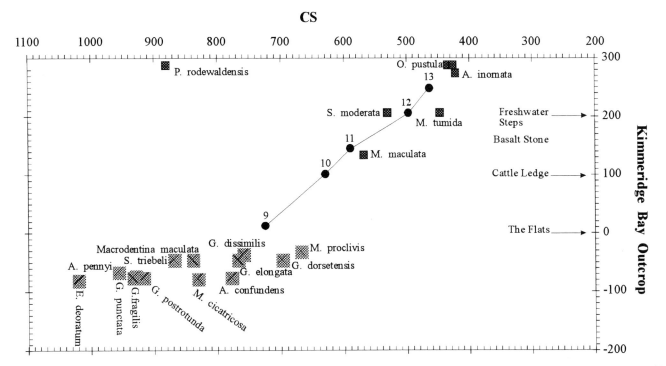

FIG. 21.—Iteration 14 in the compilation of the CS by graphic correlation. Axes of the graph represent depths in meters in the CS, and outcrops at Kimmeridge Bay, Dorset (measured up and below 'The Flats' key bed (see Fig. 6). Correlatable log and synthetic log events (see Fig. 6) are indicated by numbered black dots on the graph. Joining these points produces the hypothesized LOC. Ostracod events (squares) are taken from Kilenyi (1969). Note that the relationship of the lower ostracod events (larger squares) to the LOC is not well constrained, and are taken from a section with a much lower rock-accumulation rate.

persions can be attributed to sampling error. The ostracod *M. cicatricosa*, however, is two samples adrift of the LOC, and although it is likely that the dispersion is due to statistical fluctuations in abundance, these two samples cover most of the Kimmeridgian section and the resulting dispersion of *M. cicatricosa* is circa 300 m relative to the CS!

The FDOs at Well-7 are added to the CS via the LOC in the usual way.

Iteration 7

Well-8 is located on the west slope of the Weald Basin (Fig. 5). All log markers are present, despite the interval being less than half the thickness of the SRS, and Well-8 can be pattern correlated with confidence back to the SRS, via the intermediate wells Collendean Farm, Well-5, and Well-6. Note that the thickness of the interval encompassing log markers 14 through 18 is comparable to that of the SRS, while the thickness of the section containing events 1 to 13 is much thinner. The LOC illustrates this very well (Fig. 14), with the section of the LOC defined by log events 14 to 18 being much steeper. Seven ostracod and five foraminiferal FDOs are common to Well-8 and the CS, and as a result of the close sampling distance most of these FDO medians lie on, or close to, the LOC (Fig. 14), and dispersions can be attributed to sample spacing error.

The ostracod *Macrodentina retirugata*, present in Well-8 at 826 m but not in the CS, is projected to the LOC and added to the CS. Common FDOs at Well-8 are added to the CS via the LOC in the usual way.

Iteration 8

The ninth well to be added to the CS, Well-9, is situated on the north-west margin of the Weald Basin. The study interval is considerably thinner than at the SRS, but all the log markers are present and Well-9 can be reliably pattern correlated with the SRS via Normandy-1 (not illustrated), Farleigh Wallop (not illustrated), and Warlingham-1 (not illustrated). Only ostracods were studied at Well-9, and eight FDOs were described. Sample coverage is fair, and virtually all FDO ranges cross the LOC (Fig. 15). Most medians lie within 1 sample of the LOC, and virtually all of the dispersion can be accounted for by sampling error.

Iteration 9

Well-10 lies on the western margin of the Weald Basin and can be reliably pattern correlated with the SRS via Well-8, Well-6, Well-5, Balcombe-1 and Collendean Farm. Log markers 12 to 14 are interpreted as absent due to faulting or hiatus, and the LOC shows the characteristic 'dogleg' at this level (Fig. 16). Nine ostracod FDOs and 1 foraminiferal FDO were observed in Well-10. Most medians lie within 1 sample of the LOC, and virtually all of the dispersion can be accounted for by sampling error.

Only the foraminifer *C. canui* appears much too low in the CS. Three ostracod species, *Galliaecytheridea crendonensis, Mandelstamia rectilinea,* and *Amphicythere confundens,* and two foraminifers *Lenticulina* sp. and *Lenticulina sublata,* present at Well-10 but absent in the CS are added via the LOC.

TABLE 1.—SUMMARY OF ALL THE ITERATIONS; INDICATING THE ITERATION NUMBER, THE WELL IDENTIFICATION, AND THE DEPTH AT WHICH THE FDO WAS RECOGNISED, PLUS A RUNNING MINIMUM, MAXIMUM AND MEDIAN FOR THE FDO. ADDITIONALLY THE INTERPRETED SOURCE OF DISPERSION OR ERROR IS TABULATED.

Species and Author	Iteration Number	Well Name	Depth in Well	Depth to Sample Above	Projected Composite Depth	Minimum FDO	Median FDO	Maximum FDO	Source of Error	Figure Number
Aaleniella inornata Christensen & Kilenyi, 1970	10	Well-11	744	741	421	421	421	421		
Aaleniella inornata Christensen & Kilenyi, 1970	14	Kimmeridge outcrop	273		445	421	433	445	Within LOC Error	22
Amphicythere confundens Oertli, 1957	9	Well-10	1088	1079	919	919	919	919		
Amphicythere confundens Oertli, 1957	10	Well-11	838	796	747	919	833	747	Reworking	17
Amphicythere pennyi Kilenyi, 1969	0	Well-1	820	800	820					
Amphicythere pennyi Kilenyi, 1969	5	Well-6	1579		930	820	875	930	Within 1 Sample of LOC	12
Amphicythere pennyi Kilenyi, 1969	11	Well-12	1216	1189	972	820	930	930	Within 1 Sample of LOC	18
Cytherella fullonica Jones & Sherborn, 1888	0	Well-1	880	860	880					
Cytherella fullonica Jones & Sherborn, 1888	1	Well-2	1285	1270	921	880	901	914	Within 2 Samples of LOC	7
Cytherella fullonica Jones & Sherborn, 1888	9	Well-10	1116	1088	968	880	921	968	Within 1 Sample of LOC	16
Galliaecytheridea cf. mandelstami (Lubimova, 1955)	0	Well-1	820	800	806					
Galliaecytheridea cf. mandelstami (Lubimova, 1955)	1	Well-2	1180	1150	820	806	813	820	Within LOC Error	7
Galliaecytheridea compressa Christensen & Kilenyi, 1970	0	Well-1	360	342	360					
Galliaecytheridea compressa Christensen & Kilenyi, 1970	1	Well-2	725	705	385	360	373	385	Within 1 Sample of LOC	7
Galliaecytheridea compressa Christensen & Kilenyi, 1970	2	Well-3	320	274	380	360	380	385	Within 1 Sample of LOC	8
Galliaecytheridea compressa Christensen & Kilenyi, 1970	3	Well-4	695	671	480	360	383	480	Last Sample before Cut-Out	9
Galliaecytheridea compressa Christensen & Kilenyi, 1970	9	Well-10	933	905	411	360	385	480	Last Sample below top	16
Galliaecytheridea compressa Christensen & Kilenyi, 1970	10	Well-11	744	741	421	360	398	480	Within 2 Samples of LOC	17
Galliaecytheridea dissimilis Oertli, 1957	0	Well-1	760	740	760					
Galliaecytheridea dissimilis Oertli, 1957	1	Well-2	1130	1120	760	760	760	760	Within LOC Error	7
Galliaecytheridea dissimilis Oertli, 1957	4	Well-5	1250	1219	580	580	760	760	Reworking	11
Galliaecytheridea dissimilis Oertli, 1957	6	Well-7	808	777	820	580	760	820	Last Sample before Cut-Out	13
Galliaecytheridea dissimilis Oertli, 1957	8	Well-9	914	789	760	580	760	820	Within LOC Error	15
Galliaecytheridea dissimilis Oertli, 1957	10	Well-11	838	796	747	580	760	820	Within LOC Error	17
Galliaecytheridea dissimilis Oertli, 1957	11	Well-12	1152	1143	765	580	760	820	Within LOC Error	18
Galliaecytheridea dissimilis Oertli, 1957	12	Well-13	800	792	942	580	760	1020	Within 1 Sample of LOC	19
Galliaecytheridea dissimilis Oertli, 1957	13	98/11-1	896		719	580	760	1020	Within 2 Samples of LOC	20
Galliaecytheridea dorsetensis Christensen & Kilenyi, 1970	0	Well-1	700	680	700					
Galliaecytheridea dorsetensis Christensen & Kilenyi, 1970	1	Well-2	1090	1060	700	700	700	700	Within LOC Error	7
Galliaecytheridea dorsetensis Christensen & Kilenyi, 1970	5	Well-6	1457		765	700	700	765	Within 1 Sample of LOC	12
Galliaecytheridea dorsetensis Christensen & Kilenyi, 1970	11	Well-12	1152	1143	765	700	733	765	Within LOC Error	18
Galliaecytheridea dorsetensis Christensen & Kilenyi, 1970	13	98/11-1	890		700	700	700	765	Within 1 Sample of LOC	20
Galliaecytheridea elongata Kilenyi, 1969	0	Well-1	820	800	820					
Galliaecytheridea elongata Kilenyi, 1969	1	Well-2	1220	1210	841	820	831	841	Within 1 Sample of LOC	7
Galliaecytheridea elongata Kilenyi, 1969	3	Well-4	786	732	759	759	820	820	Within 3 Samples of LOC	10
Galliaecytheridea elongata Kilenyi, 1969	6	Well-7	808	777	820	759	820	820	Last Sample before Cut-Out	13
Galliaecytheridea elongata Kilenyi, 1969	7	Well-8	1039	972	730	730	820	820	Within 3 Samples of LOC	14
Galliaecytheridea elongata Kilenyi, 1969	9	Well-10	1042	963	769	730	795	820	Within 1 Sample of LOC	16
Galliaecytheridea elongata Kilenyi, 1969	10	Well-11	838	796	747	730	769	820	Within LOC Error	17
Galliaecytheridea postrotunda Oertli, 1957	0	Well-1	920	900	920					
Galliaecytheridea postrotunda Oertli, 1957	1	Well-2	1375	1355	1006	920	963	1006	Within 6 Samples of LOC	7
Galliaecytheridea postrotunda Oertli, 1957	5	Well-6	1637		1020	920	1006	1006	Within 1 Sample of LOC	12
Galliaecytheridea postrotunda Oertli, 1957	6	Well-7	863	710	1045	920	1013	1045	Within 2 Samples of LOC	13
Galliaecytheridea postrotunda Oertli, 1957	7	Well-8	1119	1106	955	920	1006	1045	Within 2 Samples of LOC	14
Galliaecytheridea postrotunda Oertli, 1957	8	Well-9	969	914	960	920	983	1045	Within 1 Sample of LOC	15
Galliaecytheridea postrotunda Oertli, 1957	10	Well-11	838	796	747	747	960	1045	Cut-out by faulting	17
Galliaecytheridea postrotunda Oertli, 1957	13	98/11-1	939		926	747	958	1045	Within LOC Error	20
Galliaecytheridea punctata Kilenyi, 1969	1	Well-2	1315	1295	935					
Galliaecytheridea punctata Kilenyi, 1969	0	Well-1	900	880	900	900	918	935	Within 2 Samples of LOC	7
Galliaecytheridea spinosa Kilenyi, 1969	0	well-1	381	360	381					
Galliaecytheridea spinosa Kilenyi, 1969	1	well-2	745	725	402	381	392	402	Within 1 Sample of LOC	7
Galliaecytheridea spinosa Kilenyi, 1969	5	Well-6	1183		452	381	402	452	Last Sample	12
Galliaecytheridea spinosa Kilenyi, 1969	7	Well-8	899	893	430	381	416	452	Within 2 Samples of LOC	14
Galliaecytheridea spinosa Kilenyi, 1969	8	Well-9	789	695	438	381	430	452	Within LOC Error	15
Galliaecytheridea spinosa Kilenyi, 1969	10	Well-11	741	732	421	381	426	452	Within LOC Error	17
Galliaecytheridea spinosa Kilenyi, 1969	12	Well-13	720	660	729	381	430	729	Last Sample before Cut-Out	18
Galliaecytheridea spinosa Kilenyi, 1969	13	98/11-1	811		421	381	426	729	Within LOC Error	19
Galliaecytheridea spinosa Kilenyi, 1969	14	Kimmeridge outcrop	287		425	381	425	729	Within LOC Error	22
Galliaecytheridea wolburgi Kilenyi, 1969	0	Well-1	1080	1061	1080					
Galliaecytheridea wolburgi Kilenyi, 1969	1	Well-2	1220	1210	841	841	961	1080	Within 9 Samples of LOC	7
Galliaecytheridea wolburgi Kilenyi, 1969	12	Well-13	660	660	410	410	841	1080	Within 25 Samples of LOC	19

Iteration 10

The section at Well-11, which lies on the crest of the high separating the Weald and Channel Basins, is much thinned relative to the SRS, and may be faulted in its lower part. Pattern correlation can only be reliably achieved using the sonic log, which can be used to tie the section back to the CS via Well-9. Only log events 7 through 18 were recognized. Moreover, log events 13, 14 and 15 appear to be condensed (Fig. 17). Eleven ostracod and two foraminiferal FDOs were recognized at Well-11, and the median FDOs cluster tightly about the proposed LOC, with the exception of the FDOs at the base of the section. Here they form a plateau due to faulting or hiatus (Fig. 17).

Iteration 11

Well-12 lies circa 10 km south-west of Well-10 on the high separating the Weald and Channel Basins. Only log markers 1

TABLE 1.—Continued.

Species and Author	Iteration Number	Well Name	Depth in Well	Depth to Sample Above	Projected Composite Depth	Minimum FDO	Median FDO	Maximum FDO	Source of Error	Figure Number
Macrodentina cicatricosa Malz, 1957	0	Well-1	760	740	760					
Macrodentina cicatricosa Malz, 1957	1	Well-2	1130	1120	760	760	760	760	Within LOC Error	7
Macrodentina cicatricosa Malz, 1957	5	Well-6	1469		795	760	760	795	Within 1 Sample of LOC	12
Macrodentina cicatricosa Malz, 1957	6	Well-7	863	710	1045	760	778	1045	Within 2 Samples of LOC	13
Macrodentina cicatricosa Malz, 1957	8	Well-9	924	914	780	760	780	1045	Within LOC Error	15
Macrodentina cicatricosa Malz, 1957	10	Well-11	856	838	796	760	788	1045	Within LOC Error	17
Macrodentina cicatricosa Malz, 1957	13	98/11-1	896		719	719	780	1045	Within 1 Sample of LOC	20
Macrodentina cf. proclivis Malz, 1958	5	Well-6	1326		572	572	572	572		
Macrodentina cf. proclivis Malz, 1958	9	Well-10	1042	963	769	572	671	572	Within 1 Sample of LOC	13
Macrodentina retirugata (Jones, 1885)	7	Well-8	826	820	357	357	357	357		
Macrodentina retirugata (Jones, 1885)	9	Well-10	933	905	411	357	384	411	Within 1 Sample of LOC	16
Macrodentina retirugata (Jones, 1885)	10	Well-11	698	674	365	357	365	411	Within LOC Error	17
Macrodentina transiens (Jones, 1888)	0	Well-1	321	300	321					
Macrodentina transiens (Jones, 1888)	1	Well-2	705	685	373	321	347	373	Within 1 Sample of cut-out	7
Macrodentina transiens (Jones, 1888)	3	Well-4	701	695	500	321	373	500	Within 1 Sample of cut-out	10
Macrodentina transiens (Jones, 1888)	8	Well-9	768	695	386	321	380	500	Within LOC Error	15
Macrodentina transiens (Jones, 1888)	9	Well-10	905	884	349	321	373	500	Within 2 Samples of LOC	16
Macrodentina transiens (Jones, 1888)	10	Well-11	698	674	365	321	369	500	Within LOC Error	17
Macrodentina maculata Kilenyi, 1961	0	Well-1	820	800	820					
Macrodentina maculata Kilenyi, 1961	1	Well-2	1220	1210	841	820	831	841	Within LOC Error	7
Macrodentina maculata Kilenyi, 1961	11	Well-12	1189	1152	863	820	841	841	Within 1 Sample of LOC	18
Mandelstamia maculata? Kilenyi, 1961	0	Well-1	460	440	460	460	460	460		
Mandelstamia maculata Kilenyi, 1961	3	Well-4	732	719	570	460	515	570	Within 2 Samples of LOC	10
Mandelstamia maculata Kilenyi, 1961	8	Well-9	814	780	480	480	525	570	Within 2 Samples of LOC	15
Mandelstamia maculata Kilenyi, 1961	10	Well-11	796	768	590	480	535	570	Within 1 Sample of LOC	17
Mandelstamia maculata Kilenyi, 1961	13	98/111	860		585	480	570	570	Within 1 Sample of LOC	20
Mandelstamia maculata Kilenyi, 1961	14	Kimmeridge outcrop	135		600	480	578	600	?	22
Mandelstamia rectilinea Malz, 1958	0	Well-1	720	700	720					
Mandelstamia rectilinea Malz, 1958	9	Well-10	1042	963	769		745		Within 1 Sample of LOC	16
Mandelstamia tumida Christensen & Kilenyi, 1970	1	Well-2	825	805	464	464	464	464		
Mandelstamia tumida Christensen & Kilenyi, 1970	10	Well-11	744	741	430	464	447	430	Within 1 Sample of LOC	17
Mandelstamia tumida Christensen & Kilenyi, 1970	14	Kimmeridge outcrop	206		500	500	464	430	?	22
Orthonotacythere pustulata	1	Well-2	785	765	435		435			
Orthonotacythere pustulata	14	Kimmeridge outcrop	287		435	435	435	435	Within LOC Error	22
Protocythere rodewaldensis (Klinger, 1955)	8	Well-9	951	924	881	881	881	881		
Protocythere rodewaldensis (Klinger, 1955)	14	Kimmeridge outcrop	288		430	430	656	881	Within > 20 samples of LOC	22
Schuleridea moderata Christensen & Kilenyi, 1970	0	Well-1	500	480	500					
Schuleridea moderata Christensen & Kilenyi, 1970	1	Well-2	900	880	526	500	513	526	Within 1 Sample of LOC	7
Schuleridea moderata Christensen & Kilenyi, 1970	3	Well-4	719	701	538	500	526	538	Within 1 Sample of LOC	10
Schuleridea moderata Christensen & Kilenyi, 1970	5	Well-6	1274		544	500	532	544	Within LOC Error	12
Schuleridea moderata Christensen & Kilenyi, 1970	7	Well-8	972	930	561	464	553	665	Within LOC Error	14
Schuleridea moderata Christensen & Kilenyi, 1970	10	Well-11	744	741	430	430	532	665	Within 4 Samples of LOC	17
Schuleridea moderata Christensen & Kilenyi, 1970	14	Kimmeridge outcrop	206		500	430	526	665	Within 2 Samples of LOC	22
Schuleridea triebeli (Steghaus, 1951)	0	Well-1	840	820	840					
Schuleridea triebeli (Steghaus, 1951)	1	Well-2	1220	1210	840	840	840	840	Within LOC Error	7
Schuleridea triebeli (Steghaus, 1951)	2	Well-3	671	597	870	840	847	870	Within 1 Sample of LOC	9
Schuleridea triebeli (Steghaus, 1951)	3	Well-4	823	786	854	840	847	870	Within LOC Error	10
Schuleridea triebeli (Steghaus, 1951)	4	Well-5	1219	1128	553	553	840	870	Within 10 Samples of LOC	11
Schuleridea triebeli (Steghaus, 1951)	5	Well-6	1579		930	553	847	870	Within 1 Sample of LOC	12
Schuleridea triebeli (Steghaus, 1951)	6	Well-7	844	808	982	553	854	982	Within 1 Sample of LOC	13
Schuleridea triebeli (Steghaus, 1951)	7	Well-8	1106	1070	889	553	862	982	Within 1 Sample of LOC	14
Schuleridea triebeli (Steghaus, 1951)	8	Well-9	966	951	965	553	870	982	Within 2 Samples of LOC	15
Schuleridea triebeli (Steghaus, 1951)	9	Well-10	1079	1070	861	553	866	982	Within LOC Error	16
Schuleridea triebeli (Steghaus, 1951)	11	Well-12	1216	1189	972	553	870	982	Within 1 Sample of LOC	18
Schuleridea triebeli (Steghaus, 1951)	12	Well-13	800	792	1020	553	880	1020	Within 1 Sample of LOC	19
Schuleridea triebeli (Steghaus, 1951)	13	98/11-1	920		865	553	870	1020		20
Vernoniella sequana Oertli, 1957	0	Well-1	1022	1001	1022					
Vernoniella sequana Oertli, 1957	1	Well-2	1435	1415	1045	1022	1034	1045	Within 1 Sample of LOC	7
Vernoniella sequana Oertli, 1957	3	Well-4	884	823	1029	1022	1029	1045	Within LOC Error	9
Vernoniella sequana Oertli, 1957	5	Well-6	1362		1060	1022	1037	1060	Within 1 Sample of LOC	12
Vernoniella sequana Oertli, 1957	6	Well-7	863	710	1045	1022	1045	1060	Within LOC Error	13
Vernoniella sequana Oertli, 1957	7	Well-8	1149	1143	1019	1022	1037	1060	Within LOC Error	14
Vernoniella sequana Oertli, 1957	8	Well-9	1009	969	1061	1022	1045	1061	Within LOC Error	15
Vernoniella sequana Oertli, 1957	9	Well-10	1146	1116	1065	1022	1045	1061	Within LOC Error	16
Vernoniella sequana Oertli, 1957	11	Well-12	1244	1222	1055	1022	1045	1061	Within LOC Error	18

through 9 can be recognized, but the interval is very thin. The remaining log markers having been removed by erosion. Seven ostracod FDOs and one foraminiferal FDO were recognized, all of which cluster around the proposed LOC (Fig. 18).

Iteration 12

Well-13 lies circa 40 km west of Well-9. Here, the study interval is quite thin, the log markers are relatively poorly constrained, and the sampling for micropaleontology very sparse (Fig. 19). Placement of the log-based LOC suggests considerable section is missing due to hiatus or faulting.

Iteration 13

Offshore well 98/11-1 lies on the northern edge of the Channel Basin. The study interval can be pattern correlated back to

TABLE 1.—Continued.

Species and Author	Iteration Number	Well Name	Depth in Well	Depth to Sample Above	Projected Composite Depth	Minimum FDO	Median FDO	Maximum FDO	Source of Error	Figure Number
Ammobaculites subaequalis Mjatliuk, 1839	1	Well-1	680	660	680					
Ammobaculites subaequalis Mjatliuk, 1839	2	Well-3	597	536	745	680	713	745	Within 1 Sample of LOC	9
Ammobaculites subaequalis Mjatliuk, 1839	3	Well-4	786	732	755	680	745	755	Within 1 Sample of LOC	10
Ammobaculites subaequalis Mjatliuk, 1839	5	Well-6	1469		795	680	750	795	Within 1 Sample of LOC	12
Ammobaculites subaequalis Mjatliuk, 1839	6	Well-7	808	777	820	680	755	820	Within 1 Sample of LOC	13
Ammobaculites subaequalis Mjatliuk, 1839	7	Well-8	1070	1045	795	680	775	820	Within 1 Sample of LOC	14
Ammobaculites subaequalis Mjatliuk, 1839	11	Well-12	1152	1143	765	680	765	820	Within LOC Error	18
Ammobaculites subaequalis Mjatliuk, 1839	13	98/11-1	887		700	680	760	820	Within 3 Samples of LOC	20
Citharina serratocostata (Gumbel, 1862)	0	Well-2	785	765	440					
Citharina serratocostata (Gumbel, 1862)	1	Well-1	399	381	399	399	420	440	Within 2 Samples of LOC	7
Citharina serratocostata (Gumbel, 1862)	10	Well-11	732	698	410	399	410	440	Within LOC Error	17
Citharina serratocostata (Gumbel, 1862)	12	Well-13	720	660	725	399	425	725	Within 1 Sample of LOC	19
Cribrostomoides canui (Cushman, 1910)	5	Well-6	1588		947	947	947	947		
Cribrostomoides canui (Cushman, 1910)	7	Well-8	1143	1119	1015	947	981	1015	Within 2 Samples of LOC	14
Cribrostomoides canui (Cushman, 1910)	9	Well-10	1042	963	760	760	947	1015	Within 7 Samples of LOC	16
Cribrostomoides canui (Cushman, 1910)	10	Well-11	856	838	796	760	872	1015	Within LOC Error	17
Epistomina ornata (Roemer, 1841)	5	Well-6	1579		930	930	930	930		
Epistomina ornata (Roemer, 1841)	7	Well-8	1045	1039	722	722	826	930	Within 8 Samples of LOC	14
Epistomina ornata (Roemer, 1841)	12	Well-13	792	720	990	722	930	990	Within 1 Sample of LOC	19
Epistomina parastelligera (Hofker, 1954)	0	Well-1	840	820	840	840	840	840		
Epistomina parastelligera (Hofker, 1954)	2	Well-3	671	597	868	840	854	868	Within 1 Sample of LOC	9
Epistomina parastelligera (Hofker, 1954)	12	Well-13	792	720	990	840	855	990	Within 1 Sample of LOC	19
Epistomina parastelligera (Hofker, 1954)	13	98/11-1	914		842	840	855	990	Within LOC Error	20
Haplophragmoides sp.	0	Well-1	480	460	480					
Haplophragmoides sp.	1	Well-2	870	840	505	480	493	505	Within 1 Sample of LOC	7
Haplophragmoides sp.	2	Well-3	408	320	500	480	500	505	Within LOC Error	9
Haplophragmoides sp.	5	Well-6	1228		498	480	499	505	Within LOC Error	12
Haplophragmoides sp.	7	Well-8	930	899	475	475	498	505	Within 1 Sample of LOC	14
Haplophragmoides sp.	10	Well-11	768	744	490	475	494	475	Within LOC Error	17
Haplophragmoides sp.	13	98/11-1	835		485	475	490	475	Within LOC Error	20
Lenticulina muensteri (Roemer, 1839)	0	Well-1	360	342	360					
Lenticulina muensteri (Roemer, 1839)	1	Well-2	705	685	365	360	363	365	Within LOC Error	7
Lenticulina muensteri (Roemer, 1839)	12	Well-13	660	660	415	360	365	415	last sample + fault cutout	19
Lenticulina muensteri (Roemer, 1839)	13	98/11-1	811		421	360	390	421	last sample	20
Lenticulina subalata (Reuss, 1854)	9	Well-10	963	948	500	500	500	500		
Lenticulina subalata (Reuss, 1854)	12	Well-13	660	660	415	415	458	500	Within LOC Error	19
Lenticulina uralensis (Mjatliuk, 1939)	1	Well-1	725	705	381					
Lenticulina uralensis (Mjatliuk, 1939)	2	Well-3	536	408	667	381	524	667	Within 2 Samples of LOC	9
Marginulinopsis aff. phragmites (Loeblich & Tappan, 1950)	0	Well-1	381	360	381	381	381	381		
Marginulinopsis phragmites (Loeblich & Tappan, 1950)	7	Well-8	893	826	422	381	402	422	Within 1 Sample of LOC	14
Reinholdella pseudojasanensis (Dain, 1967)	2	Well-3	536	408	667	667	667	667		
Reinholdella pseudojasanensis (Dain, 1967)	5	Well-6	1219		485	485	576	667	Within 6 Samples of LOC	12
Saracenaria oxfordiana Tappan, 1955	0	Well-1	399	381	399					
Saracenaria oxfordiana Tappan, 1955	1	Well-2	745	725	399	399	399	399	Within LOC Error	7
Saracenaria oxfordiana Tappan, 1955	13	98/11-1	823	?	453	399	399	453	Within 6 Samples of LOC	20

the SRS via Well-11 and Well-9 (Fig. 6). The section is relatively thin, and, like at Well-11, log markers 1 and 2 are missing, probably as a result of onlap onto the high separating the Weald and Channel basins. Seven ostracod FDOs and five foraminiferal FDOs were recognized here and the medians cluster around the proposed LOC (Fig. 20). The sampling interval at 98/11-1 is circa 20 m, which is much denser than at most of the wells described thus far. As a result, the median FDOs tend to lie on, or to the left of, the LOC.

Iterations 14 and 15

As a final example, we illustrate the addition of the standard European ammonite biozonal boundaries to the CS.

The Kimmeridge Clay and Portland Beds were cored in their entirety at Warlingham-1 (Fig. 4), which is located on the northern margin of the Weald Basin. These cores were examined in detail for ammonites by Callomon and Cope (1971). Only the gamma-ray curve was available for correlation back to Well-2 (Figs. 4, 22) but this was relatively straightforward, in spite of the section thinning. Loss of section occurs at the top and base of the Kimmeridge Clay due to unconformity and faulting, respectively, producing characteristic flat-spots on the LOC (Fig. 22). As in the case of the previous iterations, ammonite events were added to the CS by projecting horizontally to, and vertically from, the LOC.

As a test of the ammonite calibration at Warlingham we also attempted to tie our correlation to the surface type section at Kimmeridge Bay (Figs. 6, 21). Of course, no digital gamma-ray or sonic logs exist for this surface section, but measured sections are common. In order to facilitate the comparison of a measured surface section with the borehole logs, we digitized the measured section by scoring the lithology in a simple, systematic way. Lithologies were scored on an ordinal scale of decreasing radioactivity, as follows: black shales 2, shales 1, mudstones 0, calcareous mudstones −1, clean sandstones and limestones −2. The digitized log was then filtered using a long-wavelength gaussian filter which removed all of the frequencies

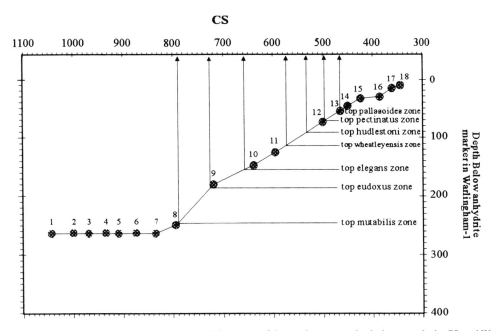

FIG. 22.—Iteration 15 in the compilation of the CS by graphic correlation. Axes of the graph represent depths in meters in the CS, and Warlingham-1 borehole. Ammonite biozone boundaries are added to the CS via the log-derived LOC.

below 20 m. This facilitates comparison of the digitised section with the gamma-ray curves of nearby wells (Fig. 6).

Pattern correlation of the filtered outcrop curve with the nearest well, 98/11-1, is fairly convincing (Fig. 6). Plotting outcrop sampled ostracod tops (Kilenyi, 1969) against our median FDOs (Fig. 21) renders the correlation more convincing, and the fact that the ammonite boundaries from Warlingham correlate exactly makes the correlation probable.

Interestingly, the ostracod FDOs appear as two distinct populations in Figure 21, with a 'dogleg' recorded in the *mutabilis* ammonite biozone, in the range of log-markers 1–6, suggesting a high-degree of condensation at this level.

CONCLUSIONS

Low bandpass filtering of gamma-ray and sonic logs covering the Upper Jurassic strata in the Wessex Basin enhances their correlatability, thereby facilitating the rapid and objective subdivision of the succession into chronostratigraphically significant packages or cycles. Correlation of these cycles, which are chronostratigraphic within the resolution of biostratigraphy and seismic stratigraphy, permits the compositing of fourteen wells and one outcrop to produce a CS in the manner outlined. All biostratigraphic events based on ostracod, foraminifers, and ammonites have been integrated in this CS (Table 1). The final results for ostracods and foraminifers are illustrated in Figures 23 and 24. The standard log response, calibrated to the ammonite biostratigraphy, is presented on the left of the diagram with the ranges of the ostracod FDOs as histograms to the right. The median for each FDO is illustrated as a diamond within the histogram, and the number of events contributing to the median is proportional to the number of columns in the histogram. A taxon combining a short range with a large number of occurrences qualifies as a good biostratigraphic index.

Within the lowermost Kimmeridgian interval, the ostracod species with the smallest dispersion of FDOs is *V. sequana*; most of the FDOs occur within 10–20 m of the LOC (referred to in Table 1 as 'LOC error'). The FDO of *V. sequana* represents an extremely good marker for the base of the *baylei* ammonite biozone and consequently the base of the Kimmeridgian Stage in the Wessex Basin.

Other good markers for the lower part of the Kimmeridgian include the ostracods *S. triebeli*, *O. postrotunda*, *O. dissimilis*, *M. cicatricosa*, and *G. elongata*. *S. triebeli* has the highest number of occurrences in the data set and is a good index for the mutabilis ammonite biozone. The anomalously high occurrence of this species is attributed to reworking, operator error, or even facies-related distribution at Well-5, while the high numbers of low FDO occurrences represent FDOs in relatively condensed stratigraphic intervals with low sampling densities. The same is probably true for the ostracod FDOs *G. postrotunda* and *G. dissimilis*.

Above the *eudoxus* ammonite biozone, within the *autissiodorensis* through *wheatleyensis* ammonite biozone interval, ostracod and foraminifer FDOs become rare. FDOs in this interval tend to be anomalously high or low occurrences of older and younger species. Possible exceptions, the foraminifer *A. subaequalis* and the rarer ostracod *G. dorsetensis*, are relatively good markers for the upper part of the lower Kimmeridge Clay. The reason for the absence of markers throughout this interval is not completely understood. However, perhaps coincidentally, this part of the Kimmeridge Clay contains the most widespread development of organic-rich black shales in the Wessex Basin (Fig. 2), testifying to hostile, anaerobic conditions for benthonic organisms.

The uppermost *wheatleyensis* to lowermost *pectinatus* ammonite biozones are characterized by reliable ostracod FDOs, *S. moderata* and *M. maculata* (Fig. 24); and the remainder of

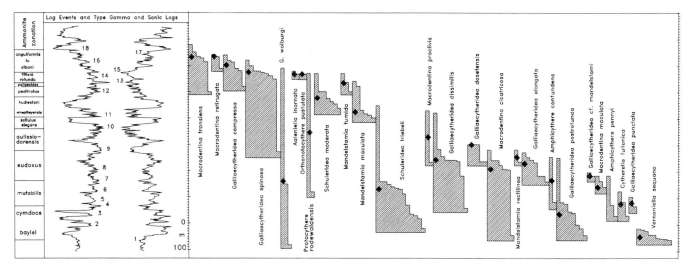

Fig. 23.—Results of graphic correlation for ostracods. Histograms represent the relative positions of all FDOs in the study, with the width of the histogram being proportional to the number of times the FDO of the species was recognised. Diamonds represent the median FDO for each species

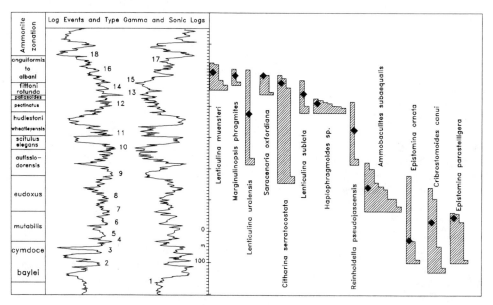

Fig. 24.—Results of graphic correlation for foraminifers. Histograms represent the relative positions of all FDOs in the study, with the width of the histogram being proportional to the number of times the FDO of the species was recognised. Diamonds represent the median FDO for each species

the Upper Kimmeridgian interval contains several events, the best of which is foraminifer *Haplophragmoides sp..* This FDO is an excellent marker for the upper pectinatus ammonite biozone, and virtually all of the dispersion associated with this event occurs within the error of the LOC (Table 1). Ostracod *G. spinosa* is also a good Upper Kimmeridgian interval marker (Fig. 24; the anomalously low occurrence of this event is attributed to stratigraphic cut-out at Well-13 (Table 1)).

The most frequent FDOs in the Portlandian interval are the ostracods *M. transiens* and *G. compressa* (Fig. 24). Other less frequent, but potentially useful FDOs, include the ostracod *M. retirugata*, and foraminifer *Lenticulina muensteri*, *Citharina serratocostata*, and *Saracenaria oxfordiana*.

The histograms in Figures 23 and 24 represent the relative stratigraphic ranking of all the FDOs studied. As such, they can be treated as a crude form of cumulative probability for each FDO. For example, consider the nine occurrences of the FDO of the ostracod *G. spinosa* (Fig. 24). In our experience, two out the nine FDOs occurred within the Lower Portlandian interval, suggesting a cumulative probability of circa 22% of encountering the FDO of *G. spinosa* in the Lower Portlandian strata. Moreover, six of the nine events occur within the *fittoni* and *rotunda* ammonite biozones, suggesting a cumulative probability of circa 88% of encountering *G. spinosa* in the uppermost Kimmeridgian or lowermost Portlandian interval. This kind of logic can be applied to all the events and demonstrates the predictive potential of the technique. One could go further with diagrams such as Figures 23 and 24, perhaps labelling or color-coding the histograms with the well of origin, or grouping according to facies, thereby taking into account the effect of facies variations on fossil distribution. This was not attempted in this study because of the lack of detailed mud-logs, cores, and a sufficient suite of logs for detailed lithology interpretation.

We are aware of published computer programs in which weightings may be assigned to selected stratigraphic events to influence the path of the LOC (Gradstein, 1990). Our method differs in that the LOC honours certain selected events (in this example those of gamma-ray and sonic log events) and that the scatter of biostratigraphic FDOs are measured relative to the LOC thus defined. Currently our calculations are carried out manually, but work is in progress to provide the method as an interactive computerised utility.

The method presented here is an extension of the graphic correlation methodology. Unique log events, enhanced by bandpass filtering of digital gamma-ray and sonic log data, can be used to derive lines of correlation between wells, and these lines of correlation can be used to map the scatter of biostratigraphic events onto a single composite standard reference section. The concepts of median FDO and range of FDOs per taxon facilitate objective biozonation. Application of the technique to quantitative biostratigraphic data holds promise for probabilistic biozonation.

ACKNOWLEDGEMENTS

The authors acknowledge the permission of The British Petroleum Company to publish this paper.

REFERENCES

CALLOMON, J. H. AND COPE, J. C. W., 1971, The stratigraphy of the Oxford and Kimmeridge Clays in the Warlingham borehole: Bulletin of the Geological Survey of Great Britain, v. 35, p. 147–178.

CHADWICK, R. A., 1986, Extension tectonics in the Wessex Basin, southern England: Journal of the Geological Society of London, v. 143, p. 465–488.

COPE, J. C. W., DUFF, K. L., PARSONS, C. F., TORRENS, H. S., WIMBLEDON, W. A. AND WRIGHT, J. K., 1980, A correlation of the Jurassic rocks of the British Isles. Part 2: London, Geological Society of London Special Report Number 15, 109 p.

COX, B. M. AND GALLOIS, R. W., 1981, The stratigraphy of the Kimmeridge Clay of the Dorset type area and its correlation with some other Kimmeridgian sequences: city, Report of the Institute of Geological Sciences United Kingdom, 80, 44 p.

EDWARDS, L. E., 1989, Supplemented graphic correlation: a powerful tool for paleontologists and nonpaleontologists: Palaios, v. 4, p. 127–143.

GRADSTEIN, F. M., 1990, Program STRATCOR (Version 1.6) for zonation and correlation of fossil events: Ottawa, Geological Survey of Canada, Report Number 2285.

HALLAM, A. AND SELLWOOD, B. W., 1976, Middle Mesozoic sedimentation in relation to tectonics in the British area: Journal of Geology, v. 84, p. 301–321.

HOLLOWAY, J. L., 1958, Smoothing and filtering of time series and space fields: Advances in Geophysics, v. 4, p. 351–388.

HOUSE, M. R., 1985, A new approach to an absolute timescale from measurements of orbital cycles and sedimentary microrhythms: Nature, v. 315, p. 712–725.

HOUSE, M. R., 1986, Are Jurassic sedimentary microrhythms due to orbital forcing?: Proceedings of the Ussher Society, v. 6, p. 299–311.

KILENYI, T. I., 1969, Ostracoda of the Dorset Kimmeridge Clay: Palaeontology, v. 12, p. 112–160.

MELNYK, D.H., ATHERSUCH, J., AND SMITH, D.G., 1992, Measuring the dispersion of biostratigraphic events in the subsurface using graphic correlation: Marine and Petroleum Geology, v. 9, p. 602–607.

MELNYK, D. H., SMITH, D. G., AND AMIRI-GARROUSSI, K., 1994, Filtering and frequency mapping as tools in subsurface cyclostratigraphy, with examples from the Wessex Basin, UK, in de Boer, P. L. and Smith, D. G., eds, Orbital Forcing and Cyclic Sedimentary Sequences: city, International Association of Sedimentologists, Special Publication Number 19, p. 35–46.

MILLER, F. X., 1977, The graphic correlation method in biostratigraphy, in Kauffman, E. G. and Hazel, J. E., eds, Concepts and Methods of Biostratigraphy: Stroudsberg, Dowden, Hutchinson and Ross, p. 165–186.

OSCHMANN, W., 1988, Kimmeridge Clay sedimentation—a new cyclic model: Paleoceanography, Paleoclimatology and Paleoecology, v. 65, p. 217–251.

PENN, I. E., COX, B. M., AND GALLOIS, R. W., 1986, Towards precision in stratigraphy: geophysical log correlation of Upper Jurassic (including Callovian) strata of the eastern England shelf: Journal of the Geological Society of London, v. 143, p. 381–410.

SCHWARZACHER, W., 1975, Sedimentation Models and Quantitative Stratigraphy: Amsterdam, Elsevier, Developments in Sedimentology 19, 382 p.

SHAW, A. B., 1964, Time in Stratigraphy: New York, McGraw-Hill, 365 p.

VAN WAGONER, J.C., MITCHUM, R.M., CAMPION, K.M., AND RAHMANIAN, V.D., 1990, Siliciclastic Sequence Stratigraphy in Well Logs, Cores, and Outcrops: Concepts for High-resolution Correlation of Time and Facies: Tulsa, American Association of Petroleum Geologists, Methods in Exploration Series 7, 55 p.

WHITTAKER, A., 1985, Atlas of the Onshore Sedimentary Basins in England and Wales: Post-Carboniferous Tectonics and Stratigraphy: Glasgow, Blackie, 71 p.

ZIEGLER, P. A., 1975, Geologic evolution of the North Sea and its tectonic framework: Bulletin of the American Association of Petroleum Geologists, v. 59, p. 1073–1097.

CORRELATION ACROSS A CLASSIC FACIES CHANGE (LATE MIDDLE THROUGH LATE CENOMANIAN, NORTHWESTERN BLACK HILLS): APPLIED SUPPLEMENTED GRAPHIC CORRELATION

CYNTHIA G. FISHER

Department of Geology, State University of New York, Cortland College, Cortland, New York 13045

ABSTRACT: An abrupt lithofacies change occurs in Cenomanian strata of northeastern Wyoming and southeastern Montana. Across a 46.75-km areal extent, calcareous rocks of the Greenhorn Formation change to non-calcareous rocks of the Belle Fourche Shale. The physical conditions that produced the facies change also restricted the lateral extension of biostratigraphically useful fossils. Historically, correlation of these strata have proven difficult. However, supplemented graphic correlation techniques generate high resolution correlations using local nonunique geologic data, such as bentonite and calcarenite beds and foraminiferal biofacies.

INTRODUCTION

Lithofacies changes typically produce correlation problems because the lateral change in rock type disrupt diagnostic beds. The change in physical conditions that produce the lithofacies also affect the distribution of the biota between the lithofacies. These lateral changes in biota can limit biostratigraphic correlation across lithofacies.

This type of local correlation problem exists in strata deposited in the mid-Cretaceous Greenhorn Seaway in the extreme southeastern corner of Montana and the northeastern corner of Wyoming (Fig. 1). An abrupt lithofacies change between calcareous shale and non-calcareous shale resulted in a dearth of laterally continuous marker beds and the discontinuity of floral and faunal species distributions across the outcrop belt.

The goal of this study is to correlate stratigraphic sections that change lithologic characteristics laterally across a distance of 46.75 km. Lithologic beds initially are used to correlate the sections, and foraminiferal biofacies refine these initial correlations.

Western Interior Seaway: Geologic Setting

Deposition of the middle and upper Cenomanian strata, comprising the facies change, occurred during the Greenhorn Cycle, the second and most extensive of five tectono-eustatic cycles (Kauffman, 1985). From the middle Cenomanian until the middle Maastrichtian, the Gulf of Mexico and the Arctic Ocean connected across the 4,800 km (3,000 miles) long seaway through the Western Interior Basin (Williams and Stelck, 1975). At maximum Greenhorn transgression, the breadth of the seaway extended from southwestern Utah to Iowa and southwestern Minnesota (greater than 1,600 km or 1,000 miles).

Deposition of the Belle Fourche Shale, Greenhorn Formation, and Pool Creek Member of the Carlile Shale in the northern Black Hills occurred as the Greenhorn Sea transgressed and then regressed across the area. The Greenhorn Formation (upper middle Cenomanian to middle Turonian) in the southern Black Hills consists of approximately 68.5 m of calcareous shale succeeded by about 18 m of alternating shale and limestone (Robinson and others, 1964). In the northern Black Hills, the calcareous shale of the Greenhorn Formation passes northwestward into black, non-calcareous shale (Robinson and others, 1964). Outcrops in southeastern Montana and northeastern Wyoming most clearly show the facies change where the Little Missouri River flows southwest across the area (Fig. 1). The incision of the river exposes calcareous clay-shale, thin limestone, and calcarenite beds of the Greenhorn Formation to the southeast and non-calcareous clay-shale and limestone and siderite concretions of the Belle Fourche Shale to the northwest (Fig. 1). Details of the facies change show that the calcareous rocks step-up section toward the northwest (Fig. 2).

The stratigraphic interval from the upper Belle Fourche Shale through the Greenhorn Formation and into the Poole Creek Member of the Carlile Formation spans nine ammonite zones. Figure 3 shows the biostratigraphic zonations of ammonites and planktonic foraminifera for the pertinent part of the Greenhorn Cycle strata.

Rubey first mapped this Cenomanian lithofacies change in southeast Montana and northeast Wyoming during work at the United States Geological Survey in the 1920's (unpublished data notebooks of Bramlette and Rubey, U.S.G.S). In their correlation of the facies change, Bramlette and Rubey apparently dealt with this difficult correlation by adjusting the thickness within their sections so that the beds are plotted within the sections as horizontal time lines (Fig. 4). Discussion in the literature does not explain the rational behind the adjustments in the thicknesses of the stratigraphic sections. Moore (1949) later used a correlation diagram constructed by Bramlette and Rubey as a classic example of a facies change in *The Meaning of Facies*. Robinson and others (1964) reproduced the facies correlation in their description of the northwestern Black Hills stratigraphy. Recent studies show the difficulty in correlating the facies change in detail (Macdonald, 1984; Macdonald and Byers, 1988; Fisher, 1991).

MATERIALS AND METHODS

The five, best-exposed and stratigraphically most-continuous sections were measured, described, and sampled: two from the calcareous facies and three from the non-calcareous facies (Figs. 1, 2). Trenching of sections exposed fresh rock surfaces, and measurements were made using a Jacob staff fitted with an adjustable Abney-level (Elder, 1989). Trenched sections were tested continuously for calcareous content using 10% hydrochloric acid. Inoceramid bivalves and ammonites were collected and their stratigraphic positions noted. Samples for foraminifera were taken at 2-m intervals, except for the lower half of the Torgerson Draw and Bull Creek sections. At Torgerson Draw, the section was sampled at irregular intervals due to intervals of poor exposure. Only the upper one-third of the Bull Creek section was sampled for foraminifera because this section had previously been thoroughly examined by Eicher and Worstell (1970). Their data, derived from continuous channel samples of approximately 3-m intervals, were incorporated into this study.

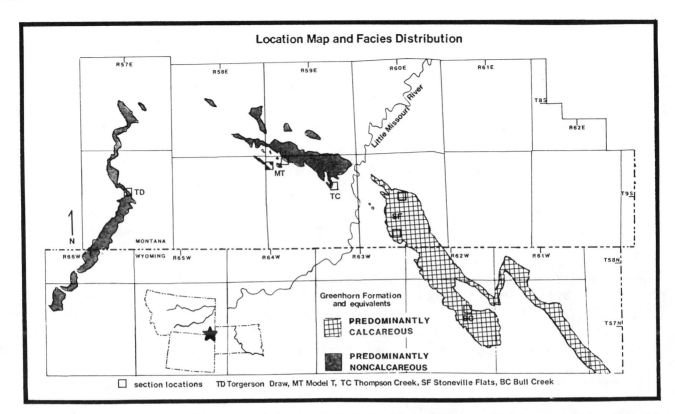

FIG. 1.—Insert shows location of the study area. Patterned areas show the distribution of Cenomanian outcrops. The calcareous facies (cross hatched) lie to the southeast of the Little Missouri River Valley; whereas, the non-calcareous facies (grey) lie to the northwest. Stratigraphic sections studied are shown as squares. Stoneville Flats and Model T are composite sections. Modified from Fisher and others (1994).

One hundred and twenty-one samples from the Stoneville Flats, Thompson Creek, Model T, and Torgerson Draw sections were selected for foraminiferal examination. Incorporation of the data from 49 samples of Eicher and Worstell (1970) from the Bull Creek section provided data from 170 samples. Eighteen planktonic species, 35 calcareous benthic species, and 26 agglutinated benthic species were identified from the study area.

LITHOSTRATIGRAPHIC CORRELATION AND
MACROFOSSILS DISTRIBUTIONS

The difficult task of correlating the strata necessitates the integration of unique and nonunique geologic data. The graphic correlation method of Shaw (1964) employ only unique events in earth history, such as first and last species datums. The limited biostratigraphic data available in this study necessitates correlation using the supplemented graphic correlation (SGC) techniques of Edwards (1989). SGC expands on Shaw's graphic correlation method by incorporating physical property data that represent recurring "events," the sequential pattern of which may or may not be uniquely recognizable. The use of nonunique data in correlation should be considered a hypothesis. Data used in correlation must pass two tests. There must be a reason to believe that the data responded to "time-significant phenomena over the areal extent of the specific problem at hand" (Edwards, 1989, p. 129), and it must be tested for geologic reasonableness before correlations are accepted.

Edwards (1989, p.127) proposes that some types of lithologic data may have temporal significance "within the geographic extent over which the conditions prevailed" to produce the bed. In this study, the nonunique data that will be tested for correlation are lithologic beds of bentonite and calcarenite, and biotic data of foraminiferal biofacies distributions. The 2-axis graph method tests the lithologic beds, and the side-by-side comparison method tests the biotic data. Edwards (1989) notes the 2-axis and side-by-side methods are equivalent.

Lithostratigraphic Correlation

Figure 5 shows the 2-axis graphing of the Model T, Thompson Creek, and Stoneville Flats sections plotted against the Torgerson Draw section using lithologic beds. Figure 6 shows the 2-axis graphing of Bull Creek plotted against Stoneville Flats. The graphs test the hypothesis that bentonite and calcarenite beds may be recognizable as temporally significant marker beds deposited in response to "time-significant phenomena" (physical event units of Kauffman, 1988; and some nonunique data of Edwards, 1989). Individual bentonite beds have obvious temporal significance and can be correlated providing they are accurately identified (Obradovich and Cobban, 1975). Calcarenites may represent geologically short duration episodes of submarine erosion, rapid deposition such as during storms, or reduced sedimentation that take place over local or regional areal extent (Kauffman, 1988).

In this study, the 2-axis graphic correlation uses only bentonites greater than 10 cm thick and calcarenite beds. The X

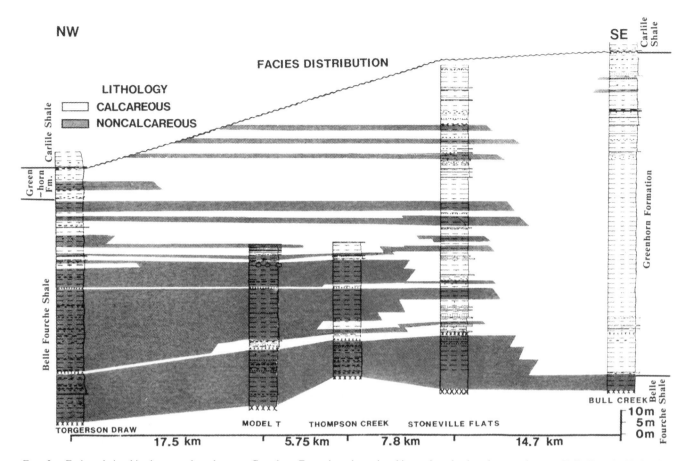

FIG. 2.—Facies relationships between the calcareous Greenhorn Formation, shown in white, and predominantly non-calcareous Belle Fourche Shale, shown in grey. Calcareous facies step-up across the area toward the northwest. Thin intervals of calcareous shale occur in the northwestern sections. Modified from Fisher and others (1994).

bentonite, a 1- to 2-m thick, grey and red unit, forms the base of each section (Figs. 5, 6). The X bentonite is widespread and easily recognizable by its color throughout the Western Interior Basin (McNiel and Caldwell, 1981). Approximately 15 to 20 m above the X bentonite lays a local bentonite known as the G bentonite (Knetchel and Patterson, 1962), which ranges in thickness from 20 to 30 cm. These two bentonites may correlate the lower third of all five stratigraphic sections (Figs. 5, 6). In the northwestern sections, a third, smaller bentonite, informally referred to as the "T" bentonite, may correlate Torgerson Draw, Model T, and Thompson Creek (Fig. 5). This 10-cm bentonite does not occur in the calcareous facies.

Kauffman (1988) considers many of the calcarenites of the Western Interior Seaway to be of temporal significance over a local to regional areal extent. Two, 50- to 100-cm thick, well-developed calcarenite horizons, informally designated as "a" and "b," occur in the top quarter of the Model T and Thompson Creek stratigraphic sections (Fig. 5). The calcarenites consist of biogenic grains of local origin, fragmented biogenic tests, and microfossils sorted in lag deposits. These units thin significantly away from Model T and Thompson Creek and may correlate with difficulty into the western Torgerson Draw section and across the facies change into the Stoneville Flats section. The units thin from 50 to 100 cm at Model T and Thompson Creek to 2 to 10 cm at Torgerson draw and Stoneville Flats (Fig. 5). A stratigraphically higher calcarenite bed, designated informally as "c," may correlate the Stoneville Flats section to the Bull Creek section (Fig. 6). Each of the data points, representing bentonites X, G, and "T" and calcarenites "a" and "b," connect by positively sloping, nearly straight lines, supporting the hypothesized correlations. Sedimentation rates may vary; however, the data do not indicate the presence of a hiatus in strata between the X bentonite and the "b" calcarenite. Figure 7 shows enlarged sections containing these lithostratigraphic correlations.

Sparse ammonite and inoceramid occurrences test these correlations (Fig. 7). Limited macrofossil data place the lithologically correlated stratigraphic sections into the five youngest Cenomanian ammonite zones of the Western Interior (Figs. 3, 7). The X, G, and "T" bentonites and "a," "b," and "c" calcarenites do not cross-cut the ammonite zones (Figs. 3, 7). The occurrence of the macrofossils supports the correlation of lithologic beds between the sections and validates them as time-significant horizons of correlation.

A disconformity occurs near the top of the Bull Creek and Torgerson Draw sections. The disconformity excludes the lower Turonian ammonite Zones of *Watinoceras devonense* Wright and Kennedy, *Pseudoaspidoceras flexuosum* Powell, *Vascoceras birchbyi* Cobban and Scott, and *Mammites nodosoides* (Schlüter) and the middle Turonian *Collignoniceras woollgari*

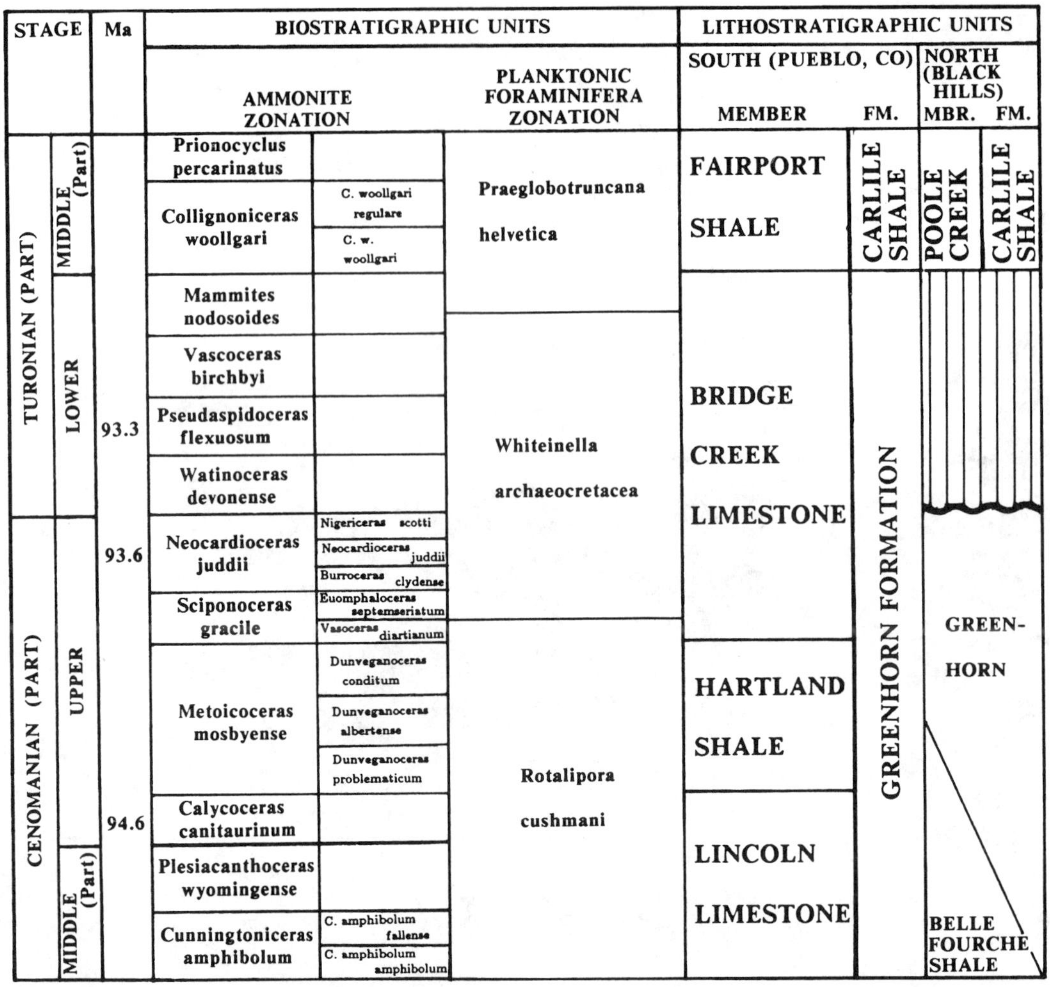

STAGE			Ma	BIOSTRATIGRAPHIC UNITS			LITHOSTRATIGRAPHIC UNITS			
				AMMONITE ZONATION		PLANKTONIC FORAMINIFERA ZONATION	SOUTH (PUEBLO, CO)		NORTH (BLACK HILLS)	
							MEMBER	FM.	MBR.	FM.
TURONIAN (PART)	MIDDLE (Part)			Prionocyclus percarinatus		Praeglobotruncana helvetica	FAIRPORT SHALE	CARLILE SHALE	POOLE CREEK	CARLILE SHALE
				Collignoniceras woollgari	C. woollgari regulare					
					C. w. woollgari					
	LOWER		93.3	Mammites nodosoides		Whiteinella archaeocretacea	BRIDGE CREEK LIMESTONE	GREENHORN FORMATION	GREEN-HORN	
				Vascoceras birchbyi						
				Pseudaspidoceras flexuosum						
				Watinoceras devonense						
CENOMANIAN (PART)	UPPER		93.6	Neocardioceras juddii	Nigericeras scotti					
					Neocardioceras juddii					
					Burroceras clydense					
				Sciponoceras gracile	Euomphaloceras septemseriatum					
					Vascoceras diartianum					
				Metoicoceras mosbyense	Dunveganoceras conditum	Rotalipora cushmani	HARTLAND SHALE			
					Dunveganoceras albertense					
					Dunveganoceras problematicum					
			94.6	Calycoceras canitaurinum			LINCOLN LIMESTONE			
	MIDDLE (Part)			Plesiacanthoceras wyomingense					BELLE FOURCHE SHALE	
				Cunningtoniceras amphibolum	C. amphibolum fallense					
					C. amphibolum amphibolum					

FIG. 3.—Relationship between ammonite and planktonic foraminiferal biostratigraphic zonations and lithostratigraphic units with the time scale of Obradovich (1993). The ammonite zonation of Cobban (1993) as used in Kauffman and others (1993) is shown. Planktonic foraminiferal zonation from Robaszynski and Caron (1979). From Fisher and others (1994).

woollgari (Mantell) Subzone (Fig. 3). At Bull Creek, the middle Turonian ammonite *Collignoniceras woollgari regulare* (Haas) occurs above rocks containing the upper Cenomanian ammonite *Neocardioceras juddii* (Guerne and Barrois), thereby marking the disconformity (Figs. 3, 7). At Torgerson Draw, the northwestern-most section, the ammonite *Collignoniceras woollgari* occurs in calcarenites overlying shales containing the calcareous nannofossils: *Rhagodiscus asper* (Stradner), *Microstaurus chiastius* (Worsley), and *Corollithion kennedyi* Crux, which have their last occurrences in the upper Cenomanian (Watkins and others, 1993). Thus at Torgerson Draw, middle Turonian strata overlay shales no younger than late Cenomanian in age, thereby excluding strata representative of the lower Turonian ammonite zones.

FORAMINIFERAL BIOFACIES CORRELATIONS

The side-by-side graphic display and the supplemented graphic correlation techniques of Edwards (1989) facilitate examination of the foraminiferal biofacies data. Expansion and compression of the vertical scale aids visualization of the correlations (Edwards, 1989). Between successive marker beds, the section with the greatest thickness is selected as the new standard thickness, and the correlative segments of the other sections are rescaled and sample positions interpolated. For example, the thickest strata occur between the top of the X bentonite and the base of the G bentonite at the Model T section (24.18 m, Fig. 7). Therefore, this thickness becomes the new thickness for all the sections from the interval from the X bentonite to the G bentonite. The samples in each section are then proportionally spaced within the standard thickness, maintaining the relative position of each sample (Fig. 8). The procedure permits more precise comparisons of samples from different sections because the effect of variable sedimentation rates between sections is reduced or eliminated.

Foraminiferal Distributions

Figures 9 and 10 show foraminiferal biofacies (planktonic/benthic and mixed agglutinated and calcareous benthic/agglutinated benthics only, respectively) plotted for the interpolated samples on the rescaled sections. Each tick mark represents a

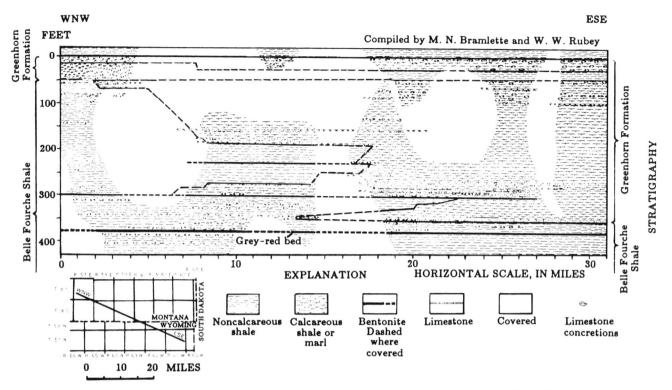

Fig. 4.—The facies change in the Greenhorn and Belle Fourche Formations in Cenomanian age strata in the northern Black Hills has been known for over half a century. This reconstruction shows the interfingering of the two facies (modified from Bramlette and Rubey, in Robinson and others, 1964). Stratigraphic distance between key bentonite and limestone beds were altered in order to correlate across the area.

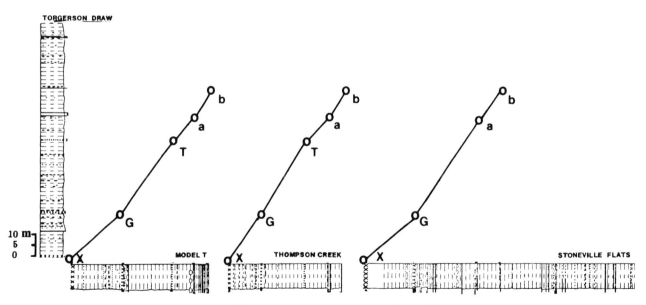

Fig. 5.—Two-axis supplemented graphic correlation of Torgerson Draw plotted against the Model T, Thompson Creek, and Stoneville Flats stratigraphic sections. The X, G, and "T" bentonites and the "a" and "b" calcarenites are the nonunique data used in the correlation. All segments of the line of correlation are positively sloped, indicating that no significant disconformities are present. Sedimentation accumulation rates appear to vary slightly.

foraminiferal sample (Fig. 8). Figure 9 shows presence/absence of planktonic foraminifera for the samples. Figure 10 shows presence/absence data of the mixed agglutinated-calcareous benthic biofacies and the agglutinated benthic biofacies for each sample. Areas between samples are interpolated according to data from the adjacent samples.

The calcareous southeastern sections (Stoneville Flats and Bull Creek) contain planktonic and calcareous benthic forami-

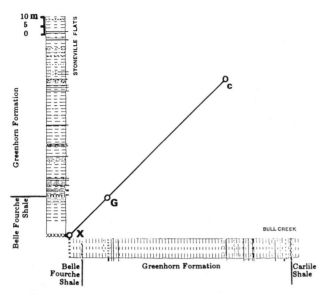

FIG 6.—Two-axis supplemented graphic correlation of Bull Creek plotted against Stoneville Flats. Only the X and G bentonites and "c" calcarenite co-occur.

nifera throughout much of the sections; however, the northwestern sections (Torgerson Draw, Model T and Thompson Creek) contain calcareous foraminifera only at specific intervals (Figs. 9, 10). The northwestern limits of the calcareous foraminiferal biofacies, across the study area, vary through time.

Foraminiferal biofacies distributions provide an example of nonunique data that may or may not have temporal significance. Edwards (1989) requires that this type of data undergo two tests: (1) that the investigator has reason to believe the data were produced in response to time-significant phenomena, and (2) that the data are tested for geologic reasonableness. The foraminiferal biofacies distributions are an inferred response to time-significant phenomena across the study area. Many previous investigators (Eicher, 1965, 1967, 1969a, b; Eicher and Worstell, 1970; Kauffman, 1977, 1985, 1988; Eicher and Diner, 1985, 1989; Fisher, 1991; Caldwell and others, 1993) identify two distinct water masses that entered the Western Interior Basin, one from the north and one from the south. These water masses advanced and retreated across the seaway and provided habitats for distinct foraminiferal biofacies (Eicher, 1965, 1967, 1969a, b; Eicher and Worstell, 1970; Kauffman, 1977, 1985, 1988; Eicher and Diner, 1985, 1989; Fisher, 1991; Caldwell and others, 1993). A mixed, agglutinated and calcareous benthic and planktonic foraminiferal biofacies represents a southern, warmer, more saline water mass; whereas, an exclusively agglutinated benthic foraminiferal biofacies represents the northern, cooler, less saline water mass.

Lateral shifts of calcareous biofacies limits within the study are hypothesized as examples of Edwards' (1989, p. 127) "response to temporally varying external condition ... within the

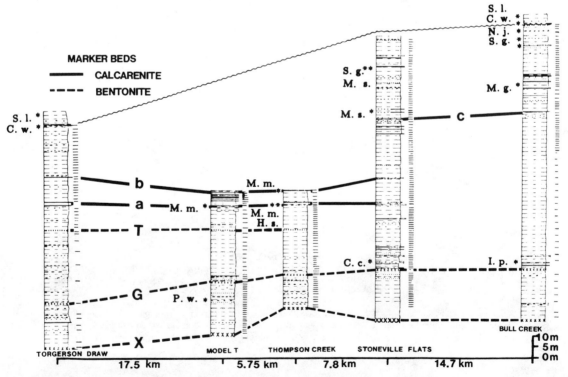

FIG. 7.—Correlations transferred to stratigraphic section enlargements. Bentonites and calcarenite "event" marker beds: X = X bentonite, G = G bentonite, T = informally named "T" bentonite, a = lower calcarenite, b = middle calcarenite, c = upper calcarenite. Macrofossils collected in the field designated by * and labeled: S.l.= *Scaphites larvaeformis*; C.w. = *Collignoniceras woollgari*; N.j. = *Neocardioceras juddii*; S.g. = *Sciponoceras gracile*; M.s. = *Moremanoceras scotti*; M.g. = *Metoicoceras geslinianum*; M.m. = *Metoicoceras mosbyense*; H.s. = *Hamites salebrosus*; C.c. = *Calycoceras canitaurinum*; P.w. = *Plesiacanthoceras wyomingense*; I. p. = *Inoceramus prefragilis*.

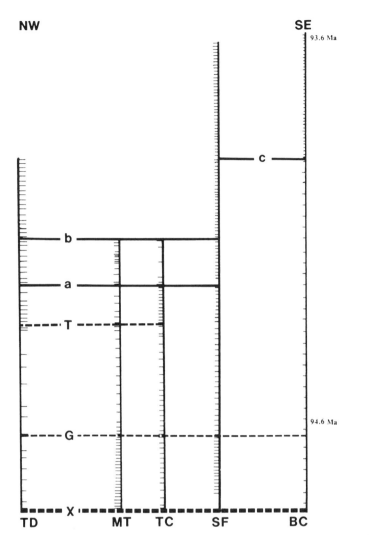

FIG. 8.—Rescaled stratigraphic sections of the northern Black Hills study area. Intervals between any two chronostratigraphic "event" beds within each of the six chronostratigraphic intervals were rescaled to the thickest section. Samples were proportionally spaced to maintain their relative positions. Modified from Fisher and others (1994).

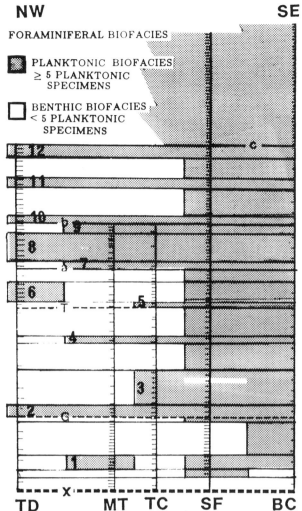

FIG. 9.—Distributions of planktonic foraminifera plotted on the rescaled sections. Grey area represents presence of planktonic biofacies (sample contains >5 planktonic individuals), 93% of these samples contain >50 planktonic specimens; white area represents absence of planktonic biofacies (sample contains <5 planktonic individuals); 89% of these samples contain only 0 to 2 planktonic specimens. "Migrations" of planktonic foraminifera into the northwestern sections (TD, MT, and TC) are numbered.

geographic extent over which the conditions prevailed." These lateral shifts of the foraminiferal biofacies may represent changes in water mass characteristics that influence the foraminiferal habitats across the area. The lateral shifts of the foraminiferal biofacies may represent nonunique "events" that have time-significance and therefore may potentially refine the lithologic correlations. Hypothesized correlation lines are drawn between sections at the stratigraphic horizons of these shifts (Figs. 9, 10). These correlation lines must be tested for their geologic reasonableness.

Planktonic foraminifera extend or "migrate" into the northwestern sections twelve times (Fig. 9). Extension or "migration" four displays a laterally continuous distribution. The planktonic biofacies is present in the southeastern sections of Bull Creek and Stoneville Flats and the northwestern sections of Thompson Creek and Model T but is absent in the northwestern Torgerson Draw section. Migration one (Fig. 9) represents a discontinuous distribution of the planktonic biofacies.

The southeastern sections, Bull Creek and Stoneville Flats, contain the planktonic biofacies, but interruption occurs at Thompson Creek and only agglutinated benthic foraminifera occur. The planktonic assemblage occurs again farther to the northwest in the Model T section but is absent at Torgerson Draw. The laterally discontinuous distribution indicates that "migration" one is probably affected by dissolution although miscorrelation or a discontinuity could produce similar results. Only two of the 12 "migrations" of planktonic foraminifera exhibit lateral discontinuities of the planktonic foraminiferal biofacies (Fig. 9). These correlations are considered valid because: (1) 83% of the intervals display continuous "migration" across the area, and the majority of lateral shifts of planktonic foraminifera biofacies appear to be unaffected by dissolution, hiatus, or miscorrelation, and (2) the planktonic biofacies "migrations" parallel the lithologic, time-significant correlation lines.

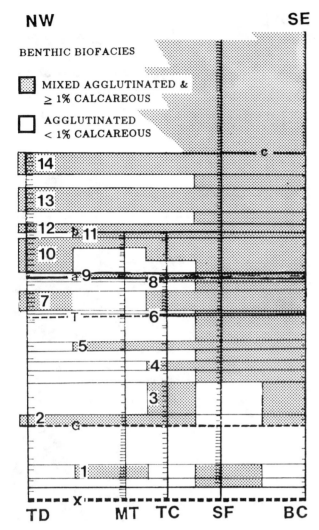

FIG. 10.—Distribution of benthic foraminiferal biofacies. Grey area represents presence of mixed agglutinated and calcareous benthic foraminiferal biofacies (sample contain >1 percent calcareous benthic foraminifera). White area represents presence of agglutinated benthic biofacies (samples contain <1 percent calcareous benthic foraminifera). "Migrations" of mixed agglutinated and calcareous benthic foraminifera into the northwestern sections (TD, MT, and TC) are numbered.

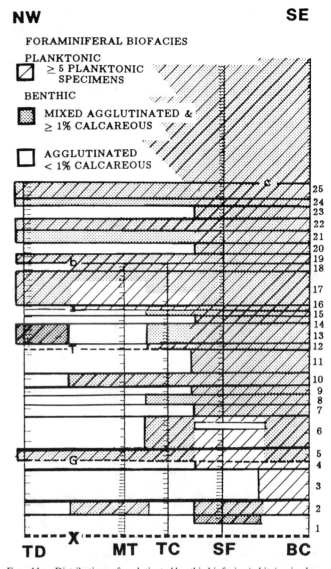

FIG. 11.—Distributions of agglutinated benthic biofacies (white), mixed agglutinated and calcareous benthic biofacies (grey), and planktonic biofacies (diagonal lines) are shown. "Time lines" are drawn at the base of lateral shifts in the distribution of these foraminiferal biofacies. "Time slices" defined by these shifts are numbered. From Fisher and others (1994).

Similar SGC analysis is conducted for benthic foraminiferal biofacies. The presence/absence of the mixed agglutinated and calcareous benthic and the agglutinated benthic biofacies are plotted on the standardized sections (Fig. 10). Agglutinated benthic foraminifera biofacies dominate the northwestern stratigraphic sections. Agglutinated benthic foraminifera dominate the benthic biofacies in all of the sections between the X and G bentonites. Lateral changes in the positions of the benthic biofacies distinguish biofacies intervals. Fourteen "migrations" of the mixed agglutinated and calcareous benthic biofacies into the northwestern sections occur (Fig. 10). Five out of fourteen intervals (36% of the intervals) display lateral discontinuities ("migrations" 1, 2, 3, 7, and 10; Fig. 10), suggesting they are affected by dissolution, minor hiatus, miscorrelation, or patchiness. Modern benthic foraminifera exhibit patchy distributions (Boltovskoy and Wright, 1976). Most of the benthic biofacies "migrations" coincide with planktonic biofacies "migrations," lending support to the hypothesis that the distributions are useful in correlation. The benthic biofacies "migrations" parallel the lithologic correlation lines and therefore appear to represent valid lines of correlation.

The hypothesis suggesting planktonic and benthic biofacies as useful in correlation appears reasonable. Migrations of planktonic and benthic biofacies parallel the lithologic correlations. The planktonic biofacies correlations and the benthic biofacies correlations are integrated and plotted in conjunction with the bentonite and calcarenite markers to produce twenty-five "time slices" (Fig. 11), creating a high resolution correlation across the facies change.

SUMMARY

Using traditional correlation techniques, lithofacies changes are generally difficult if not impossible to correlate. The use of unique and nonunique data in geologic history not only permits correlation but enhances correlation resolution across this major facies change. Despite the scant availability of biostratigraphic data, local data such as bentonites, calcarenites, and distinct foraminiferal biofacies shifts define 25 "time lines" within a period of less than approximately 1 Ma (Fig. 3). The improved resolution of correlation, obtained across this facies change, was achieved using the supplemented graphic correlation methods of Edwards (1989).

ACKNOWLEDGMENTS

I wish to thank Bill Hay, Don Eicher, Peter Roth, David Watkins, Erle Kauffman, Mark Leckie, Bill Cobban, Jim Kirkland, and Brad Sageman for their contributions and knowledgeable discussions. The comments of three anonymous reviewers enhanced the manuscript. Financial support was supplied by NSF [8618877] to Don Eicher. Financial support was also provided by the University of Colorado Museum Walker Van Riper Grant, Amoco Production Company, Shell Oil, Phillips Chemical Company, and the University of Colorado Grauberger Fund.

REFERENCES

BOLTOVSKY, E. AND WRIGHT, R., 1976, Recent Foraminifera: The Hague, Dr. W. Junk, 515 p.

CALDWELL, W. G. E., DINER, R., EICHER, D. L., FOWLER, S. P., NORTH, B. R., STELCK, C. R., AND VON HOLDT WILHELM, L., 1993, Foraminiferal biostratigraphy of Cretaceous marine cyclothems, in Caldwell, W. G. E. and Kauffman, E. G., eds., Evolution of the Western Interior Basin: St. John's, Geological Association of Canada Special Paper 39, p. 477–520.

COBBAN, W. A., 1993, Diversity and distribution of Late Cretaceous ammonites, Western Interior, United States, in Caldwell, W. G. E. and Kauffman, E. G., eds., Evolution of the Western Interior Basin: St. Johns's, Geological Association of Canada Special Paper 39, p. 435–451.

EDWARDS, L., 1989, Supplemented graphic correlation: A powerful tool for paleontologists and nonpaleontologists: Palaios, v. 4, p. 127–143.

EICHER, D. L., 1965, Foraminifera and biostratigraphy of the Graneros Shale: Journal of Paleontology, v. 39, p. 875–909.

EICHER, D. L., 1967, Foraminifera from the Belle Fourche Shale and equivalents, Wyoming and Montana: Journal of Paleontology, v. 41, p. 167–188.

EICHER, D. L., 1969a, Cenomanian and Turonian planktonic foraminifera from the Western Interior of the United States, in Bronnimann, P. and Renz, H. H., eds., Proceedings of the First Intonational Conference on Planktonic Microfossils: Leiden, E. J. Brill, v. 2, p. 163–174.

EICHER, D. L., 1969b, Paleobathymetry of Cretaceous Greenhorn Sea in eastern Colorado: American Association of Petroleum Geologists Bulletin, v. 53, p. 1075–1090.

EICHER, D. L. AND DINER, R., 1985, Foraminifera as indicators of water mass in the Cretaceous Greenhorn Sea, Western Interior, in Pratt, L. M., Kauffman, E. G., and Zelt, F. B., eds., Fine-Grained Deposits and Biofacies of the Cretaceous Western Interior Seaway: Evidence of Cyclic Sedimentary Processes: Tulsa, Society of Economic Paleontologists and Mineralogists Field Trip Guidebook 4, p. 60–71.

EICHER, D. L. AND DINER, R., 1989, Origin of the Cretaceous Bridge Creek Cycles in the Western Interior, United States: Palaeogeography, Palaeoclimatology, Palaeoecology, v. 74, p. 127–146.

EICHER, D. L. AND WORSTELL, P., 1970, Cenomanian and Turonian foraminifera from the Great Plains, United States: Micropaleontology, v. 16, p. 269–324.

ELDER, W., 1989, A simple high-precision Jacob's Staff design for the high resolution stratigrapher: Palaios, v. 4, p. 196–197.

FISHER, C., 1991, Calcareous Nannofossil and Foraminifera definition of an oceanic front in the Greenhorn Sea (late Middle through Late Cenomanian), Northern Black Hills, Montana and Wyoming: paleoceanographic implications: Unpublished Ph.D. Dissertation, University of Colorado, Boulder, 325 p.

FISHER, C., HAY, W. W., AND EICHER, D. L., 1994, Oceanic front in the Greenhorn Sea (late middle through late Cenomanian): Paleoceanography, v. 9, p. 879–892.

KAUFFMAN, E. G., 1977, Geological and biological overview: Western Interior Cretaceous Basin, in Kauffman, E. G., ed., Cretaceous Facies, Faunas and Paleoenvironments Across the Western Interior Basin: The Mountain Geologist, v. 14, p. 75–99.

KAUFFMAN, E. G., 1985, Cretaceous evolution of the Western Interior Basin of the United States, in Pratt, L. M., Kauffman, E. G., and Zelt, F. B., eds., Fine-Grained Deposits and Biofacies of the Cretaceous Western Interior Seaway: Evidence of Cyclic Sedimentary Processes: Tulsa, Society of Economic Paleontologists and Mineralogists Field Trip Guidebook 4, p. iv–xiii.

KAUFFMAN, E. G., 1988, Concepts and methods of high-resolution event stratigraphy: Annual Review Earth and Planetary Sciences, v. 16, p. 605–654.

KAUFFMAN, E. G., SAGEMAN, B. B., KIRKLAND, J. I., ELDER, W. P., HARRIES, P. J., AND VILLAMIL, T., 1993, Molluscan Biostratigraphy of the Cretaceous Western Interior Basin, North America, in Caldwell, W. G. E. and Kauffman, E. G., eds., Evolution of the Western Interior Basin: St. John's, Geological Association of Canada Special Paper 39, p. 397–434.

KNETCHEL, M. AND PATTERSON, S., 1962, Bentonite deposits of the northern Black Hills District, Wyoming, Montana, and South Dakota: United States Geological Survey Bulletin 1082-M, p. 893–1030.

MACDONALD, R. H., 1984, Depositional environments of the Greenhorn Formation (Cretaceous), Northwestern Black Hills: Unpublished Ph.D. Dissertation, University of Wisconsin, Madison, 284 p.

MACDONALD, R. H. AND BYERS, C. W., 1988, Depositional history of the Greenhorn Formation (Upper Cretaceous) northwestern Black Hills: The Mountain Geologist, v. 25, p. 71–85.

MCNIEL, D. H. AND CALDWELL, W. G. E., 1981, Cretaceous Rocks and their Foraminifera in the Manitoba Escarpment: St. John's, Geological Association of Canada Special Paper 21, 439 p.

MOORE, R., 1949, The meaning of facies, in Longwell, C.R., ed., Sedimentary Facies in Geologic History: Boulder, Geological Society of America Memoir 39, p. 1–39.

OBRADOVICH, J. D., 1993, A Cretaceous time scale, in Caldwell, W. G. E. and Kauffman, E. G., eds., Evolution of the Western Interior Basin: St. John's, Geological Association of Canada Special Paper 39, p. 379–396.

OBRADOVICH, J. D. AND COBBAN, W., 1975, A Time-scale for the Late Cretaceous of the Western Interior of North America, in Caldwell, W. G. E., ed., The Cretaceous System in the Western Interior of North America: St. John's, Geological Association of Canada Special Paper 13, p. 31–54.

ROBASZYNSK, F. AND CARON, M., eds., 1979, Cashiers de Micropalaeontologie, Atlas de Foraminiferes Planctoniques de Cretace Moyen (Mer Boreale et Tethys): Paris, Editions du Centre National de la Recherche Scientifique, Parts 1 and 2, 185 p. and 181 p.

ROBINSON, C. S., MAPEL, W. J., AND BERGENDAHL, M. H., 1964, Structure and stratigraphy of the northern and western flanks of the Black Hills uplift, Wyoming, Montana and South Dakota: Washington, D.C., United States Geological Survey Professional Paper 404, 134 p.

SHAW, A., 1964, Time in Stratigraphy: New York, McGraw-Hill, 365 p.

WATKINS, D. K., BRALOWER, T. J., COVINGTON, J. M., AND FISHER, C. G., 1993, Biostratigraphy and paleoecology of the Upper Cretaceous calcareous nannofossils in the Western Interior Basin, North Americas, in Caldwell, W. G. E. and Kauffman, E. G., eds., Evolution of the Western Interior Basin: St. John's, Geological Association of Canada Special Paper 39, p. 521–537

WILLIAMS, G. D. AND STELCK, C. R., 1975, Speculations on the Cretaceous paleogeography of North America, in Caldwell, W. G. E., ed., The Cretaceous System in the Western Interior of North America: St. John's, Geological Association of Canada Special Paper 13, p. 1–20.

GRAPHIC CORRELATION OF NEW CRETACEOUS/TERTIARY (K/T) BOUNDARY SUCCESSIONS FROM DENMARK, ALABAMA, MEXICO, AND THE SOUTHERN INDIAN OCEAN: IMPLICATIONS FOR A GLOBAL SEDIMENT ACCUMULATION MODEL

NORMAN MACLEOD

Department of Palaeontology, The Natural History Museum, Cromwell Road, London SW7 5BD, England

ABSTRACT: Since the original MacLeod and Keller (1991a, b) graphic correlation study of Cretaceous/Tertiary (K/T) boundary sections and cores, new biostratigraphic data have become available for lowermost Danian successions in high latitudes (Nye Kløv, ODP sites 690 and 738) and from sequences proximal to the proposed Chicxulub impact structure (Millers Ferry, Mimbral). Graphic analysis of these data provides an opportunity to test the prediction that rising eustatic sea level during the trans-K/T interval played a major role in controlling patterns of sediment accumulation in neritic and bathyal settings. In addition, restricting the empirical basis for development of a lowermost Danian Composite Standard (LD-CS) from all available biostratigraphic data (used in the previous study) to only datums from widely-accepted Danian and K/T survivor taxa enables determination of the extent to which previous results were biased by inclusion of data from controversial "Cretaceous" survivor taxa. Results indicate that sequences from neritic and upper bathyal settings (Nye Kløv, Mimbral) are temporally complete within their lowermost Danian intervals while sections/cores from very shallow inner neritic settings (Millers Ferry) and the deep sea (ODP Site 690, ODP Site 738) contain incomplete lower Danian stratigraphic records. These findings are consistent with results of traditional zone-based biostratigraphic analyses and with predictions of the sequence stratigraphic model. Moreover, this revised LD-CS is essentially identical to the previous K/T-CS (MacLeod and Keller, 1991b) based on total biotic data. Data from a large number of organismal groups now confirm that biotic, sedimentologic, and geochemical studies based solely on deep-sea and very shallow neritic successions are biased toward catastrophic patterns of change as a result of missing section (= time). Biotic patterns from inner neritic through upper bathyal settings have the best chance of preserving temporally complete sequences of lower Danian events. The faunal record from these sequences reveals the lower Danian succession to be characterized by progressive rather than instantaneous turnover in biotic, sedimentologic, and geochemical variables occurring over at least 500,000 years.

INTRODUCTION

For more than a decade the debate, over alternative scenarios for the Cretaceous/Tertiary (K/T) faunal turnover has captured the attention of both earth scientists and the lay public. There are a wide variety of reasons for this popularity, not the least of which includes a fascination with some of the more sensationalistic environmental changes thought to follow a bolide impact on the earth's surface (Alvarez and others, 1980, 1982). More substantially though, this debate captures much of the dynamic tension that has historically existed between (neo)uniformitarianism and (neo)catastrophism (see Gould, 1977).

The amount of physical evidence consistent with recognition of a bolide impact having occurred at or near the K/T boundary has been growing for over a decade (see Silver and Schultz, 1982; Sharpton and Ward, 1990; Ryder and others, in press). This line of research has recently culminated in identification of the Chicxulub structure beneath Yucatan as a possible K/T impact site (Hildebrand and Boynton, 1990; Hildebrand and others, 1991; Pope and others, 1991; Alvarez and others, 1992; Smit and others, 1992; Sharpton, 1993). The Chicxulub impact scenario is not without its critics (see Officer and others, 1992), however, and at least some of the preliminary data advanced in support of the Chicxulub impact are incorrect (see Jéhanno and others, 1992; Lyons and Officer, 1992; Keller and others, 1993; Stinnesbeck and others, 1993; Savrda, 1993; Keller and others, 1994). Available biostratigraphic data also suggest that the Chicxulub structure, whatever its origin, does not coincide with the internationally-recognized K/T boundary but lies well within the upper Maastrichtian (Stinnesbeck and others, 1993; Officer and others, 1992). But, even if the Chicxulub structure does eventually prove to represent a late Cretaceous impact crater, proponents of more extreme versions of the impact-mass extinction link must demonstrate that this event horizon corresponds to an instantaneous global extinction event. In other words, identification of an impact crater at or near the K/T boundary, by itself, cannot establish a causal connection between bolide impacts and mass extinctions.

Until recently, the accepted characterization of end Cretaceous faunal turnovers in ammonites, rudistid bivalves, dinosaurs, and plankton appeared to correspond well with predictions of the asteroid impact model (Alvarez and others, 1980, 1982). Recovery of iridium anomalies at K/T boundaries from many sections and deep-sea cores scattered around the world were also interpreted as confirming an association between impact debris and mass extinction (Alvarez and others, 1982). Since these initial observations were made, a large number of studies have revealed data that are inconsistent with a bolide-induced mass extinction at the K/T boundary. Foremost among these are numerous reports of "Cretaceous" taxa occurring in unquestionably Danian sediments (see Perch-Nielsen and others, 1982; Jiang and Gartner, 1986; Pospichal, 1993 [nannoplankton]; Brinkhuis and Zachariasse, 1988; [dinoflagellates]; Kitchell and others, 1986 [diatoms]; Hollis, 1993 [radiolaria]; Keller, 1988b, Keller and Lindinger, 1989; Thomas, 1990; Widmark and Malmgren, 1992 [benthic foraminifera]; Heinberg, 1979 [bivalves]; Donce and others, 1985; Peypouquet and others, 1986; Brouwers and DeDeckker, 1993 [ostracodes]; Zinsmeister and others, 1989 [ammonites]; Surlyk, 1990 [brachiopods]; Labandiera, 1992 [insects and insect-plant associations]; Sullivan, 1987 [non-dinosaurian reptiles]; Sloan and others, 1986; Charig, 1989; Clemens and Nelms, 1993 [dinosaurs]; Archibald and Clemens, 1982; Archibald and Clemens, 1984 [microvertebrates—mainly mammals]; Lerbekmo and others, 1987; Sweet and Braman, 1992; Sweet and others, 1990; McIver and others, 1991; Askin, 1990, 1992; Méon, 1990 [pollen and spores]; Johnson and Greenwood, 1993; Knobloch and others, 1993 [plant body fossils]). In the past, such reports have often been dismissed as isolated examples of reworked upper Cretaceous faunas. But, the consistency with which these anomalous faunas have now been recovered, along with the broad taxonomic scope of this biota, demands that the notion of widespread trans-K/T survivorship be taken seriously. It should also be noted that several of these studies document differentially high levels of trans-K/T survivorship in high latitudes. This suggests that the K/T extinction event, whatever its cause, exhibited a pronounced geographic as well as temporal structure.

In response to the widely perceived need for a more detailed understanding of K/T boundary stratigraphy (e.g., Feldmann, 1990), MacLeod and Keller (1991a, b) published a synoptic study of K/T boundary sections and cores based on extensive collections of trans-K/T planktic foraminiferal faunas, one of the groups thought to have been severely affected by the K/T boundary event (Bramlett and Martini, 1964; Bramlett, 1965). This study employed graphic correlation to infer the global sequence of more than 70 planktic foraminiferal and calcareous nannoplankton first and last appearance datums (FADs and LADs, respectively). After comparing boundary sections and cores representing a large suite of marine depositional settings, we concluded that sequences located in outer neritic (Israel, N. Spain, Germany) and bathyal (DSDP Sites 528 and 577) environments are typically characterized by boundary hiatuses, whereas successions located in inner and middle neritic settings (El Kef, Caravaca, Agost, Brazos River) preserve temporally complete event sequences. Many of these shallower water sections also contain evidence for one or more intervals of stratigraphic condensation or the presence of a hiatuses higher up in the lower Danian, but these do not affect completeness estimates across the boundary horizon itself. When correlated to the trans-K/T eustatic sea-level curves of Brinkhuis and Zachariasse (1988, based on El Kef) and Haq and others (1987, based on Exxon sequence stratigraphic data), this alternating pattern of sediment accumulation between shallow and deep-marine environments was found to be consistent with the predictions of sequence stratigraphy (Vail and others, 1977; Haq, 1991; Berger, 1970; Berger and Winterer, 1974; Keller and Barron, 1983, 1987; Loutit and Kennett, 1981; Goodwin and Anderson, 1985; Goodwin and others, 1986).

Graphic correlation was selected for the MacLeod and Keller (1991a, b) study in order to overcome deficiencies inherent in traditional approaches to stratigraphic analysis that allow for only very general zone-level correlations to be made and incorrectly assume the synchroneity of all biostratigraphic FADs and LADs. Results of our study are provocative in that they fail to corroborate the notion that almost all trans-K/T sections and cores contain a temporally complete record of events, an idea that has gained currency only since the proposal of the asteroid impact model (see Kauffman, 1984). Prior to publication of this hypothesis, most K/T boundary sections were thought to exhibit a pronounced boundary hiatus of unknown duration (see Harland and others, 1989; Ager, 1993; Ward and Kennedy, 1993 and references therein).

Since MacLeod and Keller's (1991) publications, new lower Danian planktic foraminiferal data have been obtained from several boundary sequences, including the comparatively high latitude localities of Nye Kløv, Denmark (Keller and others, 1993) and ODP sites 690 in the Weddell Sea and 738 on the Kerguelen Plateau (Keller, 1993), along with data from Mimbral in northern Mexico (Keller and others, 1994) and Millers Ferry in Alabama (Liu and Olsson, 1992; Olsson and Liu, 1993). The Mimbral and Millers Ferry data are especially important since these are located proximal to the proposed Chicxulub impact site. New high-resolution planktic foraminiferal data have also been obtained from the K/T boundary stratotype at El Kef, Tunisia, as a result of a 1992 Tunisian field conference on the K/T boundary (Ben Abdelkader and others, 1992; Keller and others, 1994). MacLeod (in press) documented the general nature of stratigraphic correlations between the revised boundary stratotype sequence and the high latitude sequences of Nye Kløv, ODP Site 690, and ODP Site 738. In the present study, these data along with faunal data from Mimbral and Millers Ferry are correlated with a revised Lower Danian Composite Standard (LD-CS) in order to (1) establish a globally-referenced lower Danian chronostratigraphy, (2) estimate levels of temporal completeness and identify hiatuses that may bias local records of trans-K/T event sequences, and (3) make the inter-sequence biostratigraphic comparisons necessary to develop and refine a global model of sediment accumulation in neritic-bathyal marine environments for the Mesozoic-Cenozoic transition.

METHODS

Biostratigraphic Database

This report is based on samples collected from Nye Kløv, Denmark (Keller and others, 1993); Mimbral, Mexico (Keller and others, 1994); ODP Site 690, Weddell Sea (Keller, 1993); and ODP Site 738, Kerguelen Plateau (Keller, 1993), along with new planktic foraminiferal data from the K/T boundary stratotype section at El Kef, Tunisia (Keller and others, unpubl. data). It also makes use of published faunal data from Millers Ferry (Liu and Olsson, 1992; Olsson and Liu, 1993) because actual samples or picked faunas from the Millers Ferry core were unavailable for this study (Olsson, pers. commun.). Distribution of these sections and cores plotted on a paleogeographic reconstruction of continental positions at 65 my is shown in Figure 1. Exact sample positions are provided in the primary biostratigraphic literature cited above.

All foraminiferal samples were processed according to standard micropaleontological techniques (see Keller, 1986) that included the recovery of all particulate matter $\geq$ 63 microns in diameter for all boundary sections and $\geq$ 38 microns in diameter for the boundary clay intervals of Site 738 and El Kef. Recovered planktic foraminiferal faunas were randomly split into 300–400 specimen aliquots that were identified in their entirety to the species level (see Buzas, 1990 for a justification of this procedure). Identified sample aliquots were mounted on paper slides for reference and currently reside in the planktic foraminiferal collections of Princeton University. Consistent species concepts were applied across the entire dataset. While the Danian taxonomy employed herein differs from some recent revisions (e.g., Olsson and others, 1992) the empirical data presented in those contributions primarily relates to changes in the generic assignment of particular species. These revisions have little significance in terms of species-level data summaries or biostratigraphic correlations.

Several alternative biostratigraphic zonations have been proposed for the uppermost Maastrichtian and lowermost Danian interval. These have been summarized in numerous publications, including Keller (1988a), Canudo and others (1991), Keller and Benjamini (1991), Ben Abdelkader and others (1992), and Keller (1993). Throughout this report, the zonation of Keller (1993) is employed.

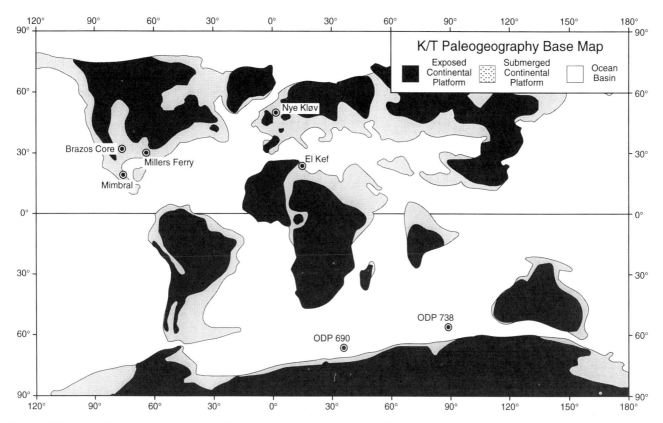

Fig. 1.—Paleogeographic map of ocean basins (white), continental shelves (stipple), and exposed continental platforms (black) during the lower Danian interval on which the positions of the sections and cores considered in this study are plotted. Reconstruction based on Denham and Scotese (1987), Stanley (1989), and Ziegler (1990).

Graphic Correlation

Graphic correlation and compositing of these biostratigraphic data into the LD-CS were carried out according to the methods described in Miller (1977) and Edwards (1984). Shaw (1964) demonstrated that graphic correlation results are consistent with those of more traditional biozone-based methods. This result has now been corroborated and extended by several subsequent investigations using both actual and simulated stratigraphic data (see Miller, 1977; Edwards, 1984, 1989; Hazel and others, 1984; Pisias and others, 1984; Pisias and others, 1985; Prell and others, 1986; Dowsett, 1988, 1989; Hazel, 1989; Johnson and others, 1989; MacLeod, 1991; MacLeod and Keller, 1991a, b).

Positions for the line of correlation (LOC) were estimated using the qualitative method of Miller (1977; see also MacLeod and Sadler, this volume) which distinguishes between the chronostratigraphic implications of FADs and LADs, and composited using multiple comparison cycles until a stable CS sequence was achieved. In contrast to MacLeod and Keller's (1991a, b) previous set of analyses, all data pertaining to Danian occurrences of "Cretaceous" taxa were excluded from consideration in order to avoid potential bias resulting from inclusion of datums some authors consider reworked (see Olsson and Liu, 1993). As outlined in several recent papers (see MacLeod and Keller, 1994; MacLeod in press a, b and references therein), there is virtually no independent evidence in favor of a reworking interpretation for these Danian occurrences, while many observations support their recognition as the remains of populations that survived the K/T boundary event. Nevertheless, sequestration of these "Cretaceous" datums from the present analysis enables an empirical test to be made of Olsson and Liu's (1993) assertion that inclusion of these datums biased our original graphic correlations. Additionally, FADs were judged to be more chronologically reliable than LADs owing to the potential for upward reworking that some authors believe to be a serious problem for the interpretation of lowermost Danian microfossil biostratigraphy (e.g., Liu and Olsson, 1992; Olsson and Liu, 1993). Since the overwhelming majority of datums used in the previous study also constituted first rather than last appearances, differential weighting of FADs would not be expected to exert a large influence on the previous correlations either in the compositing process or in the development of final correlation models. This expectation is additionally tested herein.

Elimination of controversial "Cretaceous" LADs from the original MacLeod and Keller (1991b) data leaves a total of 62 Danian biostratigraphic datums representing 43 taxa on which to base a lowermost Danian biochronostratigraphy (Appendix 1). Patterns of first and last appearances for these taxa were assessed in 11 different boundary sections/cores for CS construction. These include all boundary sequences known to be biostratigraphically complete and judged chronostratigraphically complete by the previous MacLeod and Keller (1991a, b) study (El Kef, Brazos 1, Brazos, CM4, Brazos Core, Agost, Caravaca), in addition to the five sections/cores under consideration in this report.

APPENDIX 1.—BIOSTRATIGRAPHIC DATA FOR LOWER DANIAN CORRELATIONS OF K/T BOUNDARY SECTIONS/CORES

No.	Datum	Type	LD-CS*	El Kef	Brazos 1	Brazos CM4	Brazos Core	Nye Klov	Millers Ferry**	Mimbral	ODP Site 690	ODP Site 738
1	*Chiloguembelina crinita*	LAD	479	—	—	—	—	—	—	—	—	535.5
2	*Chiloguembelina midwayensis*	FAD	137	137	—	—	—	50	207	45	15	225.5
3	*Chiloguembelina moresi*	FAD	402	—	—	—	—	—	—	65	25	146
4	*Chiloguembelina moresi*	LAD	479	—	—	—	—	—	—	—	—	535.5
5	*Eoglobigerina danica*	FAD	341	—	—	—	—	110	—	—	—	8
6	*Eoglobigerina danica*	LAD	377	—	—	—	—	690	—	—	—	15
7	*Eoglobigerina edita*	FAD	45	45	85	—	149	—	76	4	10	—
8	*Eoglobigerina edita*	LAD	950	950	170	200	200	—	1739	—	175	—
9	*Eoglobigerina eobulloides*	FAD	64	67	114	45	8	—	65	—	0	56
10	*Eoglobigerina eobulloides*	LAD	434	—	—	—	—	—	1707	—	154.5	91
11	*Eoglobigerina fringa*	FAD	0	25	20	5	38	15	—	4	—	0
12	*Eoglobigerina fringa*	LAD	377	—	—	—	—	20	—	—	—	12
13	*Eoglobigerina simplicissima*	FAD	46	50	114	55	8	490	—	45	35	6
14	*Eoglobigerina simplicissima*	LAD	472	—	—	—	—	490	—	—	85	495.5
15	*Eoglobigerina trivialis*	FAD	320	—	—	—	—	490	130	35	5	14
16	*Eoglobigerina trivialis*	LAD	487	—	—	—	—	1130	—	—	595	575.5
17	*Globanomalina pentagona*	FAD	302	302	—	125	88	77	—	40	0	8
18	*Globanomalina pentagona*	LAD	921	—	—	200	200	—	—	—	—	—
19	*Globanomalina taurica*	FAD	137	137	—	105	55	160	33	55	0	16
20	*Globanomalina tetragona*	FAD	402	—	—	—	—	670	—	65	—	—
21	*Globanomalina planocompressa*	FAD	157	157	—	—	—	—	152	—	—	—
22	*Globastica daubjergensis*	FAD	262	262	—	85	50	40	65	9	0	8
23	*Globastica sp.*	FAD	84	112	85	25	8	—	—	40	—	—
24	*Globigerinelloides monmouthensis*	LAD	479	—	—	—	—	200	—	25	—	535.5
25	*Globoconusa conusa*	FAD	0	0	5	35	20	15	65	4	10	1
26	*Globoconusa conusa*	LAD	477	400	170	115	—	80	413	70	95	91
27	*Globoconusa extensa*	FAD	420	—	—	—	—	—	—	—	45.5	225.5
28	*Globoconusa extensa*	LAD	462	—	—	—	—	—	—	—	575	—
29	*Globigerinelloides caravacaensis*	LAD	438	262	146	35	88	185	—	—	—	—
30	*Guembelitria cretacea*	LAD	459	350	—	—	—	940	402	—	—	330.5
31	*Guembelitria trifolia*	LAD	388	—	—	—	—	110	—	35	—	56
32	*Hedbergella holmdelensis*	LAD	438	55	146	35	88	—	25	60	−25	14
33	*Hedbergella monmouthensis*	LAD	411	6	114	45	−30	90	25	55	10	−1
34	*Igorina spiralis*	FAD	420	—	—	—	—	800	—	—	154.5	225.5
35	*Morozovella inconstans*	FAD	440	—	—	—	—	740	—	—	335	330.5
36	*Mucriglobigerina chasconona*	FAD	433	—	—	—	—	520	—	—	154.5	295.5
37	*Mucriglobiferina aquiensis*	FAD	420	—	—	—	—	520	—	—	154.5	225.5
38	*Parvularugoglobigerina eugubina*	FAD	57	57	100	35	20	15	65	4	—	8
39	*Parvularugoglobigerina eugubina*	LAD	500	500	170	100	68	30	228	160	—	14
40	*Parvularrugoglobigerina longiapertura*	FAD	63	—	—	—	—	9	—	4	—	—
41	*Parvularrugoglobigerina longiapertura*	LAD	479	—	—	—	—	120	—	150	—	—
42	*Planorotalities compressus*	FAD	341	—	—	—	—	110	—	30	0	8
43	*Subbotina moskvini*	FAD	378	400	—	105	148	185	—	40	0	8
44	*Subbotina moskvini*	LAD	442	—	—	—	—	270	—	—	271	146
45	*Subbotina pseudobulloides*	FAD	350	350	—	135	155	140	185	45	0	8
46	*Subbotina triangularis*	FAD	377	—	—	—	—	—	—	—	—	12
47	*Subbotina triangularis*	LAD	456	—	—	—	—	—	—	—	—	415.5
48	*Subbotina triloculinoides*	FAD	282	500	—	115	53	390	—	135	10	8
49	*Subbotina triloculinoides*	LAD	455	—	—	—	—	—	—	—	475	—
50	*Subbotina trinidadensis*	FAD	783	—	—	—	—	—	—	—	—	—
51	*Subbotina varianta*	FAD	401	—	—	—	—	390	—	—	95	146
52	*Subbotina varianta*	LAD	487	—	—	—	—	—	—	—	595	575.5
53	*Woodringina claytonensis*	FAD	273	—	—	—	—	715	65	165	35	111
54	*Woodringina claytonensis*	LAD	479	—	—	—	—	—	—	—	—	535.5
55	*Woodringina hornerstownensis*	FAD	17	20	9	5	5	9	120	4	10	91
56	*Woodringina hornerstownensis*	LAD	487	—	—	—	—	100	—	—	375.5	575.5
57	*Biantholithus sparsus*	FAD	0	0	0	—	—	—	—	—	—	—
58	*Biscutum parvulum*	FAD	538	—	—	—	—	—	—	—	—	—
59	*Biscutum romeini*	FAD	20	20	95	—	—	—	—	—	—	—
60	*Cruciplacolithus primus*	FAD	570	570	250	—	—	—	—	—	—	—
61	*Cruciplacolithus primus*	LAD	1000	1000	500	—	—	—	—	—	—	—
62	*Towiensis petalosus*	FAD	700	700	400	—	—	—	—	—	—	—
63	Ir-1	KBD	0	0	0	—	0	0	—	0	0	0
64	Red Layer	KBD	0	0	0	0	0	0	—	0	—	0

*Based on multiple compositing rounds; see text for description.
**Data from Liu and Olsson, 1992.

Initial CS estimation was achieved via designation of the El Kef boundary stratotype as the Standard Reference Section (SRS) and compositing data from the Brazos sections, Agost, and Caravaca into this sequence. As in the previous MacLeod and Keller (1991a, b) study, LOC placement during compositing utilized each section's "best-case" correlation model (= sequence most complete; sediment accumulation rate as constant as possible; simplest LOC geometry). Employment of this convention produces a conservative estimate of the true event sequence. Once a stable CS had been estimated, a separate cycle of compositing rounds was undertaken to add the information present at Nye Kløv, Millers Ferry, Mimbral, and ODP Sites 690 and 738 to the CS. Only after this second compositing cycle achieved stability were final correlations and interpretations made. The compositing process was responsible for increasing the number of datums available for stratigraphic inference from the 28 original SRS datums to 64 datums in the final CS (an increase of over 200%) and modifying the position of 23 datums within the overall composite sequence.

The purpose of CS construction is to infer the correct global sequence of biostratigraphic events. During this process, it is necessary to sequester information on datum positions provided by the consideration of individual sequences in previous compositing rounds in order to avoid self-justified LOC geometries (see Edwards, 1984). However, once a stable and independently-justified LOC has been defined, the maintenance of such a data sequestration convention into the next logical stage of a graphic correlation analysis (the comparative inference of chronostratigraphy) confuses the process of CS construction (= model creation) with the task of chronostratigraphic inference (= model application). Since the final CS represents the best available estimate of the true sequence of events, any degradation of this stratigraphic model by knowingly removing datums or altering their sequence results in a concomitant decrease in the model's information content. In order to preserve this content and utilize it to provide the best chronostratigraphic inferences possible, a fundamental distinction between CS creation and CS application should be recognized within the graphic correlation method. This distinction can be accomplished by treating the inference of final correlations between the CS and each section or core used in its creation in the same manner as a correlation model for a section or core not used in CS construction. In other words, the data sequestration convention observed during CS construction should *not* be applied when final correlation models for these sections are produced.

In most cases, the actual number of datums whose CS positions are the result of information provided by individual sequences is small. The most obvious exception to this is the SRS which, if properly chosen, may contain a larger number of biostratigraphic datums already at their respective minima or maxima. Thus, for most sections/cores, modification of this longstanding methodologic convention should not result in the production of radically different results. Moreover, if leaving the CS intact during the inference of final correlation models implies a substantively different LOC geometry, a stable CS has most likely not been achieved. The only logically consistent alternative to the revised methodology described above that also avoids degradation of the CS is to regard the CS as being unable to address the question of chronostratigraphic inference in sets of sections or cores used to create it — an alternative that would severely compromise the graphic correlation's scope and practicality.

As shown in Figure 2, development of an alternative LD-CS using only the Danian subset of biostratigraphic datums used by MacLeod and Keller (1991a, b) has comparatively little effect on correlations for previously considered sequences. Parametric (Pearson) and non-parametric (Spearman) correlation coefficients between the original (MacLeod and Keller, 1991b) and revised (Appendix 1) CS datum positions are both 0.96, indicating statistical significance well above the $\alpha = 0.999$ level. Thus, contrary to statements by Olsson and Liu (1993), results of MacLeod and Keller's (1991a, b) previous analysis do not appear to have been substantially biased by inclusion of "Cretaceous" taxa in the former study. Instead, the strong similarities observed between the LD-CS and the original MacLeod and Keller (1991a, b) K/T-CS stem from the fact that the correlations obtained in the original analysis were predominantly controlled by the Danian taxa. Moreover, these correlations result in large part from the observation that, in most of these boundary sequences, individual datums found close together in one sequence are often found far apart in another. Using traditional biozone-based methods, the chronostratigraphic implications of this observation are often ignored in favor of a model that illogically assumes all zone-defining biostratigraphic datums to be isochronous. It was the obvious limitation inherent in such traditional approaches to biostratigraphically-based chronostratigraphic inference that impelled Shaw (1964) to develop the graphic correlation method. Similarly, it is the stringent requirements that stratigraphic tests of the K/T impact/extinction model place on our ability to tell time in the fossil record that necessitate use of this method to obtain objective, high-resolution estimates of lower Danian chronostratigraphy.

RESULTS

Nye Kløv

Keller and others (1993) found the Nye Kløv section to be biostratigraphically complete at the zonal level with the lowermost Danian biozone (Zone P0). This zone is represented by the 15 cm of section that separate the thin, organic-rich boundary clay layer that coincides with a negative $\delta^{13}C$ shift (see Keller and others, 1993) from the *Parvularugoglobigerina eugubina* FAD [40]. Preliminary graphic comparison of the Nye Kløv sequence with planktic microfossil data from the El Kef boundary stratotype (MacLeod, in press a) was unable to distinguish between alternative correlation models, suggesting either continuous sediment accumulation at a uniform rate throughout the sequence or the presence of a hiatus in the lower portion of Zone P1a. This indeterminacy results from the comparatively small number of lowermost Danian taxa in common between the El Kef and Nye Kløv sequences. The expanded number of datums available as a result of compositing the El Kef SRS with data from other K/T boundary sections/cores enables some aspects of these alternative correlation models to be resolved.

Graphic correlation of the Keller and others (1994) biostratigraphic data with the revised LD-CS (Fig. 3) indicates that the Nye Kløv sequence is temporally complete but characterized by a relatively low time-averaged sediment accumulation rate in the lowermost Danian. This rate appears to change approximately mid-way through Zone P1a where the relative sediment accumulation rate increases and remains stable throughout the remaining 200 centimeters of the sampled interval. Nevertheless, the juxtaposition of lowermost Danian FADs (e.g., *Globoconusa conusa* [25], *Woodringina hornerstownensis* [55], *Eoglobigerina fringa* [11], *P. eugubina* [38]) with upper Zone P1a LADs (*E. fringa* [12], *P. eugubina* [39]) over a six centimeter interval remains a matter of concern. In other sections (e.g., El Kef, Brazos 1, Brazos CM4) these datums are separated from one another by as much as 5 m of section. Accordingly,

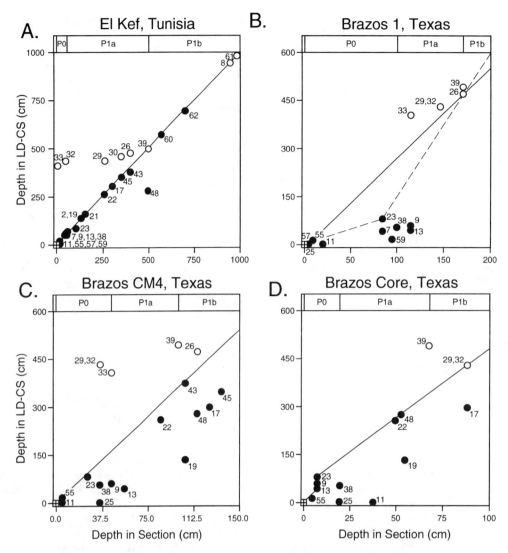

Fig. 2.—Graphic correlation diagrams for the El Kef (A), Brazos 1 (B), Brazos CM4 (C), and Brazos Core (D); biostratigraphic data (Appendix 1) plotted against corresponding datum positions in the revised LD-CS. Numbers refer to datums listed in Appendix 1. Closed circles = FADs; open circles = LADs. Box with cross indicates presence of key beds within the section. LD-CS datum positions inferred on the basis of Danian microfossil datums along with the LADs of unquestioned K/T survivor taxa. (A.) Datums present within the El Kef boundary stratotype section (x-axis = SRS) plotted against their position in the final LD-CS (y-axis). This is not a graphic correlation plot *per se* but instead serves to graphically illustrate the conservative manner in which datum adjustments were made during the compositing process. The diagonal line is not a LOC but represents the line along which datums points would fall if no adjustments to SRS datum positions occurred in production of the final CS. Note the strong similarity between these two datum sequences. This indicates that the compositing procedure resulted in the adjustment of only a small number of datums from their SRS positions. The group of adjusted datums is also dominated by LADs, reflecting the higher weight given to FADs as a result of possible upward reworking of last appearances. Despite these modifications, though, well over two-thirds of the LD-CS positions reflect datum distributions found within the El Kef boundary stratotype. (B.-C.) Graphic correlation plots of the LD-CS with data from three Brazos River K/T boundary sections (data from MacLeod and Keller, 1991b). Solid LOC = preferred (best case) correlation model; dashed LOC = alternative (worst case) correlation model. Note the correspondence between correlations of the Brazos River sections using only Danian and unquestioned survivor datums and previous graphic correlations that included datums from proposed K/T survivor taxa (MacLeod and Keller, 1991a, b). This similarity indicates that, contrary to Olsson and Liu (1993), these previous correlations models were not strongly affected by the inclusion of "Cretaceous" LADs.

their co-incidence in the Nye Kløv sequence may indicate that a hiatus spanning the lower portion of Zone P1a is still present. Since this ambiguity results from a lack of Danian datums in the interval between 15 and 40 cm, it is unlikely to be resolved until data from other fossil groups becomes available. Presence of a hiatus in this position, though, would not change the interpretation of the lowermost Danian portion of Nye Kløv as being both biostratigraphically and temporally complete.

In addition to these biostratigraphic data, the preferred chronostratigraphic correlation depicted in Figure 3 is also consistent with presence of the characteristic red clay layer within the Nye Kløv sequence. A red clay layer occurs coincident with the K/T boundary in may biostratigraphically complete neritic sequences (e.g., El Kef, Agost, Caravaca, Brazos River). Unfortunately, a lithostratigraphically similar red clay layer is also present in some biostratigraphically incomplete successions

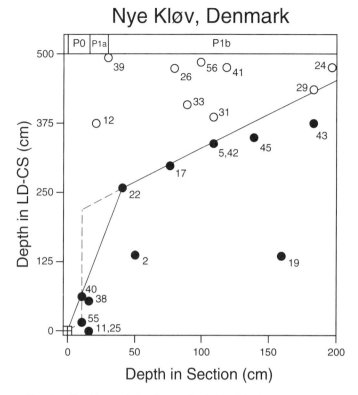

FIG. 3.—Graphic correlation diagram for the Nye Kløv biostratigraphic data (Appendix 1) plotted against corresponding datum positions in the revised LD-CS. Numbers refer to datums listed in Appendix 1. Closed circles = FADs; open circles = LADs. Box with cross indicates presence of key beds (an Ir anomaly and the red event layer) within the section. Solid line = preferred correlation model. Dashed LOC segment = alternative correlation model.

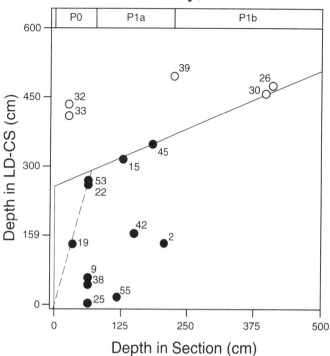

FIG. 4.—Graphic correlation diagram for the Millers Ferry biostratigraphic data (Appendix 1; from Liu and Olsson, 1992) plotted against corresponding datum positions in the revised LD-CS. Numbers refer to datums listed in Appendix 1. See Figure 2 for symbol conventions. Note lack of any of the characteristic key beds at the K/T boundary (e.g., red layer, layer of Ni-rich spinels) that are present at Brazos River (see above) and Mimbral (see below) and would be expected in a temporally complete boundary sequence.

(e.g., Gubbio, Italy). The preferred LOC geometry for the Nye Kløv correlation is well-supported by available biotic data with 67% of the datums plotting on or near the LOC.

Millers Ferry

Correlation of Liu and Olsson's (1992) Millers Ferry data with the revised LD-CS (Fig. 4) reveals patterns that are in many ways similar to the Nye Kløv correlation. Once again, a collection of datums that are spread out over some 500 cm of section in other K/T boundary sequences (e.g., the El Kef boundary stratotype) are condensed into half that interval at Millers Ferry. This suggests a relatively low sediment accumulation rate for the lowermost Danian. Also, Millers Ferry's geometry of FADs and LADs indicates that sediment accumulation rates increased at this locality by the beginning of Zone P1a. This inference is based on the much greater separation between the FADs of *Globastica daubjergensis* [22], *Woodringina claytonensis* [53], *Eoglobigerina trivialis* [15], and *Subbotina pseudobulloides* [45] and the LADs of *Globoconusa conusa* [26] and *G. cretacea* [30] at Millers Ferry section relative to other boundary sequences, including the El Kef stratotype.

Unfortunately, the precise nature of the lowermost Danian correlation at Millers Ferry remains ambiguous. Whereas the K/T boundary in the stratotype section is recognized on the basis of multiple criteria, including geochemical anomalies in Ir, $CaCO_3$, $\delta^{13}C$, Ni-rich spinels, and lithostratigraphically-defined red layer, along with the FAD of *G. conusa* [25] (see Ben Abdelkader and others, 1992; Keller and others, 1994), none of these markers has been reported from the Millers Ferry sequence. Liu and Olsson (1992) place their K/T boundary on the basis of an abundance increase in *G. cretacea* (p. 330). However, their Figure 2 shows the *G. cretacea* abundance increase at Millers Ferry to begin well below the K/T boundary in the El Kef stratotype. Keller (1989) shows the *G. cretacea* abundance increase to begin above the K/T boundary (defined by multiple geochemical, lithostratigraphic and biotic criteria) at Brazos River (see also Fig. 5 of Liu and Olsson, 1992) while Canudo and others (1991) show that this datum coincides with the Danian boundary between zones P0 and P1a—some 40,000 years into the Tertiary—in the Agost section. These results suggest that, relative to the standard K/T boundary recognition criteria (see Keller and others, 1993), the *G. cretacea* relative abundance increase is diachronous and, consequently, unable to reliably identify the K/T boundary in local successions.

Since the Millers Ferry sequence apparently contains none of the geochemical or lithostratigraphic K/T boundary markers present in all sections widely regarded as complete, and since

Liu and Olsson (1992) have used a poorly constrained biotic marker to define the boundary at Millers Ferry, the presence of a boundary hiatus in this succession must be considered possible, if not likely. Liu and Olsson's (1992) biostratigraphic data reinforce this interpretation. As shown in Figure 4 (see also their Fig. 3), the first Danian planktic foraminiferal datums at Millers Ferry include the LADs of *Hedbergella monmouthensis* [33] and *Hedbergella holmdelensis* [32] (both of which these authors regard as valid K/T survivors) and the FADs of *Praemurica* (= *Globanomalina*) *taurica* [19], *Parasubbotina* aff. *pseudobulloides*, and *Globanomalina archeocompressa* [Note the latter two datums have not been included in Figure 4 because comparable taxonomic entities have not been recognized in other boundary sequences]. The next Millers Ferry horizon containing biostratigraphic datums coincides with the Zone P0/P1a boundary and contains the FADs of five Danian taxa, including the basal Zone P1a marker *P. eugubina* [38]. The impression gained from plotting these datums against corresponding horizons in the LD-CS is one of a pronounced basal Danian discontinuity.

Unfortunately, each of these datums occurs above the boundary horizon at El Kef and a number of other boundary sections. Thus, their appearance in the lower part of the Millers Ferry sequence does little to constrain boundary placement of the chronostratigraphic interpretation of this interval. Both alternative correlation models developed in Figure 4 pass through the origin of the lithostratigraphic coordinate system (= K/T boundary) on the basis of the *Rosita contusa* LAD (see Liu and Olsson, 1992) which is known to coincide with the boundary horizon in other K/T boundary successions (e.g., El Kef, Caravaca, Agost). However, the appearance of this datum in the Miller's Ferry K/T boundary does not preclude the possibility that a boundary hiatus (extending at least into the uppermost Maastrichtian interval and possibly including a substantial lowermost Danian interval as well) is not present.

Recognition of a boundary hiatus within the Millers Ferry sequence also agrees with the lithostratigraphy. The basal (Danian) Clayton Formation sands are separated from the underlying Prairie Bluff Chalk by an undulatory surface that Mancini and others (1989) have interpreted as an erosional unconformity. Liu and Olsson (1992) argue that the boundary hiatus, if present, is very short in duration owing to the thickness of the Danian Zone P0 in the Millers Ferry core. However, the simple equation of lithostratigraphic thickness with temporal completeness must be regarded as suspect, owing to the well-known fact that lithostratigraphic thickness has no necessary chronostratigraphic implication (see Hedberg, 1976; NACSN, 1982). Moreover, in a recent ichnosedimentologic study of K/T boundary sands from Alabama, Savrda (1993) concluded that the basal Clayton sands represent a transgressive systems tract following a latest Maastrichtian sea-level lowstand (see also Baum and Vail, 1988; Donovan and others, 1988; Donovan and others, 1990; Savrda, 1993; Habib and others, 1992). This model explains the anomalous thickness of the Zone P0 interval and the absence of Danian FADs in the lowermost portion of Millers Ferry to result from facies shifts, including subareal exposure and subsequent erosion during the lowstand and grading through incised valley fill deposits to estuarine and inner neritic facies during the subsequent transgression. Savrda (1993) could find no evidence for catastrophic sedimentation patterns resulting from a K/T boundary tsunami as previously proposed by Bourgeois and others (1988) for Brazos River, Texas and by Liu and Olsson (1992; see also Olsson and Liu, 1993) for the Millers Ferry and Braggs sections in Alabama.

When taken together, these data make a compelling case for interpreting the Millers Ferry sequence to contain a pronounced lower Danian hiatus. Estimates of the duration of this hiatus remain uncertain due to the present lack of biostratigraphic control in the lowermost Danian portion of this section. The geometry of the Danian FADs shown in Figure 4, though, indicate that the entire P0 chronozone and lower half of the overlying P1a chronozone—an interval corresponding to over 100,000 years (see MacLeod and Keller, 1991b; D'Hondt and Herbert, 1991)—may be missing. It is also important to note that the preferred LOC geometry identifies a relatively small number (42%) of isochronous lower Danian datums in the Millers Ferry section. This may be due to this section's relatively shallow, estuarine to inner neritic depositional setting, coupled with variations in the species concepts applied by Liu and Olsson (1992) and other authors who contributed data to the lower Danian planktic foraminiferal database used in CS construction (see Methods section).

Mimbral

Although the Mimbral section of northeastern Mexico represents an upper bathyal depositional setting while Millers Ferry and Nye Kløv represent inner and middle neritic facies respectively (see MacLeod and Keller, 1994), many details of Mimbral's graphic correlation with the revised LD-CS are superficially similar to the much shallower Nye Kløv sequence (Fig. 5). Like Nye Kløv, but unlike Millers Ferry, Mimbral exhibits the lithostratigraphic and geochemical markers of the K/T boundary event bed. These observations provide strong support for interpreting the K/T boundary at Mimbral as complete. This interpretation must be tempered, however, with the realization that the first Danian microfossils enter the section not at the lithostratigraphically and geochemically-defined boundary, but 4 cm above this horizon. No less than seven Danian FADs occur simultaneously at this 4-cm level which together are spread over an interval of 57 cm in the El Kef stratotype and 63 cm in the revised LD-CS (Appendix 1). Consequently, two different interpretations for these data are possible. By, focusing on the geochemical and lithostratigraphic evidence, the Mimbral sequence may be regarded as complete throughout the lowermost Danian but extremely condensed with the 4-cm unfossiliferous interval representing over 50 cm of accumulation in other coeval sequences. Alternatively, by focusing on the biotic data, the coincidence of six Danian FADs representing an interval from the K/T boundary through the lower portion of Zone P1a at a single Mimbral horizon might be regarded as sufficient evidence to identify a boundary hiatus in the sequence with the basal 4 cm representing an interval in which carbonate microfossils have been removed via diagenesis. The latter interpretation explains the biotic data but not the geochemical and lithostratigraphic data while the former successfully accounts for all observations. Consequently, the 'condensed section' interpretation is preferred though the "hiatus" LOC is retained as a plausible alternative. The condensed section interpretation is also corroborated by previous findings of Keller and others (1994) based on a traditional biozone analysis.

Regardless of the interpretation for the basal Danian interval at Mimbral, beginning halfway through Zone P1a (LD-CS level 262 corresponding to the FAD of *Globastica daubjergensis* [22], Fig. 5), the sediment accumulation rate appears to have increased to several times its lowermost Danian value and then increased again in the upper portion of the zone (LD-CS level 378 corresponding to the FAD of *Subbotina moskvini* [43], Fig. 5) by a slightly lesser amount. This LOC geometry may indicate continuously increasing sediment accumulation rates throughout the interval from LD-CS level 262 to 378. In Figure 5 though, this correlation is represented as two linear LOC segments to conform to present graphic correlation conventions (see Shaw, 1964; Miller, 1977). Sediment accumulation rates appear to have remained relatively high throughout the remainder of Zone P1a deposition at Mimbral. The preferred Mimbral LOC is well supported by available biotic data with 63% of the datums plotting on or near the LOC.

ODP Site 690

Of the two deep-sea cores considered in this analysis, correlation of biostratigraphic data from Site 690, Hole 690C with the LD-CS (Fig. 6) closely matches previous results obtained by MacLeod and Keller (1991a,b) from similar deep ocean settings. Here, a series of seven different Danian planktic foraminiferal FADs occur at a horizon coincident with an Ir anomaly. Thus, in this core, the K/T boundary is recognized on the basis of biotic and geochemical criteria. None of the other typ-

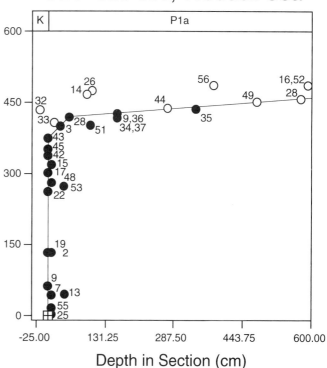

FIG. 6.—Graphic correlation diagram for the ODP Site 690, Hole 690C biostratigraphic data (Appendix 1) plotted against corresponding datum positions in the revised LD-CS. Numbers refer to datums listed in Appendix 1. See Figure 2 for symbol conventions.

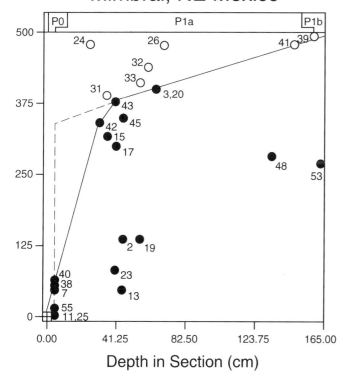

FIG. 5.—Graphic correlation diagram for the Mimbral biostratigraphic data plotted against corresponding datum positions in the revised LD-CS. Numbers refer to datums listed in Appendix 1. See Figure 2 for symbol conventions.

ical geochemical or lithostratigraphic markers of the boundary (e.g., red layer, Ni-rich spinels) have been reported to occur in the core (Barker and others, 1988). Certainly, the coincidence of seven different Danian species' FADs, which together span an interval of 302 cm in both the El Kef boundary stratotype and the revised LD-CS, provides strong evidence supporting recognition of a boundary hiatus in this core. In addition, the biostratigraphic succession within this core is incomplete with no hint of the basal Danian Zone P0. Recognition of a boundary hiatus is further corroborated by the absence of any geochemical anomaly (except that of Ir) or characteristic red lithostratigraphic layer that are, almost without exception typically found in complete boundary successions.

The only evidence in favor of recognizing continuous sediment accumulation across the Site 690C K/T boundary is an Ir anomaly that spans 2 m of section, reaches a peak abundance of 1566 ± 222 ppt, and drops off by a factor of four within 4 cm above and below peak abundance (Michel and others, 1988). Significantly, no independent evidence of impact debris (e.g., shocked quartz, microtektites) has been recovered from this core (Barker and others, 1988). In the absence of such independent evidence, the interpretation of an Ir anomaly as necessarily being the product of impact fallout and therefore evidence of continuous sediment accumulation across the boundary horizon becomes highly problematic. Plimer and Williams (1989), Colodner and others (1992), and Sawlowicz (1993) have shown that platinum-group elements (PGE's) can become significantly enriched during both syndepositional and post-depositional changes in redox conditions, while Dissanayake and Krisotakis (1984) have described similar PGE enrichments during sudden changes from aerobic to anaerobic conditions. In this context, it is important to note that Michel and others (1990, p. 161) describe the Hole 690C Ir anomaly as occurring "near the base of a dark layer in the sediment [with] tails upsection (into the dark layer) and downsection into light-colored Cretaceous sediment."

If the Site 690C Ir was derived directly from a bolide impact, ratios among PGEs should also exhibit values typical of chondritic materials. Determination of the Ir/Co, Ir/Cr, and Ir/Fe ratios for samples from the interval 4 cm above and below the Hole 690C Ir peak, though, shows these ratios to differ from chondritic values by orders of magnitude (Table 1). Similarly altered PGE ratios have recently been obtained by Wang and others (1993) in their study of the global Ir anomaly occurring at the Devonian-Carboniferous boundary. Wang and others (1993) attribute The D/C Ir anomaly to a sudden change in paleo-redox conditions during deposition and/or early diagenesis. The Site 690C PGE results do differ from the Wang and others (1993) D/C boundary values, but this is likely a function of differences in the local geochemical environment, including distinctions between the host lithologies. What is clear is that neither dataset exhibits a chondritic PGE fingerprint. In the absence of independent evidence for impact debris in this core, an Ir anomaly alone is insufficient to support an interpretation of temporal continuity across this boundary, especially in the face of available biotic data.

Of course, these results do not preclude the possibility that the PGEs present in Hole 690C were ultimately derived from a bolide impact at or near the K/T boundary. Nevertheless, taking all available biotic and geochemical data into consideration, the most parsimonious interpretation of this basal Danian record is to regard the boundary interval as missing and the Ir anomaly to be the result of either syndepositional or post-depositional concentration from ambient sources with subsequent diffusion of the Ir into underlying Cretaceous sediments. This interpretation is also consistent with the marked asymmetry of the Ir peak as well as with the recognition of a boundary hiatus in nearby Hole 689B (Michel and others, 1990; Barker and others, 1988).

As stated above, seven different Danian FADs occur coincident with the Ir anomaly in the 690C core with an additional six FADs occurring within the first 10 cm of Danian sediments. Together, these data suggest that the boundary hiatus encompasses all of Zone P0 and at least the lower two-thirds of Zone P1a. Above the boundary hiatus sediment accumulation rates rapidly change from moderate to quite high values. Datums occupying an interval of 100 cm in the El Kef boundary stratotype and revised LD-CS are spread over an interval of 600 cm in core 690C. Over its entire length, this correlation is extremely well supported by the available biotic data with over 78% of the datums plotting on or close to the LOC.

ODP Site 738

Hole 738C, drilled along the southern margin of the Kerguelen Plateau, contains what is, in many respects, an atypical lowermost Danian deep-sea biostratigraphic sequence. First, this core contains a complete sequence of biozones including the lowermost Danian Zone P0. This represents the first confirmed recovery of Zone P0 from any DSDP or ODP core [Note: D'Hondt and Keller (1991) had previously reported a 1-cm thick Zone P0 from Site 528 but this represents a sampling gap between their K/T boundary and the first Danian sample which contains the Zone P1a marker species *P. eugubina*. In addition, subsequent comparative analysis of stable isotopic data from the El Kef stratotype, Brazos Core, and Site 528 (MacLeod, in press a) failed to corroborate the existence of Zone P0 at Site 528].

Second, this core contains the first recorded, southern high-latitude occurrence of *P. eugubina* (the primary Zone P1a index). As mentioned above, the only other southern high latitude site to be studied in comparable detail (Site 690) did not produce any *P. eugubina* specimens. In the preliminary biostratigraphic analysis of Site 738, Huber (1991) also did not identify any morphotypes attributable to *P. eugubina*, though they were found subsequently by Keller (1993). These observations, among others, led Stott and Kennett (1990) and Huber (1991, 1992) to propose alternative Danian biostratigraphic zonations for the Austral Realm. The discrepancy between Huber's (1991) and Keller's (1993) biostratigraphic results with respect to *P. eugubina* lies in their differing species concepts, with Keller (1993) assigning this name to morphotypes consistent with Luterbacher and Premoli-Silva's (1964) original description and Huber (1991, along with Stott and Kennett, 1990) following Smit's (1977, 1982) substitution of a morphotypic concept de-

TABLE 1.—ELEMENTAL ABUNDANCES AND RATIOS FOR SITE 690C PGE DATA†

Sample	Ir (ppt)	Co (ppt)	Cr (ppt)	Fe (ppt)	Ir/Co	Ir/Cr	Ir/Fe
690C-15X-4, 35–36	467	1.57E + 07	5.80E + 06	2.52E + 10	2.98E − 05	8.05E − 05	1.85E − 08
690C-15X-4, 36–37	747	1.62E + 07	7.50E + 06	2.53E + 10	4.62E − 05	9.96E − 05	2.95E − 08
690C-15X-4, 37–38	1101	1.56E + 07	8.20E + 06	2.42E + 10	7.05E − 05	1.34E − 04	4.55E − 08
690C-15X-4, 38–39	1487	1.57E + 07	7.40E + 06	2.45E + 10	9.45E − 05	2.01E − 04	6.07E − 08
690C-15X-4, 39–40	1566	1.59E + 07	5.00E + 06	2.49E + 10	9.85E − 05	3.13E − 04	6.29E − 08
690C-15X-4, 40–41	983	1.18E + 07	1.09E + 07	1.74E + 10	8.34E − 05	9.02E − 05	5.65E − 08
690C-15X-4, 41–42	1237	1.46E + 07	1.20E + 07	2.40E + 10	8.50E − 05	1.03E − 04	5.15E − 08
690C-15X-4, 42–43	602	9.62E + 06	—	1.44E + 10	6.26E − 05	—	4.18E − 08
690C-15X-4, 43–44	378	1.00E + 07	—	1.51E + 10	3.78E − 05	—	2.50E − 08
Black Shale*	181				3.28E − 06	2.64E − 06	3.00E − 09
Chondrite**	473				9.30E − 04	1.80E − 04	2.60E − 06

†Data from Michel and others, 1990.
*Average of six values from Devonian/Carboniferiferous Black Shales; from Wang and others, 1993.
**Values from Anders and Ebihara, 1982.

rived from Blow's (1979) *Globorotalia longiapertura* for *P. eugubina* (see MacLeod, in press a). Recognition of *P. eugubina* s.s. in the lowermost Danian interval from Site 738 enables the standard low latitude Danian zonation to be extended into the high southern latitudes. Watkins and others (1994) demonstrate a similar extension of tropical nannofossil zones into this same region. Additionally, the coeval absence of *P. eugubina* from Site 690C also implies the existence of climatologically and oceanographically important restrictions in austral surface ocean circulation patterns during the lower Danian interval.

Third, Keller (1993) found several taxa whose FADs were previously thought to occur in the lowermost Danian in association with upper Maastrichtian faunas below the Site 738 Ir anomaly. These include *Chiloguembelina waiparaensis*, *Chiloguembelina crinita*, and *Guembelitria trifolia*. Several other Tertiary taxa whose FADs were thought to occur higher up in the Danian were also found associated with lower Danian faunas at Site 738 (e.g., *Igorina spiralis*, *Mucriglobigerina chasconona*, and *M. aquiensis*). These data are consistent with the results of a comparative biogeographic analysis for the trans-K/T planktic foraminiferal fauna as a whole (MacLeod and Keller, 1994), and suggest that the high southern (and northern) latitudes acted as centers of speciation for Danian planktic foraminifera.

Turning back to the correlation between the Site 738 biostratigraphic data and the revised LD-CS (Fig. 7), as with Site 690 (Fig. 6), an Ir anomaly exists within this core at the base of a grey shale layer, coincident with the FAD of Danian microfossils. This anomaly is not associated with the characteristic red layer present at the El Kef stratotype, Nye Kløv, Agost, Caravaca, and the Brazos River sections, shocked quartz, microtektites, Ni-rich spinels, or any other lithostratigraphic indicator of the boundary event. As was the case at Site 690, the simple presence of an Ir anomaly by itself constitutes insufficient evidence for its direct derivation from a impact event. Having made these points, however, there remains ample reason to believe that, unlike Site 690, the Ir anomaly at Site 738 occupies a position coincident with the chronostratigraphic K/T boundary at the base of a lowermost Danian interval that is temporally complete. This obtains not because of the simple presence of an Ir anomaly within the sequence, but rather because of the stratigraphic association between the Ir anomaly and available biotic data.

Both *Eoglobigerina fringa* [11] and *G. conusa* [25] occur within the first centimeter above the Ir anomaly (see Fig. 7). These observations are consistent with *G. conusa*'s FAD at the K/T boundary in the El Kef stratotype (Ben Abdelkader and others, 1992; Keller and others, 1994) and indicate that *E. fringa* FAD [11], located 25 cm above the K/T boundary at El Kef, coincides with the boundary in at least some austral habitats. A series of Danian FADs are spread between 1 and 8 cm above the Ir anomaly (e.g., *Eoglobigerina simplicissima* [13], *Eoglobigerina danica* [6]). At 8 cm, eight different FADs, including that of the Zone P0/P1a marker *P. eugubina* [38], occur simultaneously. This 8-cm interval between the Ir anomaly/FAD of Danian taxa and the *P. eugubina* FAD contains all of the necessary criteria for unambiguous recognition of a the Zone P0 chronozone.

Just as there seems little doubt that Site 738 is complete across the K/T boundary and into the lowermost Danian, there also seems little doubt that a substantial hiatus occurs 8 cm above the Ir anomaly and FAD of Danian taxa. Taxa simultaneously appearing at this horizon in the Site 738 core span an interval of almost 350 cm in the El Kef boundary stratotype and over 320 cm in the revised LD-CS (Appendix 1). The distance separating the FAD and LAD of *P. eugubina* (= duration of Zone P1a) is also dramatically truncated, dropping from 443 cm at El Kef to a mere 6 cm at Site 738. Aside from supporting the recognition of a major hiatus spanning almost all of Zone P1a in this deep-sea site, the latter observation also suggests that the hiatus extends between the uppermost portion of Zone P1a down at least to the Zone P0/P1a boundary and quite possibly well into the underlying Zone P0. Overlying the hiatus, sediment accumulation rates in the upper portion of Zone P1a and into Zone P1b appear to have been rapid and fairly constant with 200 cm of accumulation within the LD-CS being spread over an interval of more that 550 cm at Site 738. As at Site 690, this correlation is well-supported by the available biotic data with over 68% of the datums plotting on or close to the LOC.

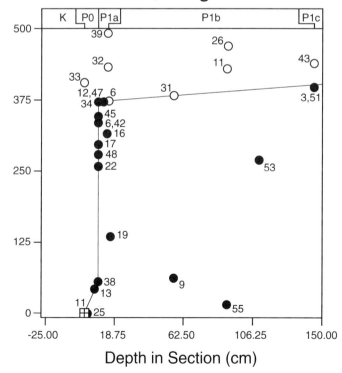

FIG. 7.—Graphic correlation diagram for the ODP Site 738, Hole 738C biostratigraphic data (Appendix 1) plotted against corresponding datum positions in the revised LD-CS. Numbers refer to datums listed in Appendix 1. See Figure 2 for symbol conventions.

Calibration of the LD-CS

Perhaps the primary advantage of graphic correlation over traditional zone-based approaches to biostratigraphic analysis is the rigor with which correlations can be studied. Unlike many quantitative stratigraphic techniques, though, graphic correlation is flexible in the sense that it allows the user a wide range of choices in terms of how the data are analyzed and what type of additional information may be used to constrain and/or calibrate the correlation. One obvious example of this flexibility lies in the method used to calibrate the CS and, via reference to individual LOC's, the correlation models. In traditional zone-based biostratigraphy, dates for zone boundaries are extrapolated from radiometrically-dated horizons under the assumptions of temporal continuity, constant sediment accumulation rate, and isochronous occurrence of the biotic datums. Much of the stratigraphic, oceanographic, and paleobiologic literature of the last several decades can be marshalled to support the contention that, in the vast majority of cases, none of these assumptions are likely to be true (see Shaw, 1964; Miller, 1977; Edwards, 1984, 1989; Hazel and others, 1984; Pisias and others, 1984; Pisias and others, 1985; Prell and others, 1986; Dowsett, 1988, 1989; Hazel, 1989; Johnson and others, 1989; MacLeod, 1991; MacLeod and Keller, 1991a, b; MacLeod, in press b; and references therein). Even when chemostratigraphy or magnetostratigraphy are used in conjunction with biozone-based biostratigraphy, the results, though often better than those based on biostratigraphy alone, are compromised by the circularity of using the biozones themselves to identify and correlate individual magnetochrons and chemical abundance peaks (Hailwood, 1989).

Graphic correlation circumvents the inherent imprecision of the zone-based approach to chronostratigraphic inference by enabling individual biotic datums to be used in the construction of an event-based biostratigraphy and then allowing this general framework to incorporate information from other stratigraphic events in order to corroborate, refute, or modify the original correlation model (see Edwards, 1989). In this way, graphic correlation can serve as a general reference system for many forms of stratigraphic analysis and all forms of stratigraphic correlation. Since radiometric techniques are used to date individual stratigraphic events (e.g., deposition of a bed containing a radiometric minerals), these events can be represented on a graphic correlation diagram and used to calibrate the spatial scale of the CS.

This calibration, of necessity, makes the same assumptions of temporal continuity, accumulation rate constancy, and datum isochrony characteristic of zone-based biostratigraphy. But, these assumptions are much more likely to be fulfilled within a composite standard reference section than within any single stratigraphic succession. If defined correctly, the standard reference section, upon which the composite is based, was selected to represent the longest and most continuous sequence available, and individual datum levels within this sequence have undergone a series of adjustments to align them with each species' global FADs and LADs. These horizons, which are isochronous by definition (see Hedberg and others, 1976) form the basis of the graphic correlations discussed herein.

Both the Harland and others (1989) and Cande and Kent (1992) time scales list the age of the K/T boundary at 65 Ma. The estimated duration of the lowermost Danian planktic foraminiferal biozone (P0) is 40,000 years (see Berggren and others, 1985; MacLeod and Keller, 1991b; D'Hondt and Herbert, 1991). Using these two datums and their assigned ages, linear interpolation allows dates to be assigned to the LD-CS datums as shown in Table 2. This method infers a duration of 350,000 years for the Zone P1a which is also in broad agreement with the previous estimates cited above. Calibration of the LD-CS with radiometrically-inferred ages for established biostratigraphic horizons enables a wide variety of sedimentologic and stratigraphic parameters to be estimated, including sediment accumulation rates, hiatus durations, and age intervals.

Naturally, these estimates are only as good as the dates upon which they are based, and no attempt is made here to provide

TABLE 2.—RANK ORDER OF LD-CS DATUMS WITH AGE ESTIMATES

Rank	No.*	Datum	Type	LD-CS	Inferred Age
1	63	Ir-1	KBD	0	65.000
2	64	Red Layer	KBD	0	65.000
3	11	Eoglobigerina fringa	FAD	0	65.000
4	25	Globoconusa conusa	FAD	0	65.000
5	57	B. sparsus	FAD	0	65.000
6	55	Woodringina hornerstownensis	FAD	17	64.988
7	59	B. romeini	FAD	20	64.986
8	7	Eoglobigerina edita	FAD	45	64.969
9	13	Eoglobigerina simplicissima	FAD	46	64.968
10	38	Parvularugoglobigerina eugubina	FAD	57	64.960
11	40	Parvularugoglobigerina longiapertura	FAD	63	64.956
12	9	Eoglobigerina eobulloides	FAD	64	64.955
13	23	Globastica sp.	FAD	84	64.941
14	2	Chiloguembelina midwayensis	FAD	137	64.904
15	19	Globanomalina taurica	FAD	137	64.904
16	21	Globanomalina planocompressa	FAD	157	64.890
17	22	Globastica daubjergensis	FAD	262	64.817
18	53	Woodringina claytonensis	FAD	273	64.809
19	48	Subbotina triloculinoides	FAD	282	64.803
20	17	Globanomalina pentagona	FAD	302	64.789
21	15	Eoglobigerina trivialis	FAD	320	64.776
22	5	Eoglobigerina danica	FAD	341	64.761
23	42	Planorotalities compressus	FAD	341	64.761
24	45	Subbotina pseudobulloides	FAD	350	64.755
25	6	Eoglobigerina danica	LAD	377	64.736
26	12	Eoglobigerina fringa	LAD	377	64.736
27	46	Subbotina triangularis	FAD	377	64.736
28	43	Subbotina moskvini	FAD	378	64.735
29	31	Guembelitria trifolia	LAD	388	64.728
30	51	Subbotina varianta	FAD	401	64.719
31	3	Chiloguembelina moresi	FAD	402	64.719
32	20	Globanomalina tetragona	FAD	402	64.719
33	33	Hedbergella monmouthensis	LAD	411	64.712
34	27	Globoconusa extensa	FAD	420	64.706
35	34	Igorina spiralis	FAD	420	64.706
37	37	Mucriglobiferina aquiensis	FAD	420	64.706
37	36	Mucriglobiferina chasconona	FAD	433	64.697
38	10	Eoglobigerina eobulloides	LAD	434	64.696
39	29	Globigerinelloides caravacaensis	LAD	438	64.693
40	32	Hedbergella holmdelensis	LAD	438	64.693
41	35	Morozovella inconstans	FAD	440	64.692
42	44	Subbotina moskvini	LAD	442	64.691
43	49	Subbotina triloculinoides	LAD	455	64.682
44	47	Subbotina triangularis	LAD	456	64.681
45	30	Guembelitria cretacea	LAD	459	64.679
46	28	Globoconusa extensa	LAD	462	64.677
47	14	Eoglobigerina simplicissima	LAD	472	64.670
48	26	Globoconusa conusa	LAD	477	64.666
49	1	Chiloguembelina crinita	LAD	479	64.665
50	4	Chiloguembelina moresi	LAD	479	64.665
51	24	Globigerinelloides monmouthensis	LAD	479	64.665
52	41	Parvularugoglobigerina longiapertura	LAD	479	64.665
53	54	Woodringina claytonensis	LAD	479	64.665
54	16	Eoglobigerina trivialis	LAD	487	64.659
55	52	Subbotina varianta	LAD	487	64.659
56	56	Woodringina hornerstownensis	LAD	487	64.659
57	39	Parvularugoglobigerina eugubina	LAD	500	64.650
58	58	B. parvulum	FAD	538	64.623
59	60	C. primus	FAD	570	64.601
60	62	T. petalosus	FAD	700	64.510
61	50	Subbotina trinidadensis	FAD	783	64.452
62	18	Globanomalina pentagona	LAD	921	64.355
63	8	Eoglobigerina edita	LAD	950	64.335
64	61	C. primus	LAD	1000	64.300

*Refers to Appendix 1.

error estimates either on the originally dated layers or on the stratigraphic horizons defined by the Danian datums under consideration. Any reasonable estimate of such error would encompass an interval many times larger than the lower Danian. Nevertheless, it is wrong to conclude that simply because the errors on these radiometrically-inferred dates are relatively large this fact in some way compromises the biostratigraphic results embodied in the LD-CS. Two of Shaw's (1964) most valuable insights were that the quality of a stratigraphic correlation depends entirely upon the quality of the data used to constrain or support it, and that the most direct evidence bearing on correlation questions will always be spatial (= observational) rather than temporal (= inferential). As such, the results discussed above, along with the correlations themselves, remain valid regardless of our confidence in the numerical ages that might be assigned to them by one method or another. Using these preliminary age estimates to scale the correlations presented in Figures 3 thru 7, a diagrammatic representation of temporal completeness with the lowermost Danian interval for Nye Kløv, Millers Ferry, Mimbral, and ODP Sites 690m and 738 is shown in Figure 8.

DISCUSSION

Application of graphic correlation to K/T boundary stratigraphy has precipitated controversy over technical aspects of the technique itself, the original database used by MacLeod and Keller (1991a, b) to make temporal completeness estimates, and the conformance of MacLeod and Keller's (1991a, b) eustatic model for global sediment accumulation during the lower Danian with other data. Several of the concerns raised by various authors have been individually addressed in previous publications (see MacLeod and Keller, 1994; MacLeod, in press a, b; MacLeod and Sadler, this volume). Instead of reiterating those arguments here, it will be more instructive to focus on general aspects of the application of graphic correlation to high-resolution stratigraphic problems in addition to tracing the implications these new correlations have for the ongoing K/T boundary controversy.

Much of the resistance to the application of graphic correlation to the high-resolution analysis of K/T chronostratigraphy stems from a basic unfamiliarity with details of the technique. In several instances, this unfamiliarity is also coupled to a lack of appreciation for assumptions regarding the nature of the stratigraphic record that are required by a traditional, zone-based approach to biostratigraphy but rendered superfluous by the graphic correlation paradigm. This volume attempts to address the former via explanation and example. However, to suggest that graphic correlation has failed to realize its potential due solely to a dearth of adequate technical descriptions and worked examples is both erroneous and misleading. Excellent descriptions of graphic correlation, beginning with Shaw's (1964) treatise, along with detailed treatments of both simulated and actual data have been repeatedly published with negligible effect. In my view, graphic correlation has been considered and rejected by many members of the stratigraphic community specifically because of the worldview it requires. If taken seriously, the assumptions made by graphic correlation challenge those upon which many stratigraphers base their analyses. Ultimate acceptance of graphic correlation then will require a Kuhnian paradigm shift within large segments of the stratigraphic community with regard to several basic concepts.

Nowhere is the clash of assumptions between practitioners and opponents of graphic correlation more evident than on the subject of the supposed isochrony of biostratigraphic datums. Despite the fact that both the ISSC and the NACSN categorically state that it is inappropriate to regard either biostratigraphically-defined or magnetostratigraphically-defined datums as isochronous (see Hedberg, 1976; NACSN, 1982), the stratigraphic literature is replete with analyses that make precisely those assumptions. In certain instances (e.g., creation of a Phanerozoic geologic time scale), this practice can be justified on the grounds that such assumptions are required to successfully complete the task and that the scope of the analysis is sufficiently long to render errors due to possible intra-datum diachrony inconsequential with respect to the whole. Such appeals, however, cannot suffice for high-resolution investigations whose temporal scope approaches the level at which intra-datum diachrony may play a decisive role in stratigraphic interpretations.

Take, for example, Olsson and Liu's (1993, p. 136) contention that any biostratigraphic datum shown to be diachronous must be axiomatically regarded as reworked because, "its last occurrence should be approximately synchronous at . . . localities [within the same climatic belt]." Certainly, post-depositional movement of fossil faunas must result in artificial range extension. To regard the mere fact of biostratigraphic datum diachrony as sufficient grounds upon which to base an interpretation of reworking, though, betrays a concept of biostratigraphy that is incompatible with the experience of those who practice graphic correlation. As was first shown by Shaw (1964), the simple exercise of constructing a two-dimensional plot of coeval biostratigraphic data from different sections should be sufficient to convince anyone that intra-datum diachrony is routinely encountered in sequences within which there is no independent evidence of reworking. This can easily be demonstrated using Liu and Olsson's (1992) own data which shows the LAD of the distinctive K/T survivor species *G. cretacea* to occur 174 cm above the LAD of *P. eugubina* at Millers Ferry whereas it occurs 41 cm below this datum in the El Kef stratotype section (see Appendix 1). According to Olsson and Liu's (1993) argument, diachrony so extreme as to place a datum in different biozones must be explained as a result of reworking, in which case their interpretation of *G. cretacea* as a K/T survivor species at Millers Ferry should be questioned. The only alternative is to admit that substantial biostratigraphic diachrony can arise as a natural consequence of many evolutionary, biogeographic, and ecologic factors. However, under this interpretation, Olsson and Liu's (1993) reason for regarding most "Cretaceous" taxa routinely recovered from Danian sediments as reworked faunal constituents loses its basis of support. So long as highly questionable assumptions such as intra-datum isochrony are recognized as acceptable stratigraphic practice, graphic correlation will not make serious inroads into stratigraphic thinking because such practices deny the rationale for preferring graphic correlation to traditional (bio)stratigraphic methods.

Aside from a failure to accept the paradigmatic aspects of graphic correlation, many traditional stratigraphers also remain unsure of the relationship between graphic and traditional zone-

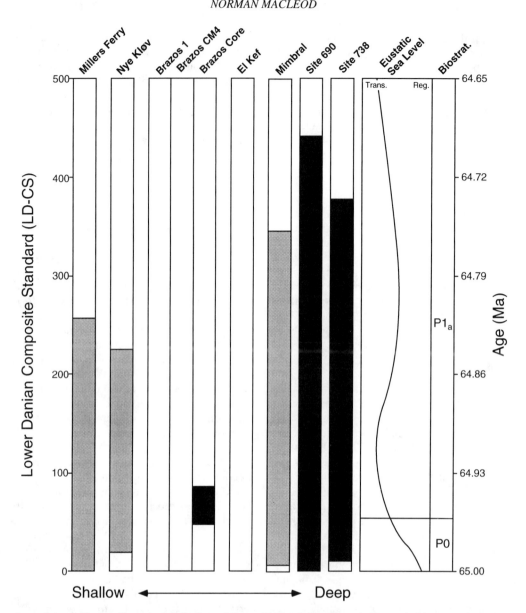

Fig. 8.—Temporal completeness diagram illustrating implications of the graphic correlations shown in Figures 2–7 in terms of local lower Danian sequences of events. White intervals = net positive sediment accumulation under both best-case and worst-case correlation models. Grey intervals = net positive sediment accumulation under best case correlation model. Black intervals = hiatus duration under both best case and worst case correlation models. Note, as predicted by the MacLeod and Keller (1991a, b) eustatic sea-level model, inner and middle sequences (El Kef, Brazos River, Nye Kløv) record complete lowermost Danian records while the very shallow successions (Millers Ferry) along with middle bathyal cores (ODP sites 690 and 738) contain extensive lower Danian hiatuses at or near the K/T boundary. These patterns appear to result from migrations of sediment deposition loci that resulted from a eustatic sea-level rise following an upper Maastrichtian sea-level lowstand. Accordingly, biotic, sedimentologic, and geochemical data based exclusively on deep-sea stratigraphic records appear biased towards instantaneous change at the K/T boundary. Boundary sequences from inner and middle neritic settings typically record a gradual pattern of biotic, sedimentological, and geochemical change extending over a considerable interval. See text for discussion.

based correlations. Within the debates surrounding estimates of K/T boundary chronostratigraphy, this is exemplified by several authors' confusion regarding the nature of evidence for widespread boundary hiatuses in deep-sea sections and cores. In a recent paper regarding breakdown in the $\delta^{13}C$ surface-to-deep gradient above the K/T boundary, Ivany and Salawitch (1993a) argue that the abruptness and magnitude of this decline in deep-sea cores indicate that 25% of the earth's biomass was burned on a time scale less than that of ocean mixing (approximately 1000 years). When, in a later comment (Keller and MacLeod, 1993), it was pointed out that all of the cores used by Ivany and Salawitch (1993a) contained evidence for a boundary hiatus, Ivany and Salawitch (1993b) responded that (1) graphic correlation, which was implicated as the basis for our recognition of a deep-sea boundary hiatus, is only precise when long periods of time spanning many complete range zones are considered, and (2) in those instances in which the lowermost Zone P0 was shown to be absent in deep-sea localities, graphic correlation would be unable to distinguish between bioturbation and a true hiatus.

With regard to the former statement, there is, to my knowledge, no publication by any experienced practitioner (or critic) of graphic correlation containing the suggestion that this techniques' accuracy is in any way linked to the temporal duration of the interval over which it is applied. Indeed being an event-based rather than a zone based, method, all previous descriptions of graphic correlation have emphasized the superior resolution graphic correlation affords over traditional zone-based approaches to biostratigraphic analysis. Without exception, the available literature on graphic correlation states that it should be preferred to zone-based methods for the development of a high-resolution biotically-based chronostratigraphy.

As for graphic correlations' ability to distinguish between a biozone's absence due to bioturbation or hiatus, Ivany and Salawitch's statement is, for the most part, meaningless, not only because no method (graphical or otherwise) based solely on biotic data can make such a distinction, but also because, in terms of stratigraphic completeness, the distinction is irrelevant. Regardless of whether or not Zone P0 is absent from the deep sea as a result of a cessation of sediment accumulation, active erosion of previously-deposited sediments, or bioturbation, all biostratigraphers agree that the zone cannot be recognized (MacLeod and Keller, 1991a, b; Olsson and Liu, 1993). Sequences lacking a complete succession of biozones are, by definition, biostratigraphically incomplete. Simple logic then forces a strong suspicion of temporal incompleteness for successions that are demonstrated to be biostratigraphically incomplete. Furthermore, even if the lack of a recognizable Zone P0 were due to bioturbation, that process itself would have disrupted or destroyed lowermost Danian biotic, sedimentologic, and geochemical patterns to the extent that deterministic interpretations of such data (e.g., those attempted by Ivany and Salawitch, 1993a) could not be made with confidence. The sharpness of the carbon isotope deflections that characterize many of the cores studied by Ivany and Salawitch (1993a) alone provides compelling evidence against recognizing elevated levels of bioturbation in these deep-sea cores, as do the sedimentological site descriptions (e.g., Barker and others, 1988), and comparative analysis of isotopic data from deep sea and neritic sequences (MacLeod, in press a). This chain of reasoning has nothing whatsoever to do with the MacLeod and Keller (1991a, b) graphic correlation results insofar are these are all qualitatively consistent (and quantitatively more accurate) with previous zone-based biostratigraphic results. Rather, recognition of the hiatuses that characterize the K/T boundary in most deep-sea successions amounts to little more than the stratigraphic equivalent of common sense.

Both Olsson and Liu (1993) and Ivany and Salawitch (1993b) obscure the issues they address by misrepresenting graphic correlation and then dismissing results they (erroneously) claim to be based on its use. More serious, though, is the latent impression, given by both sets of authors, that appears to reject the idea that biotic data of any sort are capable of high-resolution stratigraphic analysis. These are the types of objections graphic correlation can and must overcome if it is to realize its full potential.

What do these latest analyses of K/T boundary sections and cores tell us about the nature of sediment accumulation in the lowermost Danian? The sections and cores in Figure 8 are ordered from left to right by increasing water depth. A clear pattern is evident in these data showing that sequences in very shallow depositional settings (Millers Ferry, which may have undergone a period of subareal exposure) and those representing outer neritic to upper bathyal environments (Mimbral, ODP Sites 690 and 738) are largely incomplete over the lowermost Danian interval whereas sequences representing inner-middle neritic environments (Nye Kløv, Brazos River, El Kef) exhibit largely continuous Lower Danian successions. These results are consistent with predictions of the MacLeod and Keller (1991a, b) eustatic model of trans-K/T sediment accumulation patterns

As eustatic sea level rose during the very uppermost Maastrichtian and into the lower Danian (Haq and others, 1987; Donovan and others, 1988; Baum and Vail, 1988; Habib and others 1992; Donovan and others, 1990; Savrda, 1993), the locus of sediment accumulation would have shifted from outer neritic and continental slope environments to the inner and middle portions of continental shelves. This would result in elevated rates of sediment accumulation in inner and middle neritic settings with consequent stratigraphic condensation and/or hiatus formation in the deep sea. The outer neritic and upper bathyal sequences of Mimbral and Site 738 do appear to exhibit a small Zone P0 intervals, but given available biostratigraphic data, it is by no means clear just how much time is represented by these relatively thin lowermost Danian intervals. What is clear is that these successions, in addition to the Site 690C sequence, are dominated by an extensive lowermost Danian hiatus comparable in magnitude to the hiatuses previously shown to occur at DSDP sites 577 and 528 (MacLeod and Keller, 1991a, b). Neither of the trans-K/T intervals in the southern high latitudes exhibit any independent evidence of extensive bioturbation. When these data are considered as a whole, they provide strong evidence in favor of eustatic sea-level rise as a determining factor in the organization of lowermost Danian sediment accumulation patterns. Contrary to Ivany and Salawitch (1993b, p. 1151) who argue that, "the dramatic changes in biological, sedimentological, and geochemical parameters documented at numerous sites worldwide are correlative and indicative of a single catastrophic event" (see also Ivany and Salawitch, 1993a), the changes they refer to are confined solely to sequences representing middle bathyal environments where the overall absence of sediment corresponding to the lowermost Danian interval presents a catastrophically biased sequence of events. High-resolution analyses of the more temporally complete neritic K/T boundary sections have consistently documented gradual patterns of faunal turnover with many instances of trans-K/T survivorship, continuous patterns of sediment accumulation (at both biozone and event levels of resolution), and gradually decreasing surface-to-deep $\delta^{13}C$ gradients extending over hundreds of thousands of years.

SUMMARY

The twin hallmarks of major conceptual advances in our understanding of natural phenomena are synthesis and prediction. Any serious model of trans-K/T environmental and biotic change must take into consideration all available data bearing on the issue and pass the test of making conjectures regarding uncollected data that can be and indeed are later verified via empirical analysis. By these criteria, the sequence stratigraphic or eustatic model of sediment accumulation across the K/T boundary must be regarded as constituting a substantial improvement in our understanding of this important interval in earth history.

By comparing available biotic, sedimentologic, and geochemical data from a wide variety of K/T boundary sequences rather than relying solely on the deep-sea record (Ivany and Salawitch, 1993a,b) or the sequence of events observed in a single boundary section (Liu and Olsson, 1992), the eustatic model constitutes a far more integrative approach to the study of trans-K/T events. Results of both traditional biozone-level and event-level biostratigraphic analyses via graphic correlation have uncovered consistently different biotic patterns between neritic and deep-sea boundary sequences, both of which indicate that deep-sea sequences are either highly condensed or demonstrably incomplete in the lowermost Danian. These results are in accordance with biotic data from a wide variety of other neritic and terrestrial biota, all of which document widespread K/T survivorship during an extended interval of faunal turnover. The alternative scenarios advanced by Liu and Olsson (1992), Olsson and Liu (1993), and most recently by Ivany and Salawitch (1993a,b) either ignore data from neritic sequences altogether or attempt to resolve the evident contradiction between data obtained from neritic and deep-sea settings through a series of *ad hoc* explanations (e.g., reworking of Cretaceous faunas into lower Danian sediments, absence of Zone P0 from deep-sea cores as a result of bioturbation) for which no independent evidence exists. In addition, these authors grossly misrepresent many aspects of the graphic correlation theory and practice.

Ongoing interest in K/T boundary events has resulted in the biostratigraphic sampling and analysis of sequences not available at the time of the original MacLeod and Keller (1991a, b) study. These new data provide an opportunity to test the predictive abilities of the eustatic model. Use of biostratigraphic data from only fully Danian and unquestioned K/T boundary survivor taxa to constrain the resultant correlations effectively frees the analysis from bias due to differing interpretations of Danian occurrences of traditionally "Cretaceous" taxa, while at the same time providing an external check on the extent to which the inclusion of these "Cretaceous" datums influenced the results of the previous study.

Graphic correlation of Danian biostratigraphic data from Nye Kløv, Millers Ferry, Mimbral, ODP Site 690, and ODP Site 738 with a revised Danian datum-based LD-CS overwhelmingly confirms predictions of geographically alternating sediment accumulation patterns described in the original MacLeod and Keller (1991a, b) eustatic model as well as adding several important new dimensions to the understanding of K/T events. Neritic and upper bathyal sequences from Nye Kløv, Brazos River, and El Kef appear to be temporally complete throughout the lowermost Danian (Zone P0) while Mimbral and the two deep-sea cores contain evidence for substantial hiatuses during this same time period. The Millers Ferry section also appears to be incomplete across the lowermost Danian interval. This result is consistent with Millers Ferry's shallow depositional setting and the presence of a regional boundary unconformity throughout the northern Gulf of Mexico that, in turn, is tied to the uppermost Maastrichtian sea-level lowstand.

While all boundary sequences judged temporally complete in this study exhibit multiple sedimentologic and geochemical markers associated with the K/T boundary at the El Kef stratotype as well as many other (mostly neritic) sections, the Ir anomaly at Site 690 appears to be associated with a lithologic and environmental contrast across a boundary unconformity. This interpretation is supported by non-chondritic PGE abundance ratios in the boundary layer itself, as well as the association of the anomaly with a dark shale layer indicating a change in redox potential across the local K/T boundary. Recent geochemical investigations have shown that PGEs can be artificially concentrated by changes in redox potential in addition to other natural processes. Operation of natural processes that may concentrate Ir and other PGEs at contrasting lithologic/geochemical horizons in other boundary sections might additionally serve to account for multiple Ir anomalies reported from several K/T boundary sections and cores along with the presence of Ir anomalies in sections widely recognized to contain a boundary hiatus (e.g., Gubbio, Italy). In light of these recent analyses, the entire issue of a causal association between Ir anomalies and bolide impacts should be re-evaluated.

Finally, with respect to the ultimate question of the proposed causal connection between a bolide impact and biotic extinctions at the close of the Cretaceous, these data suggest that the magnitude and timing of these events cannot be properly evaluated in deep-sea settings. These results are not inconsistent with the recognition of a bolide impact as having occurred at or near the close of the Cretaceous although they do not lend any support to that scenario either. Available data are also consistent with recognition of a mass extinction having occurred across the K/T boundary, provided the term "mass extinction" refers to an interval of accelerated faunal turnover that exceeds background levels by a substantial margin. Results presented herein are clearly inconsistent, however, with the more extreme scenario that postulates the extinction of a large proportion of planktic foraminiferal species over an exceedingly small interval of time as a result of the proximal effects of a bolide impact. This scenario fails to account for a large number of physical data and biotic observations in both ancient and modern faunas and should be abandoned as a plausible model of trans-K/T events.

ACKNOWLEDGEMENTS

The author gratefully acknowledges Gerta Keller for providing most of the data upon which this analysis is based; Gilbert Klapper for constructive criticism and informative discussions on aspects of the graphic correlation technique; Steve Culver, Keith Mann, Ronald Martin, and an anonymous reviewer for reading and commenting on an earlier version of the manuscript; and Cecilia MacLeod for discussion on aspects of PGE geochemistry. This is a contribution from the Natural History Museum/University College London Global Change and the Biosphere Project.

REFERENCES

AGER, D. V., 1993, The New Catastrophism: The Importance of the Rare Event in Geological History: Cambridge, Cambridge University Press, 231 p.

ALVAREZ, L. W., ALVAREZ, F., ASARO, F., AND MICHEL, H. V., 1980, Extraterrestrial cause for the Cretaceous-Tertiary extinction: Science, v. 208, p. 1095–108.

ALVAREZ, W., ALVAREZ, L. W., ASARO, F., AND MICHEL, H., 1982, Current status of the impact theory for the terminal Cretaceous extinction, *in* Silver,

L. T. and Schultz, P. H., eds., Geological Implications of Impacts of Large Asteroids and Comets on Earth: Boulder, Geological Society of America Special Paper 190, p. 305–315.

ALVAREZ, W., SMIT, J., LOWRIE, W., ASARO, F., MARGOLIS, S. V., CLAEYS, P., KASTNER, M., AND HILDEBRAND, A. R., 1992, Proximal impact deposits at the Cretaceous-Tertiary boundary in the Gulf of Mexico: A restudy of DSDP Leg 77 Sites 536 and 540: Geology, v. 20, p. 697–700.

ANDERS, E. AND EBIHARA, M., 1982, Solar system abundances of the elements: Geochimica et Cosmochimica Acta, v. 46, p. 2263–2380.

ARCHIBALD, J. D. AND CLEMENS, W. A., 1982, Late Cretaceous extinctions: American Scientist, v. 70, p. 377–385.

ARCHIBALD, J. D. AND CLEMENS, W. A., 1984, Mammal evolution near the Cretaceous-Tertiary boundary, in Berggren, W. A. and Vancouvering, J. A., eds., Catastrophes and Earth History: The New Uniformitarianism: Princeton, Princeton University Press, p. 229–371.

ASKIN, R. A., 1990, The palynological record across the Cretaceous/Tertiary transition on Seymour Island, Antarctica, in Feldmann, R. M. and Woodburne, M. O., eds., Geology and Paleontology of Seymour Island, Antarctic Peninsula, Geological Society of America Memoir, v. 169, p. 131–156.

ASKIN, R. A., 1992, Preliminary palynology and stratigraphic implications from a new Cretaceous-Tertiary boundary section from Seymour Island: Antarctic Journal of the United States, v. 25, p. 42–44.

BARKER, P. F., KENNETT, J. P., and others, 1988, Introduction and objectives: Proceedings of the Ocean Drilling Program: Initial Reports, v. 113, p. 1–11.

BAUM, G. R. AND VAIL, P. R., 1988, Sequence stratigraphic concepts applied to Paleoogene outcrops, Gulf and Atlantic basins, in Wilgus, C, K., Hastings B. S., Kendall, C. G., Posamentier, H. W., Ros, Ch., Van Wagoner, J. C., eds., Sea Level Changes—An Integrated Approach: Tulsa, Society of Economic Paleontologists and Mineralogists Special Publication 42, p. 309–327.

BEN ABDELKADER, O., HAJ ALI, N. B., BEN SALEM, H., AND RAZGALLAH, S., 1992, International Workshop on Cretaceous-Tertiary Transitions (El Kef Section, Part II, Field Trip Guidebook: Tunisia, IUGS/GSGP/ATEIG/ Tunisian Geological Survey, 25 p.

BERGER, W. H., 1970, Planktic foraminifera: Selective solution and the lysocline: Marine Geology, v. 8, p. 111–138.

BERGER, W. H. AND WINTERER, E. L., 1974, Plate stratigraphy and the fluctuating carbonate line, in Hsü, K. J. and Jenkins, H., eds., Pelagic Sediments on Land and in the Ocean: New York, International Association of Sedimentology, p. 11–48.

BERGGREN, W. A., KENT, D. V., AND FLYNN, J. J., 1985, Jurassic to Paleogene: Part 2, Paleogene geochronology and chronostratigraphy, in Snelling, N. J., eds., The Chronology of the Geologic Record: London, Geological Society of London Memior 10, p. 141–195.

BLOW, W. H., 1979, The Cainozoic Globigerinida: Leiden, Brill, 1409 p.

BOURGEOIS, J., HANSEN, T. A., WIBERG, P. L., AND KAUFFMAN, E. G., 1988, A tsunami deposit at the Cretaceous-Tertiary boundary in Texas: Science, v. 241, p. 557–570.

BRAMLETTE, M. N., 1965, Massive extinctions in biota at the end of Mesozoic time: Science, v. 148, p. 1696–1699.

BRAMLETTE, M. N. AND MARTINI, E., 1964, The great change in calcareous nannoplankton fossils between the Maastrichtian and Danian: Micropaleontology, v. 10, p. 291–322.

BRINKHUIS, H. AND ZACHARIASSE, W. J., 1988, Dinoflagellate cysts, sea level changes and planktonic foraminifers across the Cretaceous-Tertiary boundary at El Haria, northwest Tunisia: Marine Micropaleontology, v. 13, p. 153–191.

BROUWERS, E. M. AND DE DECKKER, P., 1993, Late Maastrichtian and Danian ostracode faunas from northern Alaska: Reconstructions of environment and paleogeography: PALAIOS, v. 8, p. 140–154.

BUZAS, M. A., 1990, Another look at confidence limits for species proportions: Journal of Paleontology, v. 64, p. 842–843.

CANDE, S. C. AND KENT, D. V., 1992, A new geomagnetic polarity time scale for the Late Cretaceous and Cenozoic: Journal of Geophysical Research, v. 97, p. 13917–13951.

CANUDO, J. J., KELLER, G., AND MOLINA, E., 1991, Cretaceous/Tertiary boundary extinction pattern and faunal turnover at Agost and Caravaca, S. E. Spain: Marine Micropaleontology, v. 17, p. 319–341.

CHARIG, A. J., 1989, The Cretaceous-Tertiary boundary and the last of the dinosaurs: Philosophical Transactions of the Royal Society of London, Series B, v. 325, p. 387–400.

CLEMENS, W. A. AND NELMS, L. G., 1993, Paleoecological implications of Alaskan terrestrial vertebrate fauna in latest Cretaceous time at high paleolatitudes: Geology, v. 21, p. 503–506.

COLODNER, D. C., BOYLE, E. A., EDMOND, J. M., AND THOMSON, J., 1992, Post-depositional mobility of platinum, iridium and rhenium in marine sediments: Nature, v. 358, p. 402–404.

DENHAM, C. R. AND SCOTESE, C. R., 1987, Terra Mobilis': A plate tectonic program for the Macintosh, version 1.1: Austin, Geoimages, 26 p.

D'HONDT, S. L. AND HERBERT, T. D., 1991, Marking time at the South Atlantic Cretaceous-Tertiary boundary: Geological Society of America Abstracts with Programs, v. 17, p. A179.

D'HONDT, S. L. AND KELLER, G., 1991, Some patterns of planktic foraminiferal assemblage turnover at the Cretaceous-Tertiary boundary: Maine Micropaleontology, v. 17, p. 77–118.

DISSANAYAKE, C. B. AND KRITSOTAKIS, K., 1984, The geochemistry of Au and Pt in peat and algal mats: A case study from Sri Lanka: Chemical Geology, v. 42, p. 61–76.

DONCE, P., JARDINE, S., LEGOUX, O., MASURE, E., AND MEON, H., 1985, Les évènements à la limite Crétacé-Tertiaire: au Kef (Tunisie septentrioale), l'analyse palynoplactogique montre qu'un changement cliatique est décelable à labase du Danian: Actes du Premier Congrès National des Sciences de la Terre, Tunis, v. Sept. 1, 1981, p. 161–169.

DONOVAN, A. D., BAUM, G. D., BLECHSCHMIDT, G. L., LOUTIT, T. S., PFLUM, C. E., AND VAIL, P. R., 1988, Sequence stratigraphic setting of the Cretaceous-Tertiary boundary in Central Alabama, in Wilgus, C, K., Hastings B. S., Kendall, C. G., Posamentier, H. W., Ros, Ch., Van Wagoner, J. C., eds., Sea-Level Changes: An Integrated Approach, Society of Economic Paleontologists and Mineralogists Special Publication 42, p. 299–307.

DONOVAN, A. D., LOUTIT, T. S., AND GREENLEE, S. M., 1990, Looking at the forest as well as the trees: The sequence stratigraphic setting of the K/T boundary in the southern United States: Geological Society of America, Abstracts with Programs, v. 22, p. A279.

DOWSETT, H. J., 1988, Diachrony of Late Neogene microfossils in the southwest Pacific Ocean: Application of the Graphic Correlation method: Paleoceanography, v. 3, p. 209–222.

DOWSETT, H. J., 1989, Application of graphic correlation to Pliocene marine sequences: Marine Micropaleontology, v. 14, p. 3–32.

EDWARDS, L. E., 1984, Insights on why graphic correlation (Shaw's method) works: Journal of Geology, v. 92, p. 583–597.

EDWARDS, L. E., 1989, Supplemented graphic correlation: A powerful tool for paleontologists and nonpaleontologists: PALAIOS, v. 4, p. 127–143.

FELDMANN, R. M., 1990, On impacts and extinction: biological solutions to biological problems: Journal of Paleontology, v. 64, p. 151–154.

GOODWIN, P. W. AND ANDERSON, E. J., 1985, Punctuated aggradational cycles: A general hypothesis of episodic stratigraphic accumulation: Journal of Geology, v. 93, p. 515–533.

GOODWIN, P. W., ANDERSON, E. J., GOODMAN, W. M., AND SARAKA, L. J., 1986, Punctuated aggradational cycles: Implications for stratigraphic analysis: Paleoceanography, v. 1, p. 417–429.

GOULD, S. J., 1977, Eternal metaphors of paleontology, in Hallam, A., ed., Patterns of Evolution as Illustrated by the Fossil Record, Amsterdam, Elsevier, p. 1–26.

HABIB, D., MOSHKOVITZ, S., AND KRAMER, C., 1992, Dinoflagellate and calcareous nannoplankton fossil response to sea-level change in Cretaceous-Tertiary boundary sections: Geology, v. 20, p. 165–168.

HAILWOOD, E. A., 1989, The role of magnetostratigraphy in the development of geological times scales: Paleoceanography, v. 4, p. 1–18.

HAQ, B., 1991, Sequence stratigraphy, sea-level change and significance for the deep sea: International Association of Sedimentologists, Special Publication, v. 12, p. 3–39.

HAQ, B., HARDENBOL, J., AND VAIL, P. R., 1987, Chronology and fluctuating sea levels since the Triassic: Science, v. 235, p. 1156–1166.

HARLAND, W. B., ARMSTRONG, R. L., COX, A. V., CRAIG, L. E., SMITH, A. G., AND SMITH, D. G., 1989, A Geologic Time Scale 1989: Cambridge, Cambridge University Press, 263 p.

HAZEL, J. E., 1989, Chronostratigraphy of upper Eocene microspherules: PALAIOS, v. 4, p. 318–329.

HAZEL, J. E., EDWARDS, L. E., AND BYBELL, L. M., 1984, Significant unconformities and the hiatuses represented by them in the Paleogene of the Atlantic and Gulf Coastal province, in Schlee, J., ed., Interregional Unconformities: American Association of Petroleum Geologists Memoir, v. 36, p. 59–66.

HEDBERG, H., 1976, International Stratigraphic Guide: A Guide to Stratigraphic Classification, Terminology, and Procedure: New York, John Wiley & Sons, 200 p.

HEINBERG, C., 1979, Bivalves from the latest Maastrichtian of Stvens Klint and their stratigraphic affinities, in Birkelund, T. and Bromely, R. G., eds.,

Cretaceous-Tertiary Events Symposium, Vol. 1, The Maastrichtian and Danian of Denmark: Copenhagen, University of Copenhagen, p. 58–64.

HILDEBRAND, A. R. AND BOYNTON, W. V., 1990, Proximal Cretaceous-Tertiary boundary impact deposits in the Carribean: Science, v. 248, p. 843–846.

HILDEBRAND, A. R., PENFIELD, G. T., PILKINGTON, D. A., CAMARGO, Z. A., JACOBSEN, S. B., AND BOYNTON, W. V., 1991, Chicxulub crater: A possible Cretaceous-Tertiary boundary impact crater on the Yucatán Peninsula, Mexico: Geology, v. 19, p. 867–871.

HOLLIS, C. J., 1993, Latest Cretaceous to Late Paleocene radiolarian biostratigraphy: A new zonation from the New Zealand region: Marine Micropaleontology, v. 21, p. 295–327.

HUBER, B. T., 1991, Maestrichtian planktonic foraminiferal biostratigraphy and the Cretaceous/Tertiary boundary at ODP Hole 738C (Kerguelen Plateau): Proceedings of the Ocean Drilling Project, Scientific Results, v. 119, p. 451–465.

HUBER, B. T., 1992, Upper Cretaceous planktic foraminiferal biozonation for the Austral Realm: Marine Micropaleontology, v. 20, p. 107–128.

IVANY, L. C. AND SALAWITCH, R. J., 1993a, Carbon isotopic evidence for biomass burning at the K-T boundary: Geology, v. 21, p. 487–490.

IVANY, L. C. AND SALAWITCH, R. J., 1993b, Carbon isotopic evidence for biomass burning at the K-T boundary: Comment and Reply (Reply): Geology, v. 21, p. 1150–1151.

JÉHANNO, C., BOCLET, D., FROGET, L., LAMBERT, B., ROBIN, E., ROCCHIA, R., AND TURPIN, L., 1992, The Cretaceous-Tertiary boundary at Beloc, Haiti: No evidence for an impact in the Caribbean area: Earth and Planetary Science Letters, v. 109, p. 229–241.

JIANG, M. J. AND GARTNER, S., 1986, Calcareous nannofossil succession across the Cretaceous/Tertiary boundary in east-central Texas: Micropaleontology, v. 32, p. 232–255.

JOHNSON, D. A., SCHNEIDER, D. A., NIGRINI, C. A., CAULET, J.-P., AND KENT, D. V., 1989, Radiolarian events and magnetostratigraphic calibrations for the tropical Indian Ocean: Marine Micropaleontology, v. 14, p. 33–66.

JOHNSON, K. R. AND GREENWOOD, D., 1993, High-Latitude deciduous forests and the Cretaceous-Tertiary boundary in New Zealand: Geological Society of America, Abstracts with Programs, v. 25, p. A295.

KAUFFMAN, E. G., 1984, The fabric of Cretaceous extinctions, in Berggren, W. A. and Van Couvering, J. A., eds., Catastrophes and Earth History: The New Uniformitarianism, Princeton, Princeton University Press, p. 151–246.

KELLER, G., 1986, Stepwise mass extinctions and impact events: Late Eocene to Early Oligocene: Marine Micropaleontology, v. 10, p. 267–293.

KELLER, G., 1988a, Extinction, survivorship and evolution of planktic foraminifera across the Cretaceous/Tertiary boundary at El Kef Tunisia: Marine Micropaleontology, v. 13, p. 239–263.

KELLER, G., 1988b, Biotic turnover in benthic foraminifera across the Cretaceous-Tertiary boundary at El Kef, Tunisia: Palaeogeography, Palaeoclimatology, Palaeoecology, v. 66, p. 153–171.

KELLER, G., 1989, Extended Cretaceous/Tertiary boundary extinctions and delayed population change in planktonic foraminiferal faunas from Brazos River, Texas: Paleoceanography, v. 4, p. 287–332.

KELLER, G., 1993, The Cretaceous-Tertiary boundary transition in the Antarctic Ocean and its global implications: Marine Micropaleontology, v. 21, p. 1–46.

KELLER, G., BARRERA, E., SCHMITZ, B., AND MATTSON, E., 1993, Gradual mass extinction, species survivorship, and long-term environmental changes across the Cretaceous-Tertiary boundary in high latitudes: Geological Society of America Bulletin, v. 105, p. 979–997.

KELLER, G. AND BARRON, J. A., 1983, Paleoceanographic implications of Miocene deep-sea hiatuses: Geological Society of America Bulletin, v. 94, p. 590–613.

KELLER, G. AND BARRON, J. A., 1987, Paleodepth distribution of Neogene deep-sea hiatuses: Paleoceanography, v. 2, p. 697–713.

KELLER, G. AND BENJAMINI, C., 1991, Paleoenvironment of the eastern Tethys in the Early Paleocene: PALAIOS, v. 6, p. 439–464.

KELLER, G. AND LINDINGER, M., 1989, Stable isotope, TOC and $CaCO_3$ record across the Cretaceous/Tertiary boundary at El Kef, Tunisia: Palaeogeography, Palaeoclimatology, Palaeoecology, v. 73, p. 243–265.

KELLER, G. AND MACLEOD, N., 1993, Carbon isotopic evidence for biomass burning at the K-T boundary: Comment and Reply (Comment): Geology, v. 21, p. 1149–1150.

KELLER, G., MACLEOD, N., LYONS, J. B., AND OFFICER, C. B., 1993, Is there evidence for Cretaceous-Tertiary boundary-age deep-water deposits in the Carribean and Gulf of Mexico?: Geology, v. 21, p. 776–780.

KELLER, G., STINNESBECK, W., AND LOPEZ-OLIVA, J. G., 1994, Age, deposition, and biotic effects of the Cretaceous/Tertiary boundary event at Mimbral, NE Mexico: PALAIOS, v. 9, p. 144–157.

KITCHELL, J. A., CLARK, D. L., AND GOMBOS, A. M., 1986, Biological selectivity of extinction: A link between background and mass extinction: PALAIOS, v. 1, p. 504–511.

KNOBOCH, E., KVACEK, Z., BUZEK, C., MAI, D. H., AND BATTEN, D. J., 1993, Evolutionary significance of floristic changes in the northern hemisphere during the late Cretaceous and Palaeogene, with particular reference to central Europe: Review of Palaeobotany and Palynology, v. 78, p. 41–54.

LABANDIERA, C. C., 1992, Diets, diversity, and disparity: Determining the effect of the terminal Cretaceous extinction on insect evolution: Fifth North American Paleontological Convention, Abstracts with Programs, Special Publication of the Paleontological Society, v. 6, p. 174.

LERBEKMO, J. F., SWEET, A. R., AND ST. LOUIS, R. M., 1987, The relationship between the iridium anomaly and palynological floral events at three Cretaceous-Tertiary boundary localities in western Canada: Geological Society of America Bulletin, v. 99, p. 325–330.

LOUTTIT, T. S. AND KENNETT, J. P., 1981, Australasian Cenozoic sedimentary cycles, global sea-level change, and the deep-sea sedimentary record: Oceanologica Acta, v. Special Publication 1981, p. 46–63.

LIU, C. AND OLSSON, R. K., 1992, Evolutionary radiation of microperforate planktonic foraminifera following the K/T mass extinction: Journal of Foraminiferal Research, v. 22, p. 328–346.

LUTERBACHER, H. AND PREMOLI SILVA, I., 1964, Biostratigrafia del Limite Cretaceo-Terziario Nell'Appennino centrale: Rivista Italiana di Paleontologia e Stratigrafia, v. 70, p. 67–88.

LYONS, J. B. AND OFFICER, C. B., 1992, Mineralogy and petrology of the Haiti Cretaceous/Tertiary section: Earth and Planetary Science Letters, v. 109, p. 205–224.

MACLEOD, N., 1991, Punctuated anagenesis and the importance of stratigraphy to paleobiology: Paleobiology, v. 17, p. 167–188.

MACLEOD, N., in press a, Graphic correlation of high latitude Cretaceous-Tertiary boundary sequences at Nye Kløv (Denmark), ODP Site 690 (Weddell Sea), and ODP Site 738 (Kerguelen Plateau): Comparison with the El Kef (Tunisia) boundary stratotype: Modern Geology, v. 19.

MACLEOD, N., in press b, Cretaceous/Tertiary (K/T) biogeography of planktic foraminifera: Historical Biology, v. 8.

MACLEOD, N. AND KELLER, G., 1991a, Hiatus distributions and mass extinctions at the Cretaceous/Tertiary boundary: Geology, v. 19, p. 497–501.

MACLEOD, N. AND KELLER, G., 1991b, How complete are Cretaceous/Tertiary boundary sections? A chronostratigraphic estimate based on graphic correlation: Geological Society of America Bulletin, v. 103, p. 1439–1457.

MACLEOD, N. AND KELLER, G., 1994, Comparative biogeographic analysis of planktic foraminiferal survivorship across the Cretaceous/Tertiary (K/T) boundary: Paleobiology, v. 20, p. 143–177.

MANCINI, E. A., TEW, B. H., AND SMITH, C. C., 1989, Cretaceous-Tertiary contact, Mississippi and Alabama: Journal of Foraminiferal Research, v. 19, p. 93–104.

MCIVER, E. E., SWEET, A. R., AND BASINGER, J. F., 1991, Sixty-five-million-year-old flowers bearing pollen of the extinct triprojectate complex—a Cretaceous-Tertiary boundary survivor: Review of Paleobotany and Palynology, v. 70, p. 77–88.

MÉON, H., 1990, Palynologic studies of the Cretaceous/Tertiary boundary interval at El Kef outcrop, northwestern Tunisia: Review of Paleobotany and Palynology, v. 65, p. 85–94.

MICHEL, H. V., ASARO, F., ALVAREZ, W., AND ALVAREZ, L. W., 1990, Geochemical studies of the Cretaceous-Tertiary boundary in ODP Holes 689B and 690C: Proceedings of the Ocean Drilling Program, Scientific Results, v. 113, p. 159–168.

MILLER, F. X., 1977, The graphic correlation method in biostratigraphy, in Kauffman, E. G. and Hazel, J. E., eds., Concepts and Methods of Biostratigraphy: Stroudsburg, Dowden, Hutchinson and Ross, p. 165–186.

NACSN, 1982, North American Stratigraphic Code: American Association of Petroleum Geologists Bulletin, v. 67, p. 841–875.

OFFICER, C. B., DRAKE, C. L., PINDELL, J. L., AND MEYERHOFF, A. A., 1992, Cretaceous-Tertiary events and the Carribean caper: GSA Today, v. 2, p. 1–8.

OLSSON, R. K., HEMLEBEN, C., A., B. W., AND LIU, C., 1992, Wall texture classification of planktonic foraminifera genera in the Lower Danian: Journal of Foraminiferal Research, v. 22, p. 195–213.

OLSSON, R. K. AND LIU, C., 1993, Controversies on the placement of Cretaceous-Paleogene boundary and the K/P mass extinction of planktonic foraminifera: PALAIOS, v. 8, p. 127–139.

PERCH-NIELSEN, K., MCKENZIE, J., AND HE, Q., 1982, Biostratigraphy and isotope stratigraphy and the catastrophic extinction of calcareous nannoplankton at the Cretaceous/Tertiary boundary, in Silver, L. T. and Schultz, P. H., eds., Geological Implications of Impacts of Large Asteroids and Comets on the Earth: Boulder, Geological Society of America Special Paper 190, p. 353–371.

PEYPOUQUET, J.-P., GROUSSET, F., AND MOURGUIART, P., 1986, Paleoceanography of the Mesogean Sea based on ostracodes of the northern Tunisian continental shelf between the Late Cretaceous and Early Paleogene: Geologie Rundsdau, v. 75, p. 159–174.

PISIAS, N. G., BARRON, J. A., AND DUNN, D. A., 1985, Stratigraphic resolution of Leg 85: An initial analysis: Initial Reports of the Deep sea Drilling Project, v. 85, p. 695–708.

PISIAS, N. G., MARINSON, D. G., MOORE, T. C. J., SHACKELTON, N. J., PRELL, W., HAYS, J., AND BODEN, G., 1984, High-resolution stratigraphic correlation of benthic isotope records spanning the last 300,000 years: Marine Geology, v. 56, p. 119–136.

PLIMER, I. R. AND WILLIAMS, P. A., 1989, Salt lake concentration of platinum group elements: Terra Abstracts, v. 1, p. 20–21.

POPE, K. O., OCAMPO, A. C., AND DULLER, C. E., 1991, Mexican site for K/T impact crater: Nature, v. 351, p. 105.

POSPICHAL, J. J., 1993, Cretaceous nannofossils in Danian sediments: Survivorship or reworking?: Geological Society of America, Abstracts with Programs, v. 25, p. A-363.

PRELL, W. L., IMBRIE, J., MARTINSON, D. G., MORELY, J. J., PISIAS, N. G., SHACKELTON, N. J., AND STREETER, H. F., 1986, Graphic correlation of oxygen isotope stratigraphy: Application to the Late Quaternary: Paleoceanography, v. 1, p. 137–162.

SAWLOWICZ, Z., 1993, Iridium and other platinum-group elements as geochemical markers in sedimentary environments: Palaeogeography, Palaeoclimatology, Palaeoecology, v. 104, p. 253–270.

SAVRDA, C. E., 1993, Ichnostratigraphic evidence for non-catastrophic origin pf Cretaceous-Tertiary boundary sands in Alabama: Geology, v. 21, p. 1075–1078.

SHARPTON, V. L. AND WARD, P. D., 1990, Global Catastrophes in Earth History: An Interdisciplinary Conference on Impacts, Volcanism, and Mass Mortality: Boulder, Geological Society of America Special Paper 247, 631 p.

SHARPTON, V. L., BURKE, K., CAMARGO-ZANOGUERA, A., HALL, S. A., LEE, D. S., MARIN, L. E., SUÁREZ-REYNOSO, G., QUEZADA-MUÑETON, J. M., SPUDIS, P. D., AND URRTIA-FUCUGAUCHI, J., 1993, Chicxulub multiring impact basin: Size and other characteristics derived from gravity analysis: Science, v. 261, p. 1564–1566.

SHAW, A., 1964, Time in Stratigraphy: New York, McGraw-Hill, 365 p.

SILVER, L. T. AND SCHULTZ, P. H., 1982, Geological implications of impacts of large asteroids on the Earth: Boulder, Geological Society of America Special Paper 190, 500 p.

SLOAN, R. E., RIGBY, J. K., VAN VALEN, L. M., AND GABRIEL, D., 1986, Gradual dinosaur extinction and simultaneous ungulate radiation in the Hell Creek Formation: Science, v. 232, p. 629–633.

SMIT, J., 1977, Discovery of a planktonic foraminifera association between the Abathomphalus mayaroensis Zone and the 'Globigerina' eugubina Zone at the Cretaceous/Tertiary boundary in the Barranco del Gredero (Caravaca, SE Spain: Koninklijke Nederlandese Akademie van Wetenschappen Proceedings, Series B, v. 80, p. 280–301.

SMIT, J., 1982, Extinction and evolution of planktonic foraminifera after a major impact at the Cretaceous/Tertiary boundary, in Silver, L. T. and Schultz, P. H., eds., Geological Implications of Impacts of Large Asteroids and Comets on the Earth, Geological Society of America Special Paper, v. 190, p. 329–352.

SMIT, J., MONTANARI, A., SWINEBURNE, N. H. M., ALVAREZ, W., HILDEBRAND, A. R., MARGOLIS, S. V., CLAEYS, P., LOWRIE, W., AND ASARO, F., 1992, Tektite-bearing, deep-water clastic unit at the Cretaceous-Tertiary boundary in northeastern Mexico: Geology, v. 20, p. 99–103.

STANLEY, S., 1989, Earth and Life Through Time, Second Edition: San Francisco, W. H. Freeman, 689 p.

STINNESBECK, W., BARBARIN, J. M., KELLER, G., LOPEZ-OLIVA, J. G., PIVNIK, D. A., LYONS, J. B., OFFICER, C. B., ADATTE, T., GRAUP, G., ROCCHIA, R., AND ROBIN, E., 1993, Deposition of channel deposits near the Cretaceous-Tertiary boundary in northeastern Mexico: Catastrophic or "normal" sedimentary deposits?: Geology, v. 21, p. 797–800.

STOTT, L. D. AND KENNETT, J. P., 1990, Antarctic Paleogene planktonic foraminiferal biostratigraphy: ODP Leg 113, Sites 689 and 690: Proceedings of the Ocean Drilling Program, Scientific Results, v. 113, p. 549–569.

SULLIVAN, R. M., 1987, A reassessment of reptilian diversity across the Cretaceous-Tertiary boundary: Contributions in Science, v. 391, p. 1–26.

SURLYK, F., 1990, Mass extinction events, Section 2.13.6 Cretaceous-Tertiary (Marine), in Briggs, D. E. G. and Crowther, P. R., eds., Palaeobiology: A Synthesis: Oxford, Blackwell Publishers, p. 198–203.

SWEET, A. R. AND BRAMAN, D. R., 1992, The K-T boundary and contiguous strata in western Canada: Interactions between paleoenvironments and palynological assemblages: Cretaceous Research, v. 13, p. 31–79.

SWEET, A. R., BRAMAN, D. R., AND LERBEKMO, J. F., 1990, Palynofloral response to K/T boundary events; A transitory interruption within a dynamic system, in Sharpton, V. L. and Ward, P. D., eds., Global Catastrophes in Earth History: Boulder, Geological Society of America Special Paper 247, p. 457–469.

THOMAS, E., 1990, Late Cretaceous-Early Eocene mass extinction in the deep sea, in Sharpton, V. L. and Ward, P. D., eds., Global Catastrophes in Earth History: Boulder, Geological Society of America Special Paper 247, p. 481–495.

VAIL, P. R., MITCHUM, R. M. J., AND THOMPSON, S., 1977, Seismic stratigraphy and global changes in sea level, part three: Relative changes of sea level from coastal onlap, in Payton, C. F., ed., Seismic Stratigraphy — Applications to Hydrocarbon Exploration: American Association of Petroleum Geologists Memoir, v. 26, p. 83–98.

WANG, K., ATTREP, M., JR., AND ORTH, C. J., 1993, Global iridium anomaly, mass extinction, and redox change at the Devonian-Carboniferous boundary: Geology, v. 21, p. 1071–1074.

WATKINS, D. K., CRUX, J. A., POSPICHAL, J. J., AND WISE, S. W. JR., 1994, Upper Cretaceous calcareous nannofossils of the southern ocean and their paleocliatic implications: Programme and Abstracts, ODP and the Marine Biosphere, University of Wales, Aberystwyth, p. 32.

WARD, P. D. AND KENNEDY, W. J., 1993, Maastrichtian ammonites from the Biscay region (France, Spain): Paleontological Society Memior 34, p. 1–58.

WIDMARK, J. G. AND MALMGREN, B. A., 1992, Benthic foraminiferal changes across the Cretaceous-Tertiary boundary in the deep sea; DSDP Sites 525, 527, and 465: Journal of Foraminiferal Research, v. 22, p. 81–113.

ZIEGLER, A. M., 1990, Paleogeographic atlas project: Geotimes, v. 35, p. 22–24.

ZINSMEISTER, W. J., FELDMANN, R. M., WOODBURNE, M. O., AND ELLIOT, D. H., 1989, Latest Cretaceous/earliest Tertiary transition on Seymour Island, Antarctica: Journal of Paleontology, v. 63, p. 731–738.

GRAPHIC CORRELATION OF PLIO-PLEISTOCENE SEQUENCE BOUNDARIES, GULF OF MEXICO: OXYGEN ISOTOPES, ICE VOLUME, AND SEA LEVEL

RONALD E. MARTIN
Department of Geology, University of Delaware, Newark, Delaware 19716
AND
RUTH R. FLETCHER[1]
Department of Geology, University of Delaware, Newark, Delaware 19716

ABSTRACT: During the last two decades, the oxygen isotope curve has been used extensively as a proxy for Neogene ice volume and sea-level change. Unlike the oxygen isotope curve, however, the relation of the *Globorotalia menardii*-based Ericson and Wollin (1968) zonation to paleoclimate and sea-level change has remained obscure. Utilizing the "warm-water" *Globorotalia menardii* complex and "cool-water" *G. inflata*, we have subdivided the Pleistocene of the tropical Atlantic Ocean, Gulf of Mexico, and Caribbean Sea into 17 subzones. Graphic correlation of subzonal boundaries and oxygen isotope events reveals changes in slope of the line of correlation (changes in sediment accumulation rate) that indicate seismic sequence boundaries at the Zone Y/X (~0.09 Ma), W/V1 (~0.2 Ma), V2/V3 (~0.4 Ma), V3/U (~0.475 Ma), U/T (~0.525–0.620 Ma), T3/T4 (~0.7–0.9 Ma), T4/S1 (~1.0 Ma), S2/S3-S3/R1 (~1.2 Ma), R2/R3-R3/Q1 (~1.4–1.5 Ma), and P/Pliocene (~1.8–1.9 Ma) boundaries, and a regionally condensed section in Zone R1 (~1.3 Ma). The subzones and sequence boundaries are also recognized in an exploration well (Garden Banks Block 412 Unocal #1, Gulf of Mexico). Relative abundance of keeled globorotalids (analogous to the *G. menardii* complex) closely tracks the oxygen isotope curve in Pliocene sections of ODP Core 625B (NE Gulf of Mexico) and DSDP Core 502B (Caribbean Sea). Changes in slope of the line of correlation delineate sequence boundaries at ~2.4, 2.6, 3.0, and 3.8 Ma, which correspond to those of other workers.

Use of the graphic correlation technique not only delineates sequence boundaries and erosionally-truncated or reworked biostratigraphic markers, but also suggests further avenues of research with regard to microfossil-based zonations, paleoclimate, and sea-level change.

INTRODUCTION

Graphic correlation of fossil first and last occurrences, as well as of ecostratigraphic zones and oxygen isotope events, permits the recognition of the separate effects of paleoceanographic factors of ice volume, sea-level change, and water mass structure. During the past decade, the senior author, along with students and colleagues, has published a series of papers on ecostratigraphic zonation of the Plio-Pleistocene of the "American Mediterranean" (Gulf of Mexico, Caribbean Sea) and tropical Atlantic (Fig. 1) using a modified form of the Ericson and Wollin (1968) zonation. We recognized the zonation in both deep-sea cores and an exploration well (based on cuttings) and related it to sea-level change and resultant sequence boundaries (Martin and others, 1990a, b, 1993). Most recently, we have extended the method to Pliocene sections to produce a high-resolution stratigraphic framework (Fletcher, 1993; Martin and Fletcher, 1993). Herein we review our work to date. This paper is intended as an heuristic device, in the hope that it will guide investigators in their studies of graphic correlation. Readers interested in further details of our studies are encouraged to consult the primary literature cited herein.

PREVIOUS INVESTIGATIONS

Planktic foraminiferal biostratigraphy of the Quaternary System is problematical in that First (FADs) and Last Appearance Datums (LADs) of planktic foraminifera during the period represent primarily non-evolutionary migrational events which are only of regional significance (Thunell, 1984). In the Gulf of Mexico, only two Pleistocene planktic foraminiferal datums are routinely utilized: the FAD of *Globorotalia truncatulinoides* (~1.8–1.9 Ma; Berggren and others, 1985) and the LAD of the regional datum *Globorotalia flexuosa* (~0.09 Ma; Kennett and Huddleston, 1972a; Thunell, 1984).

Because climatic fluctuations are a fundamental characteristic of the Pleistocene, greater stratigraphic resolution can be gained by the establishment of assemblage zones (glacial versus interglacial episodes) based on the frequency variations of climatically-controlled foraminiferal assemblages ("ecostratigraphy"). This approach was pioneered by Schott (1935), who used the relative abundance of the "warm-water" *Globorotalia menardii* complex (*G. menardii*, *G. cultrata*, *G. flexuosa*) in the Atlantic Ocean, and was subsequently employed by other workers in the Atlantic, Caribbean, and Gulf of Mexico (Phleger and others 1953; Ericson and Wollin, 1956a, b, 1968; Ewing and others, 1958; Beard and others, 1982; Lamb and others, 1987). The biozones were designated Q through Z (Z representing the Holocene), and were corroborated by Beard (1969) and Poag and Valentine (1976) using the cool-water species *Globorotalia inflata*. An eleventh biozone (Zone P) was added by Briskin and Berggren (1975), which straddles the Plio-Pleistocene boundary, and which is characterized by the presence of the *Globorotalia menardii* complex. Each zone has an average duration of ~300,000 years (Thunell, 1984). The relative abundance of *G. menardii* agrees with oxygen isotope stages 1–5 (odd and even-numbered stages are interglacials and glacials, respectively). Below Zone Y, however, the relationship between the Ericson and Wollin (1968) zonation and the oxygen isotope curve is often ambiguous, which no doubt contributed to the subsequent utilization of the oxygen isotope curve by many stratigraphers and paleoceanographers (see Imbrie and Imbrie, 1979, p. 123–140, for further discussion).

Nevertheless, further refinements in the zonation were pursued by utilizing frequency variations of numerous planktic foraminiferal species (total fauna approach). Kennett and Huddleston (1972b) subdivided Ericson and Wollin zones W-Z of the western Gulf of Mexico into 18 subzones, each with an average duration of ~15,000 years (Thunell, 1984). Their study demonstrated that abundance oscillations of many species (not just the *G. menardii* complex) can be correlated from one core to another, at least within the same basin, and provided a de-

---

[1] Present address: Department of Geology, University of Hawaii, Honolulu, HI 96822.

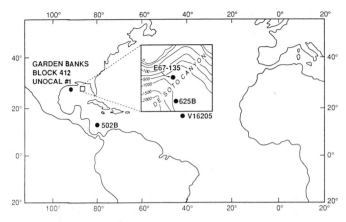

Fig. 1.—Core and well locations. Contours in meters. Published with permission of the GCSSEPM Foundation.

tailed late Pleistocene paleoclimatic history of the western Gulf of Mexico.

Neff (1985; see Thunell, 1984, for review) expanded upon Kennett and Huddleston's (1972a) approach. He utilized fluctuations in percent abundance of seven planktic foraminiferal species to subdivide zones Q-V of Eureka Core E67–135, located on the north flank of DeSoto Canyon (725-m water depth; Fig. 1), into 15 subzones, each with an average duration of ~100,000 years (Thunell, 1984; Figs. 2, 3). Although the resolution of Neff's zonation was not as high as that of Kennett and Huddleston's (1972a), it was a significant (~three-fold) improvement over the original Ericson and Wollin (1968) zonation.

Johnson (1988) and Spotz (1988) studied Ocean Drilling Program (ODP) Core 625B, which was taken on the southern flank of DeSoto Canyon (889-m water depth; Fig. 1) using an Advanced Hydraulic Piston Core (APC) system (Rabinowitz and others, 1985). Site 625 was chosen by ODP for the shakedown cruise (January, 1985) of the Joides *Resolution* because seismic data indicated that the location would provide a nearly continuous record of Early Pliocene to Holocene paleoceanographic development and sea-level fluctuation (Rabinowitz and others, 1985; Rabinowitz and Merrell, 1985). Johnson (1988; see also Martin and others, 1990a, b) was able to recognize Neff's (1985) zonation in ODP Core 625 and to correlate it to Eureka Core E67–135 using primarily the *G. menardii* complex and *G. inflata*, supplemented by left and right-coiling varieties of *G. truncatulinoides* (Fig. 2). Zonal boundaries were constrained by available biostratigraphic and magnetostratigraphic datums (Fig. 3). The salient features of the zonation have been presented elsewhere (Martin and others, 1990a, b, 1993).

GRAPHIC CORRELATION OF PLEISTOCENE DATUMS

In retrospect, it was not surprising that the zonation could be recognized in two cores located so closely to one another. The next hurdle was to extend the zonation into adjacent basins. Martin extended the zonation to Deep Sea Drilling Project Core 502B (Caribbean Sea, Colombia Basin; Fig. 4) and to Core V16–205 (tropical Atlantic; Fig. 5; data from Briskin and Berggren, 1975) via graphic correlation of ecostratigraphic datums

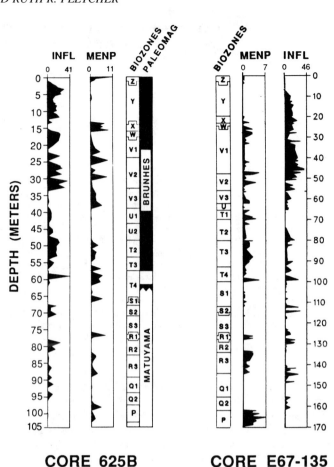

CORE 625B   CORE E67-135

Fig. 2.—Species abundances and zonation in cores 625B and E67–135. Species abundances are given as percentages of planktic foraminiferal assemblages. INFL = *Globorotalia inflata*; MENP = *Globorotalia menardii* complex.

(based on *G. menardii* complex and *G. inflata*) and bio- and magnetostratigraphic datums.

Graphic correlation was ideally suited for testing our correlations and for integrating diverse types of chronostratigraphic data. Oxygen isotope stage boundaries, which provide high stratigraphic resolution, were identified conservatively based on van Donk's (1976) analysis of Core V16–205, were incorporated into the data base by graphing, and were plotted with bio- and magnetostratigraphic datums (Martin and others, 1990a, b). Other workers have also used non-unique stratigraphic events (e.g., oxygen isotope events, well-log data) in graphic correlation (Prell and others, 1986; Edwards, 1989; Hazel, 1989; Joyce and others, 1990; Scott, 1991; Self and Scott, 1993). What is perhaps most notable about graphic correlation of the Pleistocene chronostratigraphic datums is that missing section is apparent from changes in slope of the LOC, but not necessarily from visual examination of the oxygen isotope curve alone (cf. Joyce and others, 1990).

Initially, Core 625B was adopted as the standard reference section (SRS) for several reasons (Martin and others, 1990a, b): (1) it was the longest and presumably most complete section available for the Gulf of Mexico continental slope with biostratigraphic and paleomagnetic datums (Rabinowitz and oth-

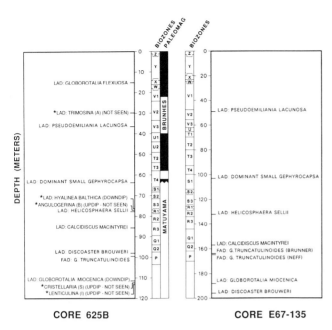

Fig. 3.—Zonation, biostratigraphic, and magnetostratigraphic datums of cores 625B and E67–135 (Brunner = Brunner and Keigwin, 1981; Neff = Neff, 1985). Calcareous nannofossil datums from Gartner and Huang (1986). Approximate positions of Gulf Coast biostratigraphic markers are indicated for Core 625B.

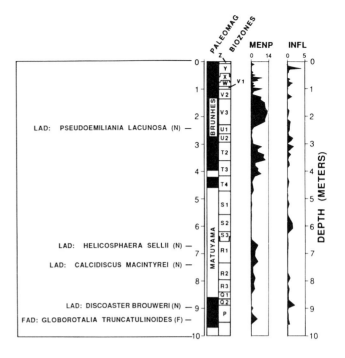

Fig. 5.—Zonation, biostratigraphic, and magnetostratigraphic datums of Core V16–205. F = Foraminifer; N = Nannofossil. Other abbreviations and species abundances as in Figure 2.

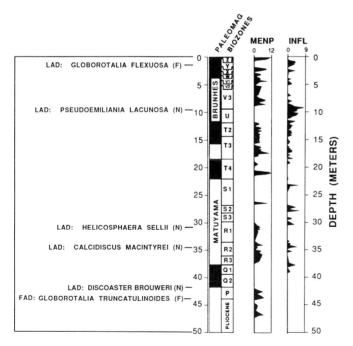

Fig. 4.—Zonation, biostratigraphic, and magnetostratigraphic datums of Core 502B. F = Foraminifer; N = Nannofossil. Other abbreviations and species abundances as in Figure 2.

ers, 1985; Rabinowitz and Merrell, 1985; see also Shaw, 1964; Miller, 1977), (2) it served as a standard reference section for the subzonal boundaries, and (3) it had been sampled at close intervals (15–20 cm) for oxygen isotope analysis (Joyce and others, 1990). Because numerous isotope stages appeared to be missing from Core E67–135 (Williams, 1984), it was not incorporated into the composite.

Rather than being a continuous column characterized by relatively constant rates of sediment accumulation, however, Core 625B represents highly episodic sedimentation. The plot of Core 502B versus the composite section based on Core 625B ("unrevised") exhibited a nearly linear relationship between zonal boundaries, oxygen isotope stage boundaries, and other datums for the Brunhes Chron. A distinct offset was observed, though, in Subzone S1 in Core 625B (Fig. 6). This offset was also observed in the plot of Core V16–205 versus the same composite (Martin and others, 1990a, Fig. 11). Composite sections were subsequently constructed using an SRS (Core 625B "revised") in which oxygen isotope stages were shifted or omitted until the best visual fit between subzonal and isotope stage boundaries was achieved (Martin and others, 1990a). In plots of Cores 502B (Fig. 7) and V16–205 (Martin and others, 1990a, Fig. 14) versus the new composite, subzonal and isotope stage boundaries better coincided in the Matuyama Chron. The T4/S1 boundary in Core 625B appeared to mark a missing or condensed section (Fig. 3). Zone S1 is ~1 m in thickness in Core 625B but is relatively expanded in the other three cores (Figs. 3, 4, 5). This zone occurred at a core break in Core 625B (Rabinowitz and others, 1985, Table 2), and was associated with

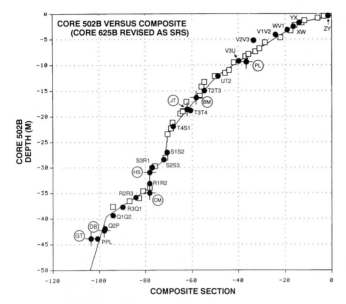

Fig. 6.—Graphic correlation plot of Core 502B datums and subzonal boundaries versus composite (composite based on Core 625B unrevised as SRS, and cores 502B and V16–205; Core E67–135 not included in composite section). Depths are sub-bottom; depths for composite section of this and subsequent plots are in composite standard units. Small open boxes indicate oxygen isotope stage boundaries. Large closed circles indicate subzone boundaries. Large cross-hairs and closed circles indicate biostratigraphic and magnetostraphic datums: GF = *Globorotalia flexuosa* LAD; PL = *Pseudoemiliania lacunosa* LAD; BM = Brunhes-Matuyama boundary; JT = Top of Jaramillo; HS = *Helicosphaera sellii* LAD; CM = *Calcidiscus macintyrei* LAD; DB = *Discoaster brouweri* LAD; GT = *Globorotalia truncatulinoides* FAD; PPL = Zone P/Pliocene boundary.

a small but distinct drop in sediment accumulation rate from ~6 to ~3 cm/kyr (Martin and others, 1990a, Fig. 8).

The position of the T3/T4 boundary in Core 625B, just above the T4/S1 boundary, is anomalous with respect to Cores 502B and V16–205. In Core 625B, it occurs at the base of the Brunhes Chron, whereas in the other cores it lies near the top of the Jaramillo Subchron (Figs. 3, 4, 5). The base of the Brunhes Chron (58.0 m) in Core 625B is associated with an abnormally high sediment accumulation rate of ~50 cm/kyr followed by an abrupt drop to ~5 cm/kyr (Martin and others, 1990a, Fig. 8). The high accumulation rate is associated with a core break at 58.5 m (Rabinowitz and others, 1985, Table 2), and was originally thought by us to have resulted from core extrusion. This level also corresponds to an unconformity between oxygen isotope stages 20 and 23 (Prell, 1982). Thus, the true T3/T4 boundary in Core 625B appears to lie in a disturbed interval, which accounts for the slight offset at this level in graphic correlation plots (Figs. 6, 7).

Stratigraphic anomalies also occur near the T/U zonal boundary. Zone U is very thin in Core E67-135 (Figs. 2, 3) and Neff (1985) therefore suggested the presence of an unconformity. Zone U can be subdivided into two subzones in Cores 625B and V16205 but not in 502B. Also, Subzone T1 is recognizable in Core E67-135 but is not present in the other cores. Johnson (1988) speculated that Subzone T1 may represent slumped (repeated) section in Core E67-135.

Subzone R1 appears to represent a regionally condensed section in Cores 625B and E67-135. Zone R1 is very narrow in both cores (Figs. 2, 3) and corresponds to a slight offset in graphic correlation plots if it is assumed that all isotope stages are present and correctly identified (Figs. 6, 7). The interval is

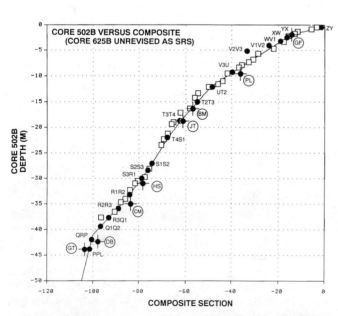

Fig. 7.—Graphic correlation plot of Core 502B datums and subzonal boundaries versus composite (composite based on Core 625B revised as SRS, and cores 502B and V16–205; Core E67–135 not included in composite section). Symbols as in Figure 6.

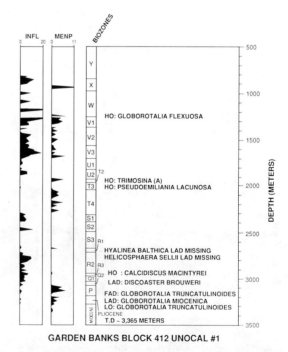

Fig. 8.—Zonation and biostratigraphic datums of Garden Banks Block 412 Unocal #1 well. Biostratigraphic datums from original Unocal in-house report. HO = Highest Occurrence (probably not true extinction point); LO = Lowest Occurrence.

characterized by a relatively low sediment accumulation rate in Core 625B (Martin and others, 1990a, Fig. 8) and corresponds to an 0.50-m interval which is heavily mottled by burrows (Rabinowitz and others, 1985). Also, the *Helicosphaera sellii* LAD is found at the base of Zone R1 in Core 625B and in close proximity to Zone R1 in Core E67-135 (Fig. 3). There are also problems with placement of the Plio-Pleistocene boundary, which may be due either to missing section or time-transgressive biostratigraphic datums (Martin and others, 1990a; see also Dowsett, 1988).

ZONATION AND GRAPHIC CORRELATION OF AN EXPLORATION WELL

Attempts at high-resolution biostratigraphic zonation of exploration wells have typically been dismissed because of presumed mixing in composite (ditch) samples (cuttings) and down-hole caving. Nevertheless, we recognized the zonation (Fig. 8) in the Garden Banks Block 412 Unocal #1 well, which was drilled on the outer continental shelf of the Gulf of Mexico (Fig. 1; Martin and others, 1993; Martin and Fletcher, 1993). Zonal boundaries in the Garden Banks well frequently coincided with abrupt changes in foraminiferal and nannofossil as-

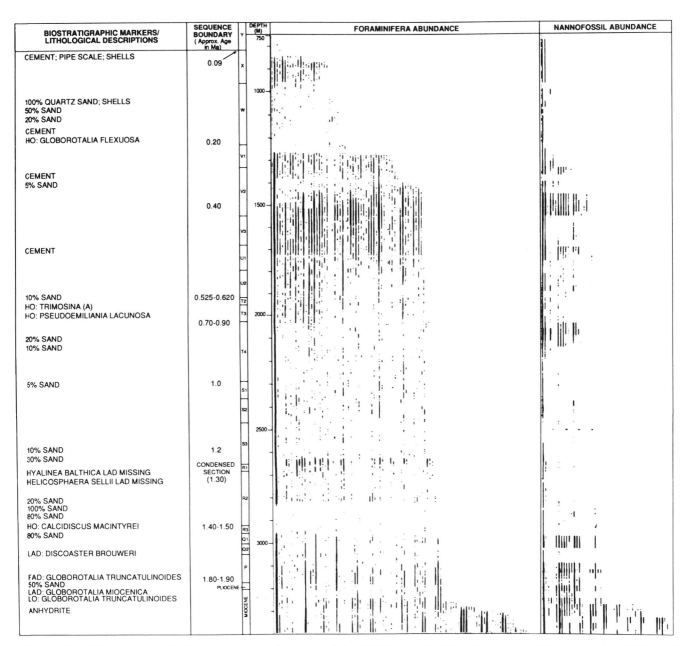

FIG. 9.—Foraminiferal and nannofossil range chart for Garden Banks well (modified from original Unocal in-house report) based on cuttings. Thickness of each line indicates relative abundance of a species, which was estimated visually by Unocal paleontologists. Lithological (sand, cement) descriptions based on visual estimates and well-log information. Subzone and sequence boundaries (based on core material) are also shown. Published with permision of the GCSSEPM Foundation.

semblages and stratigraphic markers (Fig. 9), which are either presumably truncated or depressed (Pflum and Frerichs, 1976; Armentrout, 1987, 1991; Armentrout and Clement, 1990; Armentrout and others, 1990). The standard Gulf Coast benthic foraminiferal marker *Trimosina* (A) occurred quite low in the Garden Banks well (Fig. 8) and was not employed. Two other Gulf Coast biostratigraphic markers (*Helicosphaera sellii* and *Hyalinea balthica*) were completely missing from the Garden Banks section (Figs. 8, 9).

Because of a lack of standard biostratigraphic markers (FADs, LADs), the Shaw plot of the Garden Banks well was fitted to subzonal boundaries. For these plots, we used Core 502B as SRS and incorporated previously-discussed cores to produce the composite standard. Based on our previous results with Core 625B, Core 502B appeared to be much more promising as an SRS despite its lower sediment accumulation rate (~2 cm/kyr; Prell, 1982). Abrupt changes in microfossil assemblages in the Garden Banks well (Fig. 9) frequently corresponded to changes in slope of the LOC (Fig. 10) and, presumably, sequence boundaries (Poag, 1972; Martin, 1984; Ragan and Abbott, 1989; Armentrout and Clement, 1990; Armentrout and others, 1990; Armentrout, 1991). Thus, we recognized sequence boundaries at ~0.09 Ma (Y/X boundary), 0.2 Ma (W/V1 boundary), 0.4 Ma (V2/V3 boundary), 0.525–0.620 Ma (U/T boundary), 0.7–0.9 Ma (T3/T4 boundary), 1.0 Ma (T4/S1 boundary), 1.2 Ma (S3/R1 boundary), 1.4–1.5 Ma (R2/R3-R3/Q1 boundary), and 1.8–1.9 Ma (P/Pliocene), and a regionally condensed section at approximately 1.3 Ma (Fig. 11). These sequence boundaries and condensed section were previously identified by other Gulf Coast workers (although age assignments vary somewhat) using biostratigraphic and paleoecologic (faunal discontinuities, paleodepth) information tied to well-log and seismic data (Wornardt and Vail, 1991, also recognized sequence boundaries at 0.32 and 0.02 Ma; Fig. 11; see also Wornardt and Vail, 1990). By contrast, we identified sequence boundaries primarily by graphic correlation of ecostratigraphic datums.

In several cases, the ecozonation suggested a sequence boundary where no obvious faunal or floral discontinuity existed (U/T, T3/T4, T4/S1 boundaries in Garden Banks well; Figs. 8, 9, 11). There may also be a sequence boundary at the V3/U transition (~0.475 Ma; Fig. 11). This boundary was not previously reported (Martin and others, 1993) because it corresponded to an abrupt influx of cement (casing point) in the Garden Banks well (Fig. 9), but at this level there is a slowing of sediment accumulation as reflected by a change in slope of the LOC (Fig. 10).

To further test the hypothesis of a relationship between subzonal and sequence boundaries, the occurrence of sand in the Garden Banks well was compared to sequence boundaries inferred on the basis of the zonation. Although the occurrence of sand does not absolutely confirm the presence of an unconformity (Self and Scott, 1993), in every case of a presumed sequence boundary, there was at least some influx of sand at or near the same level (Fig. 9).

With our work on the Garden Banks well came a renewed appreciation of the episodic nature of continental slope sedimentation. Comparison of the Garden Banks plot (Fig. 10) with one produced by us several years earlier (Martin and others, 1990a) for Core E67-135 versus the composite based on ODP Core 625B (Fig. 12) proves instructive. Martin and others (1990a) originally fitted the LOC for Core E67-135 using a "Least Squares" or "Principal Components" approach (Hood, 1986). Because there were no reliable oxygen isotope or paleomagnetic datums for this core, only biostratigraphic datums could be used to constrain the LOC. With the obvious exception of *Calcidiscus macintyrei*, biostratigraphic and subzonal datums fell on or very near the LOC, suggesting virtually constant sedimentation and continuous section! But Core E67-135 is located on the northwest flank of DeSoto Canyon in an essentially terrigenous province and may have had highly variable sedimentation rates (Gartner and others, 1983). A new plot of Core E67-135 versus a composite based on Core 502B indicates instead that sedimentation on the continental slope is highly episodic (Fig. 13).

Perhaps more important, however, are the positions of biostratigraphic datums with respect to the new LOC (Fig. 13). The *Pseudoemiliania lacunosa* LAD in Core E67-135 now lies above the LOC and may be reworked in Core E67-135 (compare Figs. 12, 13). This datum nearly coincides with the LOC in the Shaw plot of Core 625B (Fig. 14). The *H. sellii* LAD is associated with Subzone R1 in both Cores E67-135 and 625B (Figs. 3, 13, 14), which, as previously discussed, represents condensed section. The *C. macintyrei* LAD of Core E67-135 falls well below the LOC, just where there appears to be a slight increase in sedimentation (Q1/Q2 zonal boundary) and may, therefore, be depressed as a result of localized sediment influx (cf. Core 625B; Fig. 14).

APPLICATION TO THE PLIOCENE

We have also applied our methodology to the Pliocene Series (Martin and others, 1993; Fletcher, 1993; Martin and Fletcher,

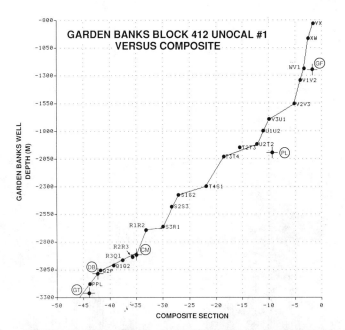

FIG. 10.—Graphic correlation plot of Garden Banks well biostratigraphic datums and subzonal boundaries versus composite (composite based on DSDP Core 502B as SRS, and cores 625B, V16-205, and E67-135, and Garden Banks well). Symbols as in Figure 6. Published with permission of the GCSSEPM Foundation.

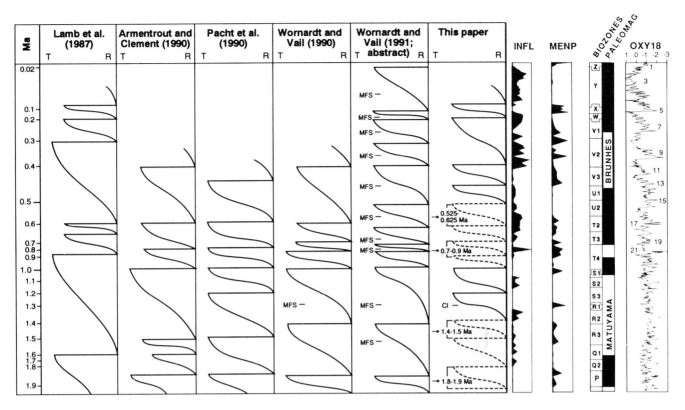

FIG. 11.—Comparison of ecozonation and sequence boundary scheme of this paper with previously-published frameworks for the Gulf Coast. Time scale, species abundances, and zonation based primarily on ODP Core 625B (NE Gulf of Mexico). Sequence boundaries of previous workers based primarily on faunal (floral) discontinuities and biostratigraphic markers tied to well-log and seismic data. Differences between our sequence boundary scheme and those of previous workers are discussed in Martin and others (1993). Dashed lines with brackets indicate uncertainty in placing sequence boundaries based on ecostratigraphic zonation. T = Transgression; R = Regression; MFS = Maximum Flooding Surface; CI = Condensed Section.

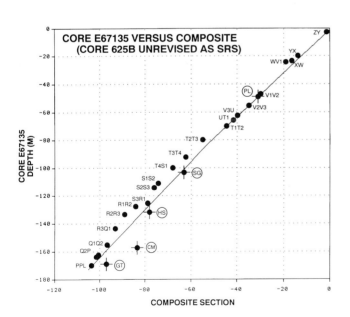

FIG. 12.—Graphic correlation plot of Core E67-135 biostratigraphic datums and subzonal boundaries versus those of composite (composite based on Core 625B unrevised as SRS, and cores 502B, V16-205, and E67-135). Position of Subzone T1 also shown. See text for further discussion. SG = Small *Gephyrocapsa* datum; otherwise, symbols as in Figure 6.

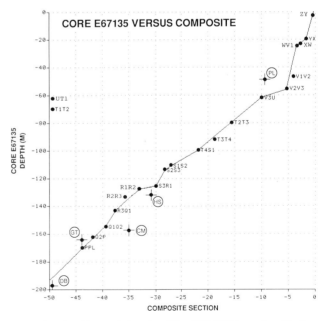

FIG. 13.—Graphic correlation plot of Core E67-135 biostratigraphic datums and subzonal boundaries versus composite (composite based on DSDP Core 502B as SRS, and cores 625B, E67-135, V16-205, and Garden Banks well). Subzone T1 not included in plot. See text for further discussion. Symbols as in Figure 6.

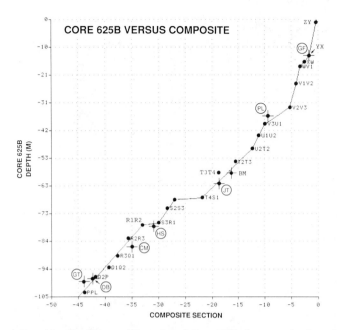

FIG. 14.—Graphic correlation plot of Core 625B datums and subzonal boundaries versus composite (composite based on DSDP Core 502B as SRS, and cores 625B, E67-135, V16-205, and Garden Banks well). Symbols as in Figure 6.

TABLE 1.—EXPLANATION OF SYMBOLS USED IN FIGURES 15–18

Paleomagnetic datums:
OB	Olduvai bottom
KT	Kaena top
KB	Kaena bottom
MT	Mammoth top
NT	Nunivak top
NB	Nunivak bottom
TB	Therva bottom

Calcareous nannoplankton datums:
CA	*Ceratolithus acutus*
CR	*Ceratolithus rugosus*
DB	*Discoaster brouweri*
DP	*Discoaster pentaradiatus*
DS	*Discoaster surculus*
DT	*Discoaster tamalis*
DQ	*Discoaster quinqueramus*
HS	*Helicosphaera sellii*
RP	*Reticulofenestra pseudoumbilica*
SA	*Sphenolithus abies*

Planktonic foraminifera datums:
DALT	*Dentoglobigerina altispira altispira*
GNP	*Globigerina nepenthes*
GFS	*Globigerinoides fistulosus*
GD	*Globoquadrina dehiscens*
GC	*Globorotalia conomiozea*
GCS	*Globorotalia crassaformis*
GE	*Globorotalia exilis*
GHR	*Globorotalia hirsuta*
GMG	*Globorotalia margaritae*
GMC	*Globorotalia miocenica*
GMT	*Globorotalia multicamerata*
GTR	*Globorotalia truncatulinoides*
GTM	*Globorotalia tumida*
SS	*Sphaeroidinellopsis spp.*

Ecostratigraphic datums:
a1	Left to right (or random) *Neogloboquadrina humerosa* + *N. acostaensis* coiling shift
a2	Right to left *N. humerosa* + *N. acostaensis* coiling shift
b1	Left to right (or random) *N. humerosa* + *N. acostaensis* coiling shift
b2	Right to left *N. humerosa* + *N. acostaensis* coiling shift
c1	Left (or random) to right *N. humerosa* + *N. acostaensis* coiling shift
c2	Right to random *N. humerosa* + *N. acostaensis* coiling shift
PPL-R	*Pulleniatina primalis* L-R coiling shift
PACHYL	Decline in abundance of left coiling *N. pachyderma*

Odd Numbers 1 to 41 refer to keeled *Globorotalia* abundance peaks

1993), utilizing Pliocene biostratigraphic and ecostratigraphic datums of ODP Core 625B and DSDP Site 502 (based on Cores 502A and B; Figs. 15, 16). The keeled *Globorotalia* curve (analogous to *G. menardii* complex) tracks the oxygen isotope curve (presumably ice volume) of Joyce and others (1990) in Core 625B; peaks and valleys of the *Globorotalia* curves of both cores (numbered in a manner analogous to the Pleistocene ox-

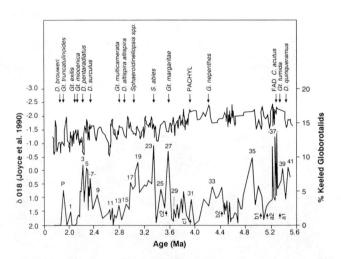

FIG. 15.—Abundance of keeled *Globorotalia* spp. (lower curve), biostratigraphic and ecostratigraphic datums, and smoothed oxygen isotope curve (upper curve) of Joyce and others (1990) for Core 625B Pliocene. Abbreviations are standard Gulf Coast biostratigraphic markers, with the exception of al-2, bl-2, and cl-2 (coiling shifts in *Neogloboquadrina* spp.) and PACHYL (coiling shift in *Neogloboquadrina pachyderma*; Fletcher, 1993; Martin and Fletcher, 1993). See Table 1 for other stratigraphic markers. Published with permission of the GCSSEPM Foundation.

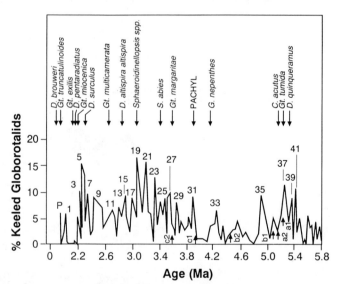

FIG. 16.—Biostratigraphic and ecostratigraphic datums for DSDP Site 502 Pliocene. See Figure 15 and Table 1 for abbreviations.

ygen isotope curve) are similar. Thus, the *Globorotalia* curves can potentially serve as a proxy for the oxygen isotope curve on an industrial basis. Where the two curves disagree, *Globorotalia* may also be responding to changes in water column stratification (productivity, temperature; Thunell and Reynolds, 1984).

Biostratigraphic (including ecostratigraphic) datums of Core 625B and Site 502 were plotted against the Pliocene composite of Dowsett (1989; based on 12 sites located in the Atlantic and Pacific oceans), who used Site 502 as the SRS for his Pliocene composite (Figs. 17, 18). A sediment accumulation rate curve for Core 625B was then constructed by transforming composite standard units into Ma (Fletcher, 1993). The accumulation rate curve again demonstrates the episodic nature of sedimentation on the continental slope of the northeast Gulf of Mexico (Fig. 19). By contrast, the accumulation rate curve of Joyce and others (1990), which is based simply on interpolation between biostratigraphic markers, is relatively smooth.

Graphic correlation of the biostratigraphic zonation for the Pliocene of ODP Core 625B shows that, in general, expanded sections (high sediment accumulation rates) represent periods

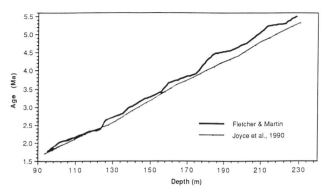

FIG. 19.—Sediment accumulation rate curve for ODP Core 625B, comparing the age model derived by Fletcher (1993) with that of Joyce and others (1990). Published with permission of the GCSSEPM Foundation.

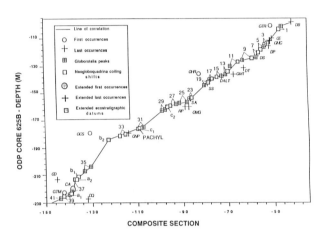

FIG. 17.—Graphic correlation plot of ODP Core 625B Pliocene versus the composite standard reference section (based on Dowsett, 1989), where appropriate biostratigraphic and ecostratigraphic datums have been extended. See Table 1 for explanation of symbols. Published with permission of the GCSSEPM Foundation.

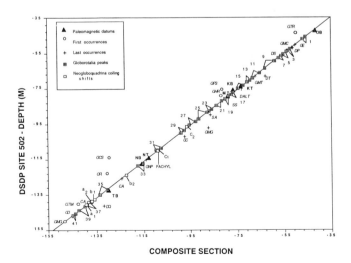

FIG. 18.—Graphic correlation plot of DSDP Site 502 Pliocene versus the composite standard (based on Dowsett, 1989), where appropriate biostratigraphic and ecostratigraphic datums have been extended. See Table 1 for explanation of symbols.

of relatively low sea level and condensed sections (low sediment accumulation rates) represent periods of high sea level (Fig. 20). Within the constraints of the graphic correlation model, high and low sediment accumulation rates in ODP Core 625B were compared to the sequence boundaries and condensed sections, respectively, of Haq and others (1987; Fig. 20), Pacht and others (1990), and Wornardt and Vail (1991) for the Gulf Coast. The 5.0-Ma, 4.0-Ma, and 2.0-Ma flooding surfaces correlated to low or decreasing sediment accumulation rates in Core 625B. The 2.45-Ma flooding surface of Pacht and others (1990) and Wornardt and Vail (1991) also appears to correlate with reduced sedimentation. Conversely, sequence boundaries at 3.8 Ma and 2.4 Ma correlated to relatively high or increasing accumulation rates. The 4.2-Ma sequence boundary was not resolved by graphic correlation in ODP Hole 625B because this part of the section exhibits no significant change in sediment accumulation rate (Figs. 19, 20). High accumulation rates in association with the 3.4-Ma and 2.7-Ma flooding surfaces and low accumulation rates at the 3.0-Ma and 2.6-Ma (Wornardt and Vail, 1991) sequence boundaries, however, violate the gross generalization between sea level and accumulation rate, and may be the result of delta-lobe switching and proximity of sediment supply (e.g., Martin and others, 1990a; see also Self and Scott, 1993). The sequence boundaries recognized by previous workers in the Gulf Coast, and by us on the basis of changes in sediment accumulation rate, are largely in agreement with those of Krantz (1991) for the Atlantic Coastal Plain.

PALEOCEANOGRAPHIC IMPLICATIONS

Because the zonation is recognizable throughout the tropical Atlantic and adjacent basins of the Caribbean and Gulf of Mexico, the Pleistocene subzones appear to primarily represent the repeated invasions (from the Indian Ocean) and regional (Atlantic) extinctions of *G. menardii* (and reciprocal contractions and expansions of *G. inflata*). The migrations appear to be a response to latitudinal shifts in water masses of the Atlantic Ocean (Kaneps, 1970). Water mass shifts, in turn, are driven by changes in northern and southern hemisphere ice volume and shifts in the position of the Subtropical Convergence and Antarctic Polar Front (Ruddiman, 1971; Stainforth and others,

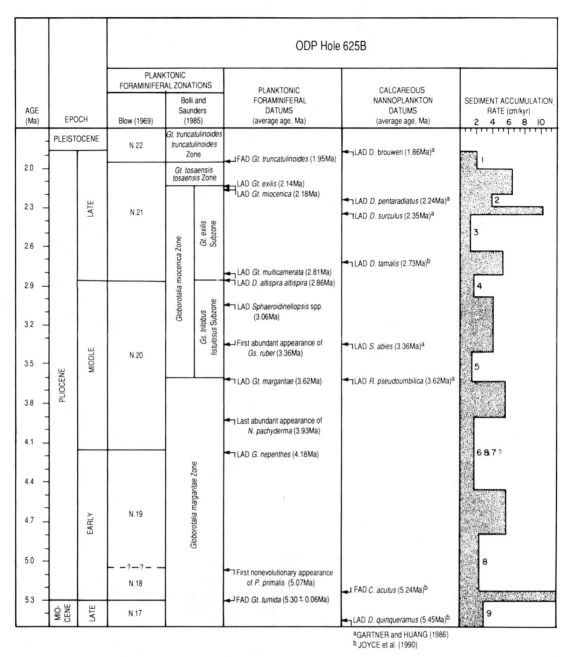

Fig. 20.—Sediment accumulation rates of ODP Core 625B correlated to the condensed sections of Shaffer (1990) and the sequence chronostratigraphy of Haq and others (1987). Published with permission of the GCSSEPM Foundation.

1975, p. 49–53; Ruddiman and McIntyre, 1976; Haq, 1982; McIntyre and others, 1989; Morley, 1989).

The behavior of the *G. menardii* complex and *G. inflata* is consistent with our recognition of major hiatuses within the Pleistocene section (Fig. 11). In most cases, there is an increase in *G. menardii* abundance associated with sea-level rise; conversely, *G. inflata* tends to increase (sometimes abruptly) in the vicinity of sequence boundaries. Ecozone-based sequence boundaries recognized herein (Fig. 11) lie near positive (even-numbered; glacial) excursions of the oxygen isotope curve (stages 1–21) or transitions between oxygen isotope stages. The sequence boundaries in the vicinity of the Y/X, W/V1, and S3/R1 zonal boundaries correspond to transitions from maximum to minimum Quaternary winter temperatures of Briskin and Berggren (1975, Fig. 10, curve T_w), and those associated with the T4/S1 and R2/R3-R3/Q1 boundaries correspond to minima of the curve. Sequence boundaries near the U/T and T3/T4 boundaries lie in cool-to-warm transitions on the winter temperature curve, although they are very close to minima. Only the V2/V3 sequence boundary lies squarely in the middle of a cool-to-warm transition on the temperature curve, although it appears to correspond to a slight decrease in temperature superimposed on the overall warming trend at this time. Martin and others (1990a, b) noted highly variable sedimentation rates

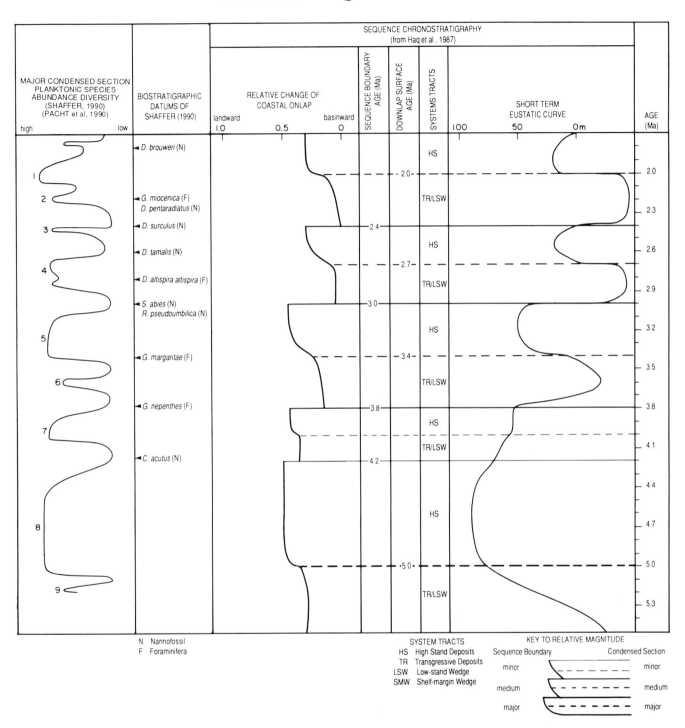

Fig. 20.—Continued

during Zone V in the northeast Gulf of Mexico, as did Ruddiman (1971) in the equatorial Atlantic, that may influence the expression of this sequence boundary. Interestingly, the coolest portions of the temperature curve occur at approximately 400,000 year intervals, which represent an eccentricity-driven portion of Milankovitch cycles (Briskin and Berggren, 1975; Pisias and Imbrie, 1986).

In some cases, the zonation suggested sequence boundaries where no obvious faunal or floral discontinuities existed (Fig. 11). Shackleton (1987), for example, concluded that sea-level fall (as reflected by the oxygen isotope curve) significantly exceeded that of the last glacial maximum during stages 12 and 16. The (presumably) eustatic curves of most Gulf Coast workers (Fig. 11) indicate significant sea-level rise from about stages 16 to 12, however. Only Pacht and others (1990) indicate a sequence boundary at about stage 12, which corresponds to the V3/U boundary discussed previously. The U/T subzonal boundary corresponds to sea-level fall of stage 16.

The occurrence of *Globorotalia* may also be related to the depth and periodic reorganization of the thermocline in the Atlantic Ocean and adjacent basins (Imbrie and others, 1992). Speciation within the *Globorotalia* complex is closely related to the position of the thermocline in the water column (Keller, 1985; Kennett and others, 1985; Stanley and others, 1988). Similarly, the occurrence of *Globorotalia* in the Pleistocene may be related to changes in water mass structure on ecological (as opposed to evolutionary) time scales (Fairbanks and Wiebe, 1980; Fairbanks and others, 1980, 1982; Mix and others, 1986a, b; McIntyre and others, 1989). Based on ^{14}C dates, Jones (1987) found that *G. tumida* repopulated the equatorial Atlantic during the Holocene at 10,200 ka, whereas *G. menardii* did not reappear until about 6700 ka. Healy-Williams (1989) determined that subtle changes in morphology of *G. inflata* and *G. hirsuta* reflected the thermal structure of the water column in the western North Atlantic. Also, Schweitzer and Lohmann (1991) and Lohmann (1992) found that changes in the oxygen isotope ratios of *G. menardii* and *G. tumida* reflected the depth (temperature) at which the species dwelled, and corresponded to changes in the pycnocline (see also Emiliani, 1971; Deuser and Ross, 1989).

The *Globorotalia* curve thus represents a composite of ice volume (sea-level change) and water mass structure (temperature, upwelling, productivity). Changes in water mass structure as opposed to temperature or ice volume (sea level) may account for the discrepancies between the ecozonation and the oxygen isotope curve below stage 5. Changes in water mass structure may also account for discrepancies between CLIMAP Sea Surface Temperatures and core top assemblages noted by Emiliani and Ericson (1991). Conceivably, the abundance and shape changes of *Globorotalia* spp. in deep sea cores could be related to high-resolution records of sea-level change from the continental margin that would allow separation of the effects of ice volume, temperature, and water mass structure that are confounded in the oxygen isotope record. Prior to the onset of major northern hemisphere ice sheets, the Pliocene *Globorotalia* curves appear to indicate changes in water mass structure in response to both the closing of the Isthmus of Panama and the fluctuations in volume of Antarctic ice sheets (Fletcher, 1993).

CONCLUSION

Graphic correlation of subzonal boundaries coupled with available oxygen isotope, biostratigraphic, and magnetostratigraphic datums demonstrates the utility of planktic foraminiferal assemblage zones in subdivision of the Pleistocene. Graphic correlation of subzonal boundaries in conjunction with biostratigraphic datums also reveals the occurrence of anomalously low (erosionally-truncated, delta-depressed) and high (apparently reworked) biostratigraphic markers, in some cases when there are no obvious changes in fossil assemblages or lithology. The zonation is recognizable in exploration wells because the much higher sediment accumulation rates of the western Gulf negate the effect of the use of rotary drill cuttings (R. Fillon, pers. commun., 1989; Fillon, 1990).

The zonation provides a rapid and consistent subdivision of the Pleistocene (based on only two easily recognizable species) that can be used to build (via graphic correlation) a sequence stratigraphic framework. Using graphic correlation, the biostratigrapher can test the synchroneity of biostratigraphic (including ecostratigraphic) datums on a local (lease block, field) and regional (parallel and perpendicular to depositional strike) basis. Bio- and ecostratigraphic, paleoecologic (e.g., faunal and floral discontinuities), lithostratigraphic (well-log), and oxygen isotope datums of cores, wells, and outcrops from the Gulf of Mexico and adjacent basins may all be integrated simultaneously into a single data base to detect potential sequence boundaries. Hence, graphic correlation is a very powerful tool; the paleontologist can build viable exploration models that can be used to predict the occurrence of reservoir sands. The geologist or geophysicist can integrate their interpretations, based primarily on well-logs and seismic sections, within the sequence framework proposed by the biostratigrapher, and vice versa.

Normally, the use of integrated data bases (paleontologic, well-log, seismic) provides higher age resolution of sequence boundaries than ecostratigraphic data alone. In some cases, however, the ecozonation presented herein has delineated sequence boundaries where no obvious faunal or floral discontinuities have been detected. Ecostratigraphically-based sequence boundaries can be tested against seismic and other data to verify (or refute) their existence. Graphic correlation of ecozone boundaries also suggests further avenues of research with regard to paleoceanographic reconstructions and paleoclimate change.

ACKNOWLEDGEMENTS

Exxon Company, U.S.A., supported study of Cores 625B and 502B. Both Unocal and Exxon U.S.A. kindly allowed us to use data from the Garden Banks well. Our thanks especially to Gregg Blake, Bob Eby, Gerald Stude, and Erich Thomas. Keith Mann and an anonymous reviewer made valuable suggestions that improved the manuscript. Thanks also to John Armentrout (Mobil Research, Dallas), Charlotte Brunner (University of Southern Mississippi), Harry Dowsett (U. S. Geological Survey, Reston, VA), Dick Fillon (Texaco, New Orleans), and Bob Scott (Amoco Research, Tulsa) for positive and encouraging critiques of previous work. Discussions with Ken Hood regarding GraphCor software have also been very helpful. Sincere appreciation is extended to Keith Mann, Rich Lane, Jeff Stein, and SEPM (Society for Sedimentary Geology) for the invitation to contribute to this volume. SEPM and the GCSSEPM Foundation allowed us to publish figures from earlier papers. Linda Smallbrook typed the manuscript and Barb Broge drafted figures.

REFERENCES

ARMENTROUT, J. M., 1987, Integration of biostratigraphy and seismic stratigraphy: Plio-Pleistocene, Gulf of Mexico, *in* Barnette, S. C. and Butler, D. M., eds., Innovative Biostratigraphic Approaches to Sequence Analysis: New Exploration Opportunities: Austin, Society of Economic Paleontologists and Mineralogists Gulf Coast Section Eighth Annual Research Conference, Papers and Abstracts, p. 6–14.

ARMENTROUT, J. M., 1991, Paleontologic constraints on depositional modeling: Examples of integration of biostratigraphy and seismic stratigraphy, Plio-Pleistocene, Gulf of Mexico, *in* Weimer, P. and Link, M. H., eds., Seismic Facies and Sedimentary Processes of Submarine Fans and Turbidite Systems: New York, Springer-Verlag, p. 137–170.

ARMENTROUT, J. M. AND CLEMENT, J. F., 1990, Biostratigraphic calibration of depositional cycles: a case study in High Island-Galveston-East Breaks areas, offshore Texas, in Armentrout J. M. and Perkins, B. F., eds., Sequence Stratigraphy as an Exploration Tool: Concepts and Practices in the Gulf Coast: Austin, Society of Economic Paleontologists and Mineralogists Gulf Coast Section Eleventh Annual Research Conference, Papers and Abstracts, p. 21–51.

ARMENTROUT, J. M., ECHOLS, R. J., AND LEE, T. D., 1990, Patterns of foraminiferal abundance and diversity: implications for sequence stratigraphic analysis, in J. M. Armentrout and B. F. Perkins, eds., Sequence Stratigraphy as an Exploration Tool: Concepts and Practices in the Gulf Coast: Austin, Society of Economic Paleontologists and Mineralogists Gulf Coast Section Eleventh Annual Research Conference, Papers and Abstracts, p. 53–58.

BEARD, J. H., 1969, Pleistocene paleotemperature record based on planktic foraminifers, Gulf of Mexico: Gulf Coast Association of Geological Societies, Transactions, v. 18, p. 174–186.

BEARD, J. H., SANGREE, J. B., AND SMITH, L. A., 1982, Quaternary chronology, paleoclimate, depositional sequences, and eustatic cycles: American Association of Petroleum Geologists Bulletin, v. 66, p. 158–169.

BERGGREN, W. A., KENT, D. V., AND VAN COUVERING, J. A., 1985, Neogene geochronology and chronostratigraphy: London, Memoir Geological Society London, v. 10, p. 211–160.

BLOW, W. H., 1969, Late Middle Eocene to Recent planktonic foraminiferal biostratigraphy: Geneva, Proceedings First International Conference on Planktonic Microfossils 1967, v. 1, p. 199–422.

BOLLI, H. M., AND SAUNDERS, J. B., 1985, Oligocene to Holocene low latitude planktic foraminifera, in Bolli, H. M., Saunders, J. B., and Perch-Nielsen, K., eds., Plankton Stratigraphy: Cambridge, Cambridge University Press, p. 155–262.

BRISKIN, M. AND BERGGREN, W. A., 1975, Pleistocene stratigraphy and quantitative paleo-oceanography of tropical North Atlantic Core V16-205, in Saito, T. and Burckle, L. H., eds., Late Neogene Epoch Boundaries: New York, American Museum of Natural History, p. 167–198.

BRUNNER, C. A. AND KEIGWIN, L. D., 1981, Late Neogene biostratigraphy and stable isotope stratigraphy of a drilled core from the Gulf of Mexico: Marine Micropaleontology, v. 6, p. 397–418.

DEUSER, W. G. AND ROSS, E. H., 1989, Seasonally abundant planktonic foraminifera of the Sargasso Sea: Succession, deep-water fluxes, isotopic compositions, and paleoceanographic implications: Journal of Foraminiferal Research, v. 19, p. 268–293.

DOWSETT, H. J., 1988, Diachrony of Late Neogene microfossils in the southwest Pacific Ocean: application of the graphic correlation method: Paleoceanography, v. 3, p. 209–222.

DOWSETT, H. J., 1989, Improved dating of the Pliocene eastern South Atlantic using graphic correlation: Implications for paleobiogeography and paleoceanography: Micropaleontology, v. 35, p. 279–292.

EDWARDS, L. E., 1989, Supplemented graphic correlation: A powerful tool for paleontologists and non-paleontologists: Palaios, v. 4, p. 127–143.

EMILIANI, C., 1971, Depth habitats of growth stages of pelagic foraminifera: Science, v. 173, p. 1122–1124.

EMILIANI, C. AND ERICSON, D. B., 1991, The glacial/interglacial temperature range of the surface water of the oceans at low latitudes, in Taylor, H. P., O'Neill, J. R., and Kaplan, I. R., eds., Stable Isotope Geochemistry: A Tribute to Samuel Epstein: Oxford, Geochemical Society Special Publication 3, p. 223–228.

ERICSON, D. B. AND WOLLIN, G., 1956a, Correlation of six cores from the equatorial Atlantic and the Caribbean: Deep-Sea Research, v. 3, p. 105–125.

ERICSON, D. B. AND WOLLIN, G., 1956b, Micropaleontological and isotopic determinations of Pleistocene climates: Micropaleontology, v. 2, p. 257–270.

ERICSON, D. B. AND G. WOLLIN, 1968, Pleistocene climates and chronology in deep-sea sediments: Science, v. 162, p. 1227–1234.

EWING, M., ERICSON, D. B., AND HEEZEN, B. C., 1958, Sediments and topography of the Gulf of Mexico, in Weeks, C. G., ed., Habitat of Oil: Tulsa, American Association of Petroleum Geologists, p. 995–1053.

FAIRBANKS, R. G., SVERDLOVE, M., FREE, R., WIEBE, P. H., AND BÉ, A. W. H., 1982, Vertical distribution and isotopic fractionation of living planktonic foraminifera from the Panama Basin: Nature, v. 298, p. 841–844.

FAIRBANKS, R. G. AND WIEBE, P. H., 1980, Foraminifera and chlorophyll maximum: Vertical distribution, seasonal succession, and paleoceanographic significance: Science, v. 209, p. 1524–1526.

FAIRBANKS, R. G., WIEBE, P. H., AND BÉ, A. W. H., 1980, Vertical distribution and isotopic composition of living planktonic foraminifera in the western North Atlantic: Science, v. 207, p. 61–63.

FILLON, R. H., 1990, Plio-Pleistocene bio-chronostratigraphy of DSDP Site 502B, Caribbean Sea: Implications for Gulf of Mexico exploration stratigraphy (abs.): American Association of Petroleum Geologists Bulletin, v. 74, p. 653.

FLETCHER, R. R., 1993, Comparative planktonic foraminiferal biostratigraphy of the Colombia Basin and northeast Gulf of Mexico: Unpublished Ph.D. Dissertation, University of Delaware, Newark, 580 p.

GARTNER, S., CHEN, M. P., AND STANTON, R. J., 1983, Late Neogene nannofossil stratigraphy and paleoceanography of the northeastern Gulf of Mexico and adjacent areas: Marine Micropaleontology, v. 8, p. 17–50.

GARTNER, S. AND HUANG, T. C., 1986, Pliocene-Pleistocene biochronology, and climate and sedimentary cycles of the northern Gulf of Mexico (abs.): Geological Society of America, Abstracts with Program, v. 18, p. 143.

HAQ, B. U., 1982, Climatic acme events in the sea and on land, in **editors?**, Climate in Earth History: Washington, D.C., National Academy Press, p. 126–132.

HAQ, B. U., HARDENBOL, J., AND VAIL, P. R., 1987, Chronology of fluctuating sea levels since the Triassic: Science, v. 235, p. 1156–1167.

HAZEL, J., 1989, Chronostratigraphy of Upper Eocene microspherules: Palaios, v. 4, p. 318–329.

HEALY-WILLIAMS, N., 1989, Morphological changes in living foraminifera and the thermal structure of the water column, western North Atlantic: Palaios, v. 4, p. 590–597.

HOOD, K. C., 1986, GraphCor: Interactive Graphic Correlation for Microcomputers: documentation for GraphCor software package, Version 2.2, Houston, 136 p.

IMBRIE, J., BOYLE, E. A., CLEMENS, S. C., DUFFY, A., HOWARD, W. R., KUKLA, G., KUTZBACH, J., MARTINSON, D. G., MCINTYRE, A., MIX, A. C., MOLFINO, B., MORLEY, J. J., PETERSON, L. C., PISIAS, N. G., PRELL, W. L., RAYMO, M. E., SHACKLETON, N. J., AND TOGGWEILER, J. R., 1992, On the structure and origin of major glaciation cycles, 1, Linear responses to Milankovitch forcing: Paleoceanography, v. 7, p. 701–738.

IMBRIE, J. AND IMBRIE, K. P., 1979, Ice Ages: Solving the Mystery: Cambridge, Harvard University Press, 224 p.

JOHNSON, G. W., 1988, Pleistocene planktic foraminiferal biostratigraphy and paleoecology: Northeast Gulf of Mexico: Unpublished Master's Thesis, University of Delaware, Newark, 256 p.

JONES, G. A., 1987, Determining the timing of repopulation of the G. menardii "Complex" in the Atlantic Ocean: A paleoceanographic application of accelerator ^{14}C dating (abs.): Eos, v. 68, p. 329.

JOYCE, J. E., TJALSMA, L. R. C., AND PRUTZMAN, J. M., 1990, High-resolution planktic stable isotope record and spectral analysis for the last 5.35 myr: ODP Site 625, northeast Gulf of Mexico: Paleoceanography, v. 5, p. 507–529.

KANEPS, A. G., 1970, Late Neogene biostratigraphy (planktonic foraminifera), biogeography, and depositional history: Unpublished Ph.D. Dissertation, Columbia University, New York, 179 p.

KELLER, G., 1985, Depth stratification of planktic foraminifers in the Miocene Ocean, in Kennett, J. P., ed., The Miocene Ocean: Paleoceanography and Biogeography: Boulder, Geological Society of America Memoir 163, p. 177–195.

KENNETT, J. P. AND HUDDLESTON, P., 1972a, Abrupt climatic change at 90,000 yr BP: faunal evidence from Gulf of Mexico cores: Quaternary Research, v. 2, p. 384–395.

KENNETT, J. P. AND HUDDLESTON, P., 1972b, Late Pleistocene paleoclimatology, foraminiferal biostratigraphy, and tephrochronology, western Gulf of Mexico: Quaternary Research, v. 2, p. 38–69.

KENNETT, J. P., KELLER, G., AND SRINIVASAN, M. S., 1985, Miocene planktic foraminiferal biogeography and paleoceanographic development of the Indo-Pacific region, in J. P. Kennett, ed., The Miocene Ocean: Paleoceanography and Biogeography: Boulder, Geological Society of America Memoir 163, p. 197–236.

KRANTZ, D. E., 1991, A chronology of Pliocene sea-level fluctuations: The U. S. Middle Atlantic Coastal Plain record: Quaternary Science Reviews, v. 10, p. 163–174.

LAMB, J. L., WORNARDT, W. W., HUANG, T. C., AND DUBE, T. E., 1987, Practical application of Pleistocene eustacy in offshore Gulf of Mexico, in Ross, R. A. and Haman, D., eds., Timing and Depositional History of Eustatic Sequences: Constraints on Seismic Stratigraphy: Lawrence, Cushman Foundation for Foraminiferal Research Special Publication 24, p. 33–39.

LOHMANN, G. P., 1992, Increasing seasonal upwelling in the subtropical South Atlantic over the past 700,000 years: Evidence from deep-living planktonic foraminifera: Marine Micropaleontology, v. 19, p. 1–12.

MARTIN, R. E., 1984, Plio-Pleistocene cyclic sedimentation at East Breaks: an integrated model for deep-water sand exploration: unpublished report, Union Oil Company of California (chart).

MARTIN, R. E. AND FLETCHER, R. R., 1993, Biostratigraphic expression of Plio-Pleistocene sequence boundaries, Gulf of Mexico, in Armentrout, J.M., Bloch, R., Olson, H. C., and Perkins, B. F., eds., Rates of Geologic Processes: Tectonics, Sedimentation, Eustasy and Climate—Implications for Hydrocarbon Exploration: Austin, Society of Economic Paleontologists and Mineralogists Gulf Coast Section Fourteenth Annual Research Conference, p. 119–126.

MARTIN, R. E., NEFF, E. D., JOHNSON, G. W., AND KRANTZ, D. E., 1990a, Quaternary planktic foraminiferal subzonation of the northeast Gulf of Mexico, Columbia Basin (Caribbean Sea), and tropical Atlantic: Graphic correlation of faunal and oxygen isotope datums: Paleoceanography, v. 5, p. 531–555.

MARTIN, R. E., NEFF, E. D., JOHNSON, G. W., AND KRANTZ, D. E., 1990b, Biostratigraphic expression of sequence boundaries in the Pleistocene: The Ericson and Wollin zonation revisited, in Armentrout, J. M., and Perkins, B. F., eds., Sequence Stratigraphy as an Exploration Tool: Concepts and Practices in the Gulf Coast: Austin, Society of Economic Paleontologists and Mineralogists Gulf Coast Section Eleventh Annual Research Conference, p. 229–236.

MARTIN, R. E., NEFF, E. D., JOHNSON, G. W., AND KRANTZ, D. E., 1993, Biostratigraphic expression of Pleistocene sequence boundaries, Gulf of Mexico: Palaios, v. 8, p. 155–171.

MCINTYRE, A., RUDDIMAN, W. F., KARLIN, K., AND MIX, A. C., 1989, Surface water response of the equatorial Atlantic Ocean to orbital forcing: Paleoceanography, v. 4, p. 19–56.

MILLER, F. X., 1977, The graphic correlation method in biostratigraphy, in Kauffman, E. G. and Hazel, J. E., eds., Concepts and Methods of Biostratigraphy: Stroudsburg, Dowden, Hutchinson, and Ross, p. 165–186.

MIX, A. C., RUDDIMAN, W. F., AND MCINTYRE, A., 1986a, Late Quaternary paleoceanography of the tropical Atlantic, 1: Spatial variability of annual ocean sea-surface temperatures, 0–20,000 years B.P.: Paleoceanography, v. 1, p. 43–66.

MIX, A. C., RUDDIMAN, W. F., AND MCINTYRE, A., 1986b, Late Quaternary paleoceanography of the tropical Atlantic, 2: The seasonal cycle of sea surface temperatures, 0–20,000 years B. P.: Paleoceanography, v. 1, p. 339–353.

MORLEY, J. J., 1989, Variation in high-latitude oceanographic fronts in the southern Indian Ocean: An estimation based on faunal changes: Paleoceanography, v. 4, p. 547–554.

NEFF, E. D., 1985, Pre-late Pleistocene paleoclimatology and planktic foraminiferal biostratigraphy of the northeastern Gulf of Mexico: Unpublished M.S. Thesis, University of South Carolina, Columbia, 123 p.

PACHT, J. A., BOWEN, B. E., SHAFFER, B. C., AND POTTER, B. R., 1990, Sequence stratigraphy of Plio-Pleistocene strata in the offshore Louisiana Gulf Coast: Applications to hydrocarbon exploration, in Armentrout, J. M. and Perkins, B. F., eds., Sequence Stratigraphy as an Exploration Tool: Concepts and Practices in the Gulf Coast: Austin, Society of Economic Paleontologists and Mineralogists Gulf Coast Section Eleventh Annual Research Conference, Papers and Abstracts, p. 269–285.

PFLUM, C. E. AND FRERICHS, W. E., 1976, Gulf of Mexico Deep-water Foraminifers: Lawrence, Cushman Foundation for Foraminiferal Research Special Publication 14, 125 p.

PHLEGER, F . B., PARKER, F. L., AND PEIRSON, J. F., 1953, North Atlantic foraminifera: Reports of the Swedish Deep-Sea Expedition, v. 7, p. 1–122.

PISIAS, N. G. AND IMBRIE, J., 1986, Orbital geometry, CO_2, and Pleistocene climate: Oceanus, v. 29, p. 43–49.

POAG, C. W., 1972, Correlation of early Quaternary events in the U. S. Gulf Coast: Quaternary Research, v. 2, p. 447–469.

POAG, C. W. AND VALENTINE, P. C., 1976, Biostratigraphy and ecostratigraphy of the Pleistocene basin, Texas-Louisiana continental shelf: Gulf Coast Association of Geological Societies, Transactions, v. 26, p. 185–254.

PRELL, W. L., 1982, Oxygen and carbon isotope stratigraphy for the Quaternary of Hole 502B: Evidence for two modes of isotopic variability: Initial Report Deep Sea Drilling Project, v. 68, p. 455–464.

PRELL, W. L., IMBRIE, J., MARTINSON, D. G., MORLEY, J. J., PISIAS, N. G., SHACKLETON, N. J., AND STREETER, H. F., 1986, Graphic correlation of oxygen isotope stratigraphy: Application to the late Quaternary: Paleoceanography, v. 1, p. 137–162.

RABINOWITZ, P. D., GARRISON, L. E., MERRELL, W. J., AND KIDD, R., 1985, Ocean Drilling Program Leg 100 report, unpublished hole summary: College Station, Ocean Drilling Program, 151 p.

RABINOWITZ, P. D. AND MERRELL, W. J., 1985, Ocean Drilling Program launches first cruise: Geotimes, v. 30, p. 12–14.

RAGAN, G. M. AND ABBOTT, W. H., 1989, High resolution correlation of Gulf of Mexico Pliocene-Pleistocene sands (abs.): American Association of Petroleum Geologists Bulletin, v. 73, p. 402.

RUDDIMAN, W. F., 1971, Pleistocene sedimentation in the equatorial Atlantic: Stratigraphy and faunal paleoclimatology: Geological Society of America Bulletin, v. 82, p. 283–302.

RUDDIMAN, W. F. AND MCINTYRE, A., 1976, Northeast Atlantic paleoclimatic changes over the past 600,000 years, in Cline, R. M. and Hays, J. D., eds., Investigation of Late Quaternary Paleoceanography and Paleoclimatology: Boulder, Geological Society of America Memoir 145, p. 111–146.

SCHOTT, W., 1935, Die foraminiferen in dem aquatorialen Teil des Atlantischen Ozeans: Wissenschaftliche Ergebnisses Deutsche Atlantische Expedition "Meteor," v. 111, p. 43–134.

SCHWEITZER, P. N. AND LOHMANN, G. P., 1991, Ontogeny and habitat of modern menardiform planktonic foraminifera: Journal of Foraminiferal Research, v. 21, p. 332–346.

SCOTT, R. W., 1991, Chronostratigraphy of Plio-Pleistocene sequences, offshore Louisiana (abs.): American Association of Petroleum Geologists Bulletin, v. 75, p. 669.

SELF, G. A. AND SCOTT, R. W., 1993, Can lithology be predicted from stratal architecture? An empirical approach to the sequence stratigraphy of the Plio-Pleistocene, Gulf of Mexico: Offshore Technology Conference, p. 419–427.

SHACKLETON, N. J., 1987, Oxygen isotopes, ice volume and sea level: Quaternary Science Reviews, v. 6, p. 183–190.

SHAFFER, B. L., 1990, The nature and significance of condensed sections in Gulf Coast late Neogene sequence stratigraphy: Gulf Coast Association of Geological Societies, Transactions, v. 40, p. 767–776.

SHAW, A. B., 1964, Time in Stratigraphy: New York, McGraw-Hill, 365 p.

SPOTZ, B., 1988, Pliocene planktic foraminiferal biostratigraphy and paleoecology, northeast Gulf of Mexico: Unpublished M.S. Thesis, University of Delaware, Newark, 168 p.

STAINFORTH, R. M., LAMB, J. L., LUTERBACHER, H., BEARD, J. H., AND JEFFORDS, R. M., 1975, Cenozoic planktonic foraminiferal zonation and characteristics of index forms, Lawrence, University of Kansas Paleontological Contributions 62, Allen Press, 425 p.

STANLEY, S. M., WETMORE, K. L., AND KENNETT, J. P., 1988, Macroevolutionary differences between the two major clades of Neogene planktonic foraminifera: Paleobiology, v. 14, p. 235–249.

THUNELL, R. C., 1984, Pleistocene planktic foraminiferal biostratigraphy and paleoclimatology of the Gulf of Mexico, in Healy-Williams, N., ed., Principles of Pleistocene Stratigraphy Applied to the Gulf of Mexico: Boston, International Human Resources Development Co., p. 25–64.

THUNELL, R. C. AND REYNOLDS, L. A., 1984, Sedimentation of planktonic foraminifera: seasonal changes in species flux in the Panama Basin: Micropaleontology, v. 30, p. 243–262.

VAN DONK, J., 1976, ^{18}O record of the Atlantic Ocean for the entire Pleistocene epoch, in Cline, R. M. and Hays, J. D., eds., Investigation of Late Quaternary Paleoceanography and Paleoclimatology: Boulder, Geological Society of America Memoir 145, p. 449–464.

WILLIAMS, D. F., 1984, Correlation of Pleistocene marine sediments of the Gulf of Mexico and other basins using oxygen isotope stratigraphy, in Healy-Williams, N., ed., Principles of Pleistocene Stratigraphy Applied to the Gulf of Mexico: Boston, International Human Resources Development Co., p. 65–118.

WORNARDT, W. W. AND VAIL, P. R., 1990, Revision of the Plio-Pleistocene cycles and their application to sequence stratigraphy of shelf and slope sediments in the Gulf of Mexico, in Armentrout, J. M. and Perkins, B. F., eds., Sequence Stratigraphy as an Exploration Tool: Concepts and Practices in the Gulf Coast: Austin, Society of Economic Paleontologists and Mineralogists Gulf Coast Section Eleventh Annual Research Conference, Papers and Abstracts, p. 391–397.

WORNARDT, W. W. AND VAIL, P. R., 1991, Tentative revision of the global Plio-Pleistocene sequences based on the sequence stratigraphy in the Gulf of Mexico (abs.): American Association of Petroleum Geologists Bulletin, v. 75, p. 1425.

APPENDIX
GRAPHIC CORRELATION AND ASSOCIATED REFERENCES

Compiled by
Keith Olin Mann and Robert W. Pierce

AMSDEN, T. W. AND SWEET, W. C., 1983, Upper Bromide Formation and Viola Group (Middle and Upper Ordovician) in eastern Oklahoma: Oklahoma Geological Survey Bulletin, 132, 76 p.

ANSTEY, R. L. AND RABBIO, S. F., 1990, Regional bryozoan biostratigraphy and taphonomy of the Edenian stratotype (Kope Foramtion, Cincinnati Area): graphic correlation and gradient analysis: Palaios, v. 4, p. 574–584.

AURISANO, R. W., 1988, Time correlation and reconstruction of geologic events in the Morondava Basin of Madagascar using the technique of graphic correlation: Abstracts of the Proceedings of the Twentieth Annual Meeting of the American Association of Stratigraphic Palynologists, Palynology, v. 12, p. 231.

AURISANO, R. W., 1994, Palynostratigraphy, graphic correlation, and the Albian-Cenomian boundary, offshore Gabon and Congo, in Lane, H. L., Blake, G., and MacLeod, N., eds., Graphic Correlation and the Composite Standard: the Methods and their Applications: Houston, Society of Economic Paleontologists and Mineralogists Research Conference, p. 2.

BAESEMANN, J. F. AND BRENCKLE, P. L., 1994, Application of composite standard methodology for local correlation of the Mississippian-Pennsylvanian Lisburne Group in northeastern Alaska, in Lane, H. L., Blake, G., and MacLeod, N., eds., Graphic Correlation and the Composite Standard: the Methods and their Applications: Houston, Society of Economic Paleontologists and Mineralogists Research Conference, p. 3.

BARON, J. A., 1975, Marine diatom biostratigraphy of the Upper Miocene-Lower Pliocene strata of southern California: Journal of Paleontology, v. 49, p. 619–632.

BARON, J. A., 1976, Revised Miocene and Pliocene diatom biostratigraphy of Newport Bay, Newport Beach, California: Marine Micropaleontology, v. 1, p. 27–63.

BARON, J. A., 1976, Middle-Miocene-Lower Pliocene marine diatom and silicoflagellate correlations in the California area: The Neogene Symposium, Society of Economic Paleontologists and Mineralogists Pacific Section, p. 117–124.

BECKER, R. T., HOUSE, M. R., AND KIRCHGASSER, W.T., 1992, Devonian goniatite biostratigraphy and timing of facies movements in the Frasnian of the Canning Basin, Western Australia, in Hailwood, E. A. and Kidd, R. B., eds., High Resolution Stratigraphy: London, Geological Society of London, Special Publication 70, p. 293–321.

BERGSTROM, S. M., 1989, Graphic correlation and event-stratigraphic assessment of Ordovician K-bentonite beds in North America and Northwestern Europe: Geological Society of America, Abstracts with Programs, v. 21, p. 133–134.

BERGSTROM, S. M., 1989, Use of graphic correlation for assessing event-stratigraphic significance and Trans-Atlantic relationships of Ordovician K-bentonites: The Proceedings of the Academy of the Estonia S.S.R., Geology, v. 38, p. 55–59.

BERGSTROM, S. M., 1991, The Baltoscandic Ordovician conodont faunal succession; a new composite standard based on sections in central Sweden: Programs and Abstracts of the Canadian Paleontology Conference I and Pander Society Meeting, University of British Columbia, Vancouver, p. 15.

BLAKE, G. H. AND GARY, A., 1994, The Application of quantitative biostratigraphy of complex tectonic settings, offshore Trinidad, in Lane, H. L., Blake, G., and MacLeod, N., eds., Graphic Correlation and the Composite Standard: the Methods and their Applications: Houston, Society of Economic Paleontologists and Mineralogists Research Conference, p. 4.

BLANK, R. G., 1979, Applications of probabilistic biostratigraphy to chronostratigraphy: Journal of Geology, v. 87, p. 647–670.

BOGGS, S., 1987, Principles of Sedimentology and Stratigraphy: Columbus, Merrill Publishing Company, 784 p.

BRENNER, R. L. AND MCHARGUE, T. R., 1988, Integrated Stratigraphy: Englewood Cliffs, Prentice Hall, 419 p.

BROWER, J. C. AND BUSSEY, D. T., 1985, A comparison of five quantitative techniques for biostratigraphy, in Gradstein, F. M., Agterberg, F. P., Brower, J. C., and Schwarzacher, W. S., eds., Quantitative Stratigraphy: Dordrecht, Reidel Publishing, p. 279–306.

BUZAS, M. A., 1990, Another look at confidence limits for species proportions: Journal of Paleontology, v. 64, p. 842–843.

DELL, R., KEMPLE, W. G., AND TOVEY, P., 1992, Heuristically solving the stratigraphic correlation problem: Institute of Industrial Engineers First Industrial Engineering Research Conference Proceedings, p. 293–297.

DENNE, R. A., 1994, Operational applications of graphic correlation, in Lane, H. L., Blake, G., and MacLeod, N., eds., Graphic Correlation and the Composite Standard: the Methods and their Applications: Houston, Society of Economic Paleontologists and Mineralogists Research Conference, p. 5.

D'HONDT, S. AND HERBERT, T. D., 1992, Comment: comment and reply on "Hiatus distributions and mass extinctions at the Cretaceous/Tertiary boundary": Geology, v. 20, p. 380–381.

DERBY, J. R., 1986, Great progress but no decision by the Cambrian-Ordovician Boundary Committee: Palaios, v. 1, p. 98–103.

DERBY, J. R., BAUER, J. A., CREATH, W. B., DRESBACH, R. I., ETHINGTON, R. L., LOCH, J. D., STITT, J. H., MCHARGUE, T. R., MILLER, J. F., MILLER, M. A., REPETSKI, J. E., SWEET, W. C., TAYLOR, J. F., AND WILLIAMS, M., 1991, Biostratigraphy of the Timbered Hills, Arbuckle and Simpson Groups, Cambrian and Ordovician Oklahoma: a review of correlation tools and techniques available to the explorationist, in Johnson, K. S., ed., Late Cambrian-Ordovician Geology of the Southern Midcontinent, 1989 Symposium, Oklahoma Geological Survey Circular 92, p. 15–41.

DOWSETT, H. J., 1986, Application of graphic correlation to Pliocene deep sea Atlantic and Pacific records of planktonic microfossils: Geological Society of America, Abstracts with Programs, v. 18, p. 587.

DOWSETT, H. J., 1988, A biochronological model for correlation of Pliocene marine sequences; application of the graphic correlation method: Unpublished Ph.D. Dissertation, Brown University, Providence, 325 p.

DOWSETT, H. J., 1988, Diachrony of Late Neogene microfossils in the Southwest Pacific Ocean: application of the graphic correlation method: Paleoceanography, v. 3, p. 209–222.

DOWSETT, H. J., 1989, Improved dating of the Pliocene of the eastern South Atlantic using graphic correlation: implications for paleobiogeography and paleoceanography: Micropaleontology, v. 35, p. 279–292.

DOWSETT, H. J., 1989, Application of the graphic correlation method to Pliocene marine sequences: Marine Micropaleontology, v. 14, p. 3–32.

EDWARDS, L. E., 1978, Range charts and no-space graphs: Computers and Geosciences, v. 4. p. 247–255.

EDWARDS, L. E., 1982, Numerical and semi-objective biostratigraphy: review and predictions: Third North American Paleontological Convention, Proceedings, v. 1, p. I47-I52.

EDWARDS, L. E., 1982, Quantitative biostratigraphy: the methods should suit the data, in Cubitt, J. M. and Reyment, R. A., eds., Quantitative Stratigraphic Correlation: Chichester, John Wiley and Sons, p. 45–60.

EDWARDS, L. E., 1984, Insights on why graphic correlation (Shaw's method) works: Journal of Geology, v. 92, p. 583–597.

EDWARDS, L. E., 1985, Insights on why graphic correlation (Shaw's method) works: a reply: Journal of Geology, v. 93, p. 507–509.

EDWARDS, L. E., 1985, Graphic correlation using various kinds of data: Abstracts of the Proceedings of the Seventeenth Annual Meeting of the American Association of Stratigraphic Palynologists, Palynology, v. 9, p. 240–241.

EDWARDS, L. E., 1989, Supplemental graphic correlation: a powerful tool for paleontologists and nonpaleontologists: Palaios, v. 4, p. 127–143.

EDWARDS, L. E., 1989, Dinoflagellate cysts from the Lower Tertiary Formations, Haynesville Cores, Richnmond County, Virginia: Washington, D. C., United States Geological Survey Professional Paper 1489-C.

EDWARDS, L. E., 1991, Quantitative biostratigraphy, in Gilinsky, N. L. and Signor, P. W., eds., Analytical Paleontology: Knoxville, Paleontological Society Short Courses in Paleontology 4, p. 39–58.

EDWARDS, L. E. AND BEAVER, R. J., 1978, The use of a paired comparison model in ordering stratigraphic events: Journal of Mathematical Geology, v. 10, p. 261–272.

EICHER, D. L., 1968, Geologic Time: Englewood Cliffs, Prentice-Hall, 149 p.

FISHER, C. G, 1993, Graphic correlation across a facies change marking the location of a Middle to Late Cenomanian ocean front: Geological Society of America Northeastern Section, Abstracts with Programs, v. 25, p. 15.

FORDHAM, B. G., 1992, Chronostratigraphic calibration of mid-Ordovician to Tournaisian conodont zones: a compilation from recent graphic correlation and isotope studies: Geological Magazine, v. 129, p. 709–721.

FREDERIKSEN, N. O., 1988, Sporomorph biostratigraphy, floral changes, and paleoclimatology, Eocene and earliest Oligocene of the Eastern Gulf Coast: Washington, D. C., United States Geological Survey Professional Paper 1448, 68 p.

FREDERIKSEN, N. O., 1991, Midwayan (Paleocene) pollen correlations in the eastern United States: Micropaleontology, v. 37, p. 101–123.

GLENISTER, B. F., 1987, Mississippian carbonates of the LeGrand area: ancienct analogs of the Bahama Banks: Iowa City, Geological Society of Iowa Guidebook 47, 37 p.

GRADSTEIN, F. M. AND AGTERBERG, F. P., 1985, Biostratigraphic correlation part III, in Gradstein, F. M., Agterberg, F. P., Bower, J. C., and Schwarzacher, W. S., eds., Quantitative Stratigraphy: Dordrecht, Reidel Publishishing Co., p. 309–357.

GRADSTEIN, F. M., KAMINSKI, M. A., BERGGREN, W. A., KRISTIANSEN, I. L., AND D'IORO, M. A., 1994, Cenozoic biostratigraphy of the North Sea and Labrador Shelf: Micropaleontology, v. 40, sup. 1–152.

HARPER, C. W. AND CROWLEY, K. D., 1985, Insights on why graphic correlation (Shaw's method) works: a discussion: Journal of Geology, v. 93, p. 503–506.

HAY, W. H. AND SOUTHAM, J. R., 1978, Quantifying biostratigraphc correlation: Annual Review of Earth Planetary Sciences, v. 6, p. 353–375.

HAZEL, J. E., 1989, Chronostratigraphy of Upper Eocene microspherules: Palaios, v. 4, p. 318–329.

HAZEL, J. E., 1994, Graphic correlation of the Eocene-Oligocene boundary stratotype at Massignano, Italy, in Lane, H. L., Blake, G., and MacLeod, N., eds., Graphic Correlation and the Composite Standard: the Methods and their Applications: Houston, Society of Economic Paleontologists and Mineralogists Research Conference, p. 18.

HAZEL, J. E., EDWARDS, L. E., AND BYBELL, L. M., 1984, Significant unconformities and the hiatuses represented by them in the Paleocene of the Atlantic and Gulf Coastal province, in Schlee, J. S., ed., Interregional Unconformities and Hydrocarbon Accumulation: Tulsa, American Association of Petroleum Geologists Memoir 36, p. 59–66.

HAZEL, J. E., MUMMA, M. D., AND HUFF, W. J., 1980, Ostracode biostratigraphy of the Lower Oligocene (Vicksburgian) of Mississippi and Alabama: Gulf Coast Association of Geological Societies, Transactions, v. 30, p. 361–401.

HOOD, K. C., 1993, Evaluating the use of average composite models in the graphic correlation technique: Geological Society of America Northeastern Section, Abstracts with Programs, v. 25, p. 24.

HOOD, K. C. AND FOSTER, D. W., 1986, Interactive graphic correlation using a microcomputer: Geological Society of America, Abstracts with Programs, v. 18, p. 639.

HUGHES, R. A., 1995, The durations of Silurian graptolote zones: Geological Magazine, v. 132, p. 113–115.

IVANY, L. AND SALAWITCH, R. J., 1993, Carbon isotopic evidence for biomass burning at the K-T boundary: comment and reply: Geology, v. 21, p. 1150–1151.

JELETZKY, J. A., 1965, Is it possible to quantify biochronological correlation?: Journal of Paleontology, v. 39, p. 135–140.

KELLER, G. AND MACLEOD, N., 1993, Carbon isotopic evidence for biomass burning at the K-T boundary: comment and reply: Geology, v. 21, p. 1149–1150.

KEMPLE, W. G., 1991, Stratigraphic correlation as a constrained optimization problem: Unpublished Ph.D. Dissertation, University of California, Riverside, 189 p.

KEMPLE, W. G., SADLER, P. M., AND STRAUSS, D. J., 1989, A prototype constrained optimization solution to the time correlation problem, in Agterberg, F. P. and Bonham-Carter, G. F., eds., Statistical Applications in the Earth Sciences: Ottawa, Geological Survey of Canada Paper 89–9, p. 417–425.

KIRCHGASSER, W. T. AND KLAPPER, G., 1992, Graphic correlation using litho- and biostratigraphic markers in the uppermost Middle and Upper Devonian (Frasnian) of New York State: Geological Society of America Northeastern Section, Abstracts with Programs, v. 24, p. 32.

KIRKLAND, J. I., 1987, Integrating high resolution event and biotic data using graphic correlation; examples from the Mid-Cretaceous of Arizona and Colorado: Geological Society of America Rocky Mountain Section, Abstracts with Programs, v. 19, p. 287.

KLAPPER, G., 1989, The effect of graphic correlation, multielement taxonomy, and shape analysis on Upper Devonian, Frasnian conodont biostratigraphy: Geological Society of America North-Central Section, Abstracts with Programs, v. 21, p. 18.

KLAPPER, G. AND FOSTER, C. T., 1993, Shape analysis of Frasnian species of the Late Devonain conodont genus *Palmatolepis*: Paleontological Society Memoir 32 (Journal of Paleontology, v. 67 supplement to no. 4), 35 p.

KLAPPER, G. AND KIRCHGASSER, W. T., 1992, Zonal and graphic correlation of the New York and Australian Upper Devonian (Frasnian) conodont and ammonoid sequences: Geological Society of America North-Central Section, Abstracts with Programs, v. 24, p. 26.

KLAPPER, G., KIRCHGASSER, W. T., AND BAESEMANN, J. F., 1993, Graphic correlation of the Frasnian Upper Devonian composite: Geological Society of America Northeastern Section, Abstracts with Programs, v. 25, p. 30.

KLEFFNER, M. A., 1989, A conodont-based Silurian chronostratigraphy: Geological Society of America Bulletin, v. 101, p. 904–912.

KLEFFNER, M. A., 1993, A revised conodont- and graptolite-based Silurian Chronostratigraphy: Geological Society of America Northeastern Section, Abstracts with Programs, v. 25, p. 30.

KLEFFNER, M. A., 1994, A conodont- and graptolite-based Silurian chronostratigraphy and development of a Silurian time scale by graphic correlation, in Lane, H. L., Blake, G., and MacLeod, N., eds., Graphic Correlation and the Composite Standard: the Methods and their Applications: Houston, Society of Economic Paleontologists and Mineralogists Research Conference, p. 8.

KREBS, W., N., 1994, Graphic correlation of the nonmarine Late Miocene Chalk Hills Formation, western Snake River Basin, Idaho, in Lane, H. L., Blake, G., and MacLeod, N., eds., Graphic Correlation and the Composite Standard: the Methods and their Applications: Houston, Society of Economic Paleontologists and Mineralogists Research Conference, p. 9.

LANE, H. R., AURISANO, R. W., AND STEIN, J. A., 1993, Worldwide and local composite standards: having your cake and eating it too: Geological Society of America Northeastern Section, Abstracts with Programs, v. 25, p. 32.

LANE, H. R., FRYE, M. W., AND COUPLES, G. D., 1994, The biothem approach: a useful biostratigraphically-based sequence-stratigraphic procedure: Willi Ziegler-Festschrift I, Sonderdruk aus CFS, Bd.168, p. 281–297.

LEMON, R. R., 1990, Principles of Stratigraphy: Columbus, Merrill Publishing Co., 559 p.

LENZ, A. C., JIN, J., MCCRACKEN, A. D., UTTING, J., AND WESTROP, S. R., 1993, Paleoscene 15, Paleozoic biostratigraphy: Geoscience Canada, v. 20, p. 41–73.

MACLEOD, N., 1991, Punctuated anagenesis and the importance of stratigraphy to paleobiology: Paleobiology, v. 17, p. 167–188

MACLEOD, N., 1993, How complete are Cretaceous/Tertiary (K/T) boundary sections II: high latitude extension of the K/T composite standard reference section: Geological Society of America Northeastern Section, Abstracts with Programs, v. 25, p. 36.

MACLEOD, N., 1994, Database Systems, in Lane, H. L., Blake, G., and MacLeod, N., eds., Graphic Correlation and the Composite Standard: the Methods and their Applications: Houston, Society of Economic Paleontologists and Mineralogists Research Conference, p. 3.

MACLEOD, N., 1994, Graphic correlation and the Cretaceous-Tertiary (K/T) boundary controversy: countering old objections with new data, in Lane, H. L., Blake, G., and MacLeod, N., eds., Graphic Correlation and the Composite Standard: the Methods and their Applications: Houston, Society of Economic Paleontologists and Mineralogists Research Conference, p. 11.

MACLEOD, N. AND KELLER, G., 1991, Hiatus distributions and mass extinctions at the Cretaceous/Tertiary boundary: Geology, v. 19, p. 497–501.

MACLEOD, N. AND KELLER, G., 1991, How complete are Cretaceous/Tertiary boundary sections? A chronostratigraphic estimate based on graphic correlation: Geological Society of America Bulletin, v. 103, p. 1439–1457.

MACLEOD, N. AND KELLER, G., 1992, Reply: comment and reply on "Hiatus distributions and mass extinctions at the Cretaceous/Tertiary boundary": Geology, v. 20, p. 381–382.

MARSHALL, C. R., 1990, Confidence intervals on stratigraphic ranges: Paleobiology, v. 16, p. 1–10.

MARTIN, R. E. AND FLETCHER, R., 1993, Graphic expression of Plio-Pleistocene sequence boundaries, Gulf of Mexico: Geological Society of America Northeastern Section, Abstracts with Programs, v. 25, p. 37.

MARTIN, R. E. AND FLETCHER, R., 1993, Biostratigraphic expression of Plio-Pleistocene sequence boundaries, Gulf of Mexico, in Armentrout, J. M., Bloch, R., Olson, H., Perkins, B., eds., Rates of Geologic Processes: Gulf Coast Section, Society of Economic Paleontologists and Mineralogists Foundation Fourteenth Annual Research Conference, p. 119–126.

MARTIN, R. E. AND FLETCHER, R. R.,1994, Graphic correlation of Plio-Pleistocene sequence boundaries, Gulf of Mexico: oxygen isotopes, ice volume and sea level, *in* Lane, H. L., Blake, G., and MacLeod, N., eds., Graphic Correlation and the Composite Standard: the Methods and their Applications: Houston, Society of Economic Paleontologists and Mineralogists Research Conference, p. 12.

MARTIN, R. E., JOHNSON, G. W., NEFF, E. D., AND KRANTZ, D. E., 1990, Subzonation of the marine Pleistocene; graphic correlation of microfossil, paleomagnetic and isotope datums of the tropical Atlantic and its marginal seas: Geological Society of America, Abstracts with Programs, v. 22, p. 43.

MARTIN, R. E., JOHNSON, G. W., NEFF, E. D., AND KRANTZ, D. E., 1990, Quaternary planktonic foriminiferal assemblage zones of the northwest Gulf of Mexico, Columbia Basin (Caribbean sea), and tropical Atlantic Ocean: graphic correlation of microfossil and oxygen isotope datums: Paleoceanography, v. 5, p. 531–555.

MARTIN, R. E. AND NEFF, E. D., 1993, Biostraigraphic expression of Pleistocene Sequence Boundaries, Gulf of Mexico: Palaios, v. 8, p. 155–171.

MARTIN, R. E., NEFF, E. D. , JOHNSON, G. W., AND KRANTZ, D. E., 1990, Graphic correlation of Quaternary bio-, magneto-, and chemostratigraphic datums of the tropical Atlantic and American Mediterranean; resolution of faunal and stable isotope signals: American Association of Petroleum Geologists Bulletin, v. 74, p. 713.

MARTIN, R. E., NEFF, E. D., JOHNSON, G. W., AND KRANTZ, D. E., 1990, Biostratigraphic expression of sequence boundaries in the Pleistocene: the Ericson and Wollin Zonation revisited: Gulf Coast Section, Society of Economic Paleontologists and Mineralogists Foundation 11th Annual Research Conference, Program and Abstracts, p. 229–236.

MARTÍNEZ, J. I., MUÑOZ, F., AND VELEZ, M. I., 1994, Biostratigraphic analysis of the Lower Magdalena Basin (northern Columbia), preliminary results using the graphic correlation technique and principal componant analysus, *in* Lane, H. L., Blake, G., and MacLeod, N., eds., Graphic Correlation and the Composite Standard: the Methods and their Applications: Houston, Society of Economic Paleontologists and Mineralogists Research Conference, p. 13.

MATTHEWS, R. K., 1974, Dynamic Stratigraphy: Englewood Cliffs, Prentice-Hall, 370 p.

MATTHEWS, R. K., 1984, Dynamic Stratigraphy (second edition): Englewood Cliffs, Prentice-Hall, 489 p.

MCLAUGHLIN, P. P. AND HOOD, K. C., 1994, Regional chronostratigraphic analysis of north Africa and the Arabian Plate—An example of graphic correlation analysis using maximum and average composite sections, *in* Lane, H. L., Blake, G., and MacLeod, N., eds., Graphic Correlation and the Composite Standard: the Methods and their Applications: Houston, Society of Economic Paleontologists and Mineralogists Research Conference, p. 10.

MCKINNEY, F. K., 1991, Exercises in Invertebrate Paleontology: Boston, Blackwell Scientific Publications, 272 p.

MELNYK, D. H., ATHERSUCH, J., AND SMITH, D. G., 1992, Estimating the dispersion of biostratigraphic events in the subsurface by graphic correlation: an example from the Late Jurassic of the Wessex Basin, UK: Marine and Petroleum Geology, v. 9, p. 602–607.

MIALL, A. D., 1984, Principles of Sedimentary Basin Analysis: New York, Springer-Verlag, 490 p.

MIALL, A. D., 1990, Principles of Sedimentary Basin Analysis (second edition): New York, Springer-Verlag, 668 p.

MIKAN, F. A. AND SWEET, W. C., 1974, Graphic correlation of key Permo-Triassic sections in Kashmir, Pakistan and Iran: North-Central Section, 8th Annual Meeting, Geological Society of America Abstracts, v. 6, p. 531.

MILLER, F. X., 1977, Biostratigraphic correlation of the Mesaverde Group in southwestern Wyoming and Northwestern Colorado: Rocky Mountain Association of Geology Symposium, p. 117–137.

MILLER, F. X., 1977, The graphic correlation method in biostratigraphy, *in* Kauffman, E. and Hazel, J., eds., Concepts and Methods of Biostratigraphy: Stroudsburg, Dowden Hutchinson and Ross, p. 165–186.

MILLER, F. X., 1980, Graphic correlation; a new concept for biostratigraphy, *in* Miall, A. D., ed., Facts and Principles of World Petroleum Occurrence: Calgary, Canadian Society of Petroleum Geologists Memior 6, p. 994–995.

MILLER, F. X., 1990, Exploration applications of the graphic correlation-composite standard methodology: Abstracts of the Proceedings of the Twenty-second Annual Meeting of the American Association of Stratigraphic Palynologists, Palynology, v. 14, p. 215.

MURPHY, M. A., 1981, The application of Shaw's method of graphic correlation to Lower Devonian conodonts from Nevada: Geological Society of America, Abstracts with Programs, v. 13, p. 516.

MURPHY, M. A., 1987, The possibility of a Lower Devonian equal-increment time scale based on lineages in Lower Devonian Conodonts, *in* Austin, R. L., ed., Conodonts: Investigative Techniques and Applications: New York, British Micropaleontological Society Series, Ellis Horwood Ltd., p. 284–293.

MURPHY, M. A. AND BERRY, W. B. N., 1983, Early Devonian conodont-graptolite colation and correlations with brachiopod and coral zones, Central Nevada: American Association of Petroleum Geologists Bulletin v. 67, p. 371–379.

MURPHY, M. A. AND EDWARDS, L. E., 1977 The Silurian-Devonian boundary in Central Nevada, *in* Murphey, M. A., Berry, W. B. N., and Sandberg, C. A., eds., Western North America: Devonian: Riverside, University of California Riverside Campus Museum Contribution 4, p. 183–189.

NEAL, J. E., STEIN, J. A., AND GAMBER, J. H., 1994, Graphic correlation and sequence stratigraphy in the Palaeogene of N.W. Europe: Journal of Micropalaeontology, v. 13, p. 55–80.

OLSSON, R. K. AND LIU, C., 1993, Controversies on the placment of Cretaceous-Paleogene boundary and the K/P mass extinction of planktonic Foraminifera: Palaios, v. 8, p. 127–139.

ORMISTON, A. R., 1973, Advantages and limitations of graphic correlation: Geological Society of America Abstracts, v. 5, p. 759.

PAK, D. N., 1982, Mathematical model for the construction of composite standards from occurrences of fossil taxa: Computers and Geosciences, v. 10, p. 107–110.

PASLEY, M. A. AND HAZEL, J. E., 1990, Use of organic petrology and graphic correlation of biostratigraphic data in sequence stratigraphic interpretations: example from the Eocene-Oligocene boundary section, St. Stephens Quarry, Alabama: Transactions—Gulf Coast Association of Geological Societies, v. XL, p. 661–684.

PHILLIPS, F. J., 1986, A review of graphic correlation: COGS Computer Contributions, v. 2, p. 73–91.

PHILLIPS, F. J., 1986, Graphic correlation on microcomputers: Fourth North American Paleontological Convention, Proceedings, v. 4, p. A35.

PHILLIPS, F. J., 1986, Graphic correlation: American Association of Petroleum Geologists Bulletin, v. 70, p.1051–1052.

PIERCE, R. W., GAMBER, J. H., AND STEIN, J. A., 1993, Composite standard graphic correlations as a tool to visualize time-stratigraphic relationships among paleontological, geological, and geophysical data, Annual Meeting American Association of Petroleum Geologists and Society of Economic Paleontologists and Mineralogists Abstracts, p. 165–166.

PIERCE, R. W. AND MILLER, F. X., 1994, An exploration application of the graphic-correlation-composite-standard methodology in Egypt, *in* Lane, H. L., Blake, G., and MacLeod, N., eds., Graphic Correlation and the Composite Standard: the Methods and their Applications: Houston, Society of Economic Paleontologists and Mineralogists Research Conference, p. 14.

PISIAS, N. G., BARRON, J. A., AND DUNN, D. A., 1985, Stratigraphic resolution of leg 85: an initial analysis: Initial Reports of the Deep Sea Drilling Project, v. 85, p. 695–708.

PISIAS, N. G., MARINSON, D. G., MOORE, T. C., SHACKELTON, N. J., PRELL, W., HAYS, J., AND BODEN, G., 1984, High-resolution stratigraphic correlation of benthic isotope records spanning the last 300,000 years: Marine Geology, v. 56, p. 119–136.

POAG, C. W. AND AUBRY, M., 1995, Upper Eocene impactites of the U.S. East Coast: depositional origins, biostratigraphic framework, and correlation: Palaios, v. 10, p. 16–43.

PRELL, W. L., IMBRIE, J., MARTINSON, D. G., MORLEY, J. J., PISIAS, N. G., SHACKLETON, N. J., AND STREETER, H. F., 1986, Graphic correlation of oxygen isotope stratigraphy application to the Late Quaternary: Paleoceanography, v. 1, p. 137–162.

PROTHERO, D. R., 1990, Interpreting the Stratigraphic Record: New York, W. H. Freeman and Co., 410 p.

REIMERS, D. D., 1990, Graphic correlation and sequence resolution: American Association of Petroleum Geologists Bulletin, v. 74, p. 747.

SADLER, P. M., 1981, Sediment accumulation rates and the completeness of stratigraphic sections: Journal of Geology, v. 89, p. 569–584.

SADLER, P. AND KEMPLE, W., 1994, Extending graphic correlation to many dimensions, *in* Lane, H. L., Blake, G., and MacLeod, N., eds., Graphic Correlation and the Composite Standard: the Methods and their Applications: Houston, Society of Economic Paleontologists and Mineralogists Research Conference, p. 15.

SCOTT, R. W., 1990, Chronostratigraphy of the Cretaceous carbonate shelf, southeastern Arabia, *in* Robertson, A. H. F., Searle, M. P., and Ries, A. C., eds., The Geology and Tectonics of the Oman Region: Boulder, Geological Society of America Special Publication 49, p. 89–108.

SCOTT, R. W., BERGEN, J. A., STEIN, J. A., EVETTS, M. J., FRANKS, P. C., AND KIDSON, E. J., 1994, Graphic correlation calibrates ages of Mid-Cretaceous depositional cycles, western interior, *in* Lane, H. L., Blake, G., and MacLeod, N., eds., Graphic Correlation and the Composite Standard: the Methods and their Applications: Houston, Society of Economic Paleontologists and Mineralogists Research Conference, p. 17.

SCOTT, R. W., EVETTS, M. J., AND STEIN, J. A., 1993, Are seismic/depostional sequences time units? Testing by SHADS cores and graphic correlation: Offshore Technology Conference, no. 25, p. 269–276.

SCOTT, R. W., FINCANNON, P. G., AND BROWN, A. L., 1994, Interface for biostratigraphic data with seismic data in the workstation, *in* Lane, H. L., Blake, G., and MacLeod, N., eds., Graphic Correlation and the Composite Standard: the Methods and their Applications: Houston, Society of Economic Paleontologists and Mineralogists Research Conference, p. 16.

SCOTT, R. W., FRANKS, P. C., STEIN, J. A., BERGEN, J. A., AND EVETTS, M. J., 1994, Graphic correlation tests the synchronous Mid-Cretaceous depositional cycles: Western Interior and Gulf Coast, *in* Dolson, J. C., Hendricks, M. L., and Wescott, W. A., eds., Unconformity-related Hydrocarbons in Sedimentary Sequences: Denver, Rocky Mountain Association of Geologists, p. 89–98.

SCOTT, R. W., FROST, S. H., AND SHAFFER, B. L., 1988, Early Cretaceous sea-level curves, Gulf Coast and Southeastern Arabia, *in* Wilgus, C. K., Hastings, B. K., Posamentier, H., Van Wagoner, J., Ross, C. A., and St. C. Kendall, C. G., eds., Sea-Level Changes — An Integrated Approach: Tulsa, Society of Economic Paleontologists and Mineralogists Special Publication 42, p. 275–284.

SELF, G. A., SCOTT, R. W., AND BUTLER, M. L., 1993, Can lithology be predicted from stratal architecture? An empirical approach to the sequence stratigraphy of the Plio-Pleistocene, Gulf of Mexico: Offshore Technology Conference, no. 25, p. 419–427.

SHAW, A. B., 1960, Quantitative fossil correlations: Geological Society of America Bulletin, v. 71, p. 1972.

SHAW, A. B., 1964, Time in Stratigraphy: New York, McGraw-Hill Book Co., 365 p.

SHAW, A. B., 1969, Presidential address: Adam and Eve, paleontology and the non-objective arts: Journal of Paleontology, v. 43, p. 1085–1098.

SHAW, A. B., 1993, The origin of graphic correlation: Geological Society of America Northeastern Section, Abstracts with Programs, v. 25, p. 78.

SMALE, J. L., THUNELL, R. C., AND SCHAMEL, S., 1987, Foraminiferal biostratigraphy, graphic correlation, and depositional history of the early Miocene, Gemsa Plain, Gulf of Suez, Egypt: American Association of Petroleum Geologists Bulletin, v. 71, p. 615.

SRINIVASAN, M. S. AND SINHA, D. K., 1991, Improved correlation of the late Neogene planktonic foraminiferal datums in the equatorial to cool subtropical DSDP sites, Southwest Pacific; application of the graphic correlation method, *in* Radhakrishna, B. P., ed., The World of Martin F. Glaessner: Calcutta, Memoir Geological Society of India, v. 20, p. 55–93.

STRAUSS, D. J. AND SADLER, P. D., 1989, Classical confidence intervals and Bayesian probability estimates for the ends of local taxon ranges: Mathematical Geology, v. 21, p. 411–427.

SWEET, W. C., 1979, Graphic correlation of Permo-Triassic rocks in Kashmir, Pakistan and Iran: Geologica et Palaeontologica, v. 13, p. 239–148.

SWEET, W. C., 1979, Late Ordovician conodonts and biostratigraphy of the western Midcontinent Province, *in* Sandberg, C. A. and Clark, D. L., eds., Conodont Biostratigraphy of the Great Basin and Rocky Mountains: Brigham Young University Geology Studies, v. 26, p.45–85.

SWEET, W. C., 1980, Conodont-based zonation of post-Chazyan Ordovician Rocks of the North American Midcontinent Province: Abhandlungen Geologisches Bundesanstalt, v. 35, p. 214.

SWEET, W. C., 1981, Fossils and time: examples from the Paleozoic micropaleontologic record: Venice, International Meeting of Paleontology, Essential of Historical Geology, Abstracts Volume, p. 18.

SWEET, W. C., 1982, Fossils and time: an example from the North American Ordovician, *in* Gallitelli, E. M., ed., Procedings of the First International Meeting on Palaeontology, Essential of Historical Geology, held in Venice Italy, Fondazione Giorgio Cini, 2–4 June 1981, p. 309–321.

SWEET, W. C., 1982, Graphic correlation of upper Middle and Upper Ordovician rocks, North American Midcontinent Province, *in* Burton, D. L. and Williams, S. H., eds., Abstracts for Meetings, IV Internatinal Symposium, Ordovician System, Paleontological Contributions from the University of Oslo, no. 280, p. 53.

SWEET, W. C., 1983, Graphic correlation of Middle and Late Ordovician rocks, North America: Appalachian Basin Industrial Associates, v. 5, p. 4–22.

SWEET, W. C., 1983, Graphic correlation of Upper Ordovician rocks, North America: Appalachian Basin Industrial Associates, Fifth Meeting, Program and Abstracts, p. 15.

SWEET, W. C., 1984, Graphic correlation of upper Middle and Upper Ordovician rocks, North American Midcontinent Province, U.S.A., *in* Burton, D. L., ed., Aspects of the Ordovician System: Oslo, Universitetsforlaget, Paleontological Contributions from the University of Oslo; no. 295, p. 23–35.

SWEET, W. C., 1984, A conodont based standard reference section in Ordovician rocks of the Cincinnati Region: Geological Society of America, Abstracts with Programs, v. 16, p. 201.

SWEET, W. C., 1984, Upper Middle and Upper Ordovician conodont biostratigraphy of the North American Midcontinent: 27th International Geological Congress, Moscow, Abstracts v. 1, p. 316

SWEET, W. C., 1984, Conodonts, conodont biostratigraphy and correlation of the Moffett Road section (Middle and Upper Ordovician), Kenton County, Kentucky: Washington, D. C., United States Geological Survey Open-File Report 84-270, 1-11, 1-18.

SWEET, W. C., 1985, A revised conodont biostratigraphy for the Lower Triassic: Geological Society of America, Abstracts with Programs, v. 17, p. 731.

SWEET, W. C., 1986, Graphic rejuvenation of Ordovician chronostratigraphic scale: American Association of Petroleum Geologists Bulletin, v. 70, p. 654.

SWEET, W. C., 1986, Graphic correlation of Lower Triassic rocks and development of a high-resolution chronostratigraphic framework: Pavia, IGCP Project No. 203 and Societa Geologica Italiana, Field Conference on Permian and Permian-Triassic Boundary in the South Alpine Segment of the Western Tethys, Abstracts Volume, p. 51.

SWEET, W. C., 1987, Futher progress on development of a high-resolution stratigraphic framework for the Upper Permian and Lower Triassic: Beijing, Final Conference on Permo-Triassic Events of East Tethys Region, Abstracts Volume, p. 17.

SWEET, W. C., 1987, Distribution and significance of conodonts in Middle and Upper Ordvian strata of the Upper Mississippii Valley Region: Geological Society of America, Abstracts with Programs, v. 19, p. 249.

SWEET, W. C., 1987, Distribution and significance of conodonts in Middle and Upper Ordvian strata of the Uppr Mississippii Valley Region, *in* Sloan, R. E., ed., Middle and Late Ordovician Lithostratigraphy and Biostratigraphy of the Upper Mississippi Valley: St. Paul, Minnesota Geological Survey, Report of Investigations 35, p. 232.

SWEET, W. C., 1988, A quantitative conodont biostratigraphy for the Lower Triassic: Seckenbergiana lethaea, v. 69, p. 253–273.

SWEET, W. C., 1988, Mohawkian and Cincinnatian chronostratigraphy: Albany, New York State Museum Bulletin 462, p. 84–90.

SWEET, W. C., 1988, High-resolution chronostratigraphy of the Permian-Triassic boundary interval eastern Tethys: Geological Society of America, Abstracts with Programs, v. 20, p. 268.

SWEET, W. C., 1989, Consequences of an Otoceras-based P/T boundary: Preliminary evidence from conodonts: Permophiles, No. 15, p. 14–16.

SWEET, W. C., 1992, A conodont-based high-resolution biostratigraphy for the Permo-Triassic boundary interval, *in* Sweet, W. C., Zunyi, Y., Dickens, J. M. and Hongfu, Y., eds. Permo-Triassic Events in the Eastern Tethys: Cambridge, Cambridge University Press, p. 120–133.

SWEET, W. C., 1992, Middle and Late Ordovician conodonts from southwestern Kansas and their biostratigraphical significance: Oklahoma Geological Survey Bulletin, v. 145, p. 181–190.

SWEET, W. C., 1993, Unembellished graphic correlation: An example from the Ordovician of North America: Geological Society of America Northeastern Section, Abstracts with Programs, v. 25, p. 82.

SWEET, W. C., 1993, Biostratigraphy of the Permo-Triassic boundary interval: Current progress and problems: Geological Society of America, Abstracts with Programs, v. 25, p. 154.

SWEET, W. C. AND BERGSTRÖM, S. M., 1986, Conodonts and biostratigraphic correlation: Annual Review of Earth and Planetary Sciences, v. 14, p. 85–112.

TIPPER, J. C., 1988, Techniques for quantitative stratigraphic correlation: a review and annotated bibliography: Geological Magazine, v. 125 p. 475–494.

UPSHAW, C. F., ARMSTRONG, W. E., CREATH, W. B., KIDSON, E. J., AND SANDERSON, G. A., 1974, Biostratigraphic framework of Grand Banks: American Association of Petroleum Geologists Bulletin, v. 58, p. 1124–1132.

VAN NIEUWENHUISE, D. S., 1994, Calibration of Jurassic maximum flooding surfaces to a regional composite standard in the North Sea, *in* Lane, H. L., Blake, G., and MacLeod, N., eds., Graphic Correlation and the Composite Standard: the Methods and their Applications: Houston, Society of Economic Paleontologists and Mineralogists Research Conference, p. 18.

WOOLLAM, R., 1994, Stratigraphic analysis of Jurassic dinoflagellate cyst distributions in the Troll Field (Norwegian sector, North Sea) using graphic correlation, *in* Lane, H. L., Blake, G., and MacLeod, N., eds., Graphic Correlation and the Composite Standard: the Methods and their Applications: Houston, Society of Economic Paleontologists and Mineralogists Research Conference, p. 19.

YUAN, D. AND BOWER, J. C., 1990, Error and error estimation for graphic correlation in biostratigraphy, *in* Agterberg, F. P. and Bonham-Carter, G. F., eds., Statistical Applications in the Earth Sciences: Ottawa, Geological Survey of Canada Paper 89-9, p. 427–438.

ZELL, M. G., 1985, Brachiopoda and graphic correlation of the late Middle Cambrian Holm Dal Formation, Peary Land, North Greenland: Unpublished M. S. Thesis, University of Kansas, Lawrence, 82 p.

INDEXES

GENERAL INDEX

A
Absolute age, 29–31, 68, 71, 74
Absolute time, 29–31, 34, 67, 102, 117, 118
Acme, 102, 104, 121,189
Alabama, 152, 153, 216, 222
Alberta Rockies, 178, 180, 182
Alleghanian, 152, 156
Antarctic Polar Front, 243
Anticosti Island, 147, 164
Appalachians, 11, 131, 134, 136, 151–153, 156, 157, 173
Arbuckle Anticline, 143
Arbuckle Mountains, 147
Arctic Ocean, 205
Assemblage zones, 12, 235, 246
Atlantic Ocean, 6, 83, 85, 86, 92, 235, 236, 243, 245, 246
Austria, 159, 162, 164, 171–173

B
Basin analysis, 45, 50, 95, 111
Bentonite, 16, 51, 54, 205–208, 212, 213
Biases, 69, 123
Biofacies, 11, 177, 182, 205, 206, 208, 210–212, 213
Biostratigraphic zonations, 3, 205, 224
Biothem, 125
Black Hills, 205
Brazos River, 216, 221, 222, 225, 229, 230

C
Calibration
   composite standard, 10, 30, 31, 102, 124, 151, 152, 159, 173, 175, 226
   individual sections, 23–25, 26, 30, 35, 227
   fossil ranges, 42, 118, 188, 191, 201
Canadian Rocky Mountains, 134
Canning Basin, 178, 180, 181
Caribbean Sea, 86, 235, 236, 243
Central Devonian Field, 178
Central North Sea, 101–103, 105, 111
Channel basin, 198–200
Chemostratigraphy, 226
Chronostratigraphic
   diagram, 41
   units, 23, 29, 33, 41, 100
Chronozones, 159, 164, 166, 167, 171–173, 175
Cincinnati, 140, 143, 146, 147
Computers, 5–10, 23, 25, 31, 42, 45, 49, 56, 66, 70, 76, 78, 79, 117, 118, 178, 203
   software, 8, 23, 25–27, 56, 117, 153, 246
Core
   see sections
Czechoslovakia, 159, 160

D
Data
   error box, 68, 71
   sample-interval boxes, 47, 49, 68, 69
   simulated, 5
   weighting, 51, 54, 55, 59–64, 68, 73, 77, 81, 97, 103, 177,191, 203, 217
Dorset coast, 185, 189

E
East Shetland Platform, 110
Ecostratigraphy, 235
Emigration, 46, 68
Endemic, 118, 123
Europe, 4, 101, 111, 174, 189

Eustasy, 95, 98, 111, 112, 117, 205, 215, 216, 227, 229, 230, 245
Evolution, 3–5, 8, 33, 65, 68, 117, 118
Extinction, 4, 5, 11, 27, 46, 65, 68, 70, 71, 117, 132, 172, 215, 219, 230
   mass, 215, 230

F
Fault, 3, 15, 16, 31, 38, 40, 41, 182

G
Germany, 111, 159, 216
Greenhorn Sea, 205
Gulf of Mexico, 23, 32, 205, 230, 235, 236, 239, 243–246
Gulf of Suez, 127

H
Hay River, 182
Hypotheses, 6, 45, 47, 49, 66
   correlation, 45, 181

I
Ice volume, 235, 242, 243, 246
Immigration, 46, 68
Index fossils, 38, 97, 103, 127
Index species, 172, 173
Indiana, 139–141, 147
Iowa, 6, 16, 18, 177, 178, 205
Ir, 221, 223–225, 230
Isotope stages, 23, 235, 237, 238, 244

K
Kentucky, 139–141, 147
Kerguelen Plateau, 216, 224
Key beds, 6, 51, 54, 63, 178, 181, 182

L
Line of correlation
   channeling, 83, 85–87, 92
   dogleg, 7, 16, 26, 31, 45, 46, 54, 55, 63, 69, 97, 162, 164, 177, 180–183, 192, 194
   economy of fit, 46, 51, 61–63, 65, 68–72, 77, 79, 80, 177
   sample-interval boxes, 47, 49, 68, 69
Lithofacies, 9, 131, 132, 180, 185, 205, 212
Llano Uplift, 77
Local composite standard, 10, 11, 117, 118, 123, 127
Logs
   chemical, 54
   electric, 17, 23, 47, 106, 125
   gamma ray, 101–108, 111, 185–191, 100–203
   geophysical, 45
   sonic, 42, 185, 187, 198, 201, 203
   wire line, 11, 101, 185, 186
Luscar Mountain, 182

M
Magnetostratigraphy, 12, 111, 226, 236, 246
   chrons, 159, 174, 175
   magnetic polarity, 61, 65
   magnetic reversal, 23, 177
   magnetozones, 51
Marker beds, 9, 111, 118, 205, 206, 208
Martin Ridge, 145
Microspherules, 45

Midcontinent, 140, 147, 178
Migration, 3, 4, 11, 85, 117, 151, 152, 155–157, 211
Mississippi Valley, 15, 131, 136
Monitor Range, 139
Montagne Noire, 178–184
Montana, 15–17, 205

N

Natural selection, 3, 18
Nevada, 139, 144, 145, 147, 159, 160, 162, 164, 173
New York, 139, 177, 178, 180–182, 205
No-space graphs, 70
Nomograph, 41
Nondeposition, 84, 96, 99, 100
Nonunique events, 45
North Dakota, 15, 16
North Sea, 99, 101–106, 110, 111, 120–124, 127
Northwest Territories, 178, 182

O

Ohio, 6, 139–141, 147, 159
Oklahoma, 6, 7, 134, 143, 145–147
Old Bohemia Valley, 181
Oxygen isotope, 12, 31, 235–238, 240, 242, 244–246

P

Pacific Ocean, 83, 85, 88, 90, 92, 243
Paleobiogeography, 3, 45, 50
Paleoecology, 3
Paleogeographic maps, 98
Paleomagnetic reversal, 45, 68
Parasequence, 97, 185–187, 189
Parsimony, 51, 55, 56, 61–63, 86, 90, 224
Pennsylvania, 11, 131
Principle of faunal succession, 3
Principle of superposition, 3
Provincialism, 117, 118

R

Radiometric ages, 23, 31
Range
  adjustments, 66, 68, 69, 71, 73, 74, 178
  composite range chart, 81, 131, 133, 135, 136, 160
  composite standard, 27, 30
  local stratigraphic, 26–29, 32, 34–38, 40, 51, 52, 65, 68, 69–73, 79, 117, 121, 124, 131
  observed, 23, 32, 65, 68, 70–74, 79, 81, 83, 85
  taxon, 54, 65, 68, 70, 72, 73, 80, 84
  total stratigraphic, 3, 27, 29, 31, 35, 36, 117
Reworking, 24, 38, 68, 71, 73, 194, 201, 217, 227, 230
Roberts Mountains, 159, 160
Rocky Mountains, 15, 134
Russian Platform, 178, 182

S

Scale
  chronometric, 29
  ordinal, 70, 74, 201
  thickness, 74
  time, 3, 31, 65, 67, 69, 70, 74, 111, 112, 118, 159, 171, 173, 174, 175, 226–228, 246
Sea level, 69, 95, 97, 99, 100, 103, 105, 107, 132, 215, 229, 243, 246
Sections
  65GV, 141, 143
  Agost, 216, 218, 221, 222, 225
  Balcombe-1, 183, 193, 196, 197
  Beards Creek, 181
  Birch Creek II, 160, 162, 164
  Braggs, 222
  Brazos 1, 217, 219, 220
  Brazos Core, 224
  Bull Creek section, 205, 207
  Calera section, 153, 155
  Caravaca, 216, 218, 222, 225
  Chicxulub, 11, 215, 216
  CM4, 219, 220
  Col du Puech de la Suque E, 177, 184
  Col du Puech de la Suque H, 178
  Collendean Farm, 193, 194, 197
  core E67-135, 236–240
  Core V16-205, 236, 237, 238
  Crossroads section, 132–134
  CZ, 143
  Denton Valley section, 153, 157
  Drab-Beavertown section, 131, 132
  DSDP core 502, 235
  DSDP core 502A, 242
  DSDP core 502B, 235
  DSDP site 502, 86
  DSDP site 528, 216, 224, 229
  DSDP site 573, 90
  DSDP site 574, 90
  DSDP site 575, 90
  DSDP site 577, 216, 229
  DSDP site 586, 88, 92
  DSDP site 606, 86
  El Kef, 120, 216–225, 227, 229, 230
  Eschelman Quarry section, 132
  Esso N16/1-1, 123
  Farleigh Wallop, 197
  Garden Banks Block 412 Unocal #1, 235, 239
  GräSfenwarth section, 159
  Gubbio, 221, 230
  Hogskin Valley, 139
  Holston Lake section, 153
  Imler quarry section, 133, 134
  Kingsport section, 156, 157
  Klonk section, 159
  L. OK77, 143
  La Serre Trench D, 178, 179
  LS, 143
  Millers Ferry, 215, 216, 219, 221, 222, 227, 229, 230
  Mimbral, 215, 216, 219, 222, 223, 227, 229, 230
  Model T section, 208, 211
  Normandy-1, 197
  Nye Kløv, 215, 216, 219–222, 225, 227, 229, 230
  ODP Core 625, 235, 236, 240, 242, 243
  ODP Core 625B, 235, 240, 242, 243
  ODP Site 690, 215, 216, 223, 230
  ODP site 738, 215, 216, 219, 224, 225, 227, 229, 230
  Ore Hill section, 135, 136
  Pete Hanson Creek IB, 173
  Pete Hanson Creek II, 160, 164
  Pete Hanson Creek IID, 160, 164
  Pic de Bissous, 179
  Potter Creek Crossroads, 131
  Sawmill Canyon, 144
  section 70ZA, 140, 147
  Shingle Pass, 144, 146–148
  Simpson Park Range I, 160, 173
  Steele Creek East section, 153, 157
  Stoneville Flats section, 206, 207
  Storrington-1, 196
  Torgerson Draw section, 206, 207, 211
  TR, 143
  Trout River, 178, 182, 183
  U. OK77, 143, 146
  Union Otter Woman Morning Gun No. 1, 15
  Upper Coumiac Auarry, 178, 180, 181, 184
  V16-205, 236–238
  Warlingham Borehole, 188

WCB 365, 181
WCB 367, 182
Well 8/27A-1, 109–111
Well 15/13-2, 108–109
Well 16/28-1, 105
Well 16/29-4, 105
Well 156/16-1, 106, 108, 109
Well N25/1-3, 110-111
Well X, 127, 129
Willow Creek I, 160
Seismic
   stratigraphy, 69, 95, 201
   synthetic seismogram, 42, 102, 104, 108
   vertical profile, 42
Sequence, 189, 235
Sequence stratigraphy, 6, 10, 61, 72, 95, 97, 98, 100, 101, 110, 111, 130, 215 216
Sevier foreland basin, 152, 155
South Viking Graben, 110
Standard time units, 5, 11, 159, 162, 164, 171, 173, 174, 175
Statistics
   analysis of variance, 60
   average composite models, 10
   bandpass filtering, 185, 186, 189, 201, 203
   confidence limits, 73
   constrained optimization, 10, 51, 61–68, 70, 77, 79, 80, 81
   correlation coefficient, 60, 219
   derived correlations, 10, 29, 83, 84, 89, 90, 92
   F-distribution, 60
   greedy search, 76
   least-squares linear regression, 31, 37, 51, 56–58, 60–63, 66, 69, 131, 132, 177
   major axis regression, 51, 57–60, 63
   moving average, 186
   multiple regression, 62
   non-parametric, 60, 63, 219
   penalty function, 65, 70–74, 77, 79
   reduced major axis regression, 51, 57, 58, 60, 63
   regression analysis, 51, 56, 59, 60, 63, 66
   sensitivity analysis, 66, 73, 80, 81
   significance tests, 51, 60, 63
   simulated annealing, 61, 65, 76–80
   standard error of estimate, 161, 177

Tabu search, 76, 78
   time series, 186
Stratigraphic cycles, 112
Subtropical convergence, 243
Sun River, 15, 17
Supplemental graphic correlation, 6, 11, 206, 212

T

Taconic orogeny, 152
Tampico Basin, 120
Tennessee, 139, 152, 156, 157
Texas, 65, 77, 134, 222
Thermocline, 246

U

Uncertainty, 47, 88, 90, 188, 191
United Kingdom, 11, 185
Utah, 139, 143, 147, 149, 205

V

Virginia, 45, 152, 156
Volcanic ash, 5, 31, 68, 45, 54, 156, 177

W

Weald Basin, 188, 192, 196–198, 200
Weddell Sea, 216
Well
   see section
Welsh basin, 171
Wessex Basin, 11, 185, 201, 202
Western Interior Basin, 205, 207, 210
Western Interior Seaway, 12, 205, 207
Whiterock Canyon, 139, 146
Williston basin, 15, 16
Worldwide composite standard, 11, 117, 120, 121, 123
Wyoming, 15, 205

Y

Yellowstone Park, 15

STRATIGRAPHIC INDEX

A

Aeronian, 159, 162, 171, 175
albani ammonite biozone, 189
Ancoradella ploeckensis Biozone, 162, 164, 172
Andrew sequence, 104
anguiformis ammonite biozone, 189, 201
Antelope Valley Limestone, 139, 145
Aphelaspis Zone, 133
autissiodorensis ammonite biozone, 201

B

Balmoral sequence, 104
baylei ammonite biozone, 189, 201
Belle Fourche Shale, 205
Blockhouse Shale, 157
bohemicus/koslowskii Chronozone, 167
bouceki/transgrediens Chronozone, 167, 173
Bromide Formation, 145
Brunhes Chron, 237, 238

C

Cambrian, 11, 16, 30, 65, 66, 73, 77, 80, 131, 144
Caradoc, 156
Carlile Shale, 205
Cashaqua Shale, 182
Cenomanian, 11, 205, 207, 208
Cenozoic, 7, 119, 120, 123, 185, 216
centrifugus/murchisoni/riccartonensis Chronozone, 169
Charles Formation, 15, 17
Chazyan, 146
Cincinnatian Series, 139–142, 144, 147
Clayton Formation, 222
Cliffia lataegenae Subzone, 131, 132, 134
Climacograptus bicornis Zone, 151–153, 155
Collignoniceras woollgari woollgari Subzone, 208
Conaspis Zone, 135, 136
Conococheague Limestone, 131
Copenhagen Formation, 145
Cretaceous, 5, 6, 7, 11, 16, 38, 205, 215, 217, 219, 224, 227, 230
cymodoce ammonite biozone, 189

D

Danian, 11, 215–230
Devonian, 7, 18, 45, 159, 162, 171, 174, 175, 177, 178, 181, 182, 226
Didymograptus murchisoni Zone, 152, 153, 156
Distomodus staurognathoides Chronozone, 164, 171

E

Edenian, 139
Egan Range, 144
Ellis Bay Formation, 147
Elvinia Zone, 131–135
Eocene, 4, 38, 95, 106
eudoxus ammonite biozone, 189, 201
Eureka Quartzite, 145

F

Fairview Formation, 141
Famennian, 177, 178, 181, 184
fittoni ammonite biozone, 202
Franconian, 131, 135, 136
Frasnian composite standard, 11, 177–182
Frasnian, 11, 177–184
Frigg sequence, 111

G

Gamachian, 147
Gatesburg Formation, 11, 131, 136
Genese Formation, 181
Givetian, 179
Glyptograptus teretiusculus Zone, 151–153, 155
Gorstian, 171, 173, 174
Greenhorn Cycle, 205
Greenhorn Formation, 205

H

High Bridge Group, 141
Holocene, 30, 235, 236, 246
Homerian, 159, 171, 173–175
House Formation, 144

I

Ibexian Series, 139, 143, 144, 147
Icriodus woschmidti Chronozone, 162, 164, 167
Irvingella major Subzone, 131, 134, 135

J

Jaramillo Subchron, 238
Joins Formation, 146, 147
Jurassic, 11, 15, 45, 185, 190, 201

K

Kanosh Formation, 144
Kimmeridge Clay Formation, 188
Kimmeridgian, 185, 188, 192, 194, 197, 201, 202, 203
Kinderhookian, 15, 16
Kockelella ranuliformis Chronozone, 172
Kockelella variabilis Chronozone, 164
Kope Formation, 141

L

Lake Valley Beds, 15
leintwardinensis Chronozone, 167, 168, 171, 173
Lexington Limestone, 141
Llandovery, 159, 167, 171, 173, 174
Llanvirn Series, 156
lochkovensis Chronozone, 167, 173
Lochkovian, 159
Lodgepole Formation, 15
Lower Sandy Member, Gatesburg Formation, 131
Ludfordian, 171, 173, 174
Ludlow, 159, 171–175
lundgreni Chronozone, 160, 167, 171, 173

M

Maastrichtian, 205, 215, 216, 222, 225, 229, 230
Madison Limestone, 15
Mammites nodosoides ammonite Zone, 207
Maureen sequence, 104
Maysvillian, 140
McLish Formation, 143, 145
Meramecian, 15
Mesozoic, 7, 119, 185, 216
Miocene, 30, 41, 127, 129
Mission Canyon Formation, 15, 17
Mississippian, 7, 15, 16
Mohawkian Series, 139, 143, 144
mutabilis ammonite biozone, 189, 201

N

nassa chronozone, 167, 173
nassa/deubeli Chronozone, 167, 173
Nemagraptus gracilis Zone, 153, 155
Neogene, 74, 235
New York regional composite, 181, 182
nilssoni chronozone, 173
nilssoni/scanicus Chronozone, 168, 171
Ninemile Shale, 139, 146

O

Oil Creek Formation, 146
Olenaspella evansi Zone, 134
Oligocene, 4, 33, 38, 41
Ordovician, 7, 11, 16, 139–147, 151–153, 156, 157, 174
Ore Hill Member, Gatesburg Formation, 11, 131–136
Osagean, 15
Ozarkodina bohemica bohemica Biozone, 172
Ozarkodina crassa Chronozone, 164, 172
Ozarkodina crispa Biozone, 164, 166, 172
Ozarkodina remscheidensis eosteinhornensis Chronozone, 162, 164, 166, 167, 172
Ozarkodina sagitta rhenana Biozone, 166, 172
Ozarkodina sagitta rhenana Chronozone, 166, 172
Ozarkodina sagitta sagitta Biozone, 164, 166, 172
Ozarkodina snajdri Biozone, 172

P

Paleogene, 23, 31, 32, 99, 101–103, 121, 123, 124
Paleozoic, 5–7, 11, 18, 119, 139
pallasoides ammonite biozone, 189
Parabolinoides Subzone, 131, 135, 136
Paralenorthis-Orthidiella brachiopod zone, 139
Parker Spring Formation, 144
parultimus chronozone, 167, 173
pectinatus ammonite biozone, 189, 201
Pedavis latialata Chronozone, 162, 164
Pelmatolepis conodont biofacies, 182
Pennsylvanian, 7, 16
Pleistocene, 12, 235, 236, 239, 242–244, 246
Pliocene, 83, 85, 92, 235, 236, 240, 243, 246
Pogonip Group, 144
Point Pleasant Formation, 141
Polygnathoides siluricus Chronozone, 164, 172
Polygnathus Frasnian conodont biofacies, 11, 177, 182
Poole Creek Member, 205
Portland Beds, 188, 200
Portlandian, 185, 188, 192, 194, 196, 202, 203
praedeubeli chronozone, 173
Prairie Bluff Chalk, 222
Pridoli, 159, 164, 167, 171–175
Pseudoaspidoceras flexuosum ammonite Zone, 207
Pterospathodus amorphognathoides Chronozone, 164, 172
Pterospathodus celloni Chronozone, 164
Purbeck Anhydrite, 190

R

Rhuddanian, 162
riccartonensis biozone, 167, 168
Richmondian, 139, 147
rigidus/ellesae Chronozone, 167, 169, 173
Riley Formation, 77, 80
Roberts Mountians Formation, 159, 160
rotunda ammonite biozone, 189, 202

S

Sauk Sequence, 143, 144
sedgwickii Biozone, 171
Sheinwoodian, 159, 164, 171, 173–175
Shingle Formation, 144
Silurian, 11, 139, 147, 159–162, 164, 167, 171–175
Sonyea Formation, 181
Sun River Dolomite, 15

T

Taenicephalus shumardi Subzone, 135, 136
Taenicephalus Zone, 131, 132, 135
Telychian, 171, 175
Tertiary, 3, 5, 7, 11, 18, 35, 38, 185, 215, 221, 225
Thanetian, 37
Tippecanoe Sequence, 143
transgrediens chronozone, 167, 173
Trenton Group, 139
Tripodus laevis conodont zone, 139
turriculatus/crispus Biozone, 171

U

U. Forties sequence, 104
ultimus chronozone, 167, 173
uniformis Chronozone, 162, 167
Upper Sandy Member, Gatesburg Formation, 136
Upper Sele sequence, 108

V

Vascoceras birchbyi ammonite Zone, 207
Vauréal Formation, 147
Viola Group, 143

W

Watinoceras devonense ammonite Zone, 207
Wenlock, 159, 164, 169, 171,–175
West Fall Formation, 181
West Spring Creek Formation, 146, 147
wheatleyensis ammonite biozone, 189, 201
Whipple Cave Formation, 144
Whiterockian Series, 139, 140, 143, 144, 146

Z

Zone P1a, 219, 221–225, 227
Zone PO, 95
Zone Q, 235, 236
Zone R1, 235, 238, 240, 244
Zone R2, 235, 240, 244
Zone R3, 235, 240, 244
Zone S1, 235, 237, 238, 240, 244
Zone S2, 235
Zone S3, 235, 240, 244
Zone T, 235, 238, 240, 244, 245
Zone T1, 238
Zone T3, 235, 238, 240, 244
Zone T4, 235, 237, 238, 240, 244
Zone U, 235, 238, 240, 244, 245, 246
Zone V1, 235, 240, 244
Zone V2, 235, 240, 244
Zone V3, 235, 240, 244, 245
Zone W, 235, 240, 244
Zone X, 235, 240, 244
Zone Y, 235, 240, 244
Zone Z, 85, 235

TAXONOMIC INDEX

A

Ammobaculites subaequalis, 192, 196
Ammomitida, 200, 201, 205, 215
Ammonoidea, 177, 178, 181
Amphicythere confundens, 197
Ancoradella ploeckensis, 162, 164–180, 181, 184
Ancoradella rotundiloba, 184
Ancoradella rugosa, 184
Ancyrodella gigas, 184
Ancyrodella lobata, 180, 182, 184
Ancyrodella rotundiloba, 179
Azgograptus incurvus, 153

B

Baltoniodus gerdae, 139
Bathysiphon taurinensis, 129
Bivalvia, 205, 215
   Inoceramid, 205, 207
   rudistid, 215
Brachiopoda, 7, 15, 136, 215
Burnetiella urania, 133
Bynumina terrenda, 132, 134

C

Calcidiscus macintyrei, 240
Camaraspis convexa, 134
Cassidulina cruysi, 129
Chiloguembelina crinita, 225
Chiloguembelina waiparaensis, 225
Citharina serratocostata, 202
Collignoniceras woollgari regulare, 208
Comanchia amplooculata, 134
Conaspis, 131, 135, 136
Corollithion kennedyi, 208
Cribrostomoides canui, 196, 198
Cryptograptus marcidus, 153
Cyrtograptus lundgreni, 173, 174
Cyrtograptus rigidus, 173
Cytherella fullonica, 190

D

Deadwoodia duris, 134
Deckera completa, 134
diatoms, 90, 215
Dicellograptus geniculatus, 153
Dicellograptus gurleyi n. ssp. A, 153
dinoflagellate, 98
dinosaurs, 215
Drabia acroccipita, 134
Drabia menusa, 134
Drumaspis occidentalis, 134

E

Elvinia roemeri, 134
Eoglobigerina fringa, 219, 225
Eoglobigerina simplicissima, 225
Eoglobigerina trivialis, 221
Eoorthis, 136
Epistomina ornata, 196
Epistomina parastelligera, 192

G

Galliaecytheridea compressa, 190, 192, 193, 202
Galliaecytheridea crendonensis, 197
Galliaecytheridea dissimilis, 193, 196, 201
Galliaecytheridea dorsetensis, 194, 201
Galliaecytheridea elongata, 190, 192, 196, 201
Galliaecytheridea fragilis, 191
Galliaecytheridea postrotunda, 194, 196, 201
Galliaecytheridea punctata, 190
Galliaecytheridea spinosa, 190, 191, 194, 202
Galliaecytheridea wolburgi, 190
Globanomalina archeocompressa, 222
Globastica daubjergensis, 221, 223
Globigerinoides bisphaericus, 129
Globoconusa conusa, 219, 221, 225
Globoconusa cretacea, 221, 227
Globorotalia cultrata, 235
Globorotalia flexuosa, 235
Globorotalia hirsuta, 246
Globorotalia inflata, 235, 236, 243, 244, 246
Globorotalia longiapertura, 225
Globorotalia menardii, 235, 236, 242–244, 246
Globorotalia truncatulinoides, 235, 236
Globorotalia tumida, 246
Globorotalia velascoensis, 35, 38
Guembelitria trifolia, 225

H

Haplophragmoides sp., 190, 192, 194, 196, 202
Hedbergella holmdelensis, 222
Hedbergella monmouthensis, 222
Helicosphaera sellii, 239, 240
Histiodella altifrons, 147
Housia vacuna, 134
Hyalinea balthica, 240

I

Icriodus woschmidti, 162, 164, 167
Igorina spiralis, 225
Irvingella major, 131, 134, 135

K

Kindbladia wichitaensis, 134
Kockelella patula, 164
Kockelella variabilis, 164

L

Lenticulina bruekmanni, 191
Lenticulina muensteri, 202
Lenticulina sp., 197
Lenticulina sublata, 197
Lenticulina uralensis, 191, 192

M

Macrodentina cf. proclivis, 196
Macrodentina cicatricosa, 194, 197, 201
Macrodentina maculata, 190, 193, 202
Macrodentina retirugata, 197, 202
Macrodentina transiens, 190, 191, 202
Mandelstamia rectilinea, 197
Mandelstamia triebeli, 190
Mandelstamia tumida, 191
Microstaurus chiastius, 208
Monograptus fritschi linearis, 167
Mucriglobigerina aquiensis, 225
Mucriglobigerina chasconona, 225

INDEX

N

nannofossils, 30, 83, 85, 100, 208
nannoplankton, 73, 127, 215, 216
Nemagraptus gracilis, 152, 153, 155
Neocardioceras juddii, 208
Neodiversograptus nilssoni, 167

O

Ostracoda, 127, 189, 191, 192, 197, 201, 202
Ozarkodina bidentatiformis, 184
Ozarkodina remscheidensis eosteinhornensis, 162, 164, 167, 172
Ozarkodina remscheidensis remscheidensis, 166, 167
Ozarkodina sagitta rhenana, 166, 172
Ozarkodina sagitta sagitta, 164, 166, 172
Ozarkodina trepta, 181, 182, 184

P

Palaeocystodinium bulliforme, 104
Palmatolepis bogartensis, 178, 181, 184
Palmatolepis domanicensis, 182, 184
Palmatolepis jamieae, 178, 184
Palmatolepis ljaschenkoae, 178, 179, 180, 184
Palmatolepis punctata, 181, 184
Palmatolepis semichatovae, 182, 184
Palmatolepis transitans, 181, 184
Palmatolepis winchelli, 178, 180, 181, 182, 184
Parasubbotina aff. pseudobulloides, 222
Parvularugoglobigerina eugubina, 219, 222, 224, 225, 227
Pedavis latialata, 162, 164
Polygnathoides siluricus, 164, 172
Polygnathus dengleri, 178, 180
Polygnathus samueli, 182
Praemurica, 222
Praeorbulina glomerosa, 34
Praeorbulina universa, 129
Probeloceras lutheri lutheri, 182
Proceratopyge rectispinata, 134
Prochorites alveolatus, 182
Pseudoclimacograptus angulatus angulatus, 153
Pseudoemiliania lacunosa, 240

Pseudosaratogia magna, 133, 134
Pterospathodus amorphognathoides, 164, 172
Pterospathodus celloni, 164
Punctospirifer solidirostris, 15

R

Radiolarian, 90
Reinholdella pseudojasanesis, 192
Rhagodiscus asper, 208

S

Saracenaria oxfordiana, 202
Schuleridea moderata, 190, 191, 194, 196, 202
Subbotina moskvini, 223
Subbotina pseudobulloides, 221
Sulcocephalus candidus, 135
Sulcocephalus granulosus, 135

T

Taenicephalus shumardi, 132, 135, 136
Trilobita, 131, 132, 134, 136
Trimosina (A), 240
Tripodus laevis, 139, 146, 147
Turborotalia peripheronda, 129

V

Vernoniella sequana, 190, 192, 194, 196, 197, 201

W

Woodringina claytonensis, 221
Woodringina hornerstownensis, 219

X

Xenocheilus spineum, 134